Plate Boundaries and Natural Hazards

Plate Boundaries and Natural Hazards

Editor:

Jaspreet Banga, Ph.D.

Delve Publishing
www.delvepublishing.com

Plate Boundaries and Natural Hazards

Editor: Jaspreet Banga, Ph.D.

708 3rd Avenue, 6th Floor
New York, NY 10017
United States of America
www.delvepublishing.com

ISBN: 978-1-68095-831-7

Library of Congress Control Number: 2016951573

About the Editor

Jaspreet Banga, Ph.D.

Jaspreet Banga has completed PhD from Central Drug Research Institute, India and Master of Technology from Indian Institute of Technology. Jaspreet's interests range from Drug development, Purification, Fermentation technology, Plant Biotechnology and Oncology.

Contents

List of Contributors

Jing-Yi Lin
Department of Earth Sciences, National Central University, 300 Jhongda Road, Jhongli City, Taoyuan County 32001, Taiwan

Jean-Claude Sibuet
Department of Earth Sciences, National Central University, 300 Jhongda Road, Jhongli City, Taoyuan County 32001, Taiwan

Shu-Kun Hsu
Department of Earth Sciences, National Central University, 300 Jhongda Road, Jhongli City, Taoyuan County 32001, Taiwan

Wen-Nan Wu
Institute of Earth Sciences, Academia Sinica, 128, Sec. 2, Academia Road, Nankang, Taipei 11529, Taiwan.

K. Mahanta
Cotton College, India

J. D. Chowdhury
Jagiroad College, India

I. Laskar
Cotton College, India

Akira Takeuchi
Graduate school of Science and Engineering for Research, University of Toyama, Japan

Passarelli L.
Section 2.1 "Physics of earthquakes and volcanoes" GFZ-German Research Centre for Geosciences in plural form, Helmholtzstr. 7, 14467, Potsdam, Germany

Rivalta E.
Section 2.1 "Physics of earthquakes and volcanoes" GFZ-German Research Centre for Geosciences in plural form, Helmholtzstr. 7, 14467, Potsdam, Germany

Shuler A.
Department of Earth and Planetary Sciences, University of California Santa Cruz, 1156 High Street, Santa Cruz, CA 95064, USA
Now at Chevron Energy Technology Company, 1500 Louisiana St., Houston, TX 77002, USA

Philip J. Heron
Department of Earth Sciences, University of Toronto, 22 Russell St. Toronto, Ontario, Canada, M5S 3B1

Russell N. Pysklywec
Department of Earth Sciences, University of Toronto, 22 Russell St. Toronto, Ontario, Canada, M5S 3B1

Randell Stephenson
School of Geosciences, University of Aberdeen, Aberdeen AB24 3UE, Scotland

Wan-Jiung Hu
Academia Sinica, Taiwan, Hong Kong University

Orlando Hernandez-Pardo
School of Earth Sciences, The Ohio State University, Columbus, OH 43210 USA, FAX 614 2927688.
Dept. of Geosciences, Universidad Nacional de Colombia, Bogotá, D.C. COLOMBIA

Ralph R. B. von Frese
School of Earth Sciences, The Ohio State University, Columbus, OH 43210 USA, FAX 614 2927688.

Jeong Woo Kim
Dept. of Geoinformation Engineering, Sejong University, Seoul, KOREA

Jiren Xu
Institute of Geology, Chinese Academy of Geological Sciences, Beijing 100037, China

Zhixin Zhao
Institute of Geology, Chinese Academy of Geological Sciences, Beijing 100037, China

R. S. M. De Plaen
School of Earth Sciences, University of Bristol, Bristol, UK
Technology and Communication, University of Luxembourg, Luxembourg City, Luxembourg

I. D. Bastow
Department of Earth Science and Engineering, Imperial College London, London, UK

E. L. Chambers
National Oceanography Centre Southampton, University of Southampton, Southampton, UK,

D. Keir
National Oceanography Centre Southampton, University of Southampton, Southampton, UK,

R. J. Gallacher
National Oceanography Centre Southampton, University of Southampton, Southampton, UK,

J. Keane
Department of Earth Science and Engineering, Imperial College London, London, UK

Naoki Uchida
Research Center for Prediction of Earthquakes and Volcanic Eruptions, Graduate School of Science, Tohoku University, Sendai 980-8578, Japan

Stephen H. Kirby
U.S. Geological Survey, Menlo Park, CA 94025, USA

Norihito Umino
Research Center for Prediction of Earthquakes and Volcanic Eruptions, Graduate School of Science, Tohoku University, Sendai 980-8578, Japan

Ryota Hino
Research Center for Prediction of Earthquakes and Volcanic Eruptions, Graduate School of Science, Tohoku University, Sendai 980-8578, Japan

Tomoe Kazakami
National Research Institute for Earth Science and Disaster Prevention, Tsukuba 305-0006, Japan

Alessandro Tibaldi
Department of Earth and Environmental Sciences, University of Milan-Bicocca, Milan, Italy

Fabio L. Bonali
Department of Earth and Environmental Sciences, University of Milan-Bicocca, Milan, Italy

Federico A. Pasquaré Mariotto
Department of Theoretical and Applied Sciences, Insubria University, Varese, Italy

Yoshiyuki Tatsumi
Department of Planetology, Kobe University, Kobe 6567-8501, Japan
Research and Development Center for Ocean Drilling Science, Japan Agency for Marine-Earth Science and Technology, Yokosuka 237-0061, Japan

Takeshi Sato
Research and Development Center for Earthquake and Tsunami, Japan Agency for Marine-Earth Science and Technology, Yokohama 236-0001, Japan.

Shuichi Kodaira
Research and Development Center for Earthquake and Tsunami, Japan Agency for Marine-Earth Science and Technology, Yokohama 236-0001, Japan.

John F. Dewey
University College, Oxford OX1 5NA, UK

Jan Golonka
AGH University of Science and Technology, Kraków, Poland

Aleksandra Gawęda
Faculty of Earth Sciences, University of Silesia, Sosnowiec, Poland

Engielle M. R. Paguican
Department of Geology, University at Buffalo, The State University of New York, Buffalo, NY, USA
Department of Geography, Earth System Sciences, Vrije Universiteit Brussel, Brussels, Belgium

Marcus I. Bursik
Department of Geology, University at Buffalo, The State University of New York, Buffalo, NY, USA

Georges Ramalanjaona
Department of Emergency Medicine, St. John's Episcopal Hospital, 327 Beach 19th Street, Far Rockaway, New York, NY 11691, USA
Department of Emergency Medicine, St. John's Episcopal Hospital, 3264 Wolfson Dive, Baldwin, New York, NY 11510, USA

Carolina Pagli
School of Earth and Environment, University of Leeds, Leeds, UK
Dipartimento di Scienze della Terra, Università di Pisa,Pisa, Italy

Hua Wang
Department of Surveying Engineering, Guangdong University of Technology, Guangzhou, China

Tim J. Wright
COMET,School of Earth and Environment, University of Leeds, Leeds, UK

Eric Calais
Ecole Normale Supérieure, PSL Research University, Paris,France

Elias Lewi
IGSSA, Addis Ababa University, Addis Ababa, Ethiopia

Kashyap Mahanta
Cotton College, India

Jyotirmoy Das Chowdhury
Jagiroad College, India

Sangita Baruah
Cotton College, India

Arun Kumar
Manipur University, India

Ashok Kumar
Tezpur University, India.

Sunil L. Singh
Manipur University, India

Prakash Barman
Tezpur University, India.

Fabien Leprieur
UMR MARBEC, (CNRS, IRD, IFREMER, UM), cc 093, Place E. Bataillon, FR-34095 Montpellier, France

Patrice Descombes
University of Fribourg, Unit of Ecology & Evolution, Ch. du Musée 10, CH-1700 Fribourg, SwitzerlandSwiss Federal Research Institute WSL, CH-8903 Birmensdorf, Switzerland
Landscape Ecology, Institute of Terrestrial Ecosystems, ETH Zürich, 8044 Zürich, Switzerland

Théo Gaboriau
UMR MARBEC, (CNRS, IRD, IFREMER, UM), cc 093, Place E. Bataillon, FR-34095 Montpellier, France

Peter F. Cowman
Department of Ecology & Evolutionary Biology, Yale University, 21 Sachem St, New Haven, Connecticut 16 06511, USA

Valeriano Parravicini
CRIOBE, USR 3278 CNRS-EPHE-UPVD, LABEX 'CORAIL', University of Perpignan, 66860 Perpignan, France

CESAB-FRB, Immeuble Henri Poincaré, Domaine du Petit Arbois, FR-13857 Aix-en-Provence, France

Michel Kulbicki
Institut pour la Recherche en Développement, UR UMR "Entropie" -Labex Corail- Université de Perpignan, 66000 Perpignan, France

Carlos J. Melián
Department of Fish Ecology and Evolution, Eawag: Swiss Federal Institute of Aquatic Science and Technology, 6047 Kastanienbaum, Switzerland

Charles N. de Santana
Department of Fish Ecology and Evolution, Eawag: Swiss Federal Institute of Aquatic Science and Technology, 6047 Kastanienbaum, Switzerland

Christian Heine
EarthByte Group, The University of Sydney, Sydney, New South Wales 2006, Australia

David Mouillot
UMR MARBEC, (CNRS, IRD, IFREMER, UM), cc 093, Place E. Bataillon, FR-34095 Montpellier, France
Australian Research Council Centre of Excellence for Coral Reef Studies, James Cook University, Townsville, Queensland 4811, Australia

David R. Bellwood
Australian Research Council Centre of Excellence for Coral Reef Studies, James Cook University, Townsville, Queensland 4811, Australia
College of Marine and Environmental Sciences, James Cook University, Townsville, Queensland 4811, Australia

Loïc Pellissier
University of Fribourg, Unit of Ecology & Evolution, Ch. du Musée 10, CH-1700 Fribourg, Switzerland
Swiss Federal Research Institute WSL, CH-8903 Birmensdorf, Switzerland
Landscape Ecology, Institute of Terrestrial Ecosystems, ETH Zürich, 8044 Zürich, Switzerland

Preface

There have been two major tsunamis, the Indian Ocean tsunami in 2004 and October 2010 Sumatra earthquake on the same fault. Luckily, the Sumatra tsunami was not as disastrous, but still there were 435 mortalities and 20,000 people were displaced in Sumatra. On the other hand, the Indian Ocean tsunami resulted in 229,866 deaths in 14 countries and many more were displaced. The undersea megathrust earthquake, of magnitude 9.1-9.3, was caused when the Indian Plate was sub ducted by the Burma Plate and triggered a series of devastating tsunamis, waves up to 30 metres, along the coasts of most landmasses bordering the Indian Ocean, creating one of the deadliest natural disasters in recorded history. There are 15 major tectonic plates, of which the majority of plates are under the ocean. The earth's continents are in motion due to the movement of these tectonic plates. These devastations in our recent history has lead us to closely understand and monitor the different types of plate boundaries

Editor
Jaspreet Banga, Ph.D.

INTRODUCTION

What are Natural Hazards?

Natural hazards are events that may pose a risk to human health, safety, and property. They can include volcanoes, fires, floods, earthquakes, and other natural events that may be difficult to prevent or control. National agencies around the world monitor ongoing hazards in their area and provide outreach, education, and support services. Internationally, countries also communicate about hazards that cross borders and could affect neighbors and allies.

Many of these events are entirely natural, although some may be influenced by human activity. For example, overuse of land with repeated crop plantings that have heavy water and nutrient demands can contribute to the creation of droughts, dust storms, and eventual desertification. Other natural hazards, like volcanoes and earthquakes, are not related to anything humans may or may not be doing on the surface of the Earth. They are unpredictable and can occur very suddenly and without any warning.

One aspect of natural hazards management is monitoring, prediction, and forecasting. Government agencies use a variety of tools to track ongoing events, take note of patterns, and issue predictions. For example, if analysis of a volcano suggests a dangerous buildup of magma that may erupt, the government can issue a warning to residents. This might be used as the basis for an evacuation order to get people out of the area in case the volcano erupts.

Government agencies can also play a role in planning and policy decisions. In an area known for natural hazards, agencies may discourage new development. At the same time, they may promote disaster preparedness and planning with things like earthquake-resistant shelters, stockpiles of materials, and routine maintenance of roads to make evacuations as smooth as possible. Hazard mitigation can also include activities like installing levees to control flooding, or building firebreaks to limit damage caused by wildfires.

Quick response to natural hazards is also part of the work of government agencies. They need to be able to get personnel on the ground quickly to monitor events, help citizens, and start the process of rebuilding and improving. This aspect of disaster planning can include the creation of mobile response units in the form of fully stocked trailers that can turn into temporary command centers at the site of a natural disaster. Ready response teams can be prepared at a moment's notice to travel to a location to provide support and assistance, ranging from respectful handling of human remains to environmental cleanup.

What is a Natural Hazard Disclosure?

A natural hazard disclosure is a document given by the seller of a real estate property to the buyer, outlining whether the property is in an area known to be prone to natural disaster, such as flooding, earthquakes, hurricanes, or tornadoes. It is designed to protect buyers from unknowingly purchasing property that might incur damages from natural disasters simply because of the property's location in a naturally hazardous area. The disclosure document is usually prepared by a third party to the transaction to prevent disclosure fraud.

In the United States, there is no federal mandate that requires sellers to provide natural hazard disclosures to buyers during a transaction. Some states have created their own disclosure regulations, however. Among these are Alaska, California, Florida, Hawaii, Idaho, and Washington — all states at high risk for varying natural hazards. Other states have no natural hazard disclosure regulations and still operate under caveat emptor, which is also known as a buyer beware sale.

California in particular has one of the most comprehensive natural hazard disclosure laws. In 1998, the state legislature created a standardized disclosure form that detailed a variety of natural hazards known to the state. Sellers are required to report information on known fire zones; seismic zones, including landslide areas; floodplains; and earthquake fault zones.

The natural hazard disclosure usually indicates whether the buyer of a real property has legal permission to develop or alter the property in any way. It might also specify whether the property falls under special insurance requirements or whether the owner has a right to federal assistance following a natural disaster. In some cases, the natural hazard disclosure may not be enough to absolve a seller's responsibility in a property. Localities that require disclosure may also specify that if a seller is aware of a natural hazard that is not present on a standardized form, he still has an obligation to report it, as well as run any special reports or obtain any relevant maps that document the hazard.

Failure to report potential disaster areas is known as natural hazard disclosure fraud. If the seller is found to have willfully withheld information about natural hazards affecting a property in order to progress a sale, he may be held liable for damages made to the property during a natural disaster. Sellers are not usually liable for mistakes or omissions on natural hazard disaster forms if the information they obtained came from a public agency or a qualified expert and was reported in good faith. The term "qualified experts" for legal real estate purposes usually includes licensed contractors, geologists, engineers, and surveyors.

What is a Natural Disaster?

A natural disaster is an event with a natural, as opposed to human, cause that results in large-scale loss of life or damage to property. It could be related to weather, geology, biology or even factors outside the Earth. Examples are earthquakes, hurricanes, droughts and flooding. Disease epidemics are sometimes considered natural disasters, but may be put into a different category. In some cases, natural and human factors may combine to produce a disaster.

Weather and Climate

The planet's weather produces disasters on a fairly regular basis. Hurricanes, also known as cyclones, are among the most prominent natural disasters, occurring quite frequently in the warmer parts of the world. They begin as low-pressure areas over warm oceans and grow into giant storms, hundreds of miles in diameter, that persist for several days. Their paths are fairly predictable, which gives forewarning to areas likely to be affected. Even so, they can cause substantial loss of life and millions of dollars of damage.=

Tornadoes cover much smaller areas and may only last for a few minutes, but during that time they can cause devastation, due to the extremely high wind speeds. In the worst category of tornado, the winds can reach 300 mph (482.8 km/h). This is enough to completely destroy brick buildings and hurl automobiles through the air. Fortunately, tornadoes of this severity are relatively rare.

Much less spectacular, but much deadlier, are droughts. Many people living in drier parts of the world often rely on seasonal rainfall to grow crops. From time to time, however, the rains fail to arrive, due to fluctuations in the Earth's climate. Prolonged lack of rainfall leads to crop failure, starvation and malnutrition, claiming millions of lives in some cases.

Excessive rainfall can cause flooding, which may result in large numbers of people losing their homes, crops being ruined or rivers bursting their banks and causing death and destruction. Flooding often results from the heavy rain that accompanies hurricanes, compounding the damage. Prolonged heavy rain can also cause disastrous landslides and mudslides.

Geological Disasters

Earthquakes are among the most destructive natural disasters. They occur at or near fault lines: cracks in the Earth's crust that mark the boundary between two different sections. When these sections move relative to one another, the resulting vibration produces an earthquake. Although old fault lines can occasionally produce small earthquakes, the most destructive ones occur in geologically active areas, near the boundaries of continental plates.

When an earthquake happens under the ocean, it may produce a tsunami, or huge wave, that travels rapidly outwards from its source. A tsunami can cause massive destruction to coastlines hundreds of miles away. These enormous waves can also be produced by landslides. There is concern that a future massive landslide on an unstable part of the island of La Palma in the Canary Isles may trigger a huge tsunami that would travel westward across the Atlantic, causing devastation to the east coast of America. This is, however, disputed by some geologists.

Volcanoes are related to earthquakes in that they occur in geologically active zones around plate boundaries. In these areas, magma, under pressure, is close to the surface and can erupt as lava. This may be of the "runny" type, which flows out relatively quietly and follows well-defined channels. Alternatively, it may be of the "sticky" type; this can solidify at the top of a volcano, causing pressure to build up until there is an explosive eruption.

Much of the death and destruction from this kind of natural disaster results from volcanic ash, which can fill the air, making breathing impossible, and accumulate on rooftops, causing buildings to collapse under the weight. Pyroclastic flow is another major hazard associated with some volcanic eruptions. This consists of a mixture of hot gas, ash and rock fragments racing at high speed from the source of the eruption and destroying everything in its path.

The biggest known volcanic natural disaster may have occurred in prehistoric times. Some scientists believe that the eruption of Mt. Toba in Indonesia over 73,000 years ago may have killed off most of the human species, leaving behind only 1,000 - 10,000 breeding pairs. This phenomenon, called a population bottleneck, has been confirmed through genetic analysis.

Diseases and other Biological Threats

Disasters that affect humans can be caused by other organisms. Historically, there have been a number of outbreaks of serious disease that have affected huge areas and claimed many lives. One such example was the "black death," a form of bubonic plague that affected much of Europe during the Middle Ages and may have reduced the population by 30-60%.

The Spanish flu epidemic of 1918-1919 is thought to have killed about 50 million people — more than World War I, which had occurred just before. The threat to humans from the emergence of a new and lethal strain of the influenza virus remains. Organisms that do not cause disease can occasionally cause disasters. Locusts, for example, can form huge swarms that can devour many acres of crops in a very short time, sometimes causing famine.

Disaster from Space

Although there are no documented cases of human fatalities resulting from meteorites or asteroids, they do pose a threat. The risk of a major impact in the near future is considered to be very low, but looking further ahead, the probability is greater. The Earth has certainly experienced such events in the past, as evidenced by obvious craters in many parts of the world. In 1908, what is thought to be a large meteorite or comet fragment devastated a vast area of the Tunguska region of Siberia. Fortunately, the area was uninhabited and there were no known human casualties.

Natural and Human Factors

Some natural disasters result from a combination of natural and human factors. For example, the primary cause of a disease epidemic may be a natural microorganism, but its spread might be encouraged by human behavior and activities, such as living in close proximity to infected animals or rapid international travel. Human activities may also have contributed in a major way to some famines. For example, bad agricultural policies are widely thought to have been at least partly to blame for the great famine of 1958-61 in China, during which 30 million people died.

Much less spectacular, but much deadlier, are droughts. Many people living in drier parts of the world often rely on seasonal rainfall to grow crops. From time to time, however, the rains fail to arrive, due to fluctuations in the Earth's climate. Prolonged lack of rainfall leads to crop failure, starvation and malnutrition, claiming millions of lives in some cases.

Excessive rainfall can cause flooding, which may result in large numbers of people losing their homes, crops being ruined or rivers bursting their banks and causing death and destruction. Flooding often results from the heavy rain that accompanies hurricanes, compounding the damage. Prolonged heavy rain can also cause disastrous landslides and mudslides.

Geological Disasters

Earthquakes are among the most destructive natural disasters. They occur at or near fault lines: cracks in the Earth's crust that mark the boundary between two different sections. When these sections move relative to one another, the resulting vibration produces an earthquake. Although old fault lines can occasionally produce small earthquakes, the most destructive ones occur in geologically active areas, near the boundaries of continental plates.

When an earthquake happens under the ocean, it may produce a tsunami, or huge wave, that travels rapidly outwards from its source. A tsunami can cause massive destruction to coastlines hundreds of miles away. These enormous waves can also be produced by landslides. There is concern that a future massive landslide on an unstable part of the island of La Palma in the Canary Isles may trigger a huge tsunami that would travel westward across the Atlantic, causing devastation to the east coast of America. This is, however, disputed by some geologists.

Volcanoes are related to earthquakes in that they occur in geologically active zones around plate boundaries. In these areas, magma, under pressure, is close to the surface and can erupt as lava. This may be of the "runny" type, which flows out relatively quietly and follows well-defined channels. Alternatively, it may be of the "sticky" type; this can solidify at the top of a volcano, causing pressure to build up until there is an explosive eruption.

Much of the death and destruction from this kind of natural disaster results from volcanic ash, which can fill the air, making breathing impossible, and accumulate on rooftops, causing buildings to collapse under the weight. Pyroclastic flow is another major hazard associated with some volcanic eruptions. This consists of a mixture of hot gas, ash and rock fragments racing at high speed from the source of the eruption and destroying everything in its path.

The biggest known volcanic natural disaster may have occurred in prehistoric times. Some scientists believe that the eruption of Mt. Toba in Indonesia over 73,000 years ago may have killed off most of the human species, leaving behind only 1,000 - 10,000 breeding pairs. This phenomenon, called a population bottleneck, has been confirmed through genetic analysis.

Diseases and other Biological Threats

Disasters that affect humans can be caused by other organisms. Historically, there have been a number of outbreaks of serious disease that have affected huge areas and claimed many lives. One such example was the "black death," a form of bubonic plague that affected much of Europe during the Middle Ages and may have reduced the population by 30-60%.

The Spanish flu epidemic of 1918-1919 is thought to have killed about 50 million people — more than World War I, which had occurred just before. The threat to humans from the emergence of a new and lethal strain of the influenza virus remains. Organisms that do not cause disease can occasionally cause disasters. Locusts, for example, can form huge swarms that can devour many acres of crops in a very short time, sometimes causing famine.

Disaster from Space

Although there are no documented cases of human fatalities resulting from meteorites or asteroids, they do pose a threat. The risk of a major impact in the near future is considered to be very low, but looking further ahead, the probability is greater. The Earth has certainly experienced such events in the past, as evidenced by obvious craters in many parts of the world. In 1908, what is thought to be a large meteorite or comet fragment devastated a vast area of the Tunguska region of Siberia. Fortunately, the area was uninhabited and there were no known human casualties.

Natural and Human Factors

Some natural disasters result from a combination of natural and human factors. For example, the primary cause of a disease epidemic may be a natural microorganism, but its spread might be encouraged by human behavior and activities, such as living in close proximity to infected animals or rapid international travel. Human activities may also have contributed in a major way to some famines. For example, bad agricultural policies are widely thought to have been at least partly to blame for the great famine of 1958-61 in China, during which 30 million people died.

How Plate Movement Affects Earthquakes, Tsunamis & Volcanic Eruptions

Earthquakes, volcanoes, and tsunamis are all dangerous natural disasters, but they also have something else in common tectonic plate movement. In this lesson, you'll see how these seemingly different events actually come from similar geological beginnings.

Moving Puzzle Pieces

The top layer of Earth is an interesting place. Also known as the 'crust,' this thin, solid layer is much more than meets the eye. If Earth were an apple, the skin of that apple could represent the crust in terms of thickness and location. But unlike an apple skin, Earth's crust isn't one large piece covering the entire planet. Instead, it's broken up into many different pieces called tectonic plates that fit together like a large puzzle.

Also unlike the apple, underneath the solid crust is not a deliciously crispy interior. Instead, directly below the crust, we have a thick liquid layer called the mantle. Because it is liquid, the mantle flows and moves around, which moves the plates sitting on top like pieces of ice on a pond.

When the plates get moved around, they wreak havoc because they crash into, and pull apart from, and rub against each other. And as you can imagine, these interactions can do some pretty serious damage. On Earth, these tectonic events result in dangerous natural disasters around the world, like earthquakes, volcanoes, and tsunamis.

Earthquakes

Earthquakes can and do happen anywhere in the world, but the majority of them occur in a region known as 'The Ring of Fire.' As you'll learn a little later in this lesson, this is also where most of the world's volcanoes are found and where the name comes from.

The reason so many earthquakes occur in these areas is because this is where many of Earth's tectonic plates come together. Earthquakes begin deep underground along plate boundaries. Tension and pressure build up as the plates slide past and bump into each other and sometimes even stick together. Although the plate boundaries themselves may be stuck, the plates keep moving and pulling. Eventually, the pulling becomes too much and the plates suddenly break free from each other, causing an earthquake!

You can think of an earthquake like a game of tug-of-war. If you and your friend are both pulling on opposite ends of the rope and suddenly your friend lets go, all of that tension quickly leaves the rope and down onto the ground you go! An earthquake is very much the same - the plates get stuck together as they move, building up tension. Suddenly, the plates slip past each other and break free, sending that built-up tension through the ground in all directions.

Volcanoes

Plate movement can also cause other natural phenomena, like volcanoes. Under the right conditions, when plates are pushed together or pulled apart, volcanoes are created, which is why they tend to occur in the same place as earthquakes.

When tectonic plates spread apart from each other, hot magma rises up and fills the space between. As it cools, it forms new land, either on the continents or on the seafloor, depending on where the plates are located.

When the plates come together, one of them may get pulled under the other one, getting recycled back into Earth's interior. During this process, called subduction, the piece of crust getting pulled under is melted and turned into magma the very magma that erupts from a volcano!

CHAPTER 1

COULD A SUMATRA-LIKE MEGATHRUST EARTHQUAKE OCCUR IN THE SOUTH RYUKYU SUBDUCTION ZONE?

Jing-Yi Lin, Jean-Claude Sibuet, Shu-Kun Hsu and Wen-Nan Wu

BACKGROUND

The Mw 9.3 Sumatra-Andaman earthquake on December 26, 2004 was the second largest earthquake recorded during the last century. The earthquake triggered a significant uplift of the seafloor and induced a series of devastating tsunamis that attacked the coasts of most landmasses bordering the Indian Ocean. The tsunami run-up was up to 30 m high and over 230,000 people were killed in 14 countries (Paris et al.2007). It is one of the deadliest natural disasters and is probably the most extensively analyzed earthquake-tsunami event in history. Since the occurrence of the Sumatra earthquake, the potential for the generation of tsunamis along subduction systems all around the world have been studied and several high risk areas have been identified (Cummins2007; Gusiakov2005; Liu et al.2007).

METHODS

Empirical laws for the possible relation between the occurrences of large subduction zone earthquakes and the related tectonic parameters, such as the convergence rate, slab age, trench sediment thickness, bathymetric features, and plate motion, have been widely investigated (Kanamori,1979; Pacheco et al.,1993; Peterson and Seno,1984; Ruff and Kanamori1983; Stein and Okal2007; Uyeda and Kanamori1979). A number of studies found that a faster convergence rate, older plate age, and higher upper plate absolute motion enhanced the triggering of the largest subduction zone events (Peterson and Seno1984; Ruff and Kanamori1983; Uyeda and Kanamori1979), even though the validity of such correlations remains controversial (Pacheco et al.1993). In less than one decade, several large subduction earthquakes occurred worldwide. Based on a more complex seismicity and subduction parameters catalog, Heuret et al. (2011) suggested that events with Mw $\geq$ 8.5 preferentially occurred in the vicinity of slab edges where the upper plate is continental and the back-arc strain neutral. Consequently, the occurrence of large subduction earthquakes seems to favor specific tectonic conditions and the comparison of the geodynamic environments of worldwide subduction zones with those of seismogenic areas where megathrust events occurred may help to identify high-risk potential zones. The Taiwan-Ryukyu margin, which is located at the westernmost corner of the western Philippine Sea plate, may present similarities

Citation: Lin et al.: Could a Sumatra-like megathrust earthquake occur in the south Ryukyu subduction zone? Earth, Planets and Space 2014 66:49. doi:10.1186/1880-5981-66-49

with the Himalaya-Sumatra zone. In this study, we discuss the geophysical parameters, including plate tectonic framework, tectonic stress pattern, bathymetric characteristics, the spatial distributions of the oceanic fracture zones, and the gravity anomalies of these two subduction systems and suggest that a Sumatra-like earthquake may occur in the south Ryukyu fore-arc area.

RESULTS

Similar Geodynamic Contexts for the Sumatra and Ryukyu Subduction Zones

Plate Tectonic Framework

In the Sumatra region, the Indo-Australian plate is subducting northwards beneath the Sunda plate at a rate of approximately 50 mm/yr (Briggs et al.2006; Gahalaut and Gahalaut2007; Paul et al.2001), and it is colliding with the Eurasian continent to the west (Curray2002; Paul et al.2001). The age of the oceanic plate in the north Sumatra-Andaman area is about 60 to 80 Ma (Müller et al.2008). The collision of the India plate with Asia results in the formation of the Himalaya Mountains and in the clockwise bending of the subduction front. From east to west along the Sunda and Andaman trenches, the plate convergence vector becomes oblique (Bock et al.2003; Michel et al.2001) (Figure 1a) and the slip is partitioned between the Sumatra trench and the right-lateral Sumatra fault system (Curray1989; Fitch1972; Genrich et al.2000; McCaffrey1991; McCaffrey et al.2000; Prawirodirdjo et al.1997). To the north, the Sumatra fault system is transformed into the spreading center of the Andaman Sea (Curray1989).

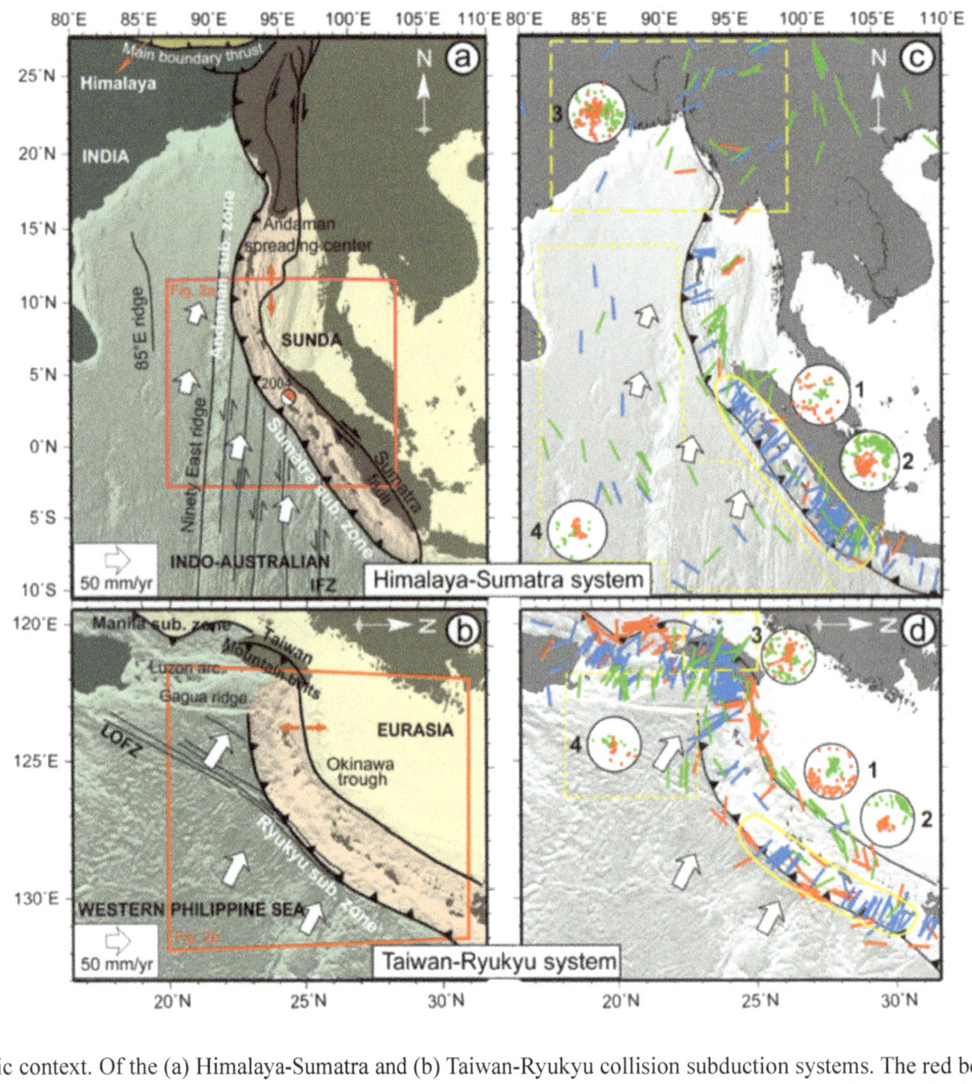

Figure 1: Tectonic context. Of the (a) Himalaya-Sumatra and (b) Taiwan-Ryukyu collision subduction systems. The red beach ball shows the location and focal mechanism of the Mw 9.3 Sumatra-Andaman earthquake on December 26, 2004. The red rectangle shows the position of Figure 2. LOFZ, Luzon-Okinawa fracture zone; IFZ, Investigator fracture zone. **(c, d)** The color bars show the stress orientations obtained from different fault types, which were retrieved from the World Stress Map (WSM) (http://dc-app3-14.gfz-potsdam.de/) (Heidbach et al.2010). Red indicates normal faulting, green indicates strike slip, and blue indicates thrust faulting. Numbered circles show the orientations of the *P*-axis (red dot) and *T*-axis (green dot) in an equal-area projection of the lower hemispheres of the focal spheres from the global centroid-moment-tensor (CMT) catalog from 1976 to December 26, 2004 for the Sumatra subduction system and from 1976 to December 31, 2011 along the Ryukyu subduction system. The patterns for groups 1 and 2 were calculated from the thrust and normal faulting earthquakes along the subduction zone shown by the yellow contour, respectively. Group 3 used the events in the collision zones (yellow dashed contour), and group 4 used the events in the oceanic domain of the subducting plate (yellow dotted contour). Relative plate motions, shown by the white arrows, are based on the MORVEL model from the plate motion calculator (http://www.unavco.org/community_science/science-support/crustal_motion/dxdt/model.html) (DeMets et al.2010).

Likewise, the Philippine Sea plate (PHS) subducts beneath the Eurasian plate (EU) along the Suruga-Nankai trough and Ryukyu trench at a rate of 80 to 85 mm/year in a 300 to 310° N direction (Yu et al.1997) (Figure 1b). The age of the west Philippine basin is roughly 30 to 60 Ma (Hall et al.1995), which is on the same order as that of the north Sumatra-Andaman area. The Luzon volcanic arc, which belongs to the PHS, collides westward with the EU margin and creates the Taiwan Mountains. Thus, the deformation of the Chinese passive margin by the Luzon arc indenter produced a clockwise bend of the southern Ryukyu arc (Letouzey and Kimura1985; Sibuet et al.1987), and the subduction of the PHS, which occurs at an oblique angle of about 40° E of Taiwan (Lallemand et al.1999). Similar to the Sumatra fault system, the median tectonic line (MTL) is a right-lateral shear fault (Kimura1996), which follows the onland arc volcanoes and is prolonged southwestward by the Okinawa trough back-arc basin.

Tectonic Stress Pattern

The stress orientations retrieved from the World Stress Map (WSM) (http://dc-app3-14.gfz-potsdam.de/) (Heidbach et al.2010) as well as the *P*- and *T*-axes distribution calculated from focal mechanisms extracted from the Global centroid-moment-tensor (CMT) catalog during the period between 1976 and December 25, 2004 for the Sumatra area and between 1976 and the end of 2011 for the Ryukyu area are shown in Figure 1c,d, respectively. On the basis of the stress regime and epicenter distribution, the focal mechanisms of the Sumatra and Ryukyu subduction zones were further divided into groups in terms of *P*- and *T*-axes orientations. Groups 1 and 2 represent the thrust and extensional events that occurred along the subduction zone (yellow contour in Figure 1c,d); group 3 illustrates the earthquakes located along the collisional front (yellow dashed contour in Figure 1c,d); and group 4 depicts the events occurring within the oceanic plate (yellow dotted contour in Figure 1c,d). The spatial pattern of the stress regime along the Himalaya-Sumatra collision-subduction zone is in good agreement with that along the Taiwan-Ryukyu collision-subduction zone. Specifically, thrust events with a *P*-axis perpendicular to the trend of the arcs dominate in the subduction areas (group 1 in Figure 1c,d), whereas those with a *P*-axis sub-parallel to the relative plate motion vector occur around the colliding areas (group 3 in Figure 1c,d). Almost all the extensional earthquakes with a trench-parallel *P*-axis occur in the seaward portion of the collision-subduction system, east of approximately 100° E for the Sumatra subduction system and east of approximately 126° E for the Ryukyu subduction system (red bars in Figure 1c,d). These extensional events are characterized by a vertical *P*-axis and a scattered *T*-axis distribution probably resulting from changes in the trench geometry (group 2 in Figure 1c,d) (Engdahl et al.2007; Tanaka et al.2006). Besides, it is noticeable that for the strike slip and thrust earthquakes occurring in the oceanic domain of the subducting plate, their *T*-axis shows a trench perpendicular direction whereas the *P*-axis is proximately sub-parallel to the relative plate motion (group 4 in Figure 1c,d). Such a stress pattern shows that large portions of the Indo-Australian and PHS are widely affected by the NW-SE compressive deformation due to the collision processes in spite of their large distance from the orogenic belts (Chamot-Rooke et al.1993; Lin et al.2013).

Bathymetric Characteristics, Oceanic Fracture Zones, and Gravity Anomalies

The large-scale features of the oceanic plate observed in the Indo-Australia plate are the Ninety East ridge and the Investigator Fracture Zone (IFZ) (98° E) (Larson et al.1978) (Figure 1a). In between these features, a set of roughly N-S sub-parallel fracture zones were identified from bathymetric, gravity, magnetic, and seismic data (Barckhausen2006; Delescluse and Chamot-Rooke2007; Deplus et al.1998; Lin et al.2009; Sibuet et al.2007) (Figures 1a and2a). Similarly, several major NE-SW-oriented fracture zones have been defined using bathymetric and magnetic data in the western PHS (e.g. Deschamps and Lallemand2002; Hsu et al.2013). Among them, the Luzon-Okinawa fracture zone (LOFZ) is the largest of these features (Figure 1b and2c) (Hsu et al.2013). Based on seismic profile interpretations and seismicity distribution, all these fracture zones subduct below the EU, which leads to kinks in the trend of the deformation front (Deplus et al.1998; Hsu et al.2013; Kopp and Kukowski2003; Lange et al.2010), and this is thought to influence the rupture behavior of major earthquakes (Abercrombie and Ekström2001; Bilek et al.2003).In both subduction systems, the free-air gravity anomaly generally displays positive values for the oceanic plates and portions of fore-arcs (Figure 2b,d). In the Ryukyu subduction zone, the entire fore-arc has a low anomaly value expect in the area where the LOFZ intersects the Ryukyu trench (approximately 125° E to 129° E; 23° N to 26.5° N) (Figure 2d). Similarly, in the Sumatra subduction system, a high gravity anomaly zone exists behind the trench (approximately 93° E to 97° E and 2.5° N to 6° N), where series of fracture zones intersect the Sumatra trench (Figure 2b). The size of the two positive free-air gravity anomalies is similar (about 450 km long).

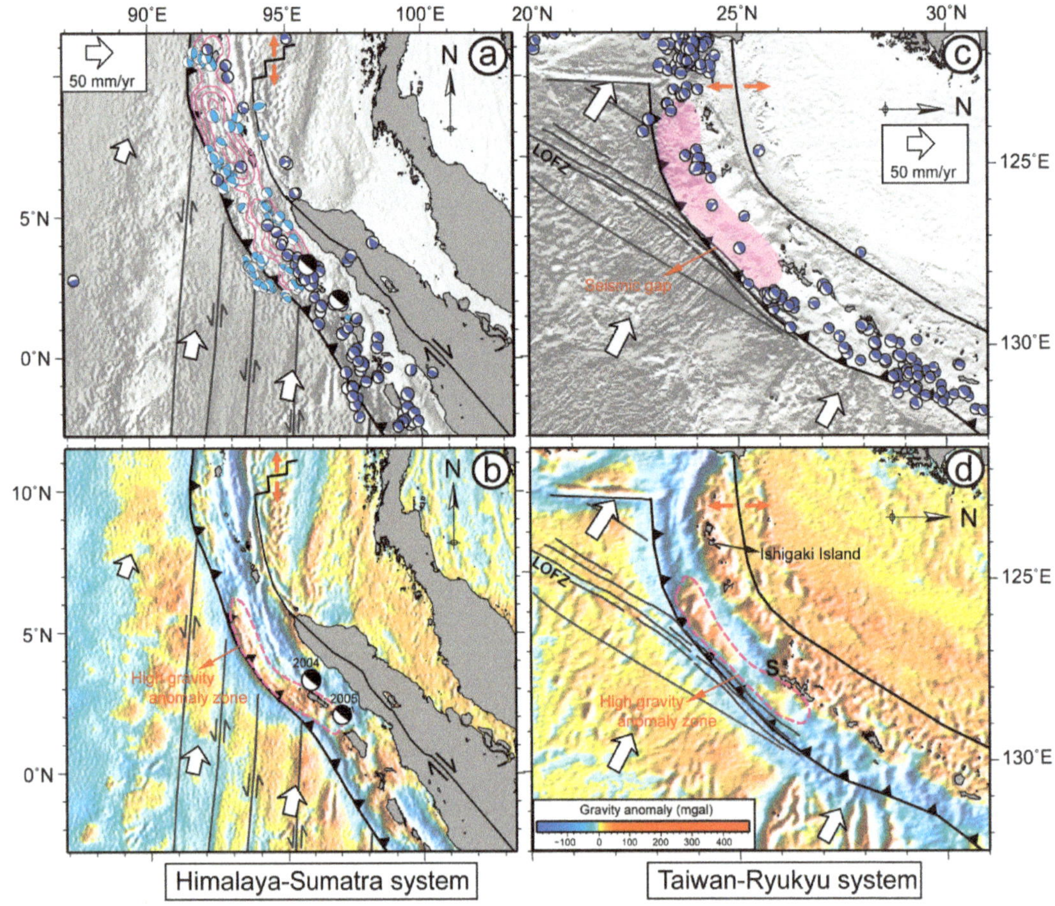

Figure 2: Focal mechanisms of the thrust-type earthquakes extracted from the Global CMT catalog. (a) Along the Sumatra subduction system from 1976 to 25 December 2004 and **(c)** along the Ryukyu subduction system from 1976 to December 31, 2011 are shown. Deep blue beach balls show focal mechanisms. Co-seismic slip contours every 5 m of the 2004 great Sumatra earthquake shown in pink in (a) are from Chlieh et al. (2007). Pink light areas show the aseismic zones. **(b, d)** The free-air gravity anomaly. Purple dashed contours show the high free-air gravity anomalies in the fore-arc areas. The letter 'S' shows the possible mainshock area if a Sumatra-like earthquake were to occur in the Ryukyu fore-arc. The black focal mechanisms in (a) and (b) show the positions of the 2004 and 2005 mainshocks.

DISCUSSION

The Role Played by the Oceanic Fracture Zones and their Orientation

Müller and Landgrebe (2012) have proposed that the fracture zones are characterized by continuous, uplifted ridges, which could be at the origin of strong, persistent coupling at the plate interface. Therefore, the occurrence of large subduction earthquakes (magnitude >8) is strongly biased towards regions at the intersections of oceanic fracture and subduction zones. This observation is supported by the occurrences of the 2004 Sumatra and 2005 Nias earthquakes, which were both located in areas where the IFZ sub-ridges enter the Sumatra trench (Figure 2a,b). As shown by Figures 1 and2 and in the previous section, the presence of fracture zones is apparent on both the Indo-Australian and PHS subducting plates. When entering in the subduction system, these fracture zones generally change the trench morphology, which is suggestive of a high resistance of the fracture zone against the subduction (Hsu et al.2013; Lin et al.2009). In addition, a trench-parallel high gravity anomaly zones are located at the intersection of the oceanic fracture ridges and the subduction zone, on the landward side of the trench wall for both subduction systems. These gravity anomaly high patterns could be the result of the presence of oceanic fracture zone material stuck beneath the accretionary wedge area (Hsu et al.2013; Lin et al.2009), which also suggests that the high resistance of fracture zones could cause a strong coupling environment (Hsu2001) and block the subduction system. It is worthnoting that epicenters of the 2004 Sumatra and 2005 Nias earthquakes were located on the eastern border of the trench-parallel high anomaly zone at the boundary between the positive and negative gravity anomalies (Figure 2b). Consequently, if this epicenter location of the Sumatra earthquakes is transferred to the Ryukyu subduction system, a possible location for a Sumatra-like earthquake in the Ryukyu area would be in area S (Figure 2d), at the junction of positive and negative gravity anomalies along the boundary of the trench-parallel high anomaly. In addition, the size of the positive free-air gravity anomaly in the both areas is similar (about 450 km long). We therefore infer that the magnitude of a potential earthquake occurring in the Ryukyu fore-arc area could be as large as the 2004 Sumatra earthquake under the premise that the whole potential source area would rupture simultaneously.

Otherwise, the angle between fracture zone directions and the plate motion are different for the two subduction zones and amount to about 120° and 0° for the Ryukyu and Sunda subduction zones, respectively (Figures 1 and2). Based on the distribution of the principle stress axes, the occurrence of earthquakes seems to be mostly controlled by the regional stress regime, i.e., the relative plate motion and the slab-pull mechanism that are caused by the plate collision and subduction processes (Chamot-Rooke et al.1993; Lin et al.2013; Yue et al.2012). It means that even the presence of fracture zones in a subduction system could result in slab coupling in the fore-arc area; their orientation seems to have a relatively small influence on the tectonic stress regime, which greatly affects the seismogenic characteristics of a subduction system. In addition, Müller and Landgrebe (2012) show that the dimension of topographic highs of the oceanic subducting plate is the main factor controlling the megathrust seismogenic potential of a subduction system. What this means is that the fracture zones characterized by laterally continuous ridges and high degrees of structural integrity could cause a strong coupling effect; small volcanic chains present a relatively fragile internal structure, which results in a weak coupling. However, the impact induced by the fracture zone orientation appears to have been neglected. Therefore, we suggest that the difference of fracture zone directions with respect to the two subduction systems may not influence the subduction processes as well as change the seismogenic characteristics.

Pre-Seismic Distribution of Earthquakes

In Figure 2a,b, the distribution of earthquakes along the Sumatra and Ryukyu subduction is extracted from the Global CMT catalog. Before the 2004 Sumatra earthquake, the thrust earthquakes with a trench-normal *P*-axis extended over the fore-arc area to the southeast of the 2004 mainshock source area (deep blue beach balls in Figure 2a). In contrast, only a few earthquakes were observed in the fore-arc region to the north, which implies that the slab interface north of the mainshock area has been locked and seismic strain has accumulated (Engdahl et al.2007). During the occurrence of the 2004 mainshock, the accumulated seismic strain was released and the rupture propagated northward (pink contours in Figure 2a) (Chlieh et al.2007) inducing a huge number of thrust-type events along the trench.

Along the Ryukyu subduction zone, thrust events with a trench-normal *P*-axis are numerous in the vicinity of the trench and fore-arc region, east of the high free-air gravity anomalies (approximately east of 128° E) (Figure 2c,d). However, the seismicity disappears west of this boundary. This spatial distribution of seismic activity is in accord with the pre-seismic pattern of the 2004 Sumatra earthquake (Figure 2a): the thrust earthquakes with a trench-normal *P*-axis occurred frequently on the oceanward side of the mainshock area, whereas few thrust-type earthquakes occurred in the cotinentward side. The area located west of 128° E with a low seismicity distribution could be locked between the interface of the PHS and EU along the high free-air gravity anomaly zone and may correspond to a potential rupture area (pink light area in Figure 2c). Therefore, if a Sumatra-like earthquake occurs in the vicinity of 128° E (area S), as was mentioned in the previous section, we would expect to see a westward rupture propagation (pink light area in Figure 2c) and the occurrence of thrust-type earthquakes along the present-day aseismic zone in the area west of 128° E.

Tsunamogenic Potential

In April 1771, a subduction earthquake generated a very large tsunami that struck the south Ryukyu Islands and killed approximately 12,000 people (Ando et al.2009; Matsumoto et al.2009; Nakamura2006,2009). Reef boulders of building size were transported by the tsunami to beaches along the east and southeast coasts of Ishigaki Island (Goto et al.2010), which suggests that the source of the tsunami was located east and southeast of Ishigaki Island (Figure 2d). As discussed in the previous section, if a Sumatra-like earthquake occurs along the Ryukyu subduction zone in the vicinity of 128° E (area S), there will be a westward propagation of the rupture. This westward rupture movement could be a possible tsunami source similar to the 2004 Sumatra earthquake. Moreover, based on seismic profile interpretations, Hsu et al. (2013) demonstrated the presence of a splay fault system within the trench-parallel high gravity anomaly area of the Ryukyu subduction zone, where a highly resistant subduction due to the integration of significantly developed fracture zones or strong plate coupling is expected (Ando et al.2009). Similar splay faults have been reported along the Sumatra subduction zone (Sibuet et al.2007). They branch at the plate interface, present steep dipping angles near the seafloor, and may have generated the devastating tsunamis. As a result, if a Sumatra-like earthquake occurs in the south Ryukyu fore-arc area, this may also trigger a tsunami similar to that during the 2004 Sumatra earthquake and induce serious damage.

Large Historical Interplate Earthquakes along the Subduction System

Although the tectonic similarity between the Sumatra and Ryukyu subduction systems was denoted, these regions have differences as regards to the occurrence of large historical interplate earthquakes, according to previous studies. Magnitude 8 to 9 class interplate earthquakes occurred frequently in the region of the Sumatra trench, such as the 1797 earthquake (Mw 8.7 to 8.9), the 1833 earthquake (Mw 8.8 to 9.2), and the 1861 earthquake (Mw 8.5). In contrast, the risk of great earthquakes and tsunamis was assumed to be low in the Ryukyu trench because the interplate coupling appeared to be weak and great interplate earthquakes (Mw >8.0) had not been recorded historically for about 300 years. However, Nakamura (2013) revealed that the 1771 Yaeyama earthquake (Mw 8.5 from the tsunami height distribution) in the south Ryukyu subduction zone and the 1911 Kikaijima earthquake (Mw 8.0) in the north-central Ryukyu subduction zone were interplate earthquakes. In addition, two historical tsunamis occurred in 1768 and 1791 on Okinawa Island, and it has been suggested that these were induced by two Mw 8 interplate earthquakes that occurred around the Ryukyu trench on the basis of the source fault model for these two tsunami events (Nakamura2013). Consequently, the lack of large historic earthquakes along the Ryukyu subduction zone may be due to incomplete records in old documents. Meanwhile, a much higher concentration of oceanic fracture zones can be observed in the Indo-Australian plate than in the PHS. As the subduction of fracture zones can cause larger earthquakes (Müller and Landgrebe2012), this suggests that there may be a higher possibility for the occurrence of large earthquakes along the Sumatra subduction area.

Comparison with Other Studies

The tectonic similarity between the Sumatra and Ryukyu subduction systems has been already raised by Hsu and Sibuet (2005), and they suggested that there was a potential megathrust earthquake in the Nankai area. Based on more detailed analyses, we found a closer tectonic similarity between the southwestern Ryukyu and Sumatra subduction zoned: the area S in Figure 2d appears to be a possible location for a Sumatra-like earthquake along the Ryukyu subduction zone. Based on a geodetic analysis, Hsu et al. (2012) suggested that the plate interface of the southernmost Ryukyu subduction zone is fully locked and a potential large earthquake (Mw 7.5 to 8.7) and tsunami might occur in the region. However, if the most southwestern end of the Ryukyu subduction zone was totally locked, the overriding EU plate should be closely contacted with the subducting PHS and largely affected by a compressional stress regime. With this hypothesis, it is difficult to explain the continued opening of the Okinawa trough and the southern migration of the Ryukyu arc already evidenced by geological, geodetic, and seismic data (Sibuet and Hsu1997; Sibuet et al.2007). Therefore, we propose that the strong coupling area is located at the intersection of the oceanic fracture ridges and the subduction zone, which block the western movement of the subducting plate resulting in low seismic activity in the southernmost Ryukyu subduction zone (approximately 125° E to 129° E; 23° N to 26.5° N).

CONCLUSIONS

In this study, we evaluated the seismic risk along the Ryukyu subduction zone by examining the similarity in tectonic environments between the Himalaya-Sumatra and Taiwan-Ryukyu collision-subduction systems. We found that both of the two systems share the four following common geodynamic contexts.

- The Indo-Australian plate is subducting beneath the Sunda plate and the Indian portion of the plate collides with the Eurasian continent resulting in the formation of the Himalaya Mountains. Similarly, the Philippine Sea plate subducts beneath the EU and the Luzon volcanic arc creating the Taiwan orogen.
- Owing to the collisional process, the subduction fronts started to bend clockwise resulting in an oblique subduction and associated partitioning.
- Oceanic fracture zones with high topographic features exist on both subducting oceanic plates. The intersections of such fracture zones with the subduction systems are associated with trench-parallel gravity highs where the foci of large earthquakes, such as the 2004 Sumatra and the 2005 Nias earthquakes, are located. The locations of these mainshocks may be related to the strong plate coupling resulting from the high resistance of the fracture zones.
- The spatial distribution of the earthquakes along the Ryukyu arc-trench system is very similar to the pre-seismic activity pattern of the 2004 Sumatra earthquake: the thrust earthquakes with a trench-normal *P*-axis occur frequently in the oceanward side of the mainshock area, whereas few thrust-type earthquakes occur in the continentward side.

From the similarities in geodynamic contexts between the Sumatra and Ryukyu subduction zones, we suggest that a potential Sumatra-like earthquake may occur along the border of the high gravity anomaly zone located in the fore-arc area of the Ryukyu subduction zone and the present low seismicity area in its western part should correspond to the possible co-seismic rupture area. Such a potential earthquake could generate a risky tsunami that would threaten many countries in Eastern Asia.

ACKNOWLEDGEMENTS

We thank two anonymous reviewers for their careful reviews which help a lot to improve this manuscript. Figures were prepared with the Generic Mapping Tool (GMT) software (Wessel and Smith1998). This research was supported by the Taiwan Earthquake Research Center (TEC) and funded through Ministry of Science and Technology (MOST) with grant numbers MOST-103-3113-M-008-001 and NSC-102-2116-M-008-024. The TEC contribution number for this article is 00104.

AUTHORS' CONTRIBUTIONS

All authors have been involved in drafting the manuscript or revising it critically for important intellectual content. JYL has made substantial contributions to conception of data. All authors read and approved the final manuscript.

REFERENCES

1. Abercrombie RE, Ekström G: Earthquake slip on oceanic transform faults. *Nature* 2001, 410: 74–77. 10.1038/35065064
2. Ando M, Nakamura M, Matsumoto T, Furukawa M, Tadokoro K, Furumoto M: Is the Ryukyu subduction zone in Japan coupled or decoupled?—the necessity of seafloor crustal deformation observation. *Earth Planets Space* 2009, 61: 1031–1039.
3. Barckhausen U: The segmentation of the subduction zone offshore Sumatra: relations between upper and lower plate. *Eos Trans Amer Geophys Union* 2006, 87: U53A-U0029A. Fall Meet. Suppl., Abstract Fall Meet. Suppl., Abstract
4. Bilek SL, Schwartz SY, DeShon HR: Control of seafloor roughness on earthquake rupture behavior. *Geology* 2003, 31: 455–458. 10.1130/0091-7613(2003)031<0455:COSROE>2.0.CO;2
5. Bock Y, Prawirodirdjo L, Genrich JF, Stevens CW, McCaffrey R, Subarya C, Puntodewo SSO, Calais E: Crustal motion in Indonesia from global positioning system measurements. *J Geophys Res* 2003, 108: 2367. doi:10.1029/2001JB000324
6. Briggs RW, Sieh K, Meltzner AJ, Natawidjaja D, Galetzka J, Suwargadi B, Hsu YJ, Simons M, Hananto N, Suprihanto I, Prayudi D, Avouac JP, Prawirodirdjo L, Bock Y: Deformation and slip along the Sunda Megathrust in the great 2005 Nias-Simeulue earthquake. *Science* 2006, 311: 1897–1901. 10.1126/science.1122602
7. Chamot-Rooke N, Jestin F, De Voogd B: Intraplate

shortening in the central Indian ocean determined from a 2100-km-long north–south deep seismic reflection profile. *Geology* 1993, 21: 1043–1046. 10.1130/0091-7613(1993)021<1043:ISITCI>2.3.CO;2

8. Chlieh M, Avouac JP, Hjorleifsdottir V, Song TRA, Ji C, Sieh K, Sladen A, Hebert H, Prawirodirdjo L, Bock Y, Galetzka J: Coseismic slip and afterslip of the great Mw 9.15 Sumatra-Andaman earthquake of 2004. *Bull Seismol Soc Am* 2007, 97: S152-S173. 10.1785/0120050631
9. Cummins PR: The potential for giant tsunamigenic earthquakes in the northern Bay of Bengal. *Nature* 2007, 449: 75–78. 10.1038/nature06088
10. Curray JR: The Sunda Arc: a model for oblique plate convergence. *Neth J Sea Res* 1989, 24: 131–140. 10.1016/0077-7579(89)90144-0
11. Curray JR: *Tectonics and history of the Andaman Sea region*. Conference on Continent-Ocean Interactions within the East Asian Marginal Seas. abstract, Chapman; 2002.
12. Delescluse M, Chamot-Rooke N: Instantaneous deformation and kinematics of the India-Australia Plate. *Geophys J Int*2007, 168: 818–842. 10.1111/j.1365-246X.2006.03181.x
13. DeMets C, Gordon RG, Argus DF: Geologically current plate motions. *Geophys J Int* 2010, 181: 1–80. 10.1111/j.1365-246X.2009.04491.x
14. Deplus C, Diament M, Hébert H, Bertrand G, Dominguez S, Dubois J, Malod J, Patriat P, Pontoise B, Sibilla J-J: Direct evidence of active deformation in the eastern Indian oceanic plate. *Geology* 1998, 26: 131–134. 10.1130/0091-7613(1998)026<0131:DEOADI>2.3.CO;2
15. Deschamps A, Lallemand S: The West Philippine Basin: an Eocene to early Oligocene back arc basin opened between two opposed subduction zones. *J Geophys Res* 2002, 107: EPM 1-1–EPM 1-24.
16. Engdahl ER, Villaseñor A, DeShon HR, Thurber CH: Teleseismic relocation and assessment of seismicity (1918–2005) in the region of the 2004 Mw 9.0 Sumatra–Andaman and 2005 Mw 8.6 Nias Island great earthquakes. *Bull Seismol Soc Am*2007, 97: S43-S61. 10.1785/0120050614
17. Fitch T: Plate convergence, transcurrent faults, and internal deformation adjacent to Southeast Asia and the western Pacific. *J Geophys Res* 1972, 77: 4432–4460. 10.1029/JB077i023p04432
18. Gahalaut VK, Gahalaut K: Burma plate motion. *J Geophys Res* 2007., 112: B10402, doi:10.1029/2007JB004928
19. Genrich J, Bock Y, McCaffrey R, Prawirodirdjo L, Stevens C, Puntodewo S, Subarya C, Wdowinski S: Distribution of slip at the northern Sumatran fault system. *J Geophys Res* 2000, 105(28):327–28,341.
20. Goto K, Kawana T, Imamura F: Historical and geological evidence of boulders deposited by tsunamis, southern Ryukyu Islands, Japan. *Earth Sci Rev* 2010, 102: 77–99. 10.1016/j.earscirev.2010.06.005
21. Gusiakov V: Tsunami generation potential of different tsunamigenic regions in the Pacific. *Mar Geol* 2005, 215: 3–9. 10.1016/j.margeo.2004.05.033
22. Hall R, Ali JR, Anderson CD: Cenozoic motion of the Philippine Sea plate: palaeomagnetic evidence from eastern Indonesia. *Tectonics* 1995, 14: 1117–1132. 10.1029/95TC01694
23. Heidbach O, Tingay M, Barth A, Reinecker J, Kurfeß D, Müller B: Global crustal stress pattern based on the World Stress Map database release 2008. *Tectonophysics* 2010, 482: 3–15. 10.1016/j.tecto.2009.07.023
24. Heuret A, Lallemand S, Funiciello F, Piromallo C, Faccenna C: Physical characteristics of subduction interface type seismogenic zones revisited. *Geochem Geophys Geosyst* 2011., 12: Q01004, doi:10.1029/2010GC003230
25. Hsu S-K: Lithospheric structure, buoyancy and coupling across the southernmost Ryukyu subduction zone: an example of decreasing plate coupling. *Earth Planet Sci Lett* 2001, 186: 471–478. 10.1016/S0012-821X(01)00261-8
26. Hsu S-K, Sibuet J-C: Earthquake off Japan could generate strong tsunami arrays. *Eos, Trans Amer Geophys Union* 2005, 86: 169–170.
27. Hsu YJ, Ando M, Yu SB, Simons M: The potential for a great earthquake along the southernmost Ryukyu subduction zone. *Geophys Res Lett* 2012., 39: L14302 L14302
28. Hsu S-K, Yeh Y-C, Sibuet J-C, Doo W-B, Tsai C-H: A mega-splay fault system and tsunami hazard in the southern Ryukyu subduction zone. *Earth Planet Sci Lett* 2013, 362: 99–107.
29. Kanamori H: A semi-empirical approach to prediction of long-period ground motions from great earthquakes. *Bull Seismol Soc Am* 1979, 69: 1645–1670.
30. Kimura M: Active rift system in the Okinawa Trough and its northeastern continuation. *Bull Disas Prev Res Inst* 1996, 45: 27–38.
31. Kopp H, Kukowski N: Backstop geometry and accretionary mechanics of the Sunda margin. *Tectonics* 2003, 22: 1072.
32. Lallemand S, Liu C-S, Dominguez S, Schnürle P, Malavieille J: Trench-parallel stretching and folding of forearc basins and lateral migration of the accretionary wedge in the southern Ryukyus: a case of strain partition caused by oblique convergence. *Tectonics* 1999, 18: 231–247. 10.1029/1998TC900011
33. Lange D, Tilmann F, Rietbrock A, Collings R, Natawidjaja DH, Suwargadi BW, Barton P, Henstock T, Ryberg T: The fine structure of the subducted investigator fracture zone in western Sumatra as seen by local seismicity. *Earth Planet Sci Lett*2010, 298: 47–56. 10.1016/j.epsl.2010.07.020
34. Larson RL, Carpenter GB, Diebold JB: A geophysical study of the Wharton Basin near the Investigator fracture zone. *J Geophys Res* 1978, 83: 773–782. 10.1029/JB083iB02p00773
35. Letouzey J, Kimura M: Okinawa Trough genesis: structure and evolution of a backarc basin developed in a continent. *Mar Petrol Geol* 1985, 2: 111–130.

10.1016/0264-8172(85)90002-9

36. Lin J-Y, Le Pichon X, Rangin C, Sibuet J-C, Maury T: Spatial aftershock distribution of the 26 December 2004 great Sumatra-Andaman earthquake in the northern Sumatra area. *Geochem Geophys Geosyst* 2009., 10: doi:10.1029/2009GC002454

37. Lin J-Y, Chen Y-F, Lee C-S, Hsu S-K, Liang C-W, Lin Y-C, Hsieh H-S: Strike-slip intraplate earthquakes in the Western Philippine Sea Plate. *Tectonophysics* 2013, 608: 499–504.

38. Liu N, Chen QF, Niu FL, Chen Y: Rupture of the 2004 Sumatra-Andaman earthquake inferred from direct P-wave imaging.*Chin Sci Bull* 2007, 52: 1986–1991. 10.1007/s11434-007-0286-x

39. Matsumoto T, Shinjo R, Nakamura M, Doi A, Kimura M, Ono T, Kubo A: Did the submarine, across-arc normal fault system in the southwest RyuKyu arc trigger the 1771 Tsunami? Field evidence from multibeam survey and in-situ observation. *J Environ Stud* 2009, 18: 123–129.

40. McCaffrey R: Slip vectors and stretching of the Sumatran forearc. *Geology* 1991, 19: 881–884. 10.1130/0091-7613(1991)019<0881:SVASOT>2.3.CO;2

41. McCaffrey R, Zwick PC, Bock Y, Prawirodirdjo L, Genrich JF, Stevens CW, Puntodewo S, Subarya C: Strain partitioning during oblique plate convergence in northern Sumatra: geodetic and seismologic constraints and numerical modeling. *J Geophys Res* 2000, 105(28):363–28,376.

42. Michel GW, Yu YQ, Zhu SY, Reigber C, Becker M, Reinhart E, Simons W, Ambrosius B, Vigny C, Chamot-Rooke N, Le Pichon X, Morgan P, Matheussen S: Crustal motion and block behaviour in SE-Asia from GPS measurements. *Earth Planet Sci Lett* 2001, 187: 239–244. 10.1016/S0012-821X(01)00298-9

43. Müller RD, Landgrebe T: The link between great earthquakes and the subduction of oceanic fracture zones. *Solid Earth Discuss* 2012, 4: 1229–1280. 10.5194/sed-4-1229-2012

44. Müller RD, Sdrolias M, Gaina C, Roest WR: Age, spreading rates, and spreading asymmetry of the world's ocean crust.*Geochem Geophys Geosyst* 2008., 9: doi:10.1029/2007GC001743

45. Nakamura M: Source fault model of the 1771 Yaeyama tsunami, southern Ryukyu Islands, Japan, inferred from numerical simulation. *Pure Appl Geophys* 2006, 163: 41–54. 10.1007/s00024-005-0007-9

46. Nakamura M: Fault model of the 1771 Yaeyama earthquake along the Ryukyu Trench estimated from the devastating tsunami. *Geophys Res Lett* 2009., 36: L19307 L19307

47. Nakamura M: The 1768 and 1791 Okinawa tsunamis in the Ryukyu Trench region. *Eos Trans Amer Geophys Union* 2013, 87: 2574. Fall Meet, Abstract Fall Meet, Abstract

48. Pacheco JF, Sykes LR, Scholz CH: Nature of seismic coupling along simple plate boundaries of the subduction type. *J Geophys Res* 1993, 98(14):133–14,159.

49. Paris R, Lavigne F, Wassmer P, Sartohadi J: Coastal sedimentation associated with the December 26, 2004 tsunami in Lhok Nga, west Banda Aceh (Sumatra, Indonesia). *Mar Geol* 2007, 238: 93–106. 10.1016/j.margeo.2006.12.009

50. Paul J, Bürgmann R, Gaur VK, Bilham R, Larson KM, Ananda MB, Jade S, Mukal M, Anupama TS, Satyal G, Kumar D: The motion and active deformation of India. *Geophys Res Lett* 2001, 28: 647–650. 10.1029/2000GL011832

51. Peterson ET, Seno T: Factors affecting seismic moment release rates. *J Geophys Res* 1984, 89(10):233–10,248.

52. Prawirodirdjo L, Bocl Y, McCaffrey R, Genrich J, Calais E, Stevens C, Puntodewo S, Subarya C, Rais J, Zwick P: Geodetic observations of interseismic strain segmentation at the Sumatra subduction zone. *Geophys Res Lett* 1997, 24: 2601–2604. 10.1029/97GL52691

53. Ruff L, Kanamori H: Seismic coupling and uncoupling at subduction zones. *Tectonophysics* 1983, 99: 99–117. 10.1016/0040-1951(83)90097-5

54. Sibuet J-C, Hsu S-K: Geodynamics of the Taiwan arc-arc collision. *Tectonophysics* 1997, 274: 221–251. 10.1016/S0040-1951(96)00305-8

55. Sibuet J-C, Letouzey J, Barbier F, Charvet J, Foucher JP, Hilde TW, Kimura M, Chiao LY, Marsset B, Muller C: Back arc extension in the Okinawa Trough. *J Geophys Res* 1987, 92(14):041–14,063.

56. Sibuet J-C, Rangin C, Le Pichon X, Singh S, Cattaneo A, Graindorge D, Klingelhoefer F, Lin J-Y, Malod J, Maury T, Schneider J-L, Sultan N, Umber M, Yamaguchi H, Team SA: 26th December 2004 great Sumatra–Andaman earthquake: Co-seismic and post-seismic motions in northern Sumatra. *Earth Planet Sci Lett* 2007, 263: 88–103. 10.1016/j.epsl.2007.09.005

57. Stein S, Okal EA: Ultralong period seismic study of the December 2004 Indian Ocean earthquake and implications for regional tectonics and the subduction process. *Bull Seismol Soc Am* 2007, 97: S279-S295. 10.1785/0120050617

58. Tanaka Y, Okuno J, Okubo S: A new method for the computation of global viscoelastic post-seismic deformation in a realistic earth model (I)—vertical displacement and gravity variation. *Geophys J Int* 2006, 164: 273–289. 10.1111/j.1365-246X.2005.02821.x

59. Uyeda S, Kanamori H: Backarc opening and the mode of subduction. *J Geophys Res* 1979, 84: 1049–1061. 10.1029/JB084iB03p01049

60. Wessel P, Smith WHF: New, improved version of generic mapping tools released. *Eos Trans Amer Geophys Union* 1998, 79: 579. doi:10.1029/98EO00426 10.1029/98EO00426

61. Yu S-B, Chen H-Y, Kuo L-C: Velocity field of GPS stations in the Taiwan area. *Tectonophysics* 1997, 274: 41–59. 10.1016/S0040-1951(96)00297-1

62. Yue H, Lay T, Koper KD: En echelon and orthogonal fault ruptures of the 11 April 2012 great intraplate earthquakes. *Nature*2012, 490: 245–249. 10.1038/nature11492

CHAPTER 2

EARTHQUAKES AND CRUSTAL DEFORMATION STUDY USING GLOBAL POSITIONING SYSTEM (GPS)

K. Mahanta, J. D. Chowdhury, A. Kumar, A. Kumar, I. Laskar, Sunil L. Singh, P. Barman

Keywords: Crustal deformation, Kopili Fault, Global Positioning System.

INTRODUCTION

Tectonically, the whole North-East Region can be divided into six physiographic cum tectonic domains. These units are Eastern Himalayan Collision Zone, Mishmi Block of the Syntaxial Zone of Himalayan Arc and Myanmar Arc, Indo-Myanmar Subduction Zone or Mobile Belt, the Brahmaputra Valley, the Meghalaya Plateau and the Bengal Basin. The Main Boundary Fault (MBF) and the Main Central Thrust (MCT) are the two major structural element of the Eastern Himalaya which extends upto the strategic southward bend of the Himalaya at Namche Barwa. The MBF has developed during the Pliocene and represents the main central boundary between the Indian lithospheric plate and the Eurasian Plate. The Mishmi Block tectonically separated the Eastern Himalaya from the mobile Myanmar Arc, where a number of lineaments in NW-SE and N-S direction are active. The North-East Region lies at the junction of Himalayan Arc to the North and Myanmar Arc to the East. The collision of Indian Plate and Eurasian Plate involves the large scale of active continental deformation and therefore a rather diffuse seismicity prevails in this region (Das Chowdhury, J. 2005).

Citation: K. Mahanta, J. D. Chowdhury, A. Kumar, A. Kumar, I. Laskar, Sunil L. Singh, P. Barman Earthquakes and crustal deformation study using Global Positioning System (GPS) ISSN 2277-937X.

Tectonic History

Millions of years ago, the whole landmass of the earth was combined, as there was only one continent and one ocean. The Pangaea theory states that all present continents were once together and collectively known as a 'supercontinent' or Pangaea. Pangaea was the large landmass that was disintegrated to form different continents. The process of separation of this landmass was started in 225 million years ago and it is still continuing. In 100 million years ago, India was surrounded by a vast ocean. The Indian landmark was moving towards North – North East direction and collided with the Eurasian Plate, forming the mountain Himalaya. Geologically, the Tethys Sea separated India from Eurasian Landmass. With continuous plate movement, Indian plate collided with Eurasian plate face to face and the mighty Himalayas evolved from the Tethys Sea. This collision began about 45 million years ago. The stresses of the Indian plate on the Eurasian Plate is still continuing, the height of the Himalaya is still rising, which also causes the various earthquake, landslides and all other physical events on the region stretching from Pamir to the Indo-Myanmar border (Mahanta, K. 2011). So, this region is seismically very sensitive and active.

Previous Work on this Region

The first scientific study of earthquake in the North East region is done by T. Oldham. In 1889, the discovery of oil in Assam also added another chapter in the scientific geological exploration of this region. R.K. Verma and M. Mukhopadhaya has extended the gravity field measurement covering the whole N-E India. R.P. Singh has done research work on geomorphology of Shillong plateau; M.N. Kulkarni has done extensive work on application of geodesy to monitor earthquake hazards. B.C Bhatia, M.R. Kumar and H.K. Gupta has also done a elaborate work on drawing seismic hazard map of India and adjoining region. But the most important investigative research work was done by D.R. Nandy. He has done a brief work of geodynamics of North East India. But in the most active Kopili Fault region, nobody has done any attempt of scientific study. So within this tectonic setup of the North East, this research work relating to study the dynamic role played by the Kopili Fault in generating any major Earthquake events in this region has great importance.

Earthquakes History in North-East India

The North-East part of the country is known to be one of the six most seismically active regions of the World. The other five regions are Maxico, Taiwan, California, Japan and Turkey. This region lies between the northern collision and eastern subduction margins of the Indian plate(Das Chowdhury, J., Barman, P. 2008). Out of the 5 great earthquakes of magnitude greater than 8 in Richter Scale (RS) experienced by India, two occurred in this region. These are the 1897 'Great Assam Earthquake' and '1950 Assam Earthquake'. The Great Assam Earthquake occurred in June 12, 1897 with epicenters at 26° N 91° E has devastated Shillong, which developed Chidrang and Sami faults and fractured the whole North-East. The magnitude of this earthquake was 8.7 in RS. The next one occurred with the almost same magnitude (ie. 8.7 in RS) in August 15, 1950 with epicenters at 28.5° N 95.5° E at the depth of 29 km (Tiwari. R.P.) has destroyed the whole Assam. Total numbers of about 1520 people were killed. This earthquake was one of the World's largest known earthquakes. Apart from these two, if we analyze the history of the major earthquakes in the North-East India and its neighborhood, we see that the Kopili fault is seismically very active. Two major earthquakes of magnitude more than 7 in RS occurred in this region. In January 10, 1869, an earthquake with magnitude 7.8 in RS occurred at a depth of 60 km along the Kopili Fault Zone. This earthquake has severely damaged the entire Nagaland. In October 23, 1943, there was an earthquake with epicenters at 27.5° N 93.5° E originated at Kopili Fault with magnitude 7.3 in RS. It shook the whole North-East and damaged a lot.

Geological Setting

Geologically, in North-East India the Brahmaputra Valley is built by the deposition of about more than 1000 m thick alluvium upon sag formed during the period of the rise of the Himalaya. The Brahmaputra Valley is a relatively narrow valley bounded by two mobile young mountain belts, Mishmi block to the northeast and Meghalaya Plateau to the south. Another notable formation of this region is the discontinuous exposure of metamorphic rocks along the northern most part of the ridge in Nagaland and adjacent Myanmar area (Mahanta, K. 2011). This metamorphic complex is thrusted over from east and south east and may be of pre-cambrian age.

Seismicity in N-E India

The North-East region lies at a junction of Himalayan arc to the north and the mobile Myanmar arc in the east.Zone wise, the Eastern Himalayan Collision Zone has experienced 2 large earthquakes, one in 1941 (M =7.1) and another one in 1947 (M =7.5). The Indo-Myanmar subduction zone is seismically very active, where 10 large earthquakes of M >7.0 have occurred during the last 100 years. The Mishmi Block of the syntaxial zone of the Himalaya and Myanmar Arc is the zone of the Great Assam Earthquake of 1950 (M = 8.7). A number of thrusts and faults make this region seismically more active. A cluster of earthquake events is evident in Namcha Barwa area at the interaction of Tethyan Himalayan domain and NW Mishmi Block (Bryant, H. Richard 1999). The Shillong Plateau and its plate boundary zone is active all throughout. It was the seat of Shillong earthquake of 1897 (M = 8.7). Moreover, it is the Kopili Fault that constitutes the tectonic boundary between the two stress domains of the Meghalaya and Mikir Hills Massif. It is the Kopili Fault Zone that triggered the 1869 (M = 7.8) and 1943 (M = 7.3) events and three of (M = 6.0 to 7.0) and several of magnitude 4.5 to 6 (Chugh, R.S. & Valdiya, K.S. 1989) . This seismically active Kopili fault cuts across the Brahmaputra about 35 km NW of Nagaon between Mikir Hills and the prolongatition of Meghalaya massif (Das Chowdhury, J. 2002).

Moreover, it is also observed that the hidden basement ridge of Assam Shelf including the block from Guwahati - Dergaon have been consistently rising, the vertical velocity increasing progressively from 0.3 mm/yr to 4.5 to 31 mm/yr (Das Chowdhury, J. 2001) at Dergaon a part of Mikir Block. North of the Dauki Fault, the Meghalaya block has been rising at the rate of 0.3 to 0.4 mm/yr.(Das Gupta, S., Mukhopadhaya, M. & Nandy, D.R. 1987).

METHODOLOGY

To understand the crustal deformation in North-East India, the work has been carried out on the latest technology using Global Positioning System (GPS). In this study, twelve (12) points in different locations of Kopili Fault region have been selected for campaign study. The first field study was initiated in October, 2006 and then after it was continued for sequent four years up to February, 2011. These GPS campaign stations are studied with Trimble 5700, Leica500 and Leica1200 receivers with choke ring and Zephyr geodetic antennae. The selection of twelve sites was based on granite and feldspar rock formations or on hard stable rocks. All the points were selected in an open area without having any disturbance to the antenna. The machines were installed in all the campaign sites at least for continuous 72 hours, so that we can collect 48 hour data without break. The GPS data so collected has been converted into RINEX observation files and quality check has been conducted using TEQC (Translations, Editing and Quality Checking) software. All these processes of quality check plots were carefully examined and data with high cycle slips, multi-path were removed from the analysis (Sunil & Kumar 2008). The data collected for the last four epochs (2006-11) processed along with the selected IGS (International GPS Service) stations KUNM, HYDE, IISC, POL2, KIT3. In the processing, we have used GAMIT/GLOBK software developed by Massachusetts Institute of Technology (MIT), USA and a reference frame ITRF 2005 is used for all time series. Ambiguity free and ambiguity fixed solutions were performed with ionosphere free linear combination to account for carrier phase ambiguities and signal delay due to ionosphere (Sunil & Kumar 2009). The horizontal errors ellipses are calculated by GAMIT/GLOBK software from the uncertainties in the North and East coordinates/velocities using the correlation between the baseline component estimates.

RESULTS

GPS velocity vectors of all twelve sites with the IGS sites have been estimated in ITRF05 (GLOBK version 5.16 & 5.19). The analysis of data gives a clear picture that the North-East Region is moving in North-North-East (NNE) direction with a average velocity of 42.36E(2.20mm) and 29.00N(1.66mm). A remarkable convergence is seen between the points NATU-KUNM, NATU-POL2, UMRA-IISC, UMRA-POL2, UMRA-KIT3, PANI-IISC, PANI-KIT3, RAJA-POL2 and KHER-POL2. Between the campaign points KHER and RAJA, a clear divergence is seen at a rate of 2.245 mm/yr. (Table1 & 3).

Table 1: Estimated velocities of campaign sites of North-East India (using GLOBK Version 5.16)

Sl. No.	Name of the Stations	Station Code	Long (E)	Lat (N)	East (mm/yr)	North (mm/yr)
1	Natun Bazar	NATU	92.602	25.521	44.50(2.36)	28.17(1.90)
2	Umrangsho	UMRA	92.725	25.526	41.97(2.47)	27.26(1.70)
3	Sokra Pam	SOKR	92.736	25.596	44.58(2.34)	27.07(1.80)
4	Panimura	PANI	92.825	25.717	42.03(2.47)	30.66(1.92)
5	Raja Gaon	RAJA	92.629	26.073	43.98(2.27)	29.96(1.88)
6	Kheroni	KHER	92.857	25.812	49.44(2.45)	30.85(1.98)
7	Kumoi	KUMO	92.248	26.196	41.09(.04)	28.22(0.92)
8	Jagiroad	JAGI	92.204	26.117	39.51(1.37)	27.98(1.24)
9	Bura Mayong	BURA	92.013	26.246	39.59(1.26)	28.69(1.12)
10	Amsoi	AMSO	92.484	26.156	36.74(2.29)	30.01(2.08)
11	Kampur	KAMP	92.650	26.163	43.36(2.72)	30.93(2.34)
12	Khetri	KHET	92.080	26.126	41.54(1.21)	28.28(1.07)

The average velocity of IGS stations KIT3 and POL2 in Eurasian plate is 5.19N(0.395mm) and 26.96E(0.385mm). The IGS sites in India IISC and HYDE have the average velocity of 35.28N(0.47mm) and 41.125E(0.48mm); means the Southern part of the Eurasian plate is more stable than the Indian plate(Table 1 & 2 and figure1)

Table 2: Estimated velocities of IGS sites of different plates (using GLOBK Version 5.16)

Sl. No.	Name of the Stations	Station Code	Long (E)	Lat (N)	East (mm/yr)	North (mm/yr)
1	Kunming	KUNM	102.797	25.030	29.11 (.43)	-18.51 (.43)
2	Lhasa	LHAZ	91.104	29.657	47.87 (.36)	15.61 (.34)
3	Hyderabad	HYDE	78.551	17.417	41.03 (.50)	34.61 (.50)
4	Bangalore	IISC	77.570	13.021	41.22 (.46)	35.95 (.44)
5	Bishkek	POL2	74.694	42.680	26.77 (.38)	5.13 (.38)
6	Kitab	KIT3	66.885	39.135	27.15 (.39)	5.25 (.41)

We have estimated the deformation rate between the sites by dividing the convergence rate by their average baseline length. A remarkable deformation is seen between the campaign points SOKR-POL2, UMRA-POL2, KHER-POL2 and PANI-POL2. The deformation rate of SOKR-POL2 is -8.978 10^{-10} per year where as UMRA-POL2 is -6.399 10^{-09}. But the both UMRA and SOKR campaign points are situated very nearby. The baseline length between them is 7897.33 m and is separated by the river Kopili. The point SOKR which is in the North bank of the river Kopili is more stable than UMRA in the South bank of the river. The convergence rate of SOKR and UMRA are -2.24 mm/yr and -16.0025 mm/yr respectively; which means about 13.7625 mm/yr is accommodated in the Kopili Fault that is passing through the river. A notable deformation between KHER and PANI is also seen. The difference of the both points from POL2 is almost 6377.06 m and the convergence rate of PANI-POL2 and KHER-POL2 are -20.34 mm and -13.80 mm respectively; means about 6.5375 mm/yr is accommodated in between the Kopili Fault. The points KHER and RAJA are separated from each other by a rate of 2.245 mm/yr, with the rate of deformation 6.10050 10^{-08}. If we carefully see the deformation rate of all points of Kopili Fault in the North-East India, a clear picture of deformation on the fault line is seen (Table 3).

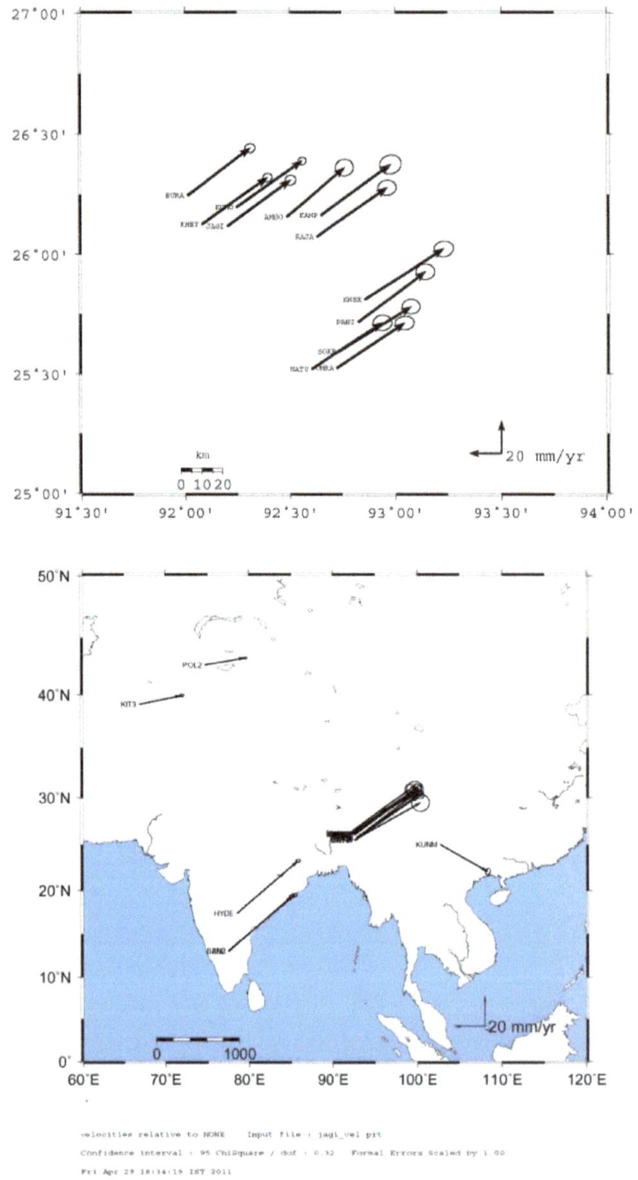

Figure 1: Velocity of various campaign sites along the IGS sites.

Table 3: Estimated Baseline lengths of campaign sites and IGS sites for four years and deformation rate

From Station	To Station	Baseline Length 2007-08 (m)	Baseline Length 2008-09 (m)	Baseline Length 2009-10(m)	Baseline Length 2010-11(m)	Convergence Rate between first and fourth epoch (mm/yr)	Deformation Rate (yr^{-1})
UMRA	NATU	12387.01187	12387.06905	12387.07388	12386.99608	-3.9475	-3.18681 10^{-07}
KHER	RAJA	36800.27833	36800.28107	36800.28156	36800.28731	2.2450	6.10050 10^{-08}
KUMO	JAGI	9818.19435	9818.18257	9818.22700	9818.19298	-0.3425	-3.48842 10^{-08}
AMSO	BURA	48178.25627	48178.13204	48178.00854	48178.25224	-1.0075	-2.09119 10^{-08}
KAMP	KHET	57092.91790	57092.91671	57092.94546	57092.91462	-0.8200	-1.43626 10^{08}
NATU	KUNM	1027193.80016	1027193.80893	1027193.79702	1027193.75797	-10.5475	-1.02683 10^{-08}

NATU	IISC	2087536.92697	2087536.89292	2087536.97656	2087536.90626	-5.1775	-2.4802 10^{-09}
NATU	POL2	2493764.20584	2493764.17029	2493764.17200	2493764.15385	-12.9975	-5.212 10^{-09}
NATU	KIT3	2814617.20855	2814617.17598	2814617.19552	2814617.18252	-6.5075	-2.31204 10^{-09}
UMRA	IISC	2097424.50216	2097424.15828	2097423.95525	2097424.45460	-11.8900	-5.66886 10^{-09}
UMRA	POL2	2500513.13754	2500512.72713	2500512.39583	2500513.07353	-16.0025	-6.39969 10^{-09}
UMRA	KIT3	2823738.88858	2823738.43554	2823738.07702	2823738.84443	-11.0375	-3.90882 10^{-09}
SOKR	IISC	2102999.61881	2102999.28951	2102999.10269	2102999.60584	-3.2425	-1.54185 10^{-09}
SOKR	POL2	2494925.91432	2494925.52098	2494925.20530	2494925.90536	-2.2400	-8.97822 10^{-10}
PANI	IISC	2118075.89568	2118075.54774	2118075.34069	2118075.84903	-11.6625	-5.50618 10^{-09}
PANI	POL2	2489562.35615	2489561.93409	2489561.59432	2489562.27479	-20.3400	-8.17011 10^{-09}
PANI	KIT3	2818531.99147	2818531.52476	2818531.15727	2818531.93020	-15.3175	-5.43457 10^{-09}
RAJA	IISC	2127503.74338	2127503.72510	2127503.79341	2127503.71831	6.2675	-2.94594 10^{-09}
RAJA	POL2	2447084.38356	2447084.35363	2447084.32492	2447084.32029	-15.8175	-6.46381 10^{-09}
RAJA	KIT3	2779851.52491	2779851.49582	2779851.48331	2779851.48115	-10.9400	-3.93546 10^{-09}
KHER	IISC	2127037.22916	2127037.21977	2127037.29566	2127037.20774	-5.3550	-2.51759 10^{-09}
KHER	POL2	2483185.2694	2483185.24962	2483185.22491	2483185.21419	-13.8025	-5.55839 $1^{0-0}9$

CONCLUSION

The whole North-Eastern Region is seismically very active due to the various faults present in this region. The NNW-SSE trending Kopili Fault which extends from western part of Manipur upto the tri-junction of Bhutan, Arunachal Pradesh and Assam covers a distance of about 400km. The Kopili fault bisects the Meghalaya Plateau and isolated the Mishmi block from the main part of the plateau. In the last two years more than four notable earthquakes (M>5.0) occurred in the two endpoint of the Kopili Fault (Manipur and Bhutan), which signify that present day the Kopili Fault is very active. Our results also show the extreme deformation in the fault area as well as the North-East India.

ACKNOWLEDGEMENT

We acknowledge Ministry of Earth Sciences (MoES), Government of India for providing funds vide Project no. MoES/P.O.(Seismo)/GPS/48/2005 to Department of Geography, Jagiroad College for carrying out the GPS surveys in the North-East Region.

REFERENCES

1. Bryant, H. Richard: Physical Geography, Rupa & Co., New Delhi 1999
2. Chugh, R.S. & Valdiya, K.S.: India J. of Geol. 1989 Vol. 61, P 1-13
3. Das Chowdhury, J., Barman, P. (2008.: "Micro-seismic Study of the Kopili Fault of North-East India"
4. Das Chowdhury, J.: Plate Tectonics: Tha latest Findings, paper on National Seminar at Mongaldoi College, Assam Dt. 21st & 22nd Sept. 2002
5. Das Chowdhury, J.:Earthquake and Related Hazard Management; National Seminar Organised by Science Trial Organisation, Silchar, Oct. 5-7, 2001
6. Das Gupta, S., Mukhopadhaya, M.& Nandy, D.R.: Active tectonic feature in the Central part of Himalaya Techonophysics 1987, Vol. 136 p 255-264
7. Das Gupta, S.& Nandy, D.R.: Physics, Chemistry of the Earth (1991) Vol. 18 P 1147-1163
8. Das Chowdhury, J.(2005): A feasibility report on "A delineation of the seismically active
9. Kopili Fault Zone of North-East India"
10. Mahanta, K. (2011): "Impact of climate change on crustal movement in the North-East India: a scientific field study using Global Positioning System (GPS)" a submitted article to the International Seminar on Bioresourses and Human Sustenance organised by department of zoology, Cotton College & Zoological Society of Assam
11. Sunil, S.L.,Kumar,A.(2008): "Present day crustal deformation along Indo-Myanmar Arc Region"
12. Sunil S. L., Kumar, A(2009).: "Establishment of a GPS permanent Reference station along Indo-Myanmar region at Manipur University, Imphal, Manipur, India".

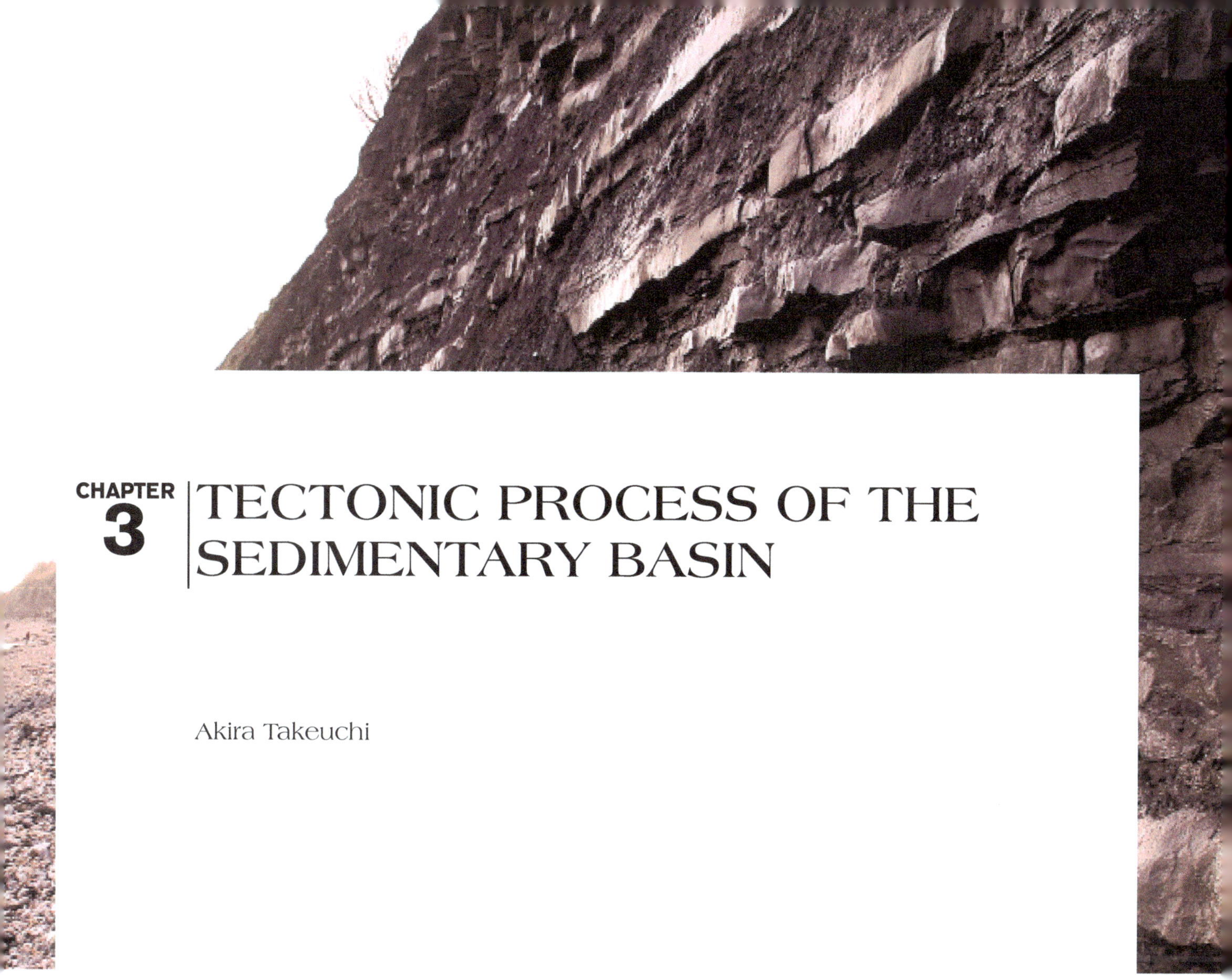

CHAPTER 3 TECTONIC PROCESS OF THE SEDIMENTARY BASIN

Akira Takeuchi

INTRODUCTION

The Japanese archipelagoes in the northwest Pacific consist of southwest Japan arc, Izu - Ogasawara arc, Ryukyu arc and Northeast Japan arc and Krile arc, and the Japan Sea is a marginal sea of northeast Asian continent separated from the northwest Pacific by three islands of Sakhalin, Hokkaido, and Honshu. The thrust-and-fold zones of the Neogene and Quaternary systems in the coastal tectonic belt along the eastern margin of Japan Sea continue their activity until now. Particularly, the seismogenic zone continued from Sakhalin is arrested with the convergent boundary between Amur Plate and the Okhotsk Plate because reverse-fault type of earthquakes with magnitudes larger than 7.4 in Richter's scale have been occurred along this heteromorphic belt every several hundred years. From this point of view to the plate tectonic condition around Japan, the central Japan acts as a multiple junction area unique in the earth, where four pieces of plates, such as the Amur, the Okhotsk, the Philippine Sea, and the Pacific plates, gather and converge together in and around the Japanese archipelagoes (Figure 1). Although the GPS geodetic observations confirmed the presence of the micro-plates [3-5], the structural features of the incipient boundary between the Okhotsk and Amur Plates seemed still immature but recognized as strain accumulated zone along the eastern margin of Japan Sea [6-11].

The Pacific plate have subducted beneath the Okhotsk, the Philippine Sea, and Amur plates which are converging together. In particular, central Japan is the place where the Southwest Japan, the Northeast Japan, the Izu-Ogasawara arcs mutually collide, and therefore the process and mechanism of the development of geologic structures here is quite complicated and hard to be interpreted. Lately, in such situation, the damaging earthquakes of the middle scale have occurred in the Japan Sea side of the central Japan in succession. In 2004, the Chuetsu earthquake (M_W6.6) occurred in the inland Chuetsu area, Niigata Prefecture. Two years and nine months later, in 2007, the Noto Peninsula earthquake (M_W6.7) occurred at a 11km deep hypocenter beneath the west coast of northern Noto Peninsula. Then, only 3.7 months later, the Niigata Prefecture Chuetsu-oki earthquake (M_W6.7) occurred approximately 15km in depth, 32km distant from the epicenter of the 2004 earthquake, attracting an attention to the relations of three earthquakes from a time-space point of view. Furthermore, the

Citation: Akira Takeuchi (2013). Tectonic Process of the Sedimentary Basin InTech, DOI: 10.5772/56805.

Naganoken Hokubu earthquake (M_W 6.35) was generated by the hypocenter 8km in depth at a moment 13 hours 13 minutes after the 2011 off the Pacific coast of Tohoku Earthquake (M_W 9.0) occurred in the Japan Trench on March 11, 2011.

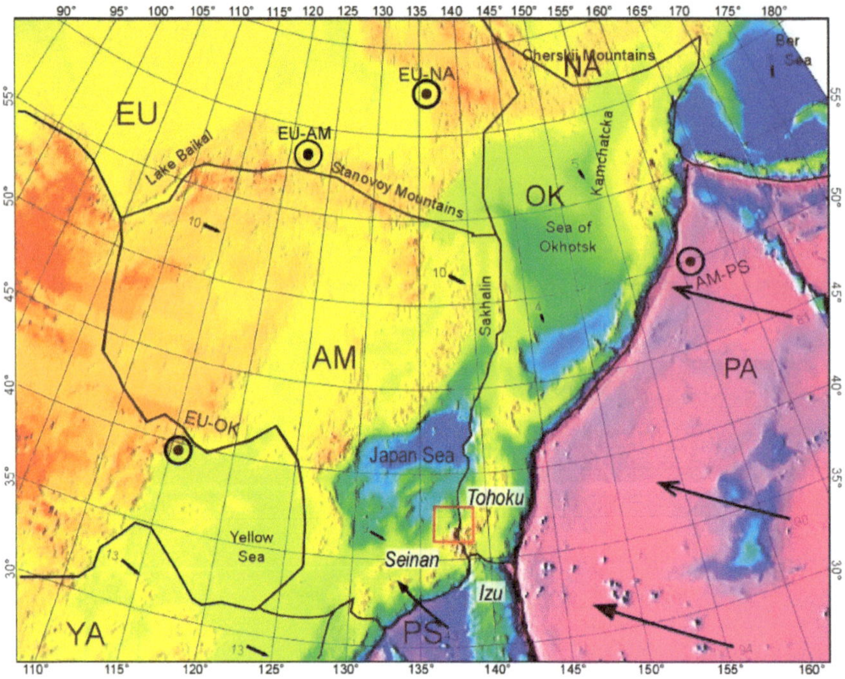

Figure 1. Index Map of Plate Framework in the Northeast Asia. Study area is depicted by open red box. Boundaries of the Okhotsk (OK) and Amur (AM) plates are shown. Surrounding plates include Eurasia (EU), North America (NA), Pacific (PA), Philippine Sea (PS), and Yangtze (YA). Black vectors give model velocities (with numbers in mm/a) relative to plate whose identifier is underlined. Black circles are locations of Euler poles. Simplified from [1] with an addition of Euler pole EU-AM [2].

As for the generation style of middle scale and larger earthquakes occurring along the eastern margin of Japan Sea, including the above mentioned earthquakes in central Japan, major listric faults contributed with back arc spreading during the Miocene has been explained as inversion tectonics which inverted the sense of faulting from normal dip-slip into the reverse slip since the Pliocene. However, the western Noto Peninsula earthquake and the Naganoken Hokubu earthquake were the focal mechanisms that they could not be explained by means of simple inversion tectonics concerning the geologic structure of shallower layers so far. Main shocks of these earthquakes are commonly occuured around 15km in depth, which is almost equivalent to the basal depth of the seismogenic layer.

Since the thick sedimentary cover of the seismogenic layer has been remarkably deformed into faulted and folded structures, the character of tectonics in progress at the Present is questionable whether thin-skinned tectonics or basement-involved tectonics, being concerned with the presence of the detachment surface between the seismogenic layer and the sedimentary cover [12-14]. Thus, this paper discusses the formation process of sedimentary basins in the collision zone between island arcs from the viewpoint of earthquake tectonics about the recent crustal earthquakes. For the purpose of elucidating the active tectonics (crustal movement in progress) along the Japan Sea coast in central Japan, this paper focuses on the specificity of the geologic structure and geomorphology of Eastern Hokuriku district and Fossa Magna - Toyama Trough region, and traces the history of geomorphologic and geologic development in order to propose a comprehensive relation with earthquake occurrence and crustal movement.

THE FEATURES AND DEVELOPMENT PROCESS OF GEOMORPHOLOGY AND GEOLOGIC STRUCTURE

Target Area

The target area for this paper, the Hokuriku-Shin'etsu district, is composed of three adjacent Neogene sedimentary basins located in the Japan Sea side of central Japan including the seabed area (Toyama Trough) between Noto Peninsula and Sado Island (Figure 2). The trough is administratively enclosed by Ishikawa, Toyama, and Niigata Prefectures. During the early to middle Miocene periods the Hokuriku, Shin'etsu, and Niigata basins had developed obliquely upon the basement geologic zones geotectonically belonging to the inner belt of pre-Cenozoic Southwest Japan. Although Shin'etsu, and Niigata basins tends to be treated as a single sedimentary basin, herein, the most part of Niigata basin is excluded from the Fossa Magna area as long as geomorphology and geologic history are concerned [17, 18]. In addition, the north-south trending, narrow basin in the central part of continental slope offing the Japan Sea side of Honshu, the Toyama Trough, borders Northeast Japan and Southwest Japan in the seabed area.

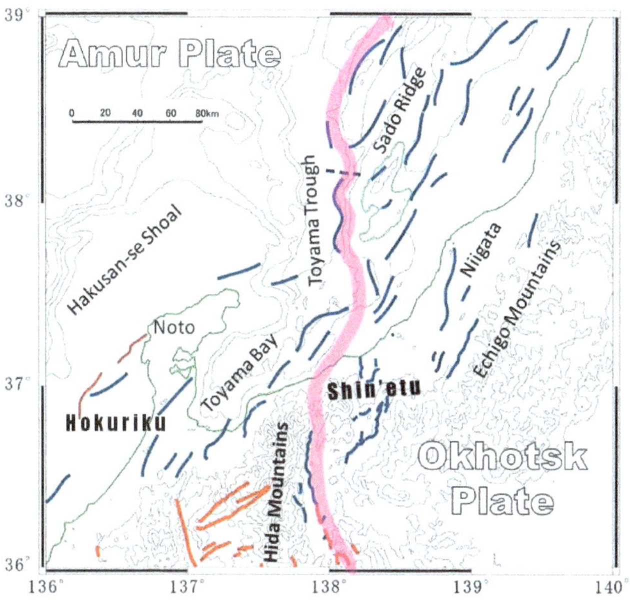

Figure 2. Index Map of Active Faults for the central Japan. Blue line denotes reverse fault, red does strike-slip fault. Thick pink line indicates plate boundary between Amur and Okhotsk plates. Place names are also indicated. Simplified and compiled from [15, 16].

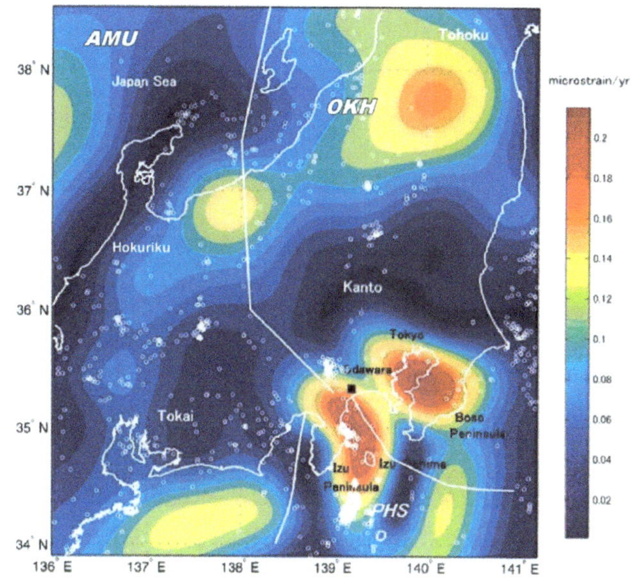

Figure 3. Maximum shear strain rates in central Japan. Estimated from the two-year improved time series data from April 1996 to March 1998 [25]. White circles indicate epicenters of the earthquakes with depths shallower than 30 km and magnitudes greater than 3.0 during the period from January 1996 to March 1998. Note that the strain distribution belt intersects the Itoigawa-Shizuoka tectonic line bounding Amur and Okhotsk plates. This belt corresponds to the Shinanogawa seismic zone and Atotsugawa fault zone in the Niigata Kobe tectonic zone [19].

Tectonic Provinces of Target Area

When a zonal division is available on the basis of regional characteristics of fault distribution such as fault length and orientation, inclination, type of displacement sense (normal, reverse, or strike-slip), and the density of fault distribution in a certain geological age, a tectonic unit in this paper is defined as a fault province. A fault province composed of active faults is called an active fault province which reflects regional characteristics of the seismogenic stress field. From this point of view to the recent crustal movement regionally in a geodetic to geological time scale, a strain concentration zone denotes a geodetic zone where a pattern of displacement field demonstrates a belt of larger strain rate, and geologically it corresponds to a zone where deformation structures such as faults and/or folds develop intensively [6, 9, 19, 20].

Including Mizuho-Fossa Magna folded belt [21], concentrated deformation belts were known in many places in the Cenozoic Japan, but existence of the Niigata - Kobe tectonic belt [19] becomes recognized by the GPS precise geodetic observation network of Geospacial Information Authority of Japan (GSI) having been maintained in and after 1995.

Figure 2 shows active fault distribution [16] and the active fault province [15, 22] of the inner Chubu District. The reverse fault province occupies the inner Tohoku arc and the strike-slip fault province is located inland area adjacent to the Hida mountain range. Concerning generation of earthquakes in the inland crust, the strong shortening in the Niigata - Kobe tectonic belt lately has attracted attention by intersecting the active fault provinces and the plate boundary between the Amur and Okhotsk plates, Itoigawa - Shizuoka tectonic line as is recognized from Figure 3.

Plate Tectonic Framework

From the plate tectonic point of view, the central Japan acts as a multiple junction area unique in the earth where four pieces of plates, such as the Amur, the Okhotsk, the Philippine Sea, and the Pacific plates, gather and converge together in and around the Japanese archipelagoes. The border of Amur plate and the Okhotsk plate has just jumped from the west margin of the Hidaka Mountain Range into the eastern margin of Japan Sea at about 0.5Ma. The former plate boundary between the North American plate and the Eurasian plate had been situated in the central Hokkaido where another collision between the Kurile and the Tohoku arcs had performed. As for the Seinan fore-arc, the commencement of subduction with the northing of Philippine Sea plate was represented by the 15Ma intrusions of outer-zone granite and the bended structure of the earlier Nankai trough caused by the paleo-Izu indentation at 15-14Ma. This remarkable transition might have affected the convergent boundary between the Eurasia plate and the North America plate and the both continental plates would be put together in the collision state. Contrastingly, the Pacific plate has continued almost steady subduction along the Japan Trench for the past 40 million years without significant change in the northwestward motion, despite tectonic episodes of back-arc spreading in Japan Sea, Okhotsk Sea and Shikoku Basin.

In the eastern margin of Japan Sea and the Fossa Magna region, the environment of the crustal movement switched totally from the calm period in the late half of Miocene to the Pliocene contraction tectonics. The start of folding in the northern Fossa Magna region dates up by evidence of the paleomagnetism in at least 4Ma [24]. However, the start of folding was much older because of the sedimentological fact that turbidite flowed down the trough-like basins of syncline and the stratigraphic fact that the base of Pliocene andesites (5.4Ma) covered obliquely the anticline which has already begun growth [25-27].

By the way, due to the migration of trench triple junction, the moving direction of the Philippine Sea plate switched at 3Ma from the north direction to northwest [28], and, therefore, the colliding force against the border area between east and west Japan as well as the southern Fossa Magna should have weakened in comparison with the past. The contraction tectonics in the Japan Sea side could be attributed to starting of eastward motion of the Amur plate, because the start of the contractinal tectonics in the eastern margin of Japan Sea was significantly older than 3Ma.

Time Scale Setting

As for the upper Cenozoic system distributed over the Hokuriku and Shin'etsu areas, the biochronological stratigraphy was almost established in the 1980s [29-31]. A complicated stratigraphy on terrestrial sediments of the lower Cenozoic system widely distributed in Noto Peninsula has become elucidated based on age-determination data of volcanic rocks [31].

In addition, in late years for the purpose of analysis of the marine paleoenvironment, high precision chronostratigraphy is performed by means of age-marker for the period after the Pleistocene in particular.

Based on the recent advance in the Pliocene stratigraphic correlation and age determination of tephra distributed widely over the central Japan [32, 33], there was large progress for historical studies on the fault activities and upheaval of Hida Mountain ranges [34-36]. In conformity with these results, this paper also obeys a new definition of the Quaternary period recently revised by the International Stratigraphic Committee (http://quaternary.stratigraphy.org/definitions/).

FORMATION AND DEVELOPMENT OF SEDIMENTARY BASINS

The Present Hida Plateau and Noto Peninsula are upheaval zones which expose the basement rocks of pre-Cenozoic system, and are different from the Present coastal plains and near shore waters which comprise the thick sedimentary layers. As for the approximately 5 million years period previous than 1 million years ago, it is thought that the area of Noto Peninsula is a large terriginous flat or is a very shallow archipelago [30], and that this area formed a peninsula after 0.5 Ma [38]. This paper considers the geomorphplogical development of the seabed and coastal places of Japan Sea mainly for block structures of Honshu Island since the Oligocene, with paying attention to the following five stages of crustal movement concerned with a geological development of Southwest Japan west of the Itoigawa-Shizuoka tectonic line. Besides, Northeast Japan saying in this paper includes the northern Fossa Magna region for convenience and excludes the southern Fossa Magna region.

Rifting Phase [32 Ma — 28 Ma]

The drillings into Japan Basin and Yamato Basin conducted in 1989 by International Ocean Drilling Program (ODP) provided an important data related to the timing of formation of the Japan Sea area. According to [39], the formation of the Japan Basin began by thinning of the continental crust in the early Oligocene (32 Ma), and such a tectonic style changed into expanding of the sea floor in the late Oligocene (28 Ma).

Sea Floor Spreading Phase [28ma — 18ma]

The tectonic domain of the sea floor spreading in the Japan Sea area had moved from the widened Japan Basin area to the southwest, and formed both Yamato Basin and Tsushima Basin by crustal expansion, but it ceased in 18 million years ago [39-42]. The rifted structures with trends of north-south direction or northwest-southeast were formed in Toyama Trough and the Hokuriku and Niigata areas in the period from the end of Oligocene to the early Miocene [43-45]. In the Hokuriku district in the middle Early Miocene (20Ma-18Ma), submarine volcanic activities occurred and tearing of the basement, i.e. intra-arc rifting, formed graben-like depressions. According to[46], tectonically distinct boundary between Tohoku and Seinan Honshu arcs had been formed or activated at the end of this phase. Toyama Bay was originally an embayment that branched off the Toyama Trough into the Hokuriku area, and the sea-bottom faults along the coastal line were activated for the period of opening of Japan Sea [9].

Sedimentary-Basin Forming Stage [18ma — 15ma]

Marine sedimentary basins were formed by sudden subsidence to reach 3,000m during this period in the Hokuriku district [30, 47, 48]. These sedimentary basins are located in the almost same places of the Present coastal alluvial plains including Kaga and Toyama Plains.

Intense volcanic activities occurred in the inner belt of Honshu arc on the Japan Sea side of Tohoku region, and submarine volcaniclastics known as 'green tuff' deposited in the period from 24Ma to 14Ma. In detail, around the time of 17Ma, southerly warm current water called 'paleo-Kuroshio' had become to emerge the back arc area, and then in substitution relatively calm subsidence commenced at 16Ma. During the period from 15Ma to 14Ma, marine transgression enlarged the entire intra-arc area, and the bathypelagic black mudstones (Nanatani Formation and its correlatives) were deposited in seabed area where the previous morphologic ups and downs would be buried.

In the Hokuriku district, the depth of the bedrock is at least 2,000m - 3,000m for last Neogene of Kaga - Toyama plains sandwiched between Noto Peninsula and the Hida Highlands [50]. In addition, the zone of relatively high-density rocks such as andesite and basalt lavas occupying the graben-like depressions of the basement is expressed as the narrow zone of highly positive features of Bouguer gravity anomaly [51-54, 35]. These kinds of volcanism are not product of the pervious syn-rift phase of back-arc spreading but of the intra-arc rifting due to commencement of arc volcanism of the Honshu arc [29, 31, 35].

Basin Differentiation Phase [15ma — 5ma]

Lateral variations in thickness of the middle to upper Miocene strata among the sedimentary basins became remarkable in this period. Spatial variety in sedimentary thickness of individual deposition centers were well documented in the Shin'etsu sedimentary basin, suggesting a syn-sedimentary fault-block movement [25]. In the Hokuriku district, however, the Present-day hilly and mountainous countries including Iouzen-Hodatsu Hill, Imizu Hill, and Yatsuo area and Noto Peninsula bordered the sedimentation basins, where the rates of sedimentation reduced during the time from 15 Ma until 13 Ma[55].

As for the Shimane Peninsula in the San-in district of the western Seinan arc, [56] mentioned that the sedimentary basin had begun its inversion tectonics under the crustal stress field of the north-south compression in 14 million years ago, and the formation of the Shinji folded zone was completed in 6 million years ago [57-59]. The expanse of such the north-south compression field became broader, and the concentrated zone of east-west trending reverse faults and related folds parallel to the Southwest Japan arc developed from San-in to Hokuriku districts in the Japan Sea side from 8 million years ago [60]. The late-Miocene east-westerly deformation zone in the Hokuriku district includes Houdatsu-san Kita fault zone in the southern part of Noto peninsula and Wakayama-gawa fault zone and in northern Noto Peninsula. Landfill underwent ahead through the Hokuriku sedimentary basin from the side of Hida area towards the former Toyama Bay. In the Noto Peninsula, however, nanofossil chronostratigraphy detected several times of hiatuses when glauconites produced on the seabed after the Middle Miocene [55, 61]. Therefore the Present Toyama Bay is the remains of the Miocene graben half of which had been filled with the sediment [9].

Evolving Stage [5ma — Present]

Upheaval and subsidence (i.e. undulation of the basement) with axes striking north-south began in the later Pliocene in the eastern part of Southwest Japan, but another tectonic regime of northwest-southeast compression has superposed by the collision with the Izu arc, and the structural trend in the northeast-southwest direction reaches the expression of remarkable current active structure [22, 59, 62,63]. The fault block movement of this stage is a process of modification where the existing geologic structure change into new one, and is deeply participated in the geomorphology development such as Hida Mountain Ranges, Noto Peninsula, and Toyama Bay in the Present period. This process did not begin at the same time in all the areas of Southwest Japan, but a tendency to migrate from the southeast (the Tokai district) to the northwest (the Hokuriku district and the Kinki district) and to the northeast (the northern Fossa Magna) is recognized [64, 65]. In addition, the inversion process included locally the one of fault-slip sense where former normal faults trending north-south to northeast-southwest directions became re-activated as reverse faults (e.g., Kureha-yama fault: [66]). In the Hokuriku district of the Neogene sedimentary basins, however, alluvial plains continue their sedimentation without performing "basin inversion" like the Miocene Shin'etsu sedimentary basin, where the whole area of subsidence with thick sedimentary layers had changed reversely into the upheaval zone [27-29].

TECTONIC INVERSION OF SEDIMENTARY BASINS AND RELATED FAULTS

The style and degree of basin inversion varies from the Tohoku arc to the Seinan arc. Although any collisional inversion of continental margin rift complexes did not occur in the Japan margin of Amur Plate, the whole basin uplift and major structural inversions with substantial thrust reactivation of earlier extensional structures have performed along the coastal belt of Japan Sea, such as the Shin'etsu basin and Sado ridge on the Tohoku side and Noto and San'in districts on the Seinan side. A gentle inversion of intra-arc rifts has occurred in the Niigata basin. While, in the Hokuriku district, the Miocene sedimentary basins remains as coastal plains or relatively low-lying area where remobilization of earlier master faults is not clear.

Some inversion mechanisms are intrinsic to the existence and lithospheric structure of the basin, and the likelihood of fault reactivation depends on the attitude of the existing fault plane such as the dip and strike to the principal stress axes [67]. If the existing fault were too steep, antithetic accessary faults might develop as new reverse faults in the footwall of the earlier extensional fault.

Figure 4 illustrates the spatial variation of inversion tectonics. The section is obtained by a tomographic inversion method in the analytical line from the western Fukushima through Echigo Plain, Sado Island and Toyama Trough to Yamato Trough [68]. The earlier normal faults are distinctively distributed in the lower part of Toyama Trough to the northwest on the Hakusan-se Shoal and in the Yamato Trough, while the later inverted reverse faults developed on Sado Island and Niigata sedimentary basin. The latter corresponds to the strain concentration zone along the eastern margin of Japan Sea [9].

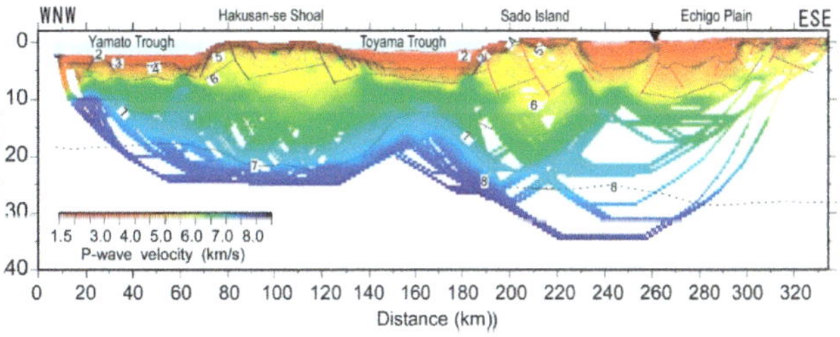

Figure 4. P wave velocity image for the crustal structure. The section is obtained by a tomographic inversion method in the analytical line from the western Fukushima through Echigo Plain, Sado Island and Toyama Trough to Yamato Trough. Faults are distinguished into earlier normal faults (blue) and later inverted reverse faults (red). Note their distinctive distribution. Compiled from [63].

As for the Tohoku arc, the start of eastward motion of Amur Plate at around 5 Ma [69] might have resulted in a new plate boundary along the strain concentration zone since 0.5 Ma. Moreover, [36] examined the U-Pb age data of Kurobegawa granite in the Hida mountain range and concluded that the granites were emplaced incrementally through the amalgamation of many intrusions since the late Miocene up to the latest intrusion event at 0.8 Ma, and that such magmatic intrusions caused rapid uplift and erosion of the Hida mountain range in the Quaternary.

As mentioned already, Japanese archipelagoes forming marginal seas between the northeastern Eurasian Continent and the northwestern Pacific Ocean comprise five island arcs (Kurile, Northeast Japan, Izu-Ogasawara, Southwest Japan, and Ryukyu arcs) which perform collisions each other in their adjacent terminations. Especially in central Japan three arcs (Northeast Japan, Izu-Ogasawara, and southwest Japan arcs) are mutually colliding, where deformed structures and active faults associated with inland crustal earthquakes are concentrated along the fringing zone east and south of Japan Sea. The mobile belt along the Japan margin of Amur Plate runs from Sakhalin - Hokkaido on the Okhotsk plate side, through the volcanic inner zones of the Northeast Japan arc, to the Southwest Japan arc on the Amur plate side [1, 2, 70, 71]. In detail, this belt includes the tectonic zone along eastern margin of Japan Sea, the Noto – San'in tectonic zone, and the Niigata - Kobe tectonic zone. Therefore, such a tectonic phenomenon could not be attributed to back arc compression of a single island-arc due simply to subduction of the oceanic plate on the Pacific side. The belt is situated in a circumference equivalent to the outer margin of the domain of back arc spreading of the Honshu arc.

Such characteristics of deformations and active faults in the inland crust as remarkable along the Japan Sea east margin is not seen at the epicentral and adjacent areas of Mw9.0 class trench-type earthquakes such as the Chile earthquake (on May 22, 1960 Mw 9.5), the Alaska earthquake (on March 28, 1964 Mw9.2), the Sumatra earthquake (on December 26, 2004 Mw9.0) and others. Conformably, the inland crustal strain accumulation and deformation directly by northwest motion of the Philippine Sea plate is not admitted conspicuous in the inner zone of southwest Japan either. After all the above-mentioned thing is the reason that cannot let the cause of inland East-West compression and the crustal earthquake generation in northeast Japan belong to northwest motion of the Pacific plate directly.

A feature of the tectonic stress field to produce crustal earthquakes can explain this most clearly. As shown in Figure 5, it is significant that a uniform compressional stress field wide spreads over the Japan Sea side of Hokkaido and Honshu, whereas variety of regionality is remarkable as for the Pacific coasts from Hokkaido to Kyushu [72].

SUMMARY

Basin Formation

Based on chronostratigraphy of Hokuriku established in late years, geomorphology, geologic structures and history of Hokuriku-Shin'etsu area were briefly summarized as follows.

After the marginal sea, i.e. Japan Sea, had been formed in the back arc area of the Honshu arc during the period from the Oligocene to Miocene time, there occurred broad transgression associated with calming of magma activity followed by cooling in central Japan. As the northward motion of the Philippine Sea plate commenced at around 15Ma, the western half of Honshu arc rotated clockwise with a decrease in area of the Japan Sea, while a buoyant subduction of Izu Arc into Honshu Arc had started.

Consequently, the mega-chasm from Fossa Magna to Toyama Trough was formed above the subducted paleo-Izu arc and the northern extension, and then the single Honshu arc differentiated into the Seinan arc and the Tohoku arc. During this process until 13Ma, the Hokuriku sedimentary basin in the Seinan arc and Shin'etsu and Niigata sedimentary basins in the Tohoku arc were developed in the Japan Sea side in the short term.

First Tectonic Inversion

According to [67], it is possible that the tectonic inversion was attributed to temporal variations in stress patterns within plates, resulting from forces caused by changes in plate boundary configuration. The sedimentary basins mentioned above had evolved individually in the period from the late half of middle Miocene to the beginning of the Pliocene. Namely, across-arc contraction tectonics with the E-W trending reverse faults and folds proceeded in the inner zone of Seinan arc, while along-arc subsiding piled up the thick sedimentary sequence in the inner zone of Tohoku arc. The start of buoyant subduction or collision of the Izu arc against the Seinan arc would have changed the configuration and relative motion at nearby plate boundaries as shown in Figure 6.

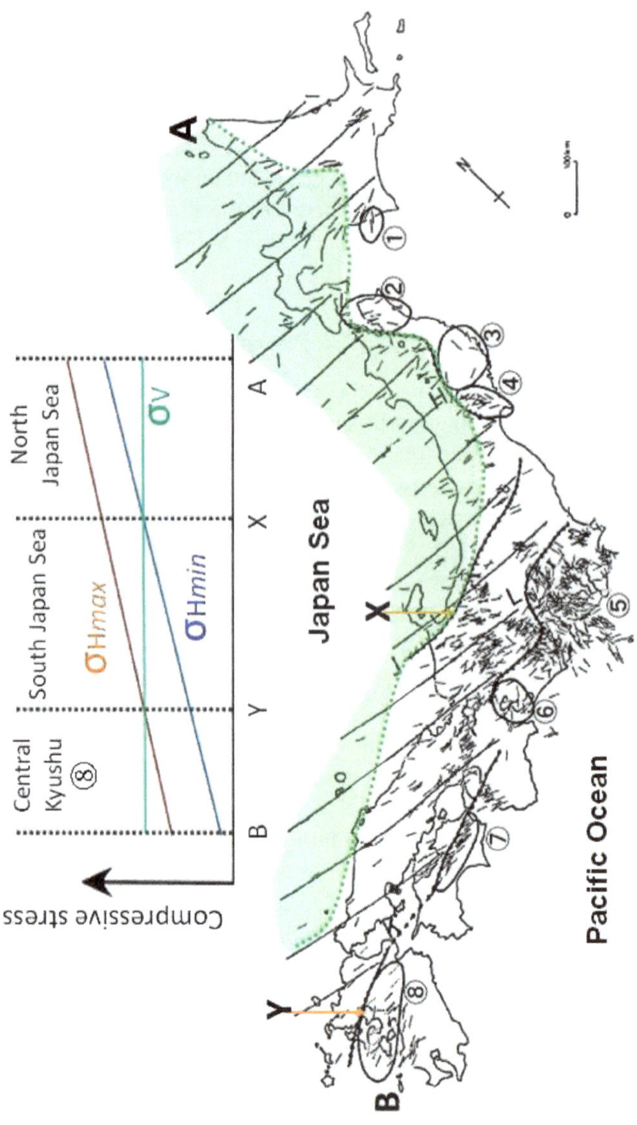

Figure 5. Regional seismogenic stress provinces in Japanese Islands. Inset is a simplified model for variation of fault types due to the along-arc stress gradient of horizontal compression. After [72] with a slight addition.

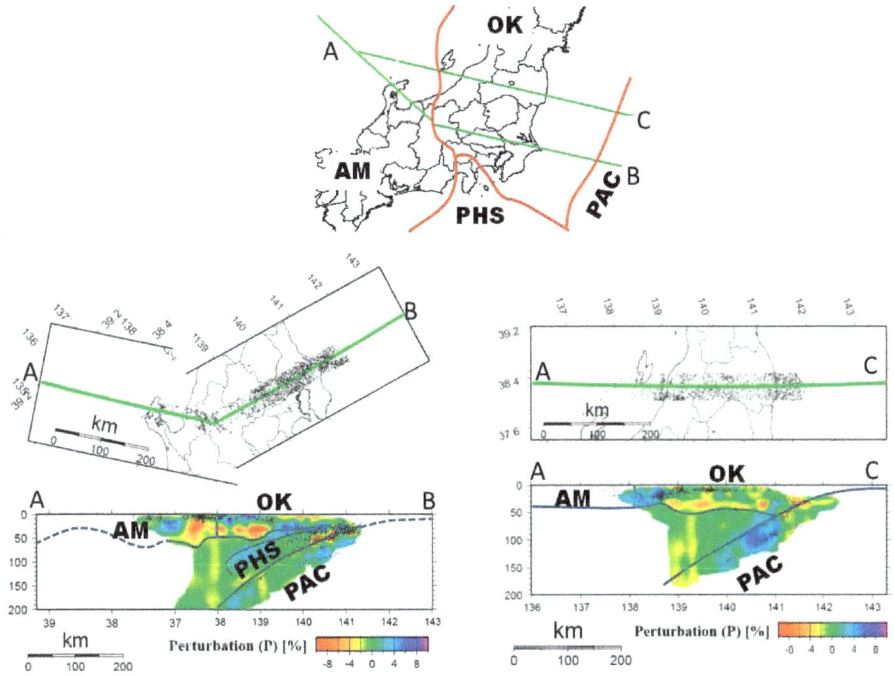

Figure 6. P-wave perturbation structures beneath the central Japan. This section image obtained by high-density seismic stations by seismic tomography using a viewer software developed by NIED to estimate the 3D seismic velocity structure typical of under Japan. Data are quoted from [73]. Tohoku and Seinan arcs are distinctive in rheology structure due to difference in plate configuration.

Superposition of Second Inversion

An eastward motion of the Amur plate happened to commence in association with the structural development of Himalaya - Tibetan plateau since the end of Miocene at around 7 Ma. Then the Tohoku, Izu and Seinan arcs started collisions mutually in central Japan, forming a core site of orogeny from where compressional tectonics has gradually spread to *far from the central Fossa Magna to the Niigata*. Therefore, sedimentary basins in the Hokuriku-Shin'etsu area developed deformation structures with reducing the deposition area of clastic sediments and being restricted into the present coastal plains and inland fault-basins. In the Hokuriku-Shin'etsu area after the late Miocene, overlapping of faults and folds in the three structural trends of north-south, east-west, and northeast-southwest is well recognizable and such a superposed structure has been illustrated also by current geomorphology. This intersecting feature has a broad expanse throughout the eastern margin of Japan Sea and is displayed in the seafloor topography conspicuously in particular along the continental slope of Okushiri and the Sado ridges. In addition, Present crustal earthquakes of moderate magnitudes occur by reverse faulting with a sense to promote the geomorphology development of the northeast - southwest direction of hills and mountains surrounding coastal plains evolved from the Miocene sedimentary basin.

Lateral Variation in Modes of Active Tectonics

This paper also noticed the present-day deformed structures and the spatial variety about the existence of basin inversion is also recognized. Namely, in the reverse fault province of Hokuriku, the inversion structure by earlier normal fault which formed the Miocene sedimentary basin is not seen, but typical basin inversion structures are seen in the reverse fault province such as the faults along the western margin of Nagano basin and the middle and northern segment of ISTL in the northern Fossa Magna area.

This cause can be considered as area characteristics of the principal stress axes arrangement by the stress gradient in the seismogenic upper crust. In the former province, stress field is in a state of strong horizontal compression ($\sigma 2 >> \sigma 3 = \sigma V$) and the latter state is somehow neutral ($\sigma 2 \approx \sigma 3$) where a strike-slip faulting is easily exchanged into a reverse faulting ([72]; see Figure 5). Moreover, [35, 48, 74, 75] presented a possible model for the deeper geologic structure, where high-angled block faults among tectonic provinces originated as transform faults and rooted in vertical weak zones in the lower crust beneath the basement of the sedimentary basin.

Based on the sedimentary basin evolution discussed in this paper and in accordance to the results of GPS geodesy and related studies [3-5, 76], the hypothesis of tectonic belt along the eastern margin of Amur Plate [70] is promising for the origin of strain concentration belt running oblique through the zone. This hypothesis includes an eastward motion of the Amur plate with convergence along the east Japan margin and transpression along the west Japan margin as well as its collision in central Japan [1,2, 71].

CONCLUSION

The development of the thrust/fold belt is attributed not only to horizontal compression but also to vertical block movements as a basement-involved tectonics. In response to the Pliocene and later compression regime, not only master fault but also secondary antithetic faults of the earlier fault-block boundaries are reactivated, and continued differential block movement such as subsiding of the sedimentary basin and uplifting of the igneous provinces.

The Neogene thrust-fault and folded belts in the Tohoku arc comprises the present-day tectonic zone of strain concentration in the sedimentary cover along the eastern margin of Japan Sea and Fossa Magna, while the stress regime of strike-slip faulting occupies the basement as inferred from focal mechanism solutions for small events. In order to account for the tectonic environment, the existence of subducted slab of the Philippine Sea plate, i.e. the paleo-forearc sliver of Izu arc, and related mechanism of rheological accommodation are possibly appreciated to have been worked in the asthenosphere mantle of the late Cenozoic arc-arc collision zone.

In the present study we conclude that an understanding of the tectonics of central Japan arc system provides useful insight into basin formation and evolution in general. The arc-to-arc colliding system in central Japan thus provides one of typical example for understanding how the development of a sedimentary basin is related to plate tectonics, because the GPS geodesy, seismicity, and active fault distribution are constraining the present process better than elsewhere.

ACKNOWLEDGEMENTS

The author is deeply grateful to Dr. Yasuto Itoh and Dr. Shigekazu Kusumoto for their encouraging comments and suggestions.

REFERENCES

1. Bird P. An updated digital model of plate boundaries. *Geochemistry Geophysics Geosystems* 2003; 1027 doi: 10.1029/2001GC000252.
2. Ashurkova S, San›kova V, Miroshnichenkoa A, Lukhneva A, Sorokinb A, Serovb M, Byzov L. *Russian Geology and Geophysics* 2011; 52(2) 239–249. doi:10.1016/j.rgg.2010.12.017.
3. Heki K. Vertical and horizontal crustal movements from three dimensional VLBI kinematic reference frame: implication for the geomagnetic reversal timescale revision. *Journal of Geophysical Research.* 1996;101(B2) 3187-3198.
4. Heki K, Miyazaki S, Takahashi H, Kasahara M, Kimata F, Miura S, Vasilenco N, Ivashchenco A, An K. The Amurian plate motion and current plate kinematics in Eastern Asia. *Journal of Geophysical Research.* 1999;104(B12) 29147-29155.
5. Heki K, Miyazaki S. Plate convergence and long-term crustal deformation in Central Japan. *Geophysical Research Letters* 2001;28 2313-2316.
6. Okamura Y, Watanabe M, Morijiri R, Satoh M. Rifting and basin inversion in the eastern margin of the Japan Sea. *The Island Arc* 1995;4(3) 166–81.
7. Takeuchi A, Shipboard Scientific Party of R/V Yokosuka Japan Sea Cruise. Submersible observations in the epicenter area of the 1993 earthquake off southwestern Hokkaido Sea of Japan. *Journal of Geophysical Research* 1998;103(B10) 24109–24125 doi: 10.1029/98JB00572.
8. Okamura Y. Inversion tectonics along the eastern margin of the Japan Sea. *Journal of the Japanese Association for Petroleum Technology*2000;65(1) 40-47. (In Japanese with English abstract)
9. Okamura Y. The Neogene and later strain concentration zone. In: Otake M, Ota Y, Taira A. (ed). *Earthquake tectonics of the active faults along the eastern margin of Japan Sea.* Tokyo: University of Tokyo Press; 2002 Chapter 7 p111-121.
10. Okamura Y. Fault-related folds and an imbricate thrust system on the northwestern margin of the northern Fossa Magna region central Japan. *TheIsland Arc* 2003;12(1) 61–73 doi: 10.1046/j.1440-1738. 2003.00379.x.
11. Taira A. Tectonic Evolution of the Japanese Island Arc System. *Annual Review of Earth and Planetary Sciences* 2001;29 109–34 doi: 10.1146/annurev.earth.29.1.109.
12. Coward MP. Thrust tectonics, thin skinned or thick skinned, and the continuation of thrusts to deep in the crust. *Journal of Strutural Geology*1983;5 113–123.
13. Tozer R, Butler R, Corrado S. Comparing thin- and thick-skinned thrust tectonic models of the Central Apennines, Italy. EGU Stephan Mueller Special Publication Series 2002;1 181–194.
14. Lacombe O, Mouthereau F. Basement-involved shortening and deep detachment tectonics in forelands of orogens: Insights from recent collision belts (Taiwan, Western Alps, Pyrenees). *Tectonics* 2002;21(4), 1030, doi:10.1029/2001TC901018.
15. Research Group for Active Faults in Japan. Active Faults in Japan Sheet Maps and Inventories. Tokyo: University of Tokyo Press; 1991.
16. Nakata T, Imaizumi T. (eds) [DVD-ROM] *Digital active fault map of Japan* DVD-ROM2. Tokyo: University of Tokyo Press; 2002.
17. Uemura T, Yamada T. (eds.) Regional Geology of Japan Part 4 (Chubu I). Tokyo: Kyoritsu Publishing; 1988.
18. Uemura T. Late Cenozoic Folding Mechanism and Crustal Dynamics in the Southern Back Arc Area of Northeast Honshu, Japan. In: *Development of Sedimentary Basins and Its Relation to Folding.* Memoirs of the Geological Society of Japan 34 p199-209. Geological Society of Japan; 1990.
19. Sagiya T, Miyazaki S, Tada T. Continuous GPS Array and Present-day Crustal Deformation of Japan. *Pure and Applied Geophysics* 2000;157(11-12) 2303-2322.
20. Okamura Y. Ishiyama T. and Yanagisawa Y. Fault-related folds above the source fault of the 2004 mid-Niigata Prefecture earthquake in a foled-and-thrust belt caused by basin inversion along the eastern margin of Japan Sea. *Journal of Geophysical Research* 2007;112(B3) B03S08 doi:10.1029/2006JB004320.
21. Otuka,Y. Active folded structures. *Jishin (Series 1)* 1942;14, 46-63.
22. Takeuchi A. *Vicissitude of the stress field and tectonic evolution in the Hokushin'etsu area since the Pliocene. The Earth Monthly* 1999;21, 583-588.
23. Li J, Miyashita K, Kato T, Miyazaki S. GPS time series modeling by autoregressive moving average method: Application to the crustal deformation in central Japan. *Earth Planets Space* 2000;52 155–162.

24. Niitsuma S, Niitsuma N, Saito K. Evolution of the Komiji Syncline in the North Fossa Magna central Japan: Paleomagnetic and K-Ar age insights.*The Island Arc* 2003;12(3) 310-323.

25. Takeuchi A. Temporal changes of regional stress field and tectonics of sedimentary basin. *Journal of the Geological Society of Japan* 1981;87(11) 737-751.

26. Takano O. Changes in depositional systems and sequences in response to basin evolution in a rifted and inverted basin: an example from the Neogene Niigata–Shin'etsu basin Northern Fossa Magna central Japan. *Sedimentary Geology* 2002;152(1-2) 79–97 doi: 10.1016/S0037-0738(01)00286-X

27. Takano O, Tateishi M, Endo M. Tectonic controls of a backarc trough-fill turbidite system: The Pliocene Tamugigawa Formation in the Niigata–Shin'etsu inverted rift basin Northern Fossa Magna central Japan. *Sedimentary Geology* 2005;176(3-4) 247–279 doi: 10.1016/j.sedgeo.2005.01.004

28. Takahashi M. Tectonic Development of the Japanese Islands Controlled by Philippine Sea Plate Motion. *Journal of Geography* 2006;115(1) 116-123. (In Japanese with English abstract)

29. Kano K, Kato H, Yanagisawa Y, Yoshida F. (ed.) Stratigraphy and Geologic History of the Cenozoic of Japan. *Report of Geological Survey of Japan*no.274 Tsukuba: Geological Survey of Japan; 1991. (In Japanese with English abstract)

30. Fujii S, Kaseno Y, Nakagawa T. Neogene paleogeography in the Hokuriku region, Central Japan, based on the revised stratigraphic correlation.*Memoirs of Geological Society of Japan* 1992;(37) 85-95. (In Japanese with English abstract)

31. Geological Society of Japan (ed.) *Regional Geology of Japan 4-Chubu.* Tokyo: Asakura Publishing; 2006.

32. Tamura I, Yamazaki H, Nakamura Y. The wide-spread tephra in the Hokuriku Group and the Quaternary tectonics of the Toyama Basin, Japan. Journal of the Geological Society of Japan 2010;116 (Sup.) S1-S20. doi: 10.5575/geosoc.116.S1

33. Kurokawa K. Studies of subaqueous tephra beds stratigraphy sedimentology volcanology -stratigraphy, sedimentology and volcanology-. *Chikyu Kagaku* 2005;59(1) 62-67.

34. Harayama S, Wada H, Yamaguchi Y. Quaternary and Pliocene granites in the Northern Japan Alps. In: *Hutton Symposium V field Guidebook for TripA1*. Interim-Report 28 p3-21. Tsukuba: Geological Survey of Japan; 2003.

35. Takeuchi A. Duplex Stress Regime in the North Fossa Magna, Central Japan. *Bulletin of the Earthquake Research Institute* 2008;83 1-8. Tokyo: University of Tokyo; 2008.

36. Ito H, Yamada R, Tamura A, Arai S, Horie K, Hokada T. Earth›s youngest exposed granite and its tectonic implications: the 10–0.8 Ma Kurobegawa Granite. *Scientific Reports* 2013;3 doi:10.1038/srep01306.

37. International Stratigraphic Committee. ISC: Formal stratigraphical definitions. http://quaternary.stratigraphy.org/ definitions/ (accessed 28 June 2013)

38. Ota Y, Hirakawa K. Marine Terraces and their Deformation in Noto Peninsula, Japan Sea Side of Central Japan. *Geographical Review of Japan*1979;52(4) 169-189. (In Japanese with English abstract)

39. Tamaki K, Suyehiro K, Allan J, Ingle J, Pisciotto K. Tectonic synthesis and implications of Japan Sea ODP drilling program. *Proceedings of the Ocean Drilling Program, Scientific Results,* 1992; 127/128(Part 2) p1333–1348.

40. Jolivet L, Tamaki K. Neogene kinematics in the Japan Sea region and the volcanic activity of the Northeast Japan arc. In: *Proceedings of the Ocean Drilling Program, Scientific Results.* 1992;127/128(Part 2) 1311–1331.

41. Jolivet L, Tamaki K, Fournier M. Japan Sea opening history and mechanism: A synthesis. *Journal of Geophysical Research* 1994; 99(B11) 22237-22259.

42. Itoh Y, Uno K, Arato H. Seismic evidence of divergent rifting and subsequent deformation in the southern Japan Sea and a Cenozoic tectonic synthesis of the eastern Eurasian margin. *Journal of Asian Earth Sciences* 2006;27 933–942.

43. Shiki T, Tateishi M. On the hypothesis of aulacogen for the Fossa Magna. In: Proceedings Memorial retired Professor Seiya Ueda -Active Margin-. *The Earth Monthly* 1991;*extra* 3 106-112.

44. Ishida H. Characteristics of the basement structure and formation of oil folding: examples of Toyama Bay through Off-Niigata-Kubiki. Annual Report of Technology and Research Center (TRC)'s Activities for the Year 1995. p17–24. Chiba: Japan Oil, Gas, Metal National Corporation (JOGMEC); 1995.

45. Kobayashi I. (ed) Chubu District I In: Editorial Committee for the Supplement Edition of Geology of Japan (ed.) Geology of Japan – Supplement. Tokyo: Kyoritsu Shuppan; 2005 p129-166.

46. Sakamoto T. Cenozoic Strata and Structural Development in the Southern Half of the Toyama Basin, Central Japan. *Report of Geological Survey of Japan* 1966; no.213, 1-28. (In Japanese with English abstract)

47. Kaseno Y. *Geology of Ishikawa Prefecture –Supplements*. Kanazawa: Hokuriku Geological Institute; 2001.

48. Takeuchi A. Basement-involved tectonics in the North Fossa Magna central Japan: The significance of the northern Itoigawa-Shizuoka Tectonic Line. *Earth Planets Space* 2004;56 1261-1269.

49. Takahashi M. Tectonic Boundary between Northeast and Southwest Japan Arcs during Japan Sea Opening. *Journal of the Geological Society of Japan* 2006;112(1) 14-32.

50. Furuse N, Kono Y. A Three Dimensional Gravity Inversion Method for Layered Structures with Lateral Density Variation. An Application to the Hokuriku District, Central Japan. *Jishin 2*, 1990;43, 1-11. (In Japanese with English abstract)

51. Senna S, Fujiwara H, Kawai S, Aoi S, Kunugi T, Ishii T, Hayakawa M, Morikawa N, Honda R, Kobayashi K, Oi M, Yasojima Y, Okumura N. A Study on Strong-Motion

Maps for Scenario Earthquakes in Morimoto Togashi Fault Zone. *Technical Note* no.255. Tsukuba: Institute of Disaster and Earth Sciences (NIED); 2004.

52. Senna S, Fujiwara H, Kawai S, Aoi S, Kunugi T, Ishii T, Hayakawa M, Morikawa N, Honda R, Kobayashi K, Oi M, Yasojima Y, Okumura N. A study on strong-motion maps for scenario earthquakes in Tonami plain fault zone. [CD-ROM] Tsukuba: *Technical Note* no.263. Tsukuba: Institute of Disaster and Earth Sciences (NIED); 2005.
53. Senna S, Fujiwara H, Kawai S, Aoi S, Kunugi T, Ishii T, Hayakawa M, Morikawa N, Honda R, Kobayashi K, Oi M, Yasojima Y, Okumura N. A Study on Strong-Motion Maps for Scenario Earthquakes in Takayama-Oppara Fault Zone. [CD-ROM] Tsukuba: *Technical Note* no.282. Tsukuba: Institute of Disaster and Earth Sciences (NIED); 2005.
54. Ishiwatari A. Diverse lava and pyroclastic rocks of continental arc: the Oligocene-Miocene volcanic rocks in Noto. In: Geological Society of Japan (ed.) *Regional Geology of Japan 4-Chubu.* Tokyo: Asakura Publishing; 2006 p336-337.
55. Yanagisawa Y, Yoshikawa T. Diatomaceous mudstone and Glauconite sandstone in the Suzu district. In: Geological Society of Japan (ed.) *Regional Geology of Japan 4-Chubu*. Tokyo: Asakura Publishing; 2006 p346-347.
56. Kano K, Yamauchi Y, Miyake Y. Miocene subaqueous lavas and volcaniclastic rocks in the Shimane Peninsula, SW Japan. In: *Field excursion guidebook for 107th annual meeting of Geological Society of Japan.* Tokyo: Geological Society of Japan; 2000 p23-34.
57. Tai Y. On the «Shinji Folded Zone». *Memoir of Geological Society of Japan* 1973;no.9 137-146. (In Japanese with English abstract)
58. Kano K, Yoshida F. Geology of the Sakaiminato District. 1:50000 Quadrangle Series 12-Okayama-7-57. Tokyo: Geological Society of Japan; 1985.
59. Nomura R. Geology of the central part in the Shiimane Peninsula - Part 1 Miocene stratigraphy -. *Journal of Geological Society of Japan* 1986;92(6) 405-420. (In Japanese with English abstract)
60. Tokuyama H, Honza E, Kimura M, Kuramoto S, Ashi J, Okamura Y, Arato H, Ito Y, Soh W, Hino R, Nohara T, Abe H, Sakai S, Mukaiyama K. Tectonic development in the regions around Japan since latest Miocene. Marine Surveys and Technology 13(1). [CD-ROM] Japan Society for Marine Surveys and Technology; 2001.
61. Watanabe M. Glauconite sandstone along the boundary-layer between Yabuta and Sugata formations. In: Geological Society of Japan (ed.)*Regional Geology of Japan 4-Chubu*. Tokyo: Asakura Publishing; 2006 p348-349.
62. Kaizuka S. Quaternary morphogenesis and tectogenesis of Japan. *Zeitschrift für Geomorphologie* 1987;(Supplement Bd63) 61-73.
63. Takeuchi A. Recent crustal movements and strains along the eastern margin of Japan Sea Floor. In: Isezaki N (ed). *Geology and Geophysics of the Japan Sea.* Volume 1 of *Japan-USSR Monograph*. Tokyo: Terra Scientific Pub Co; 1996. p385-398.
64. Takeuchi A, Okada A. Crustal movement. In: Chapter 7.1 of *Geology of Japan 5 – Chubu Area.* Tokyo: Kyoritsu Publishing; 1988 p192-194. (in Japanease)
65. Takeuchi A. Geologic structure and late Cenozoic evolution of the Hokuriku and Shin'etsu regions, north-central Japan. *Journal of the Geological Society of Japan* 2010;116(11) 624-635.
66. Toyama Prefecture. Survey on the Kurehayama Fault. In: *Result Report FY 1995 grant for Earthquake Research.* Science and Technology Agency of Japan; 1996 p235.
67. Dewey JF. *Kinematics and dynamics of basin inversion.* Geological Society London Special Publications. 1989;44(1) 352.
68. Japan Agency for Marine-Earth Science and Technology (JAMSTEC). Multichannel Seismic survey of strain concentration zone along the eastern margin of Japan Sea. In: *FY2009 Report of Observation and survey research project focused on strain concentration zone.* Tsukuba: National Research Institute for Earth Science and Disaster Prevention (NIED);2010. p236-250.
69. Zonenshain LP, Savostin LA. Geodynamics of Baikal rift zone and plate tectonics of Asia, *Tectonophysics* 1981;76 1–45, 1981 doi:10.1016/0040-1951(81)90251-1
70. Ishibashi K. The 1995 Kobe Japan earthquake (M7.1) in the "Mobile Belt along the Eastern Margin of Amur Plate" and its implication to the regional seismic activity (preliminary report). *Chishitsu Nyusu* 1995;490 14–21 Geological Survey of Japan .
71. Wei D, Seno T. Determination of the Amurian plate motion. In: Flower M, Chung SL, Lo CH, Lee TY (ed.) *Mantle dynamics and plate interactions in East Asia.* Geodynamics Series 27 AGU;1998 p337-346 doi: 10.1029/GD027p0337.
72. Tsukahara H. Regional seismogenic stress provinces in Japanese Islands. *The Earth Monthly* 2003;25 302-309.
73. Matsubara M, Obara K. The 2011 Off the Pacific Coast of Tohoku earthquake related to a strong velocity gradient with the Pacific plate. *Earth Planets Space* 2011;63 663-667.
74. Iio Y, Sagiya T, Kobayashi Y, Shiozaki I. Water-weakened lower crust and its role in the concentrated deformation in the Japanese Islands. *Earth Planetary Science Letters* 2002;203(1) 25-253.
75. Shibazaki B, Garatani K, Iwasaki T, Tanaka A, Iio Y. Faulting processes controlled by the nonlinear flow in the deeper crust and upper mantle beneath the northeastern Japanese island arc. *Journal of Geophysical Research* 2008;113(B8) B08415 doi:10.1029/2007JB005361.
76. Hyodo M. and K. Hirahara. A viscoelastic model of interseismic strain concentration in Niigata-Kobe Tectonic Zone of central Japan. *Earth Planets Space* 2003;55 667–675.

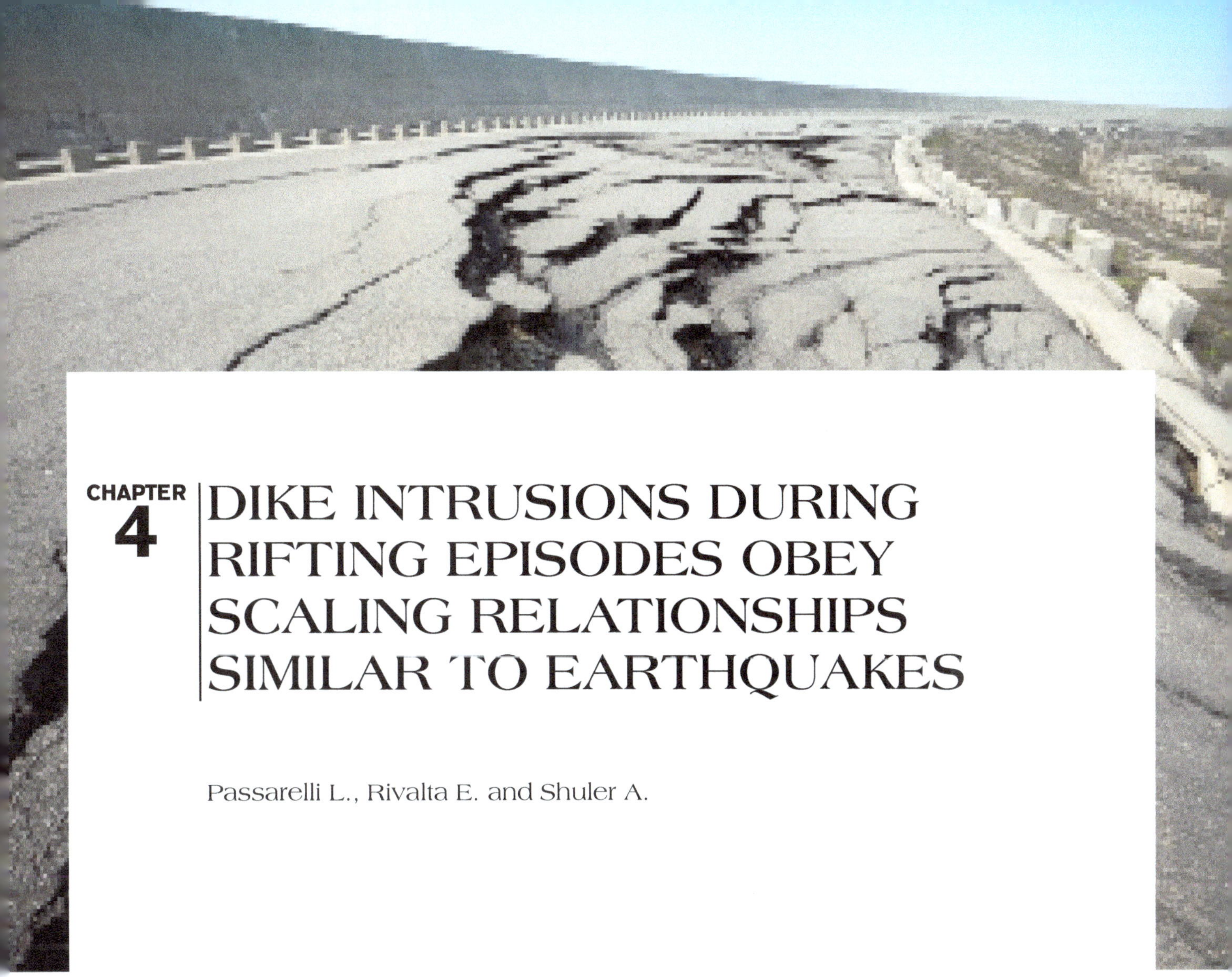

CHAPTER 4

DIKE INTRUSIONS DURING RIFTING EPISODES OBEY SCALING RELATIONSHIPS SIMILAR TO EARTHQUAKES

Passarelli L., Rivalta E. and Shuler A.

INTRODUCTION

There is a general consensus that earthquakes exhibit self-similar behavior over a wide range of magnitudes on both local and global scales. Tectonic earthquakes satisfy two power-law scaling relationships to a reasonable approximation: the Gutenberg-Richter relation, which describes the frequency-magnitude distribution, and the modified Omori law, which describes the temporal decay of aftershocks. Driven by the accumulation of tectonic stresses, the earthquake rupture process is controlled by local stress conditions and the frictional and material properties of faults.

Self-similarity is also observed in volcanic systems. Both the global rate of volcanic activity and the frequency-volume distributions of major volcanic eruptions have been shown to follow power laws. Most volcanic eruptions are initiated by dike intrusions, which nucleate when the pressure inside an inflating magma reservoir exceeds a threshold value. Once a dike is nucleated, propagation is controlled by the reservoir pressure as well as viscous, elastic and thermal stresses. Magma replenishment is necessary for additional intrusions to be sourced from the same reservoir.

Along late-stage continental rifts, plate boundary separation is accommodated primarily by dike intrusions, and to a lesser extent by faults, during discrete rifting episodes. Tensile stresses accumulated over decades to centuries exert the dominant controls on the deformation cycle, although crustal magma accumulation influences the timing of individual dike intrusions. Discrete rifting episodes have been observed along two subaerial divergent plate boundaries - the Krafla segment of the Northern Volcanic Rift Zone in Iceland (1975–1984) and the Manda-Hararo segment of the Red Sea Rift in Ethiopia (2005–2010). In both episodes, the spatial and temporal patterns of dike intrusions share many similarities with tectonic earthquakes in mainshock-aftershock sequences. The initial and largest dike intrusions were followed by clusters of smaller dike intrusions. Additionally, in a process that closely resembles the triggering of aftershocks by static stress changes, later dike intrusions were emplaced in areas where local tectonic stresses had been either increased or not completely relieved by earlier dike intrusions.

In the East African Rift System, the spatial distributions of earthquakes and volcanic vents are known to follow power laws. Here, we examine the scaling relationships for dike intrusions associated with the Krafla and Manda-Hararo discrete rifting episodes. We perform a detailed statistical

Citation: Passarelli, L., Rivalta, E. & Shuler, A. Dike intrusions during rifting episodes obey scaling relationships similar to earthquakes. Sci. Rep. 4, 3886; DOI:10.1038/ srep03886 (2014).

analysis of the dimensions and timings of individual dike intrusions and assess whether their behavior follows power laws analogous to the Gutenberg-Richter relation and the modified Omori law.

Data

Our analysis is based on published records of dike intrusions associated with the Krafla and Manda-Hararo rifting episodes (Fig. 1).

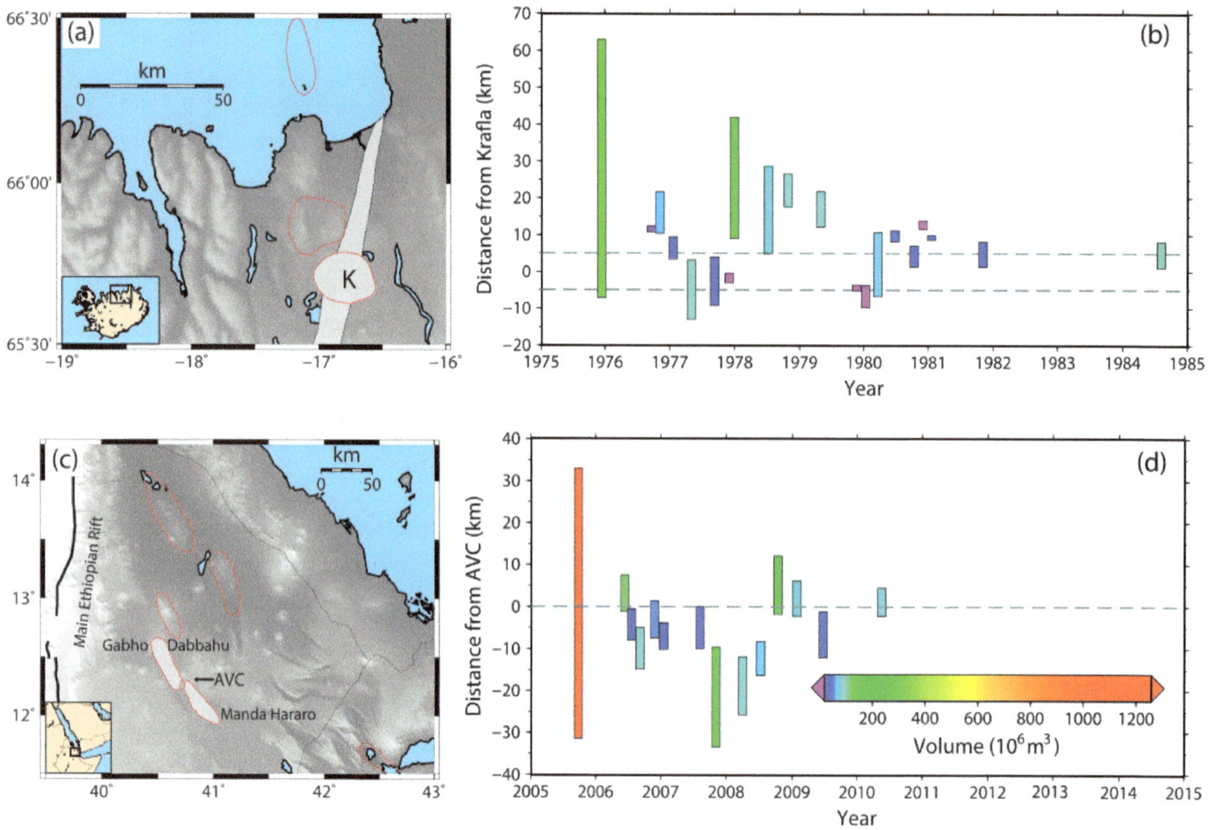

Figure 1: Summary of the Krafla and Manda-Hararo rifting episodes.

The Krafla rifting episode is composed of 20 major dike intrusions associated with subsidence of the Krafla caldera. Accounting for the fact that reservoirs expand as their internal pressures decrease, the volume of each dike intrusion is estimated to be twice the subsidence volume measured from elevation and gravity data (Fig. 1d)[17,18]. The lengths of the Krafla dike intrusions are constrained by field measurements of fractures and the hypocenters of triggered earthquakes.

The dimensions of the 14 dike intrusions of the Manda-Hararo rifting episode are constrained well by seismic and geodetic observations. We use the MHA (and I. Hamling personal communication 2011), and MHB datasets for estimates of the lengths and volumes of each dike intrusion. The MHB dataset does not include information for the most recent intrusion, but does provide estimates of the dislocated areas of the first thirteen dike intrusions. For the volume of the first intrusion in the MHB dataset, we exclude contributions from magma chambers beneath Dabbahu and Gabho volcanoes, and only consider the volume sourced from the magma reservoir associated with the Ado'Ale Volcanic Complex (Fig. 1a), which was active during the entire rifting episode.

In addition to examining the dimensions of the dikes associated with each rifting episode separately, we also combine the volume information to produce two global rifting datasets - KMHA (Krafla and MHA) and KMHB (Krafla and MHB).

A Gutenberg-Richter relation for rifting episodes

Empirically, tectonic earthquakes follow the Gutenberg-Richter relation[1], $N(M) = 10^{a - bM}$, where N(M) is the number of earthquakes with magnitudes greater than or equal to M occurring in a given time, and a and bare constants describing earthquake productivity and the relative distributions of small and large earthquakes, respectively. Rewritten in terms of seismic moment, M_0, the Gutenberg-Richter relation can be transformed into a probability density function $f(M_0) = \beta M_C M_0^{-1-\beta}$ where M_C is the seismic moment corresponding to the magnitude of completeness for a given seismicity catalog and $\beta = 2/3$ for the typical situation where the b-value is one. The probability that an earthquake meets or exceeds a given size is therefore described by the survivor function $F(M_0) = (M_C/M_0)^{\beta}$. Because seismic moment is proportional to rupture area, if earthquakes can be approximated by slip on circular fault patches, similar power-law relationships can be defined for rupture area and length.

The size of a dike intrusion is often described by the geodetic moment, which is proportional to the volume of magma intruded in an opening tensile crack. If dike intrusions along divergent plate boundaries obey self-similar scaling relationships like earthquakes, and if dike intrusions can be approximated by circular cracks, then the lengths, dislocated areas and volumes of dike intrusions should all follow power law distributions.

Here, we examine each of our rifting datasets for power-law behavior. Assuming that dike intrusions obey self-similar scaling relationships above some threshold size, x_{min}, we model each survivor function as $N(X \geq x) = (x_{min}/x)^{\beta}$, where $N(X \geq x)$ is the probability that a dike intrusion meets or exceeds a given size and x represents either the volumes, dislocated areas, or lengths of dike intrusions in a given rifting dataset. We estimate the parameters β and x_{min} following Clauset et al and assess the goodness-of-fit using the Kolmogorov-Smirnov statistic. Using the value of x_{min} determined from the power-law parameterization, we also model each rifting dataset with exponential and log-normal distributions and use the likelihood ratio test to determine which right-skewed distribution provides the best fit to the data (see Method and Tab. 1).

Table 1: Parameterizations of rifting datasets. For each of the nine rifting datasets, we report maximum likelihood estimates of power-law, exponential and log-normal parameterizations. p_{KS} **values greater than 0.10 are reported in bold and indicate statistically** ...

Rifting Dataset	Power Law			Exponential		Log-Normal			Power Law vs Exponential		Power Law vs Log-Normal	
	β	x_{min}	p_{KS}-value	λ	p_{KS}-value	μ	σ	p_{KS}-value	LR	p_{LR}-value	LR	p_{LR}-value
KMHA volumes	1.46 ± 0.49	58 ± 16	**0.35**	0.048	0.00	4.73	0.77	0.06	–	–	–	–
KMHB volumes	1.28 ± 0.64	37 ± 20	**0.83**	0.008	0.00	4.39	0.76	**0.15**	–	–	5.36	**0.09**
MHA volumes	1.06 ± 0.53	37 ± 11	**0.84**	0.048	0.08	4.54	0.96	**0.21**	–	–	3.65	**0.03**
MHB volumes	1.08 ± 0.57	48 ± 19	**0.91**	0.004	**0.19**	4.80	0.95	**0.33**	4.3	**0.03**	2.21	**0.10**
Krafla volumes	1.62 ± 0.52	40 ± 14	**0.53**	0.011	0.00	4.29	0.56	**0.43**	–	–	2.23	0.22
MHB areas	1.59 ± 0.94	62 ± 23	**0.52**	0.008	**0.34**	4.76	0.48	**0.68**	0.30	0.93	−0.63	0.77
MHA lengths	1.84 ± 0.50	9 ± 1	**0.57**	0.048	0.00	2.69	0.58	**0.12**	–	–	4.13	**0.00**
MHB lengths	2.11 ± 0.33	13 ± 2	**0.90**	0.045	0.00	3.00	0.43	**0.41**	–	–	1.74	0.11
Krafla lengths	1.17 ± 0.47	6 ± 3	**0.97**	0.056	0.00	2.56	0.72	**0.40**	–	–	2.55	0.13

For the two global rifting datasets (KMHA and KMHB), we find that intrusion volumes are modeled well by power laws and poorly by exponential and log-normal distributions (Fig. 2). In most cases, power laws also outperform the alternative distributions for individual volume, length and area datasets (Tab. 1 and Fig. 3), although the small sample sizes preclude a statistically robust comparison. Nonetheless, our observation that a power law consistently performs well across the various rifting datasets suggests that dike intrusions have self-similar scaling relationships.

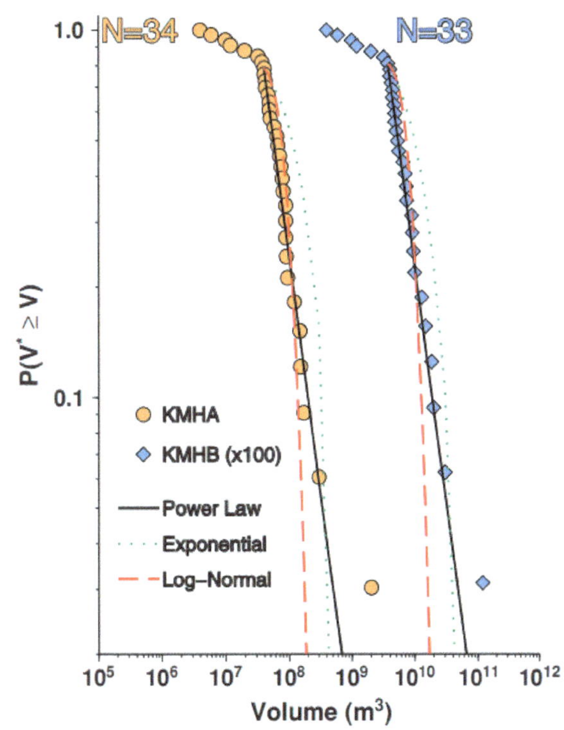

Figure 2: Survivor functions for the global rifting datasets.

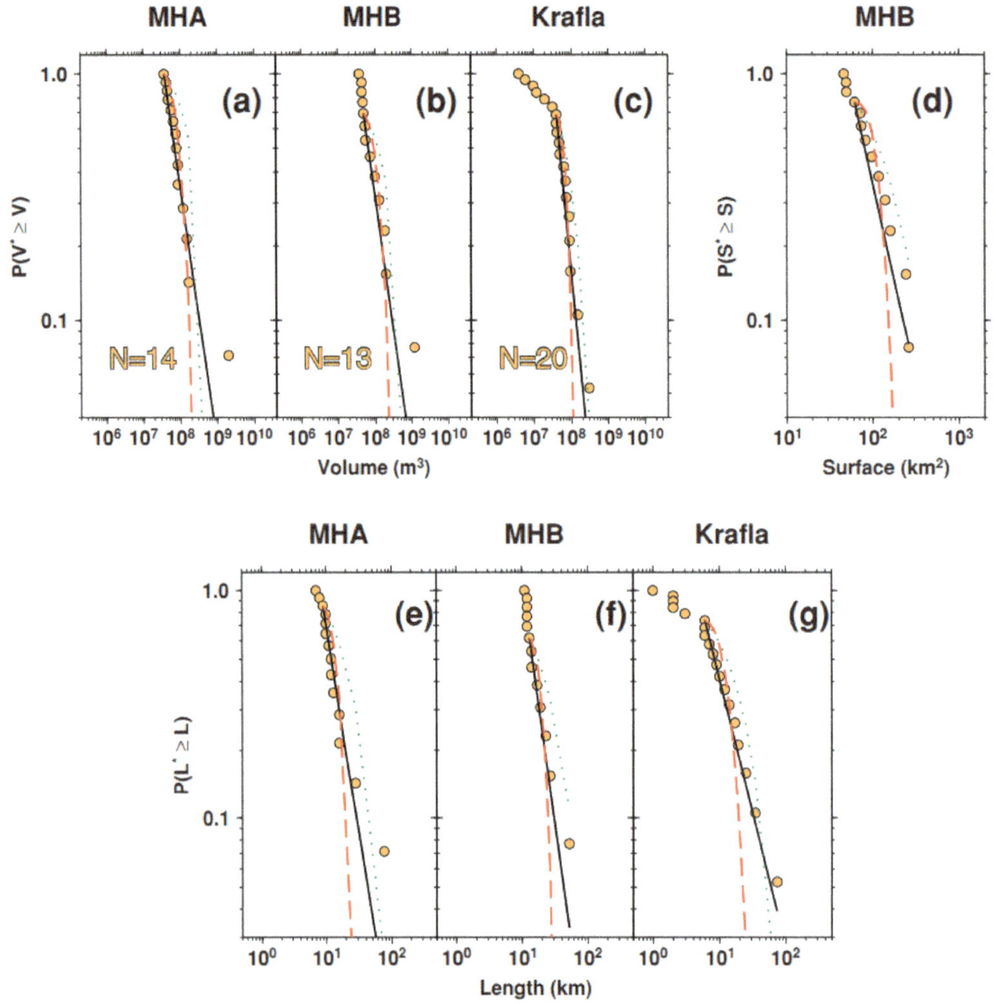

Figure 3: Survivor functions for individual rifting datasets.

Power-law parameterizations for the global rifting datasets are characterized by β-values of ~1.3 to ~1.5, which are significantly higher than the theoretical value of $\beta = 2/3$ for earthquakes as well as the empirical value of β found for most tectonic earthquake sequences. Earthquake sequences with elevated β-values (i.e. $1 < \beta < 2$), which have a larger proportion of small events, have been observed along divergent plate boundaries as well as in volcanic and geothermal environments, and have been attributed to low values of confining stress30, material heterogeneity, large thermal gradients, and the presence of fluids.

We note that the value of x_{min} for both global rifting datasets is ~40 ×10⁶ m³ to ~60 ×10⁶ m³. Due to differences in how dike intrusions were detected at Krafla and Manda-Hararo, however, it is difficult to determine if x_{min} results from catalog incompleteness or whether this quantity is associated with some threshold property of the source magma chambers, such as a minimum volume or pressure required for dike nucleation.

Individual dike intrusions of the Krafla rifting episode were identified by rapid subsidence of Krafla caldera caused by the deflation of a 3–5 km deep magma chamber. There, we note that the volume rifting datasets (Fig. 3c) could be fit by two power laws with different values of β corresponding to intrusions smaller and larger than x_{min}. The smallest intrusions occurred later in the sequence within 10 km of the center of the caldera (Fig. 1d), and were accompanied by fissure eruptions. Our observation of a potential piecewise power-law relationship may be due to the fact that lateral propagation played a lesser role in the later dike intrusions, or because these events were controlled to a greater degree by local stresses rather than tectonic extension.

At Manda-Hararo, individual dike intrusions were identified by local seismic and satellite geodetic data rather than monitoring the 8–10 km deep source magma chamber directly, and we find that the volume, length and surface area rifting datasets are well modeled by a single power-law relationship. The values of x_{min} for these datasets are close to the smallest dimensions observed, although we cannot exclude the possibility that additional small dike intrusions are missing from the catalog since the resolution of geodetic data decreases with depth and the maximum magnitude of earthquakes induced by dike intrusions is correlated with the volume of magma injected in the rift.

A Modified Omori law for Rifting Episodes

For earthquakes, the decay of aftershock activity with time follows an empirical relationship known as the modified Omori law, $n(t) = \frac{K}{(t+c)^p}$, where n(t) is the rate of aftershocks in a given magnitude range occurring at time t since the mainshock, and K, c, and p are constants. Assuming that aftershocks follow the Gutenberg-Richter relation and the modified Omori law, the cumulative seismic moment released in a time interval can be expressed as the product of the average seismic moment released per event and the total number of events in the time interval, the integral form of the modified Omori law. It follows therefore that the seismic moment release rate also exhibits power-law time dependence. These simple relationships result from self-similarity in the processes associated with static and dynamic earthquake triggering, and deviations are often linked to the occurrence of strong aftershocks that generate secondary sequences of dependent events.

A modified Omori law has also been identified for earthquakes preceding and following volcanic eruptions, which suggests that the brittle crust may respond similarly to tectonic and volcanic stresses. Here, we evaluate whether the rate of volume emplaced in the crust during the rifting sequences can be modeled with power laws similar to the modified Omori law for earthquakes.

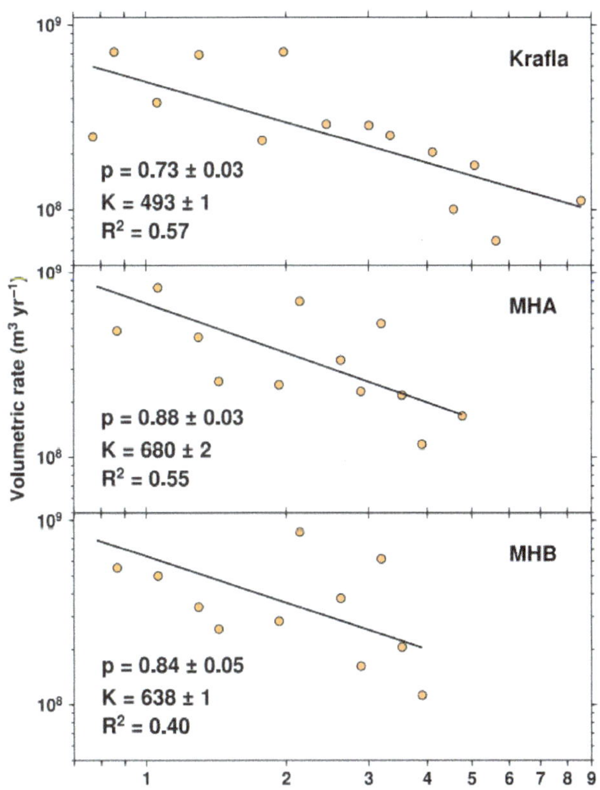

Figure 4: Omori-type decay of dike intrusion volume.

Analogous to earthquakes in a mainshock-aftershock sequence, the long-term volumetric rate during rifting episodes decays with time whereas the time between consecutive dike intrusions increases with time (Fig. 1). Secondary diking sequences are also observed following the largest dike intrusions. We quantify this behavior by calculating rifting volumetric rates using the equation $v_r = K/t^p$, where t is the time from the onset of a rifting episode, p is the power-law exponent and K is a constant (Fig. 4). From a simple linear regression in logarithmic space, we find that ~40–60% of the variance in the datasets can be explained by power-law relationships with p-values of 0.73 for Krafla and 0.84–0.89 for Manda-Hararo. These parameterizations closely resemble the form of the modified Omori law, which is predicted by rate-state theory for any impulsive stress change. The relatively short emplacement times of the dike intrusions, which were on the order of a few hours to days, compared to interevent times of a few months to years may explain why we find consistency between the temporal behavior of earthquakes and dike intrusions in rifting episodes.

For earthquake sequences that follow the Gutenberg-Richter relation and the modified Omori law, it has been shown that aftershock productivity is a function of mainshock magnitude. If rifting episodes are governed by similar physics, this could explain why the initial dike intrusions of the Krafla and Manda-Hararo sequences, which had volumes larger than $\sim 100 \times 10^6$ m³, were followed by several smaller diking events. Sequences with multiple dike intrusions and large cumulative volumes have also been recognized during past rifting episodes in the Northern Iceland Volcanic Zone. Conversely, single dike intrusions are associated with rifting episodes at Dallol in Ethiopia in 2004 and Lake Natron in Tanzania in 2007. The volumes emplaced during the Dallol and Lake Natron episodes, 60×10^6 m³ and 90×10^6 m³respectively, are comparable to the values of x_{min} for the Krafla and Manda-Hararo rifting datasets.

At volcanoes, the interevent repose time, or the time between magmatic eruptions, is correlated with erupted volume. In open conduit systems characterized by short repose periods, volcanic eruptions generally follow the time-dependent model, where interevent time is dependent on both the size of the last eruption and the magma recharge rate. Here, we investigate the role of the magma recharge rate in controlling the timing of individual dike intrusions at Krafla and Manda-Hararo by examining the time intervals between successive dike intrusions in each rifting episode. We use the Bayesian hierarchical time-predictable model of Passarelli et al, assuming that the interevent time, $r_i = t_{i+1}$ t_i, follows the empirical equation $r_i = \delta v_i^{\gamma}$, where v_i is the volume of the ith dike intrusion, and δ and γ are constants. If $\gamma = 0$ the system is not time-predictable, whereas if $\gamma = 1$ the magma recharge rate is constant and the interevent times and volumes of dike intrusions scale linearly. We perform 10,000 simulations of the statistical model to determine the value of γ for each rifting dataset.

We find that the average values of γ are 0.84 for the Krafla dataset and 0.63 and 0.69 for the Manda-Hararo datasets (Fig. 5). These values of $\gamma < 1$ demonstrate a non-linear relationship between dike intrusion volume and interevent time, such that the interevent times following large dike intrusions are longer than those following small dike intrusions, but shorter than those predicted by the classical time-predictable model. Because the inflation of magma reservoirs is observed to decay exponentially after dike intrusions, this observation suggests that the shallow magma chambers beneath Krafla and Manda-Hararo are hydraulically connected to larger and deeper reservoirs that allow magma recharge to occur at faster rates for larger pressure drops. The residuals about the best-fit regression lines for the Krafla and Manda-Hararo datasets are larger than those obtained for similar analysis completed for Kilauea and Etna volcanoes, which may suggest that magmatic stresses play a lesser role in controlling the timing of individual dike intrusions during rifting episodes relative to purely volcanic systems.

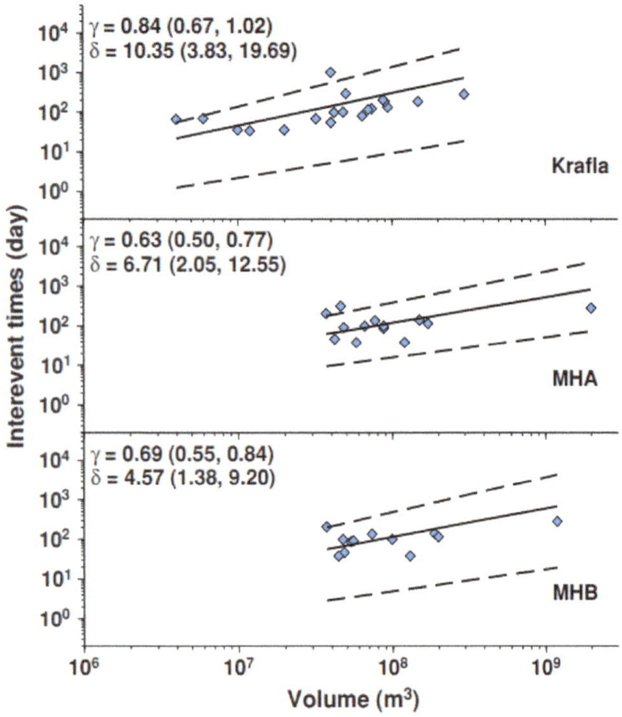

Figure 5: Time-predictable model for volume rifting datasets.

Influenced by the magma recharge rate, the time-predictable behavior of dike intrusions at Krafla and Manda-Hararo seems to be in direct contrast to earthquakes, where the time interval between two events is inversely correlated with the magnitude of the first event, such that shorter interevent times are observed following larger earthquakes.

Discussion

We have demonstrated that the Krafla and Manda-Hararo diking episodes follow self-similar scaling relationships analogous to the Gutenberg-Richter relation and the modified Omori law for tectonic earthquakes. High p_{KS}-values demonstrate the goodness of fit of the best-fit power-laws at high significance levels, and low p_{LR}-values prove that power-laws reliably outperform the best-fit log-normal and exponential distributions even for our small sample sizes. Our statistical analysis demonstrates that the dimensions of dike intrusions follow a power-law model with small levels of variability around the best-fit parameters. With the possible existence of a pressure or volume threshold for dike nucleation, the major processes influencing the dimensions and timings of individual dike intrusions appear to be the accumulation of tectonic stress caused by long-term plate motion, static stress transfer between intrusions, and magma recharge.

Along continental rifts as well as mid-ocean ridges, dike emplacement is controlled by tectonic stresses and the magma supply rate. In areas with limited magma supply, where the pressure inside a crustal magma chamber is reduced during dike propagation, a sequence of dike intrusions rather than a single diking event may be necessary to fully release the accumulated tectonic stress.

If the first dike intrusion encounters a structural barrier, it may continue to widen rather than lengthen until the driving pressure reaches a critical level, resulting in an aspect ratio that differs from later, smaller dikes intruded at lower magma pressures. Analogously, the thickness of the seismogenic zone limits the downdip rupture width, resulting in different aspect ratios for small and large earthquakes. The largest earthquakes are known to deviate substantially from the typical Gutenberg-Richter relation54, which is similar to our observation that the largest dike intrusions from the Krafla and Manda-Hararo rifting episodes do not obey the scaling laws that fit smaller dike intrusions well. We suggest that this discrepancy may be the result of factors such as the limited magma supply rate, threshold pressure conditions for dike propagation that change over time, and the structural constraints of lithospheric thickness and rift segment length.

Given that we are currently limited to studying rifting episodes along subaerial divergent plate boundaries, and since rifting episodes at a given location have recurrence intervals on the order of hundreds of years, we are restricted to this relatively small dataset. Nonetheless, the power-law behavior identified here is surprisingly robust, and with time and improvements in geophysical instrumentation, it may be possible to evaluate whether these observations apply more broadly to divergent plate boundaries in general.

Methods

In order to estimate the parameters β and x_{min} we follow the methodology presented by Clauset et al and select values that jointly minimize the Kolmogorov-Smirnov (KS) statistics. Errors for β and x_{min} are estimated from 1000 bootstrap iterations. Given the best-fitting values of β and x_{min}, we calculate 1000 synthetic power-law distributed datasets and their KS statistics. The p_{KS}-value associated with each rifting dataset is obtained by counting the fraction of the time that the KS statistics for the synthetic datasets are larger than the KS statistic for the empirical dataset. We rule out a power law distribution for datasets with p_{KS}-value ≤ 0.125.

We also model each rifting dataset using exponential and log-normal models left-truncated at x_{min}, with the functional forms $f(x) = \lambda \exp(\lambda x_{min}) \exp(-\lambda x)$ and $f(x) = \sqrt{2/x^2\pi 2\sigma^2}\left[erfc\left(\ln(x_{min} - \mu)/\sqrt{2}\sigma\right)\right]^{-1} \exp\left[-(\ln(x) - \mu)^2/2\sigma^2\right]$, respectively. Once the best-fit distributions are found and the fits are assessed by the KS test as for the power law model, we use the likelihood ratio test for model selection. The logarithm of the likelihood ratio, LR, has a positive sign if the power-law distribution provides a better fit to the data than the alternative distribution. The associated p_{LR}-value describes the statistical significance of LR, that is whether it is robustly far from zero. For large p_{LR}-values, the likelihood ratio test cannot discriminate between the two distributions, however, for p_{LR}-value ≤ 0.1, the sign of LR is a reliable indicator of which model provides a better fit to the data23.

Analogous to the classical Omori-decay fitting procedure, we bin the time intervals of each dataset (40 bins for MHA and MHB and 45 bins for Krafla) in order to obtain the rate of volume emplacement in each time interval. Bins with zero events are discarded. We use classical linear regression analysis to solve the equation $v_r = K/t^p$ for the best-fit parameters and their associated errors, which are reported in Fig. 4 along with the R^2 value. We perform an F-test on the slope of the regression line where the null hypothesis is that the slope $p = 0$. The p-values are less than 0.02 in all three cases, which means we can reject the null hypothesis at the 95% confidence level. We evaluate a time-predictable model for dike intrusions using the Bayesian hierarchical model developed in Passarelli et al. We assume that the errors for the interevent times and volumes are one day and 25%, respectively. The posterior distributions of the γ and δ parameters are obtained by performing 10,000 simulations using the Metropolis-Hastings integration technique. Using these posterior distributions, we select the average values of the parameters as the best-fit values of the model and plot the 5th and 95thpercentiles to demonstrate model variability.

AUTHOR CONTRIBUTIONS

L.P. and E.R. were involved in all stages of this study. A.S. contributed to the interpretation of the results and writing.

ACKNOWLEDGMENTS

L.P. and E.R. were funded by the ERC Starting Grant project CCMP-POMPEI, Grant N. 240583. A.S. was funded by an NSF Earth Science Postdoctoral Fellowship. Fruitful discussions with Torsten Dahm, Sebastian Hainzl and Emily Brodsky helped improve the manuscript. The power-law fitting was performed using the Matlab scripts of Clauset et al, which are freely available athttp://tuvalu.santafe.edu/~aaronc/powerlaws/. All figures were drawn using the Generic Mapping Tools.

REFERENCES

1. Gutenberg B. & Richter C. F. Frequency of earthquakes in California. Bull. Seismol. Soc. Am. 34, 185–188 (1944).
2. Frohlich C. & Davis S. D. Teleseismic b values; Or, much ado about 1.0. J. Geophys. Res. 98, 631–644 (1993).
3. Utsu T., Ogata Y. & Matsu'ura R. S. The Centenary of the Omori formula for a decay law of aftershock activity. J. Phys. Earth 43, 1–33 (1995).
4. Shcherbakov R. D., Turcotte D. L. & Rundle J. B. A generalized Omori's law for earthquake aftershock decay. Geophys. Res. Lett. 31, L11613 (2004).
5. Gusev A. A. Temporal structure of the global sequence of volcanic eruptions: Order clustering and intermittent discharge rate. Phys. Earth Planet. Inter. 166, 203–218 (2008).
6. Sachs M. K., Yoder M. R., Turcotte D. L., Rundle J. B. & Malamud B. D. Black swans, power laws, and dragon-kings: Earthquakes, volcanic eruptions, landslides, wildfires, floods, and SOC models.Europ. Phys. J. – Spec. Top. 205, 167–182 (2012).
7. Rubin A. M. Propagation of magma-filled cracks. Annu. Rev. Earth Planet. Sci. 23, 287–336 (1995).
8. Segall P. Volcano deformation and eruption forecasting. Geol. Soc. London Spec. Publ. 380, (2013).
9. Wright T. J. et al. Geophysical constraints on the dynamics of spreading centres from rifting episodes on land. Nature Geosci. 5, 242–250 (2012).
10. Rubin A. M. A comparison of rift-zone tectonics in Iceland and Hawaii. Bull. Volcanol. 52, 302–319 (1990).
11. Buck W., Einarsson P. & Brandsdottir B. Tectonics stress and magma chamber size controls on dike propagation: constraints from the 1975–1984 Krafla rifting episode. J. Geophys. Res. 111, B12404 (2006).
12. Qin R. & Buck W. R. Why meter-wide dikes at oceanic spreading centers? Earth Planet. Sci. Lett.265, 466–474 (2008).
13. Grandin R. et al. Sequence of rifting in Afar, Manda-Hararo rift, Ethiopia, 2005–2009: Time-space evolution and interactions between dikes from interferometric synthetic aperture radar and static stress change modeling. J. Geophys. Res. 115, B10413 (2010).
14. Hamling I. J., Wright T. J., Calais E., Bennati L. & Lewi E. Stress transfer between thirteen successive dyke intrusions in Ethiopia. Nature Geosci. 3, 713–717 (2010).
15. Mazzarini F. & Isola I. Monogenetic vent self-similar clustering in extending continental crust: Examples from the East African Rift System. Geosphere 6(5), 567–582 (2010).
16. Mazzarini F., Keir D. & Isola I. Spatial relationship between earthquakes and volcanic vents in the central-northern Main Ethiopian Rift. Journal of Volcanology and Geothermal Research (2013).
17. Tryggvason E. Widening of the Krafla fissure swarm during the 1975–1981 volcano-tectonic episode. Bull. Volcanol. 47, 47–69 (1984).
18. Arnadottir T., Sigmundsson F. & Delaney P. T. Sources of crustal deformation associated with the Krafla, Iceland, eruption of September 1984. Geophys. Res. Lett. 25, 1043–1046 (1998).
19. Ebinger C. et al. Length and timescales of rift faulting and magma intrusion: the Afar rifting cycle from 2005 to present. Annu. Rev. Earth. Planet. Sci. 38, 439–466 (2010).
20. Grandin R. et al. Transient rift opening in response to multiple dike injections in the Manda-Hararo rift (Afar, Ethiopia) imaged by time-dependent elastic inversion of interferometric synthetic aperture radar data. J. Geophys. Res. 115, B09403 (2010).
21. Sornette D. & Sornette A. General theory of the modified Gutenberg-Richter law for large seismic moments. Bull. Seismol. Soc. Am. 89, 1121–1130 (1999).
22. Kagan Y. Seismic moment-frequency relation for shallow earthquakes: Regional comparison. J. Geophys. Res. 102, 2835–2852 (1997).
23. Kagan Y. Y. Seismic moment distribution revisited: I. Statistical results. Geophys. J. Int. 148, 520–541 (2002).
24. Kanamori H. & Anderson D. L. Theoretical basis of some empirical relations in seismology. Bull. Seismol. Soc. Am. 65, 1073–1095 (1975).
25. Clauset A., Shalizi C. R. & Newman M. E. J. Power-law distributions in empirical data. SIAM Rev. 51, 661–703 (2009).
26. Bird P. & Kagan Y. Y. Plate-tectonic analysis of shallow seismicity: Apparent boundary width, beta, corner magnitude, coupled lithosphere thickness, and coupling in seven tectonic settings. Bull. Seismol. Soc. Am. 94,

2380–2399 (2004).

27. Boettcher M. S. & Jordan T. H. Earthquake scaling relations for mid-ocean ridge transform faults. J. Geophys. Res. 109, B12302 (2004).
28. Farrell J., Husen S. & Smith R. B. Earthquake swarm and b-value characterization of the Yellowstone volcano-tectonic system. J. Volcanol. Geotherm. Res. 188, 260–276 (2009).
29. Shuler A. & Nettles M. Earthquake source parameters for the 2010 western Gulf of Aden rifting episode. Geophys. J. Int. 190, 1111–1122 (2012).
30. Schorlemmer D., Wiemer S. & Wyss M. Variations in earthquake-size distribution across different stress regimes. Nature 437, 539–542 (2005).
31. Mogi K. Magnitude-frequency relation for elastic shocks accompanying fractures of various materials and some related problems in earthquakes. Bull. Earthquake Res. Inst. Univ. Tokyo 40, 831–853 (1962).
32. Warren N. W. & Latham G. V. An experimental study of thermally induced microfracturing and its relation to volcanic seismicity. J. Geophys. Res. 75, 4455–4464 (1970).
33. Hill D. P. et al. The 1989 earthquake swarm beneath Mammoth Mountain, California: An initial look at the 4 May through 30 September activity. Bull. Seismol. Soc. Am. 80, 325–339 (1990).
34. Woessner J. & Wiemer S. Assessing the quality of earthquake catalogues: Estimating the magnitude of completeness and its uncertainty. Bull. Seismol. Soc. Am. 95, 684–698 (2005).
35. Nakashima Y. Static stability and propagation of a fluid-filled edge crack in rock: Implication for fluid transport in magmatism and metamorphism. J. Phys. Earth 41, 189–202 (1993).
36. Grandin R. et al. Seismicity during lateral dike propagation: Insights from new data in the recent Manda Hararo–Dabbahu rifting episode (Afar, Ethiopia). Geochem. Geophys. Geosys. 12, Q0AB08 (2011).
37. Kagan Y. Y. & Houston H. Relation between mainshock rupture process and Omori›s law for aftershock moment release rate. Geophys. J. Int. 163, 1039–1048 (2005).
38. Enescu B., Mori J. & Miyazawa M. Quantifying early aftershock activity of the 2004 mid-Niigata prefecture earthquake (Mw6. 6). J. Geophys. Res. 112, B04310 (2007).
39. King G. C. P., Stein R. S. & Lin J. Static stress changes and the triggering of earthquakes. Bull. Seismol. Soc. Am. 84, 935–953 (1994).
40. Hill D. P. & Prejean S. G. in Treatise on Geophysics Vol. 4, (ed. Schubert G., 257–292 (Earthquake Seismology (ed. Kanamori H.), Elsevier, 2007).
41. Schmid A. & Grasso J.-R. Omori law for eruption foreshocks and aftershocks. J. Geophys. Res. 117, B07302 (2012).
42. Dieterich J. A constitutive law for rate of earthquake production and its application to earthquake clustering. J. Geophys. Res. 99, 2601–2618 (1994).
43. Reasenberg P. A. & Jones L. M. Earthquake hazard after a mainshock in California. Science 243, 1173–1176 (1989).
44. Nobile A., Pagli C., Keir D., Wright T. J., Ayele A., Ruch J. & Acocella V. Dike-fault interaction during the 2004 Dallol intrusion at the northern edge of the Erta Ale Ridge (Afar, Ethiopia). Geophys. Res. Lett. (2012).
45. Calais E. et al. 2008. Strain accommodation by slow slip and dyking in a youthful continental rift, East Africa. Nature 456, 783–787.
46. Passarelli L. & Brodsky E. E. The correlation between run-up and repose times of volcanic eruptions. Geophys. J. Int. 188, 1025–1045 (2012).
47. Marzocchi W. & Zaccarelli L. A quantitative model for the time-size distribution of eruptions. J. Geophys. Res. 111, B04204 (2006).
48. Passarelli L., Sandri L., Bonazzi A. & Marzocchi W. Bayesian hierarchical time predictable model for eruption occurrence: An application to Kilauea Volcano. Geophys. J. Int. 181, 1525–1538 (2010).
49. Passarelli L., Sansò B., Sandri L. & Marzocchi W. Testing forecasts of a new Bayesian time-predictable model of eruption occurrence. J. Volcanol. Geotherm. Res. 198, 57–75 (2010).
50. Burt M. L., Wadge G. & Scott W. A. Simple stochastic modelling of the eruption history of a basaltic volcano: Nyamuragira. Zaire Bull. Volcanol. 56, 87–97 (1994).
51. Rivalta E. Evidence that coupling to magma chambers controls the volume history and velocity of laterally propagating intrusions. J. Geophys. Res. 115, B07203 (2010).
52. Corral A. Time-decreasing hazard and increasing time until the next earthquake. Phys. Rev. E 71, 017101 (2005).
53. Corral A. Universal earthquake-occurrence jumps, correlations with time, and anomalous diffusion.Phys. Rev. Lett. 97, 178501 (2006).
54. Scholz H. The mechanics of earthquakes and faulting. Cambridge university press (2002).
55. Wessel P. & Smith W. H. F. New, improved version of generic mapping tools released. Eos Trans, AGU, 79(47), 579–579, 10.1029/98EO00426 (1998).

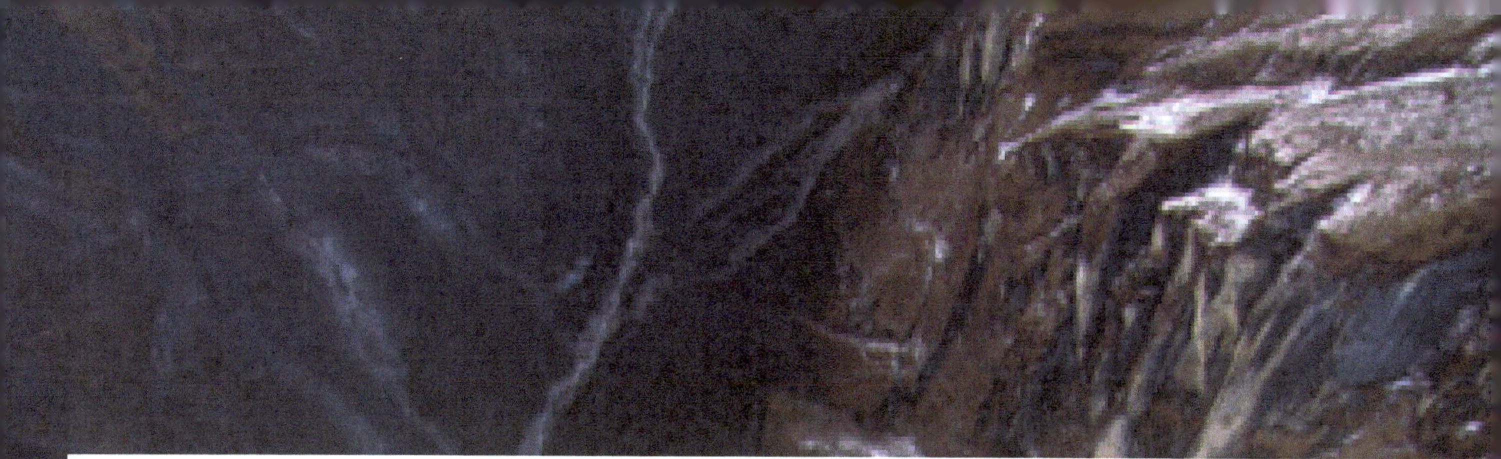

CHAPTER 5 PERENNIAL PLATE TECTONICS

Philip J. Heron, Russell N. Pysklywec and Randell Stephenson

INTRODUCTION

The expanded regions of future (and past) orogenesis may be determined by a 'perennial plate tectonics' map (Fig. 1), which highlights active and latent plate boundaries and deformation zones. Seismic reflection data from Canada, the Alps, the Pyrenees, the North Sea, and the Skagerrak and Hebrides shelf have identified tectonic structures as scarred lithosphere extending deep into Earth's interior[1,2,3] (yellow crosses, Fig. 1). Such seismic reflectors in the mantle lithosphere (ML) have been interpreted to be related to relict subduction zones linked to the closing of oceanic basins (that is, 'subduction scars' generated at an ancient plate boundary) (for example, refs [4, 5, 6]).

ML scars are well established in the geological record (for example, refs [1, 4, 6]) but they have not received the same attention as, for example, the reactivation of faults within the Earth's crust[7] for localizing intraplate deformation. Experiments on rock properties find that deformation generates weak zones that, over time, can be dormant (or be reactivated) depending on how the material strength is affected by changes in ambient stresses. A reduction in grain size is a characteristic of this lithospheric damage[8], which can be abundant at tectonic margins in the form of ML peridotite mylonites[9,10]. As deep seismic reflector anomalies are generally inferred to represent deformation relating to ancient tectonic margin processes (for example, refs [4,11]), these ML scars can be classified as weak zones (that is, a region of reduced grain size at mid-lithospheric depth) via the continuum theory of damage mechanics[8,12].

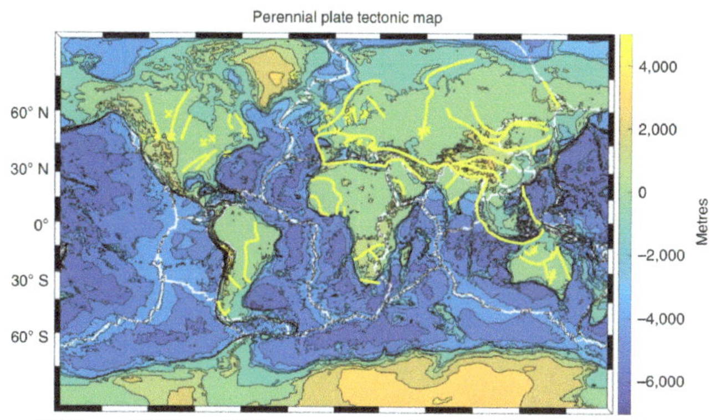

Figure 1: A perennial plate tectonic map.

Citation: Philip J. Heron, Russell N. Pysklywec and Randell Stephenson. perennial plate tectonics. Nat. Commun. 7:11834 doi: 10.1038/ncomms11834 (2016).

Here we show the results of thermo-mechanical finite-element experiments with continental parameters to understand further the role of ML heterogeneities by comparing weak zones at different depths and of different geometries. We implemented a strong ML to model the conditions required for a normal continent (for example, (ref. [13])), with a lower crustal rheology that permits crust–mantle interaction (for example, ref.[14]). We model pre-existing zones of weakness (for example, (refs [15, 16, 17])) by specifying a region with a low angle of internal friction (to mirror the processes of lithospheric damage (for example, ref. [8])), and simulate inherited structures in the upper crust (UC), lower crust (LC) and ML. We have chosen a two-dimensional (2D) geometry (over three-dimensional) so that the experiments track the evolution of fault zones in the crust–mantle tectonic models to high resolution (1-km vertical resolution in the crust).

RESULTS

Analysing the Role of ML Scars on Tectonics

Figure 2a shows 11 differently angled weak zones in both UC and LC, and one ML scar. We then compare convergent (1 cm per year) models after a significant amount of shortening (350–450 km) that feature only UC scars (Fig. 2b), LC scars (Fig. 2c), UC and LC scars (Fig. 3a), and all the available weak zones (Fig. 3b).

The reactivation of crustal scars is time-dependent, with shifting strain patterns occurring once a fault has localized over time[18]. The time-dependence of the UC scars is shown in Fig. 2b by markers A, B and C, with these faults reactivating in alphabetical sequence as strain is accumulated and transferred. The UC fault reactivation produces folding in the UC, but acts to thicken the LC and produce subcrustal subduction within the ML. When only LC scars are present (Fig. 2c), a time-dependent strain pattern change remains (markers D–G) but the thickening of the LC produces widespread folding in the UC.

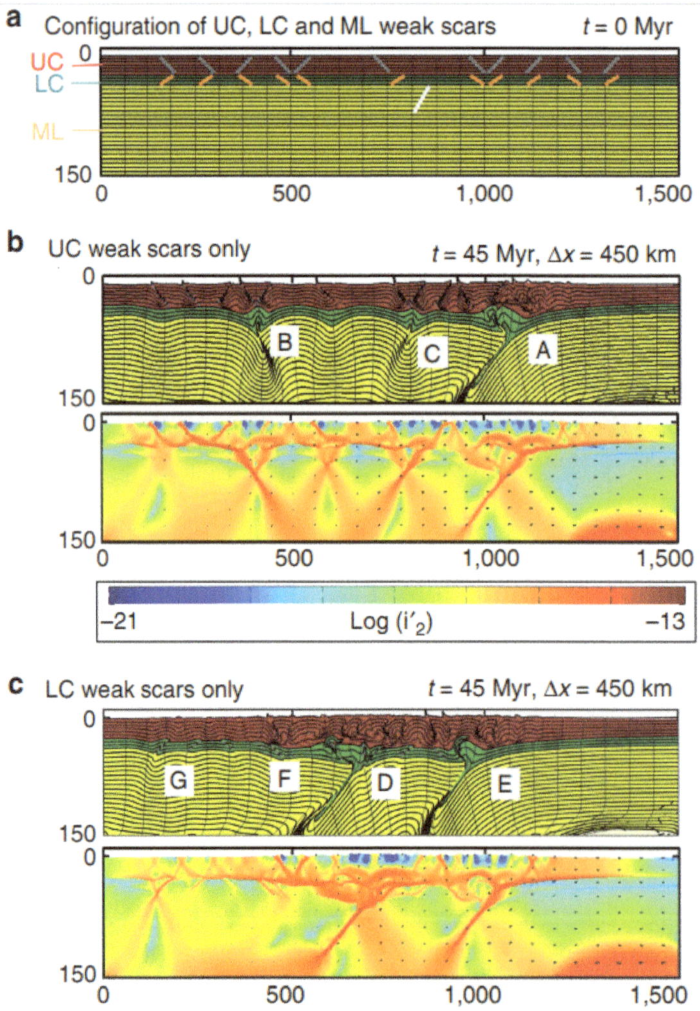

Figure 2: Model set-up and results with UC and LC weak zones.

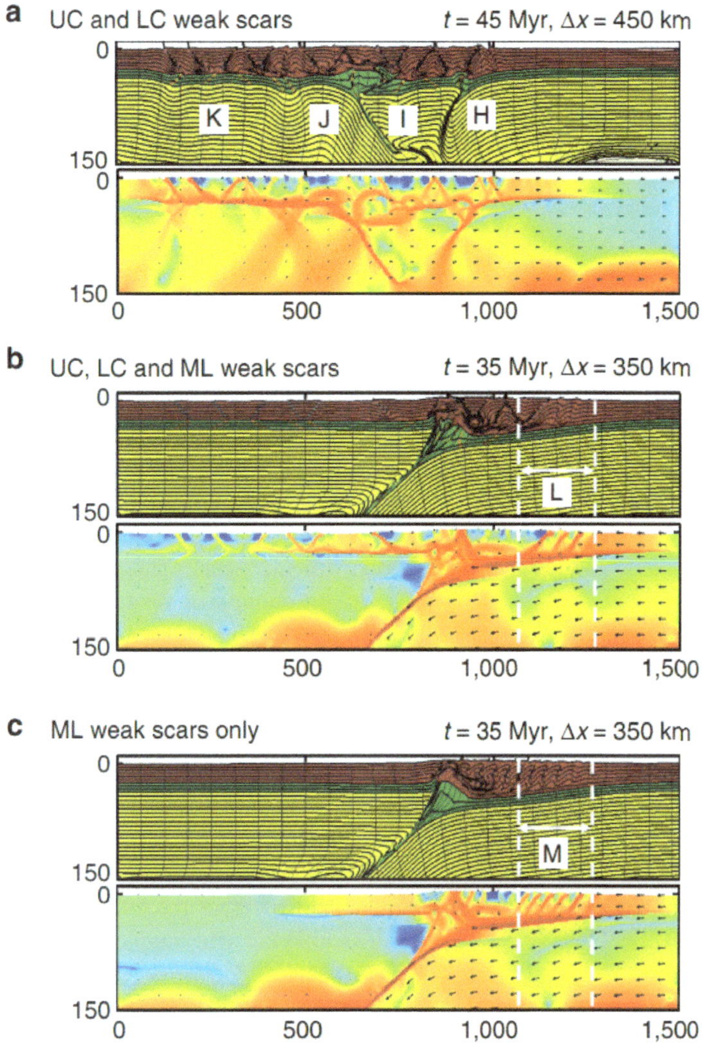

Figure 3: Model results for combinations of weak scars.

The combination of both LC and UC scars (Fig. 3a) produces a shifting pattern of strain that appears to be controlled by the lower crustal heterogeneities rather than the shallower features. Timing of crustal fault ruptures (H–K) coordinate with those of LC scars only (Fig. 2c). Furthermore, the combination of LC and UC faults produces a strong decoupling of the LC from the UC, as shown by the high strain rate along the boundary.

The introduction of a ML scar to convergent models featuring 22 remnant crustal features produces a marked focusing of deformation (Fig. 3b). The ML scar supersedes the shallower pre-existing faults to produce 'pseudo-subduction' (that is, under thrusting of the ML). Lower crustal thickening impinges on the UC to produce folding. Over time, the folding propagates as the convergence continues. However, the deformation produced by the ML scar is localized (despite the numerous crustal features present) with little time-dependence. In fact, the pattern of deformation produced by UC, LC and ML weak scars (Fig. 3b) is almost identical to that of a convergent model featuring a ML scar only (Fig. 3c). Subsequently, the evolution of intraplate deformation follows the later stages of a continent collision (for example, ref. [19]).

Model Validation

Additional experiments were performed to test how robust the model set-up is to changes in rheological parameters. Supplementary Fig. 2 shows the strength profiles for the rheology used in Figs 2 and 3(Supplementary Fig. 2a), as well as for the additional experiments (Supplementary Fig. 2b–d). The models presented in Supplementary Figs 3 and 4 show ML scars to be important in activating tectonics, while the reduction in crust and mantle strength may also determine the dominant factor in tectonic evolution. The models presented in Figs 2 and 3 are the most relevant as we apply a strong LC and ML to simulate relatively stable, strong continental interiors that feature long-lived ML scars. We present additional models here to understand the limits of our study.

Jelly Sandwich or Crème Brûlée

Supplementary Fig. 3 shows results from using a crust and ML rheology that is not as strong as those experiments presented in Figs 2 and 3. We implement a different configuration of weak zones for this suite of models (Supplementary Fig. 3a) as compared with Fig. 2a, with 12 weak zones in the LC and UC and 2 in the ML. We acknowledge that previous studies have highlighted the possible 'jelly sandwich' or 'crème brûlée' strength profiles of the mantle (for example, ref. [13]), and present experiments taking this into account in Supplementary Fig. 3. Our jelly sandwich experiments have a weaker LC than shown in Figs 2 and 3(strength profile as given in Supplementary Fig. 2b). For models just featuring UC scars, we find that the

UC dominates tectonic evolution in this jelly sandwich setting (Supplementary Fig. 3b). The time markers A–C show the evolution of strain within the UC.

The introduction of weak zones into the LC and ML (Supplementary Fig. 3c) generates more localized deformation. Although the UC weak zones are reactivated, their influence on regional tectonics and the deformation of the LC is minimal. ML scars again show to be dominant in activating tectonics. In applying a crème brûlée rheology, we detach the LC and mantle from the UC leading to décollment tectonics (for example, ref. [13]). Our results indicate that regions where a crème brûlée rheology is present, perennial plate tectonics may not persist. However, in order for stable continents to be preserved over time, Burov and Watts[13] found that a jelly sandwich rheology would be required over a crème brûlée rheology. As a result, the crème brûlée' rheology is not applicable to our study of intraplate tectonics.

LC Strength

The strength of the LC and ML is important in activating tectonics (for example, refs [14, 18, 20, 21]). To show that our results are not dependent on the rheological parameters, we present a suite of models that use material values that are similar to a previously published paper[22]. In these models we use the weak zone configuration as shown in Supplementary Fig. 3a, and implement flow laws for wet quartz for the UC[23], wet anorthite for the LC[24] and dry olivine for the upper mantle[25] (Supplementary Table 2). These rheological parameters produce strength profiles that are intermediate in the range of published flow laws for crustal and mantle materials (Supplementary Fig. 2d).

Supplementary Fig. 4 shows the material deformation after 270 km of shortening for the rheological parameters given in Supplementary Table 2. The UC weak zones in Supplementary Fig. 4a generate strong folding of the crust. The changing patterns of stress build-up shifts to produce pseudo-subduction of the crust into the ML. The implementation of wet anorthite for the LC generates a layer that is not primed for failure (even with regions of designated weakness). The wet anorthite flow law generates a weak LC that plays no role in the deformation pattern. This is shown with the introduction of UC and LC weak zones (Supplementary Fig. 4b) and the dominance of the upper crustal weak zones in the deformation pattern.

The ML weak zones again dominate the deformation pattern when introduced in Supplementary Fig. 4c. Although the UC weak zones reactivate, the overall tectonic pattern is controlled by the ML scars (with the LC weak zones playing no role). The time-dependent nature of the UC deformation pattern is replaced by the ML scarring's ability to accommodate shortening through sub-crustal subduction. In an experiment featuring only ML scars (Supplementary Fig. 3d), the deformation is localized. However, the pattern of crustal deformation with all weak zones present (Supplementary Fig. 4c) follows closely with that of the simulation with just ML weak zones (Supplementary Fig. 4d), rather than that of UC only (Supplementary Fig. 4a), implying that the deeper scars control tectonic evolution.

Plate Velocity

By increasing the convergence rate of the model from $\upsilon=1$ to 2.5 cm per year, we can show the difference in deformation due to plate forcing. Here we compare simulations that have been shortened by 450 km using the rheological parameters of Supplementary Fig. 2a and Supplementary Table 1 (and the weak zone configuration shown in Fig. 2a). The only major difference that occurs from increasing the convergence velocity (when comparing with Fig. 2) happens for the model featuring UC scars only (Fig. 2b and Supplementary Fig. 5a). At a lower convergence velocity the stress pattern changes over time reactivating faults over time (Fig. 2b and Supplementary Fig. 2b). However, at $\upsilon=2.5$ cm per year the stress pattern does not change and the time-dependence of the model disappears. In the presence of a higher convergence velocity, the result that lower crustal weak scars control deformation over shallower features is highlighted due to this lack of time-dependence with UC scars. The deformation pattern from UC and LC scars (Supplementary Fig. 5c) is more similar to LC tectonics (Supplementary Fig. 5b) than UC tectonics (Supplementary Fig. 5a). Overall, increasing the convergence velocity allows for the same conclusions: deep lithospheric heterogeneities can control deformation over shallower features.

ML Weak Zone Orientation

Supplementary Fig. 6 give an indication of how the orientation of a ML weak scar would affect the pattern of deformation. In all previous models we applied similar angles to the scars within the ML. Supplementary Fig. 6a gives four different ML scar configurations: ML(i) two scars with angles as previously shown; ML(ii) two steeper ML scars; ML(iii) two vertical ML scars; and ML(iv) two smaller ML scars (of the same size as the lower crustal scars). All models are realized with UC and LC scars as in Supplementary Fig. 3a. Models ML(i), ML(ii) and ML(iv) exhibit very similar deformation patterns despite having different configurations. In a result shown here, we found that if a ML scar becomes small enough, the crustal scars can control deformation.

The vertical nature of the weak scars in model ML(iii) inverses the deformation pattern, which indicates that there appears to be a preferred angle in which the ML can pseudo-subduct. However, the overall result from changing the orientation of the weak scars in the ML is that deeper heterogeneities often controls the pattern of deformation (despite changing the angles of scars).

DISCUSSION

Our results demonstrate that deformation patterns in continent interiors are influenced by the depth of lithospheric heterogeneities, and that ML scars can guide deformation over other pre-existing features in the UC and LC. The range of deformation patterns exhibited from the models presented here indicates the complex nature of lithospheric heterogeneities and plate tectonics. The activation of dormant lithospheric structures depends on changes in the regional stress regime[18].

The tectonic history of China provides one reference to understand plate tectonics beyond plate boundaries with regards to our results. Although there are many regions where continents have been accreted by closure of paleo-oceans between micro-plates, China is a unique reference as the far-field convergent stress from the Indian–Eurasian collision is relatively recent and ongoing (Fig. 4a).

The Altyn Tagh Fault (ATF), on the northern margin of the Tibetan Plateau, has a distinct present-day ML heterogeneity linked to a continent–continent suture[26]. The ATF accommodates some of the convergence between the Indian and Eurasian plates[27], and is characterized by localized deformation that has produced ~475±70 km of staggered displacement since the mid-Oligocene[26]. Although focal mechanisms of earthquakes close to the ATF show strike–slip motion, compressional processes account for earthquakes to the south[27], with numerous thrust faults also inhabiting the area (Fig. 4b). Geophysical studies of the ATF show deformation that penetrates the entire crust to link to heterogeneous structures in the ML[27,28,29] (Fig. 4c).

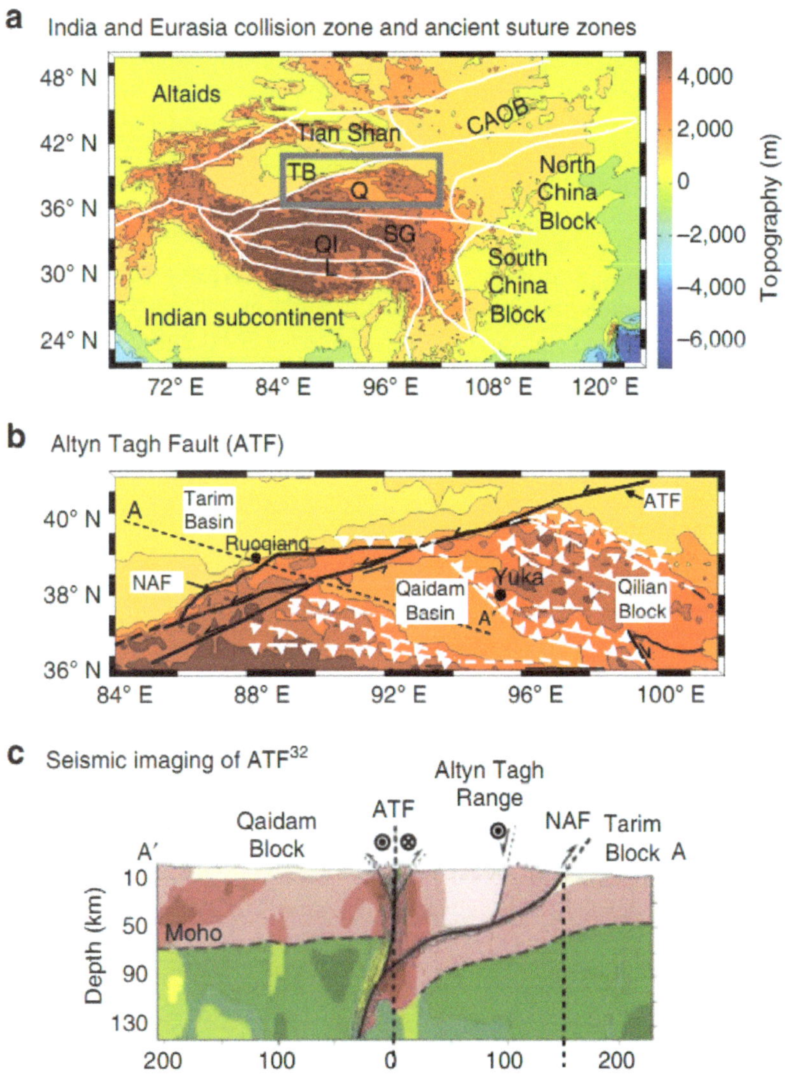

Figure 4: The suture zones of Chinese tectonics and the ATF.

Supporting our findings, the ATF can be interpreted as a ML scar originating as a continent–continent collision in the Palaeozoic[30] that controls intraplate deformation during periods of compression (with the most recent episode starting in the Oligocene by the India–Eurasia collision). A similar process could be considered for the crustal deformation in the presence of ML heterogeneities beneath the Alps, the Laramide, the Pyrenees and the Eurekan orogens[31,32,33], as well as deformation in the Tien Shan, the Caucasus and the southeastern Ukraine (for example, refs [34, 35] and references therein). However, geological examples exist where lithospheric folding in relatively strong oceanic lithosphere has been generated due to age-dependent plate tectonic forces. An example in particular being buckling during the recent Eurasian convergence in the northeastern Indian Ocean[36], results of which are supported by numerical modelling of net plate tectonic forces[37]. The role of plate tectonic forces in concentrating stresses inside plates is a mechanism that may need to be incorporated into this ML scarring method of developing intraplate deformation.

Even though the choice of crust and ML rheology is indicative of a stable continent[13,14], it may not be applicable to all geological settings. Supplementary Fig. 3 shows that continental lithosphere rheology is important in the generation of crustal deformation (through coupling dynamics) in agreement with recent analogue models (for example, refs [21, 38, 39, 40]). Our results offer an explanation as to why preserved structures in the continental crust do not always reactivate during periods of compression; the localization of deformation from strong ML heterogeneities can change strain patterns during periods of shortening and dominate over the crustal response. Furthermore, our result that ML dominates crust in controlling tectonic processes is consistent with findings from geological studies (for example, refs [11, 20, 41]) and recent physical scaled analogue modelling of heterogeneous lithosphere (for example, refs [21, 38, 39, 40]) (although no previous studies have incorporated both crustal and mantle heterogeneities as done here).

ML scars related to oceanic subduction have been the most documented in the geological record (for example, refs [2, 3, 4, 5, 6]). However, processes that do not feature oceanic subduction may also account for seismically inferred heterogeneities. Rayleigh–Taylor instability, lithospheric underplating, lithospheric delamination and lithospheric subduction are all related to continental collisions (that is, suture zone tectonics) and may produce lasting (seismically visible) ML structures[11,42]. An increase in the number of high-resolution imaging studies of the upper ML (at sites of previous continental collisions in particular) would be able to test this extended theory through the discovery of further ML heterogeneities.

The ability of deep lithospheric heterogeneous structures to exist over long periods in stable continental settings (Fig. 1) allows for a new mechanism for intraplate evolution (following external forcing). If, as an example, the ATF has a long-lasting ML scar from a continental collision that is controlling the crustal evolution, then plate tectonics may indeed display 'perennial' processes. As such, an increase in intraplate orogenesis would be observed during future (and past) periods of global compression and extension (that is, supercontinent formation and dispersal).

A tenet of the conventional theory of plate tectonics is that crustal deformation is confined to near the boundaries of plates. Our work implies that this remains true for general planetary deformation as ML scars (that can control tectonic evolution) in a continent interior may originate from an ancient plate boundary. In this way, ancient and present-day plate boundaries could be represented together as latent and active boundaries. A global map of perennial plate tectonics (Fig. 1) presents a redefined illustration of tectonic activity and modifies the conventional theory of plate tectonics.

METHODS

Governing Equations

The experiments are modelled using the 2D, thermal–mechanical finite element numerical code SOPALE[43], which implements an Arbitrary Lagrangian Eulerian method to solve for the deformation of high Prandtl number incompressible viscous–plastic media. The governing hydrodynamic equations for the numerical models include the conservation equations of mass, momentum and internal energy:

$$\nabla \cdot \mathbf{u} = 0 \quad (1)$$

$$\nabla \cdot \sigma_{ij} + \rho g = 0 \quad (2)$$

$$\rho c_p\left(\frac{\partial T}{\partial t} + \mathbf{u}\cdot\nabla T\right) = k\nabla^2 T + H \quad (3)$$

respectively. In the equations above, $\mathbf{u}$ ($m s^{-1}$), σ (Pa), ρ ($kg m^{-3}$), g ($m s^{-2}$), c_p ($J kg^{-1} K^{-1}$), T (K), k ($W m^{-1} K^{-1}$), H ($W m^{-3}$) and t (s) are the velocity, stress tensor, density, gravitational acceleration, specific heat capacity, temperature, thermal conductivity, volumetric rate of internal heat production and time, respectively. The system is completed by an associated linearized equation of state:

$$\rho = \rho_o(1 - \alpha(T - T_o)) \quad (4)$$

where α is the coefficient of thermal expansion, ρ_o is the reference material density and T_o is the reference temperature.

The stress tensor in equation (2) may be divided into the deviatoric stress tensor, σ', and a pressure term,

$$\sigma_{ij} = \sigma'_{ij} - \delta_{ij}P \quad (5)$$

where for an incompressible fluid, δ is the Kronecker delta and P is the dynamic pressure (which is given as $-\frac{1}{3}\sigma_{ii}$). The deviatoric stress is determined at each computational node as the lesser value of a yield stress, σ_y, or viscous stress, σ_v. In the numerical code, the frictional plastic yield stress is given by a pressure-dependent incompressible Drucker–Prager yield criterion

$$\sigma_y = P\sin(\phi_e) + C_o \quad (6)$$

where φ_e is the angle of internal friction and C_o the cohesion. The viscous stress is given by

$$\sigma_v = 2\eta_e \dot{I}'_2 \quad (7)$$

where $\dot{I}'_2$ is the second invariant of the deviatoric strain rate tensor, and η_e the effective viscosity. When the thermally activated power law creep is used, the effective viscosity is given as

$$\eta_e^v = \left(3^{\frac{-(1+n)}{2n}} 2^{\frac{1-n}{n}}\right) f A^{\frac{-1}{n}} \left(\dot{I}'_2\right)^{\frac{(1+n)}{2n}} e^{\frac{Q}{nRT}} \quad (8)$$

(where A is the material constant, f is a scaling parameter, n is the power law exponent, Q is the thermal activation energy, R is gas constant and T is the temperature).

The Arbitrary Lagrangian Eulerian method calculates creeping flow on a Eulerian grid (resolution here 401 × 217) that is restricted to small vertical dilations (corresponding to the evolution of topography on the free upper surface). The fully deforming Lagrangian grid (resolution 801 × 649) is advected with the computed velocity field and is used to updated the evolving thermal and material properties onto the Eulerian grid (for example, ref. [43]).

Model Set-Up and Material Parameters

Supplementary Fig. 1 shows the set-up of the reference case for modelling intraplate deformation. We model the upper and LC, the ML and portion of the sub-lithospheric mantle in 1,500-km width and 600-km depth 2D numerical experiments. The models show convergence in a continental setting, with material properties given in Supplementary Table

1. The geological motivation for this configuration is an old continental lithosphere where ML heterogeneities are found away from plate boundaries (for example, strong crust and ML). The model set-up allows for a heterogeneous lithosphere, with a number of different configurations of weak zones (that is, Supplementary Fig. 3a). As discussed in the manuscript, we model pre-existing zones of weakness by specifying a region with a low angle of internal friction (to mirror the processes of lithospheric damage), and simulate inherited structures in the UC, LC and ML. All weak zones have the same material properties as the layer they occupy, with the specification of a zero internal angle of friction (φ_e). By this method, all faults are equally primed to fail.

An initial (laterally uniform) temperature field (for example, a geotherm) is prescribed to all models. The geotherm has the surface and basal temperature fixed throughout the duration of the model runs (at 20 °C and 1,570 °C, respectively). However, as both the mechanical and thermal calculations are carried out for each time step, the interior temperature field can evolve over space and time, and is not fixed to the preliminary geotherm. Initially, a linear temperature increase of 20–550 °C from the surface to the Moho depth (36 km), with a further linear temperature increase to 1,350 °C at the base of the lithosphere. Radiogenic heat production is applied to the crust, with the majority of internal heat generation coming from the UC (2.1 µW m^{-3}) (for example, ref. [44]) rather than the LC (0.7 µW m^{-3}).

Creeping motion in the model is driven by internal buoyancy forces and an imposed plate motion. In this study, plate motion is modelled by introduced new lithosphere into the domain at the horizontal velocity of υ_c = 1 cm per year (Supplementary Fig. 1). The left margin of all models is held fixed, while a small outward flux, υ_0, maintains the mass balance of the system (Supplementary Fig. 1).

All calculations were performed on a supercomputer at the SciNet HPC consortium[45].

Supplementary Fig. 7 shows a recent comparison between two different seismic surveys of the central ATF, specifically a schematic based on the seismic data set used in Wittlinger et al.[28] (Supplementary Fig. 7a) and the more recent study of Zhao et al.[29] (Supplementary Fig. 7b). A ML structure can be seen in both models. Further details of the interpretation of these structures can be found, alongside new magneto-telluric data, in Zhang et al.[27].

Data Availability

The data that support the findings of this study are available from the corresponding author upon request.

SUPPLEMENTARY MATERIAL

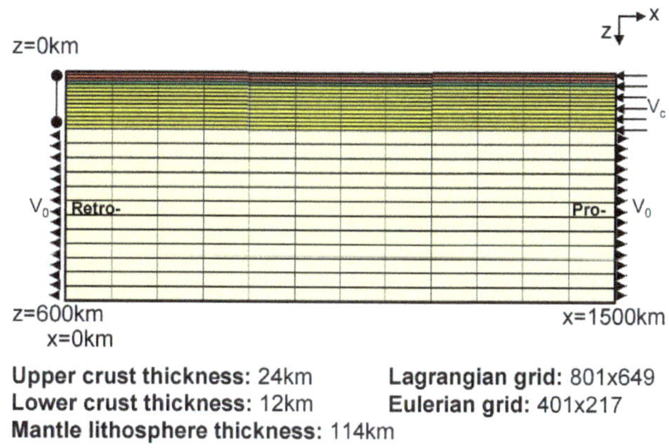

Supplementary Figure 1: The initial configuration for the numerical experiments, showing the configuration of weak zones. Upper crust (denoted by red and grey colours and has a thickness of 24 km), lower crust (green and blue, thickness 12 km), mantle lithosphere (yellow and white, thickness 114 km), and sub-lithospheric mantle (beige, the bottom 450km). The corresponding physical parameters are given in Supplementary Table 1. Temperature increases linearly throughout the solution domain; surface temperature is 20°C, the Moho temperature is 550°C, the base of the lithosphere is 1350°C, and the base of the model is 1570°C. The Lagrangian and Eulerian grid resolutions are 801×649 and 401×217, respectively. The Lagrangian grid (black mesh) is only partially represented here and shows 17% and 39% of the grid occupies the crust and mantle lithosphere, respectively (the same ratios apply to the Eulerian grid). Continental convergence is incorporated by introducing new lithosphere at the right boundary of box with velocity vc=1 cm yr^{-1}.

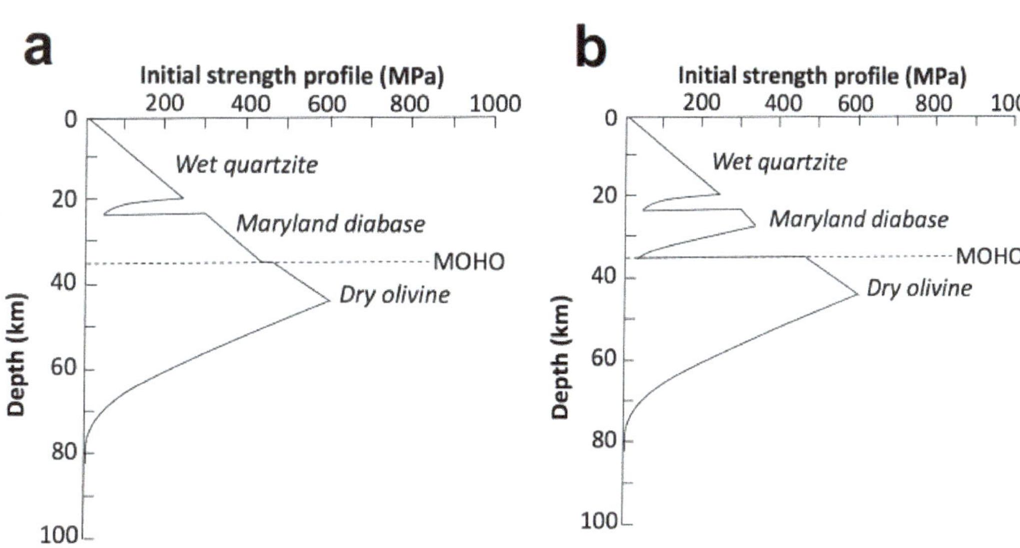

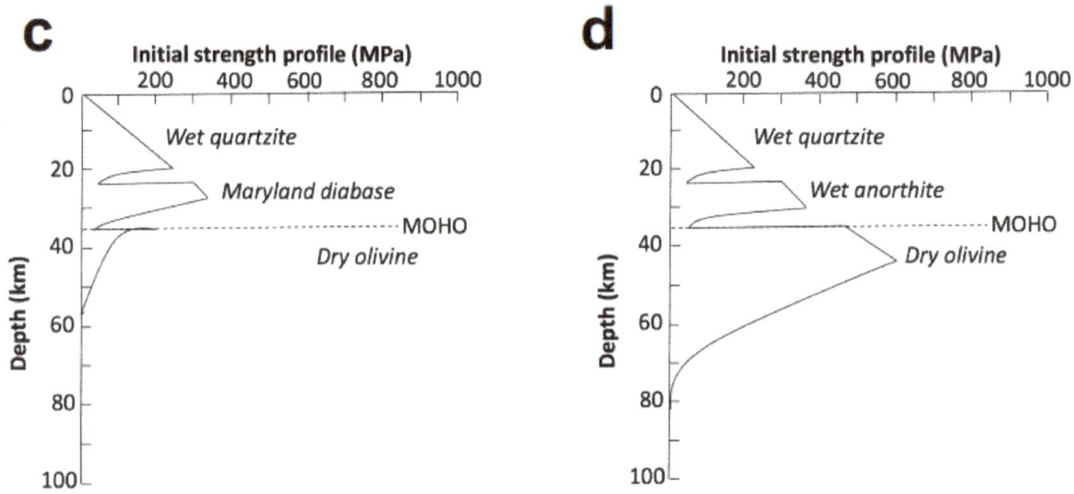

Supplementary Figure 2: Strength envelopes of differential stress versus depth for different rheologies used in the study. The strength of the materials is determined by the minimum of the brittle and viscous strengths. Parameter values are given in Supplementary Table 1 and 2, with the initial geotherm as discussed in Supplementary Table 1 and Supplementary Figure 1. We show a number of different mantle lithosphere scenarios (at a modest representative strain rate of $10^{-15}s^{-1}$) to highlight the role of crustal and mantle lithosphere strength in activating tectonics. We present the following rheologic scenarios: a. a strong continental crust and mantle lithosphere (as used in the manuscript and Supplementary Table 1); b. a weaker lower crust ('jelly sandwich') continental rheology of (a) (Q = 356 kJ mol^{-1} in lower crust (LC)); c. a weaker lower crust and mantle lithosphere ('creme brulee') continental rheology of (b) (Q = 420 kJ mol^{-1} in mantle lithosphere (ML)); d. a wet anorthite lower crust (as Supplementary Table 2).

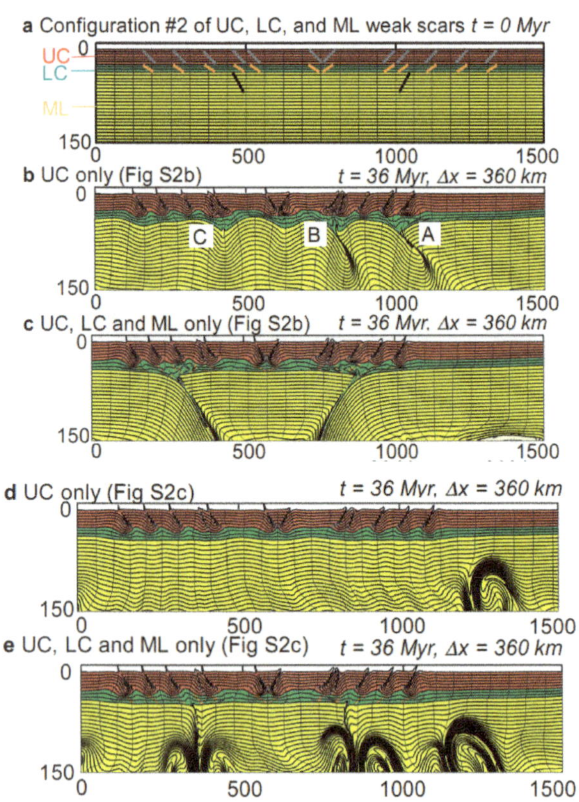

Supplementary Figure 3: Model setup and results from rheological profiles in Supplementary Figure 2b-c. a. Model set-up (red: upper crust (wet quartzite); green: lower crust (Maryland diabase); yellow: mantle lithosphere (dry olivine)). The top 150 km (2x vertical exaggeration) of the 600 km deep model is shown at the initial condition (t = 0 Myr) with the configuration of upper crust (UC), lower crust (LC), and mantle lithosphere (ML) weak scars. The full width of the model is shown. Continental convergence is incorporated by introducing new lithosphere at the right boundary of box with velocity v=1 cm yr^{-1}; b. Material deformation for the weaker lower crust ('jelly sandwich') strength profile (Supplementary Figure 2b) with UC only weak zones. The alphabetical markers (A-C) show the progression of the deformation throughout the full model simulation and track the evolution of tectonics; c. as b with UC, LC, and ML scars; d. Material deformation for weaker lower crust and mantle lithosphere ('creme brulee') strength profile (Supplementary Figure 2c) with UC only weak ′ zones; e. as d with UC, LC, and ML scars.

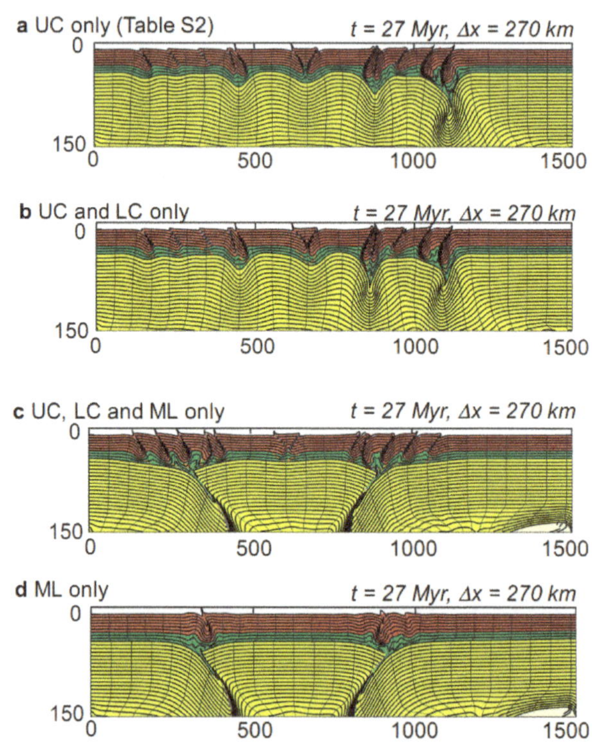

Supplementary Figure 4: Model results from rheological profiles in Supplementary Figure 2d. The material fields for shortening models featuring configuration #2 (Supplementary Figure 3a) with upper crust (UC), lower crust (LC), and mantle lithosphere (ML) flow laws as Supplementary Table 2 and Supplementary Figure 2d. a. Upper crust weak scars only; b. Upper and lower crust weak scars only; c. Upper and lower crust weak scars and two mantle lithosphere scars; d. Two mantle lithosphere scars only.

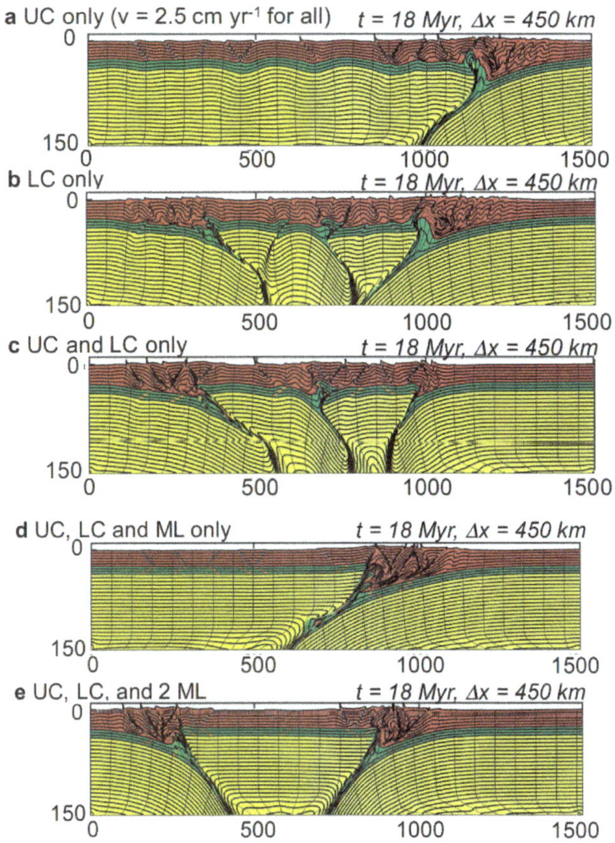

Supplementary Figure 5: Changing the convergence velocity. The material fields for shortening models featuring configuration #2 (Supplementary Figure 3a) with an increased convergence velocity from v = 1 cm yr^{-1} to v = 2.5 cm yr^{-1} . a. Upper crust weak scars only. b. Lower crust weak scars only. c. Upper and lower crust weak scars only. d. Upper and lower crust weak scars and one mantle lithosphere scar (as given in by white marker in Figure 2a). e. Upper and lower crust weak scars and two mantle lithosphere scars (as given in by black markers in Supplementary Figure 3a).

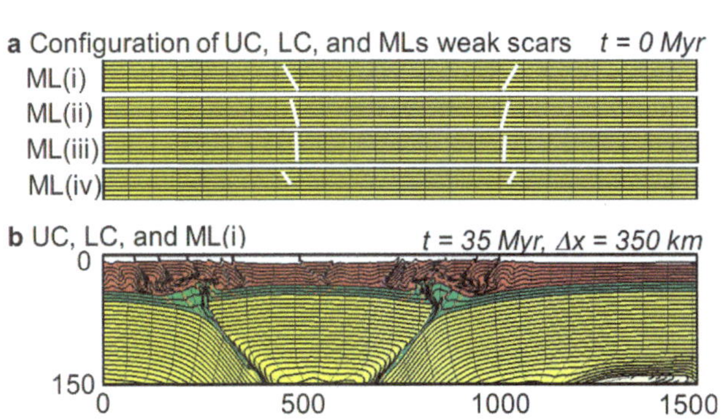

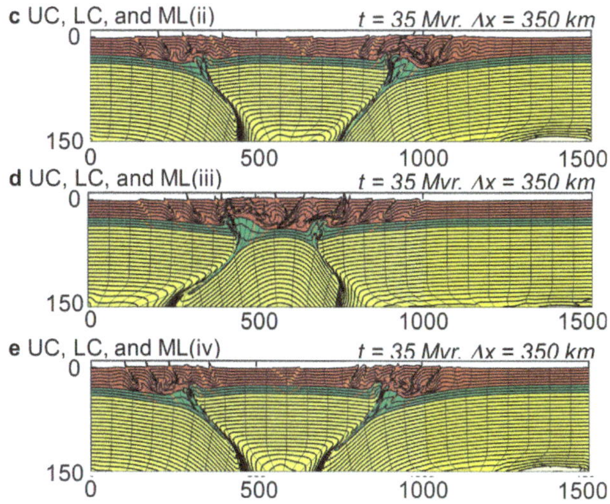

Supplementary Figure 6: Changing the orientation of mantle lithosphere scars. Shortening models are set-up with UC and LC scars as shown in Supplementary Figure 3a. a. Configuration of the different orientation of ML scars (for models ML(i - iv)). The deformation pattern after 350 km shortening for ML(i) b, ML(ii) c, ML(iii) d, and ML(iv) e.

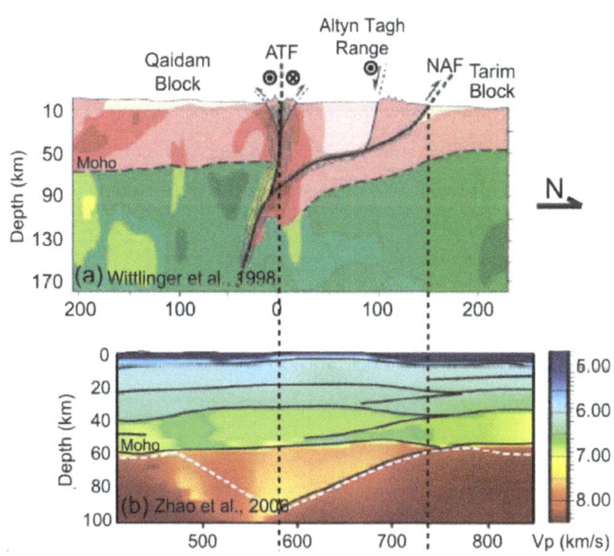

Supplementary Figure 7: Comparison between different geophysical models/datasets across the central Altyn Tagh Fault (ATF). a. Schematic seismic model of Wittlinger et al. [7]. Red and green regions indicate the crust and mantle, respectively. Regions that are more yellow or red in the model are low velocity zones. b. Seismic model of Zhao et al. [9]. Black lines are inferred discontinuities from the original paper. Regions above the white dashed line are of relatively high resolution due to dense ray coverage. Horizontal distances in (a) - (b) are using coordinates from the original papers. This figure is taken from Zhang et al. [8].

Supplementary Table 1: Rheological parameters used for the continental collisions in the manuscript. Upper crust, lower crust, mantle lithosphere, the sub-lithospheric mantle, and a weak zone are denoted by UC, LC, ML, SM, and WZ, respectively. Symbols are as parameters given in the text (units for C_0, A, Q, ρ_o, and T_0 are Pa, Pa^{-n} s^{-1}, kJ mol^{-1}, K kg m^{-3}, and K). $(\dot{I}_2')^{\frac{1}{2}}$ is accumulated strain. Reference list for Ref: (1) Gleason and Tullis [1]; (2) Ranalli [2], Mackwell et al. [3]; (3) Hirth and Kohlstedt [4], Kawazoe et al. [5]. The flow laws that represent each region are: wet quartzite for UC; Maryland diabase for LC; and dry olivine for ML and SM. For WZ the flow law is governed by the host material (with a prescribed $f_e = 0$). Physical parameters that remain constant across all regions are $\alpha=3\times10^{-5}K^{-1}$, $k=2.25\ W\ m^{-1}K^{-1}$, and $c_p=750\ J\ kg^{-1}\ K^{-1}$. The upper crust and lower crust have radioactive heat production values of 2.1 $\mu W\ m^{-3}$ and 0.7 $\mu W\ m^{-3}$, respectively

	C_o	ϕ_2	ϕ_2	$(\dot{I}'_2)_1^{\frac{1}{2}}$	$(\dot{I}'_2)_2^{\frac{1}{2}}$	A	f	Q	n	Ref	ρ_o	T_0
UC	10^6	15°	2°	0.5	1.5	1.4×10^{-28}	0.3	223	4	1	2700	293
LC	10^6	15°	2°	0.5	1.5	4.4×10^{-29}	0.05	485	4.7	2	2900	293
ML	-	15°	2°	0.5	1.5	8.3×10^{-18}	0.3	535	3.5	3	3250	1609
SM	-	15°	2°	0.5	1.5	8.3×10^{-18}	0.3	535	3.5	3	3250	1609
WZ	*	0	0	*	*	*	*	*		*	*	*

Supplementary Table 2: A different set of rheological parameters used for the continental collisions. Upper crust, lower crust, mantle lithosphere, the sub-lithospheric mantle, and a weak zone are denoted by UC, LC, ML, SM, and WZ, respectively. * denotes that values are given as the values for which material the weak zone is within. Symbols are as parameters given in the text (units for C_o, A, Q, ρ_o, and T_0 are $Pa^{-n}\ s^{-1}$, kJ mol^{-1}, K kg m^{-3}, an K). $(\dot{I}'_2)^{\frac{1}{2}}$ is accumulated strain. Reference list for Ref: (1) Gleason and Tullis [1]; (2) Rybacki and Dresen [6]; (3) Hirth and Kohlstedt [4]. The flow laws that represent each region are: wet quartzite for UC; wet anorthite for LC; and dry olivine for ML and SM. For WZ the flow law is governed by the host material (with a prescribed $f_e = 0$). Physical parameters that remain constant across all regions are $\alpha=3\times10^{-5}\ K^{-1}$, k = 2.25 W m−1 K^{-1}, and cp = 750 J kg−1 K^{-1}. The upper crust and lower crust have radioactive heat production values of 2.1 $\mu W\ m^{-3}$ and 0.7 $\mu W\ m^{-3}$, respectively

	C_o	ϕ_2	ϕ_2	$(\dot{I}'_2)_1^{\frac{1}{2}}$	$(\dot{I}'_2)_2^{\frac{1}{2}}$	A	f	Q	n	Ref	ρ_o	T_0
UC	10^6	30°	15°	0.5	1.5	8.6×10^{-28}	0.3	223	4	1	2800	0
LC	10^6	30°	15°	0.5	1.5	1.8×10^{-16}	0.05	356	3	2	2900	0
ML	10^6	30°	15°	0.5	1.5	2.9×10^{-16}	0.3	535	3.5	3	3250	602
SM	10^6	30°	15°	0.5	1.5	2.9×10^{-16}	0.3	535	3.5	3	3300	602
WZ	*	0	0	*	*	*	*	*	*	*	*	*

ACKNOWLEDGMENTS

R.N.P. and P.J.H. are grateful for funding from an NSERC Discovery Grant. Computations were performed on the GPC supercomputer at the SciNet HPC Consortium. SciNet is funded by the Canada Foundation for Innovation under the auspices of Compute Canada, the Government of Ontario, Ontario Research Fund—Research Excellence and the University of Toronto. Numerical calculations were done using a modified version of the SOPALE (2000) software. The SOPALE modelling code was originally developed by Philippe Fullsack at Dalhousie University with Chris Beaumont and his Geodynamics group.

AUTHOR CONTRIBUTIONS

The design of the study was conducted by all authors. P.J.H. analysed data and wrote the paper, and all authors discussed the results, and commented on the manuscript. R.N.P. and R.S. contributed equally to the study.

REFERENCES

1. Steer D. N., Knapp J. H. & Brown D. L. Super-deep reflection profiling: exploring the continental mantle lid. Tectonophysics 286, 111–121 (1998).
2. Flack C. & Warner M. Three-dimensional mapping of seismic reflections from the crust and upper mantle, northwest of Scotland. Tectonophysics 173, 469–481 (1990).
3. Morgan J. V., Hadwin M., Warner M. R., Barton P. J. & Morgan R. P. L. The polarity of deep seismic reflections from the lithospheric mantle: evidence for a relict subduction zone. Tectonophysics 232, 319–328 (1994).
4. Calvert A. J., Sawyer E. W., Davis W. J. & Ludden J. N. Archean subduction inferred from seismic images of a mantle suture in the Superior Province. Nature 375, 670–674 (1995).
5. van der Velden A. J. & Cook F. A. Relict subduction zones in Canada. J. Geophys. Res. 110, B08403 (2005).
6. Schiffer C. et al. . A sub-crustal piercing point for North Atlantic reconstructions and tectonic implications. Geology 43, 1087–1090 (2015).
7. Sibson R. H. Implications of fault-valve behaviour for rupture nucleation and recurrence.Tectonophysics 211, 283–293 (1992).
8. Bercovici D. & Ricard Y. Plate tectonics, damage and inheritance. Nature 508, 513–516 (2014).
9. Warren J. M. & Hirth G. Grain size sensitive deformation mechanisms in naturally deformed peridotites. Earth Planet. Sci. Lett. 248, 438–450 (2006).
10. Skemer P., Warren J. M., Kelemen P. B. & Hirth G. Microstructural and rheological evolution of a mantle shear zone. J. Petrol. 51, 43–53 (2010).
11. Cloetingh S. et al. . Lithospheric memory, state of stress and rheology: neotectonic controls on Europe's intraplate continental topography. Quat. Sci. Rev. 24, 241–304 (2005).
12. Krajcinovic D. Damage Mechanics North-Holland (1996).
13. Burov E. & Watts A. B. The long-term strength of the continental lithosphere: 'jelly sandwich' or 'crème brûée'? GSA Today 16, 4–10 (2006).
14. Copley A., Avouac J.-P. & Wernicke B. P. Evidence for mechanical coupling and strong Indian lower crust beneath southern Tibet. Nature 472, 79–81 (2011).
15. Zhong S. & Gurnis M. Viscous flow model of a subduction zone with a faulted lithosphere: long and short wavelength topography, gravity and geoid. Geophys. Res. Lett. 19, 1891–1894 (1992).
16. Zhong S. & Gurnis M. Mantle convection with plates and mobile faulted plate margins. Science 267, 838–843 (1995).
17. Zhong S. & Gurnis M. Interaction of weak faults and non-newtonian rheology produces plate tectonics in a 3D model of mantle flow. Nature 383, 245–247 (1996).
18. Crone A. J., De Martini P. M., Machette M. N., Okumura K. & Prescott J. R. Paleoseismicity of two historically quiescent faults in Australia: implications for fault behavior in stable continental regions.Bull. Seismol. Soc. Am. 93, 1913–1934 (2003).
19. Gray R. & Pysklywec R. N. Geodynamic models of mature continental collision: evolution of an orogen

from lithospheric subduction to continental retreat/ delamination. J. Geophys. Res. 117, B03408 (2012).

20. Buck W. R. Modes of continental Lithospheric Extension. J. Geophys. Res. 96, 20161–20178 (1991).
21. Calignano E., Sokoutis D., Willingshofer E., Gueydan F. & Cloetingh S. Strain localization at the margins of strong lithospheric domains: insights from analog models. Tectonics 34, 396–412 (2015).
22. Buiter S. J. H., Pfiffner O. A. & Beaumont C. Inversion of extensional sedimentary basins: a numerical evaluation of the localisation of shortening. Earth Planet. Sci. Lett. 288, 492–504 (2009).
23. Gleason G. C. & Tullis J. A flow law for dislocation creep of quartz aggregates determined with the molten salt cell. Tectonophysics 247, 1–23 (1995).
24. Rybacki E. & Dresen G. Dislocation and diffusion creep of synthetic anorthite aggregates. J. Geophys. Res. 105, 26017–26036 (2000).
25. Hirth G. & Kohlstedt D. L. Water in the oceanic upper mantle: Implications for rheology, melt extraction and the evolution of the lithosphere. Earth and Planetary Science Letters 144, 93–108 (1996).
26. Cowgill E., Yin A., Harrison T. M. & Xiao-Feng W. Reconstruction of the Altyn Tagh fault based on U-Pb geochronology: role of back thrusts, mantle sutures, and heterogeneous crustal strength in forming the Tibetan Plateau. J. Geophys. Res. 108, 2346 (2003).
27. Zhang L. et al. . Structure of the Central Altyn Tagh Fault revealed by magnetotelluric data: new insights into the structure of the northern margin of the India-Asia collision. Earth Planet. Sci. Lett.415, 67–79 (2015).
28. Wittlinger G. et al. . Tomographic evidence for localized lithospheric shear along the Altyn Tagh Fault.Science 282, 74–76 (1998).
29. Zhao J. et al. . Crustal structure across the Altyn Tagh Range at the northern margin of the Tibetan Plateau and tectonic implications. Earth Planet. Sci. Lett. 241, 804–814 (2006).
30. Sobel E. R. & Arnaud N. A possible middle Paleozoic suture in the Altyn Tagh. NW China. Tectonics18, 64–74 (1999).
31. Dèzes P., Schmid S. M. & Ziegler P. A. Evolution of the European Cenozoic Rift System: interaction of the Alpine and Pyrenean orogens with their foreland lithosphere. Tectonophysics 389, 1–33 (2004).
32. English J. M. & Johnston S. T. The Laramide orogeny: what were the driving forces? Int. Geol. Rev.46, 833–838 (2004).
33. Heron P. J., Pysklywec R. N. & Stephenson R. Intraplate orogenesis within accreted and scarred lithosphere: example of the Eurekan Orogeny, Ellesmere Island. Tectonophysics 664, 202–213 (2015).
34. Avouac J. P., Tapponnier P., Bai M., You H. & Wang G. Active thrusting and folding along the Northern Tien Shan and late Cenozoic rotation of the Tarim relative to Dzungaria and Kazakhstan. J. Geophys. Res. 98, 6755–6804 (1993).
35. Stephenson R., Egholm D. L., Nielsen S. B. & Stovba S. M. Role of thermal refraction in localizing intraplate deformation in southeastern Ukraine. Nat. Geosci. 2, 290–293 (2009).
36. Stein C. A., Cloetingh S. & Wortel R. Seasat-derived gravity constraints on stress and deformation in the northeastern Indian Ocean. Geophys. Res. Lett. 16, 823–826 (1989).
37. Cloetingh S. & Wortel R. Stress in the Indo-Australian plate. Tectonophysics 132, 49–67 (1986).
38. Sokoutis D. & Willingshofer E. Decoupling during continental collision and intra-plate deformation.Earth Planet. Sci. Lett. 305, 435–444 (2011).
39. Willingshofer E., Sokoutis D., Luth S. W., Beekman S. W. & Cloetingh S. Subduction and deformation of the continental lithosphere in response to plate and crust-mantle coupling. Geology 41, (12): 1239–1242 (2013).
40. Calignano E., Sokoutis D., Willingshofer E., Gueydan F. & Cloetingh S. Asymmetric vs. symmetric deep lithospheric architecture of intra-plate continental orogens. Earth Planet. Sci. Lett. 424, 38–50 (2015).
41. Bird P. & Gratz A. J. A theory for buckling of the mantle lithosphere and Moho during compressive detachments in continents. Tectonophysics 177, 325–336 (1990).
42. Dewey J.F. Suture zone complexities: a review. Tectonophysics 40, 53–67 (1977).
43. Fullsack P. An arbitrary Lagrangian-Eulerian formulation for creeping flows and its application in tectonic models. Geophys. J. Int. 120, 1–23 (1995).
44. Beaumont C., Jamieson R. A., Nguyen M. H. & Medvedev S. Crustal channel flows: 1. Numerical models with applications to the tectonics of the Himalayan-Tibetan orogen. J. Geophys. Res. 109, B06406 (2004).
45. Loken C. et al. . SciNet: lessons learned from building a power-efficient top-20 system and data centre.J. Phys. Conf. Ser. 256, 012026 (2010).
46. Burke K., Dewey J. F. & Kidd W. S. F. World distribution of sutures-the sites of former oceans.Tectonophysics 40, 69–99 (1977).
47. Hoffman P.F. United plates of America, the birth of a craton: early Proterozoic assembly and growth of Laurentia. Annu. Rev. Earth Planet. Sci. 16, 543–603 (1988).
48. Schiffer C., Balling N., Jacobsen B. H., Stephenson R. A & Nielsen S. B. Seismological evidence for a fossil subduction zone in the East Greenland Caledonides. Geology 42, 311–314 (2014).
49. Bird P. An updated digital model of plate boundaries. Geochem. Geophys. Geosyst. 4, 1027 (2003).
50. Watson M. P., Hayward D. N., Parkinson D. N. & Zhang Zh. M. Plate tectonic history, basin development and petroleum source rock deposition onshore China. Mar. Petrol. Geol. 4, 205–225 (1987).

CHAPTER 6

EARTHQUAKE IS ABRUPT RELEASE OF RADIATION FROM CRUST FISSURE OR HOLE

Wan-Jiung Hu

MAINTEXT

Professor Alfred Wegener proposed continental drift theory in 1912. He suggested that all the continents were joined together in old times and they separated by some unknown forces. In the beginning, his idea was thought to be absurd and was difficult to be accepted by people. However, further researches on ocean crust revealed that the ocean crust is actually expanding. This phenomenon happens in the mid-Atlantic ridge. Ocean crust is generated from the mid-Atlantic ridge and spread to both east and west side. We can find the symmetrical alternative geomagnetism reversal pattern in the two sides from mid-Atlantic ridge. The continental coasts of west Africa and east South America can match perfectly. In addition, fossils of the same species can be found in both South America and Africa. Thus, it suggested that South America and Africa was joined together in old era. Thus, continental drift theory was accepted, and then plate tectonics theory was developed to explain the happening of earthquakes and volcano eruptions. Plate tectonics theory says that earth crust can be divided into several "plates", and these "plates" crush each other to induce earthquakes. The plate crush movement is caused by ocean crust expansion and continental drift. Currently, it is a well-accepted theory for earthquake happening.

However, plate tectonics theory has several fetal defects. First of all, the sudden happening of earthquake in certain point should not be due to the slow process of continental drift. The continental drift is at a speed around 1cm per year. Thus, it only accumulates 30cm per 30 years. It is a very slow process and it cannot account for the huge energy release during a major earthquake. How can a 30cm plate movement per 30 year cause an energy release which is greater than 100x atom bomb energy? The slow continental drift cannot explain the sudden movement of rocks of plates during a major earthquake. In addition, plate tectonics movement or continental drift is a large scale movement and it cannot be accounted for earthquake happening in a single point. According to Young's modulus formula, the pressure is P=F/A=E*(DL/L). Since the palate tectonic is wide ranged (huge L), the pressure caused by plate should be very small. Thus, plate tectonics theory cannot well explain the great force of earthquake.

Second, the continental drift is only proved between South America and Africa. The continental coast of South America and Africa matches only mean that there is ocean expansion in the Atlantic Ocean. In addition, the fossils which are similar in South America and Africa mean that only South America

Citation: Hu, W. (2011). Earthquake is abrupt release of radiation from crust fissure or hole.PHILICA.COM Article number 278. ISSN 1751-3030.

and Africa were joined together. For other major continents, the evidences for all continents were joined together are weak. Thus, we cannot say all the continents are moving. We cannot say all the continents are moving which causes earthquake. Where the evidence is, where the conclusion is.

Third, plate tectonics theory cannot explain intrapalate earthquakes. There are some "hotspots" which have frequent earthquake. For example, Hawaii suffers from a lot of earthquakes per year. However, Hawaii is not located in the boundary of plates. In addition, there are several detrimental intraplate earthquakes. For example, the Sichuan earthquake currently is a huge intraplate earthquake. Plate tectonics theory cannot explain this. How can an earthquake happen when the location is not in the boundary of plates? Based on the plate tectonics theory, the intraplate earthquakes should be small or trivial. How can it cause such a huge and detrimental earthquake such as Sichuan earthquake?

Here, we propose a new mechanism for earthquake. Earthquake is actually a radiation release from the inside earth. Radiation energy acting on faults causes earthquake. Radiation energy is important in our earth. The earth core and mantle has an estimated temperature around 4000 to 5000 absolute temperature. Radiation needs to be released from inside earth to outside space because it needs to diffuse out to outer space to gain maximal entropy. We can view these heat particles are photons with momentum, and they will collide each other to expel each other to diffuse out. The formula for Stefan-Boltzmann's law is:

Power=σAT^4 ($A=4\pi R^2$)

The unit of radiation power is joule per second. Earth is covered by crusts to prevent radiation release. It means that radiation will reflex if it touches the inside crust. Thus, it means that in every second, radiation energy is accumulating when radiation is not released to outside space. After 30 years, we can see huge amount of radiation energy achieves. Because the earth core and mantle have very high temperature, we can see that the accumulated radiation energy can cause very detrimental effect. Normally, the radiation energy of inside earth is under the suppression of outer earth crust. After accumulation for decades, the accumulated radiation becomes greater and greater and it finally needs to be released.

We hypothesize that the radiation is released from earth inner core with its 5700 absolute degree and its 1200 Km in diameter. It can travel through the viscous-liquid mantle:

Thus, power=$1.1*10^{21}$ watt

Even in one second, the total energy is huge.

Energy=$1.1*10^{21}$ joule

We know the energy of a typical atomic bomb is 10^{14} joule.

Thus, the energy released from inner core is 10^7 larger than an atomic bomb.

We assume the heat energy can be transformed into work:

W=F*S (if S=1km=1000meter up-moved crust by the released inner energy)

The total force=10^{18} Newton->very great force

It is also possible that the radiation is released from mantle with average temperature 1000K. We assume the radiation is from a 1meter*1meter area in mantle. When it is emitted and touched the inner side of crust, it will go back and accumulate. However, when it reflected back, it may change its direction a little bit. After 30 years, the group of reflected radiation beam can have a lot of energy. If it finds a direction (with a refraction angle) which can emit out from the crust fissure and hole. Then, the radiation can be released out and cause earthquake.

Power=$7.16*10^5$ Watt(1000K and 1meter*1meter mantle area)

After 30 years, the total energy:

Energy=$7*10^{14}$ joule which is equal to 7 atomic bombs

If the crust is up-moved by 1000 meter, there will be:

W=F*S

Force=$7*10^{11}$ Newton, still very large

Frequently, the radiation is released from crust fissures and holes. Earth crust fissures include mid-Atlantic ridge and peri-Pacific troughs. Crust holes are the hot spots such as Hawaii. Molten lava from mantle also can be released from these fissures and holes. That is why «Pacific ring of fire» (peri-Pacific troughs) is both for volcano eruptions and earthquakes. That is why the Hawaii «Hotspot» is both for volcano eruptions and earthquakes. Radiation theory can explain why volcano happens in the same places as earthquake happens. When the radiation is released from inside earth and is passing through the earth crust, radiation energy can cause physical movement of faults of crust. When radiation pushes the crust to move upward, then the crust can move downward to return due to gravity. However, another radiation flush pushes the crust to move upward again. The back and forth phenomenon causes the shock wave. Then, earthquake happens and seismic wave occurs. If there is a fault A\B, the radiation releasing between A layer and B layer can push B layer upward or push A layer upward or downward. That explains why some P wave is upward and the other is downward. Touching A layer or B layer is depending on the angle of radiation beam.

The new theory can explain intraplate earthquakes. If there is fault in the earth crust, there is possibility that radiation pressure can be released via these faults. They are naturally gaps which allow radiation pressure to pass through. Thus, it can explain why some intraplate earthquakes are also very huge and detrimental. For examples, Sichuan or Tangshan earthquakes are both huge and detrimental intraplate earthquakes. Plate tectonic theory cannot well explain this.

We can see similar effect on our sun. When solar wind or solar flare is released from the sun surface, it also causes huge amount of seismic wave release.[1] It is called «sunshock» or «sunquake». We think the basic mechanism of sunquake and earthquake is the same: both are due to the release of radiation from inside to outside. Because the sun has much greater absolute temperature, the sunquake caused by sun flare is even much greater than major earthquakes.

In this new theory, we can also explain many earthquake phenomenon which cannot be explained by traditional plate tectonics theory. Earthquake light before or during earthquake is an important phenomenon related to earthquake happening.[2-5] For example, many people observed white to blue light shining in the sky for all night during the Peru earthquake. Plate tectonics theory says that it is caused by suddenly transformation of rocks into a semi-conductor (p-hole) like substance. Thus, rocks can emit light. This mechanism explained by plate tectonics theory is very strange and cannot be accepted. Rocks are electrically neutral and are simply mass not charge. Even the rock mass is accelerated, it cannot emit electromagnetic radiation based on fundamental Maxwell's classical electromagnetism. Why can an earthquake transform rocks into semi-conductors? In addition, radio disturbance and ionosphere anomaly are usually observed during or before earthquake happening[6-11]. Electromagnetism field anomaly or infrared light anomaly is also observed by satellite or other instruments during a major earthquake[12-21]. Some researchers also noted some changes of animal behaviors. If the earthquake is the release of radiation pressure, we can easily explain visible earthquake light, radio disturbance, EMW/ionosphere anomaly, and infrared light anomaly. Animal behavior change may be due to that some animals can detect the change of radiation wave.[22]

In majority of earthquake observations, infrared light and radio frequency disturbance are frequently seen before earthquake. However, UV abnormal signals are seldom seen. We will deduct the reason here. When a radiation beam is emitted from earth mantle or inner core to outside space, it must suffer from a phenomenon called refraction. It is like light is emitted from water to air. Because the radiation energy is accumulated, it is not going straight from the mantle just below the crust fissure or hole. There is a refraction angle. The refraction index is:

Snell›s law: N=C/v

(C: light speed in air or vacuum; v: light speed in medium such as mantle)

During the refraction, there is a physical phenomenon called dispersion. That is different wave lengths have different refraction indices. Red light which has longer wavelength has less refraction indices; blue light which has shorter wavelength has larger refraction indices. According to the above formula, red light has smaller N and faster v. (lightspeed C is constant). Thus, it means that radiation with longer wavelength transmit faster in the medium such as earth mantle. We assume that the radiations from earth compose a wide spectrum of radiation wavelengths since the radiation substances inside earth is complicated and varied. Thus, radiations with longer wavelength will be released out from inner earth faster. That is why infrared or radio signal which have longer wavelengths are seen before the main phase of earthquakes. This situation is important for earthquake prediction. Because radiations with shorter wavelengths are associated with higher energy, the short wavelength radiations are more possible to cause earthquake detrimental effects. (E=hf). However, it radiations with longer wavelengths and smaller energy are emitted out faster, we can use them as warning signals. Besides, we might expect to see a radiation spectrum change during earthquake. It should be infrared signal first, then visible light such as earthquake light, and finally UV spectrum light. This serial change of radiation spectrum can also help us for earthquake prediction. A recent review study pointed out there is really EM frequency increase pattern before earthquakes.[23]

Based on the cause-effect relationship, we can examine the following criteria for the link of earthquake and EM radiation.[24252627] First of all, we check the strength association. Stronger earthquakes are accompanied by strong EM radiation signals and more intense ionospheric anomalies. Second, we look at the consistency. All measured earthquakes can be detected the release of EM radiation. Third, we can see the temporality. Infrared light anomalies and earthquake light are detected before or during the onset of earthquakes. Fourth, we examine the plausibility. That earthquake caused by the release of electromagnetic radiation energy is more plausible than the current plate tectonic theory. There is no driving force for such an abrupt and vigorous movement in a single point for the current plate tectonic theory. Some say that tectonic plate movement is driven by mantle convection. However, how can this slowly and continuously creeping heat motion cause an abrupt and vigorous signal point disaster? Based on the Unruh-Hawking equation, heat radiation can be mutually transformed into acceleration. You may argue the heat radiation could be from the crust acceleration. However, the truth is the opposite. The heat radiation from the inside earth was transformed into crust acceleration during earthquakes. There is no plausible reason and driving force for the abrupt happened crust acceleration based on plate tectonic theory. If we look at the formula of F=ma, the net acceleration caused by earthquake will be very small with the large mass of the whole tectonic plate due to the plate tectonic theory. Thus, the heat radiation transformed by the tiny acceleration should be neglected due to the plate tectonic theory. Tectonic plates always move continuously and slowly, and we can find out the temporal association between the plate movement and the occurrence of earthquakes. In addition, there is a lack of strength association. We never observe, detect, or confirm that a more vigorous plate movement has a direct link to a more detrimental earthquake. Plate tectonics theory also has a defect in the consistency. In the detrimental and huge Sichuan earthquake, this earthquake didn›t even happen in the boundary of plate-and-plate. How can plate tectonic movement cause such a huge earthquake?

In summary, we think this new earthquake theory is the actual mechanism of earthquake happening. Plate tectonics theory fails to help to do earthquake prediction. This new theory can shed a light for earthquake prediction. If we can monitor or detect radiation anomaly (earthquake light, radio disturbance, EMW anomaly, infrared light anomaly) before earthquake, we might be able to know when and where the earthquake will happen. Earthquake can be very huge and detrimental. Around 100 thousand people might be killed in current Sichuan earthquake. Current continental drift/plate tectonics theory is wrong, and this wrong theory prevents us from doing earthquake prediction. We sincerely hope this new theory can be helpful in earthquake prediction to avoid huge amount of human life loss.

REFERENCES

1. Martinez-Oliveros J.C. et al. Helioseismic analysis of the solar flare-induced sunquake of 2005 January 15 II. A magnetoseismic study. Monthly Notices of the Royal Astronomical Society 389,1905-1910 (2008)
2. Hedervari P. Earthquake light phenomenon Nature 301,368 (1983)
3. Lloyd J.J. Earthquake light Science 193,1070 (1976)
4. Richter C.F. Earthquake light in focus Science 194,259 (1976)
5. Kamogawa M. et al. Earthquake light: 1995 Kobe earthquake in Japan Atmospheric Research 76,438-444

(2005)
6. Biagi P.F. et al. LF radio anomalies revealed in Italy by the wavelet analysis: possible preseismic effect during 1997-1998. Physics and Chemistry of the Earth, Parts A/B/C, 31, 403-408 (2006)
7. Biagi P.F. et al. Possible earthquake precursors revealed by LF radio signals Nature Hazards and Earth System Sciences 1,99-104 (2001)
8. Liu J.Y. et al. Seismoionospheric GPS total electron content anomalies observed before the May 2008 Mw7.9 Wenchuan earthquake Journal of Geophysical Research 114,1-10 (2009)
9. Ruzhin Y.Y. et al. Earthquake precursors in magnetically conjugated ionosphere regions Advance in Space Research 21,525-528 (1998)
10. Liu J. Y. et al. Seismo-ionospheric signatures prior to M>6.0 Taiwan earthquakes Geophysical Research Letters 27,3113-3116 (2000)
11. Fujiwara H. et al. Atmospheric anomalies observed during earthquake occurrences Geophysical Research Letters 31,L17110 (2004)
12. Ni S.D. et al. Energy radiation from the Sumatra earthquake Nature 434,582 (2005)
13. Uyeda S. et al. Electric and magnetic phenomena observed before the volcano-seismic activity in 2000 in the Izu island religion, Japan PNAS 99,7352-7355(2002)
14. Uyeda S. et al. Geoelectric potential changes: possible precursors to earthquakes in Japan PNAS 97,4561-4566 (2000)
15. Varotsos P.A. et al. Electric fields that arrive before the time derivative of the magnetic field prior to major earthquakes Physical Review Letters 91, 148501(2003)
16. Ouzounov D. and Freund F. Mid-infrared emission prior to strong earthquakes analyzed by remote sensing data Advances in Space Research 33,268-273 (2004)
17. Tronin A. A. et al. Thermal IR satellite data application for earthquake research in Japan and China Journal of Geodynamics 33,519-534 (2002)
18. Fujinawa Y. et al. Emission of electromagnetic radiation preceding the Ito seismic swarm of 1989 Nature 347,376-378 (1990)
19. Fujinawa Y. et al. Experiments to locate sources of earthquake-related VLF electromagnetic signals Proc. Jpn. Acad 73,33-38 (1997)
20. Oike K. et al. Electromagnetic radiations from rocks J. Geophys. Res. 90,6245-6249 (1985)
21. Varatos P. et al. Physical properties of the variations of the electric field of the earth preceding earthquake Tectonophysics 110,73-98 (1984)
22. Ikeya M. et al. Electric shocks resulting in seismic animal anomalous behaviors J. Phys. Soc. Jpn. 65,710-712 (1996)
23. Chen ZY. Et al. The observation and research on pre-earthquake electromagnetic radiation information Acta Seismologica Sinica 6,485-494 (1993)
24. Lin JW Principal component analysis method in the detection of total electron content anomalies in the 24hrs prior to large earthquakes Arab J Geosci online 2010 October
25. Biagi RF Possible earthquake precursors revealed by LF radio signals Natural Hazards & Earth System Science 1,99-104 (2001)
26. Fujinawa Y et al. Electromagnetic radiations associated with major earthquakes Physics of the Earth & Planetary Interiors 105, 249-259 (1998)
27. Kamogawa M Preseismic lithosphere-atmosphere ionosphere coupling Eos 87,417-424 (2006)

CHAPTER 7 CRUSTAL THICKNESS VARIATIONS AND SEISMICITY

Orlando Hernandez-Pardo, Ralph R. B. von Frese, Jeong Woo Kim

Keywords: Crustal thickness, Gravity, Seismicity, Northwestern South America.

INTRODUCTION

The increased number of seismological networks established during the last century led to the discovery that earthquakes are not randomly distributed, but tend to occur along well defined earthquake belts (Shearer, 1999). These belts are largely concentrated along the margins of tectonic plates that shift slowly over geologic time. The relative motions between adjacent plates give rise to earthquakes along the plate boundaries that include spreading oceanic ridges, converging subduction zones, collisional continental plate boundaries, and transform faults along which they shear past each other (Bird, 2003; Turcotte and Schubert, 2002; Cediel *et al.*,2003). For northwestern South America, the improvement in instrumentation and expansion of seismological networks has led to the production of relatively complete and accurate catalogs of earthquake locations and ground motions.

Citation: Orlando Hernandez-Pardo, Ralph R. B. von Frese, Jeong Woo Kim, "Crustal Thickness Variations and Seismicity," ISSN 1794-6190.

SPECTRALLY CORRELATED TERRAIN AND FREE AIR GRAVITY ANOMALIES

The isostasy of northwestern South America was investigated considering the topography/ bathymetry data from National Imagery and Mapping Agency (NIMA) from −8°S to 23.5°N latitude and from −90°W to −58.5°W longitude. Surface and bathymetry elevations from the JGP95E terrain data base (Smith and Sandwell, 1994; 1997) were processed to produce the Digital Elevation Model for the water and rock terrain gravity components at 0.5° nodal spacing. Free-air gravity anomalies (FAGA) were estimated from the EGM96 spherical harmonic Earth Gravity Model to degree and order 360 (Lemoine *et al.*, 1998) at 20 km altitude over the 32° x 32° area at 0.5° nodal spacing. The altitude of 20 km was chosen to help minimize the effects of local density errors in the terrain gravity modeling (Leftwich *et al.*, 2005). The terrain gravity effects were modeled in spherical coordinates at 20 km altitude by Gauss-Legendre Quadrature integration (von Frese, 1980). The terrain gravity modeling used densities of 2.8 gm/cm3 for the crust and 1.03 gm/cm3 for oceanic water. Spectral correlation theory was used to analyze the co-registered FAGA and TGE for their anomaly correlations using MatLab (MATHWORKS, 2005). Specifically, the Fourier transforms T and F of TGE and FAGA, respectively, were used to obtain their correlation spectrum (Davis, 1986; von Frese et al., 1997a; Kim *et al.*, 2000) given by:

$$CC(k) = \cos(\Delta\theta k) = Re\ \frac{F(k)\ T(k)}{T(k)||F(k)||} \qquad (1)$$

where CC(k) is the correlation coefficient between the kth wavenumber components F(k) and T(k), and denotes taking the real parts of the wavenumber components. Usually, CC(k) is evaluated from the cosine of the phase difference ($\Delta\theta k$) between the two kth wavenumber components. Using the correlation spectrum between FAGA and TGE, spectral correlation filters were designed to extract terrain-correlated free-air gravity signals. Those wavenumber components showing intermediate to high positive ($CCp(k) \geq 0.3$) and negative ($CCn(k) \leq 0.3$) correlations were identified. The cut off values for the correlation filter were determined to minimize correlative features between the terrain-decorrelated free air and compensating terrain gravity components. Inversely transforming positively and negatively correlated free-air wavenumber components according to the selected cut off values yielded the terrain- correlated free air gravity anomalies (TCFAGA).

The residual terrain-decorrelated free-air gravity anomalies (TDFAGA) were calculated by subtracting TCFAGA from FAGA, so that

$$FAGA = TCFAGA + TDFAGA \qquad (2)$$

TCFAGA are explained by anomalies associated with the topography while TDFAGA include the gravity effects of sources within the crust (e.g., local bodies) and the subcrust.

CRUSTAL MODELLING

A new crustal thickness model for northwestern South America was developed using the compensated terrain gravity effects (CTGE) that resulted when the TGE were subtracted from the TCFAGA. The CTGE represent isostatically adjusted complete Bouguer anomalies that correspond to the gravity effects of the terrain in isostatic equilibrium. This approach is feasible because 90% of the earth is in equilibrium with the mean global free-air gravity anomaly being zero (Heiskanen and Moritz, 1967). MOHO and related crustal thickness variations were modeled from the CTGE by inversion, assuming the constant nominal density contrast of 0.4 gm/cm3 of the mantle relative to the crust. This methodology has been successfully applied instudies of the mantle-crust interface for East Asia (Tan and von Frese, 1997), Antarctica (von Frese *et al.*, 1999), Greenland (Roman, 1999), Ohio (Kim *et al.*, 2000), and Iceland and the North Atlantic (Leftwich *et al.*, 2005; Leftwich, 2006). Negative CTGE values are located along the eastern Andes Mountains suggesting some degree of partial compensation and thickening of the crust. The CC between the TCFAGA and the CTGE is -0.3377 showing that most of the topography/bathymetry is isostatically disturbed.

Areas where the TCFAGA values are excessively negative or positive are more prone to seismic activity than areas where the TCFAGA values are closer to zero (Song and Simons, 2003). This paper analyzes and compares seismic data from the Advanced National Seismic System (ANSS) and the Red Sismológica Nacional de Colombia (RSNC) catalogs with crustal thickness estimates from gravity anomalies by understanding improved large-scale dynamic models of earthquakes and tectonics. Gravity anomalies along trenches, continental converging margins, crustal discontinuities, mountain ranges, and cratonic areas are compared with their seismic signatures from earthquake data collected over the last 363 years.

ZERO CURVATURE OF CRUSTAL THICKNESS VARIATION

The terrain-correlated free-air anomalies (TCFAGA) in Figure 1 highlight regions with isostatically disturbed crustal features. The zero anomalies mark areas of crustal equilibrium so that the positive and negative anomalies reflect the compression and tension of the crust, respectively. Positive and negative TCFAGA values can mark crust that is isostatically too thin (under-compensated) or thick (overcompensated), respectively, and hence under pressure to equilibrate by the compressive inflow or expansive out-flow of crustal material (Artyushkov, 1973). Thus, these anomalies can reflect lithosphere in subsidence or uplift (Kim et al., 2000; von Frese *et al.*, 1999a), or alternatively dynamic surface topography that is too high or too low, respectively, to be in isostatic equilibrium.

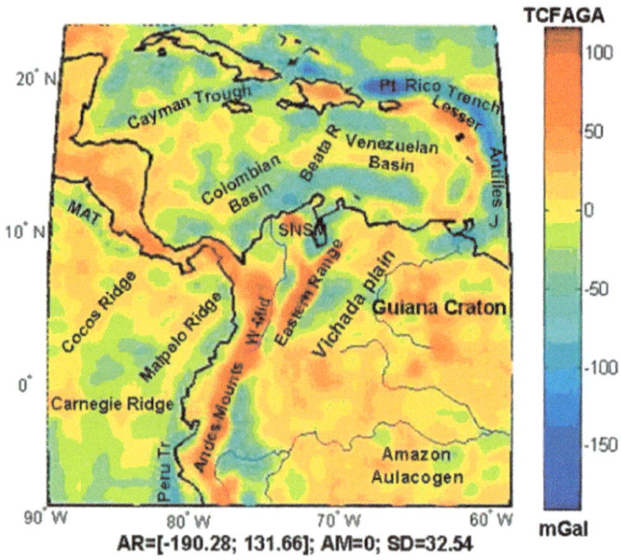

Figure 1. Terrain-correlated FAGA (TCFAGA) at 20 km elevation for the study area. Map annotations include the amplitude range (AR) of (min; max) values, the amplitude mean (AM), and standard deviation (SD). SNSM = Sierra Nevada of Santa Martha, W-Mid = Western-Central ranges. This map was produced using the Albers equal-area conic projection.

Figure 2 gives the crustal thickness variations for northwestern South America obtained by adding the gravity MOHO estimates (Hernandez, 2006). The zero curvature contour that estimates the edges of thickness variations was obtained by calculating the second vertical derivative of the crustal thickness (Figure 2). The use of vertical derivatives has been always a standard method of enhancing high frequency features in potential field data. Second order vertical derivatives were computed using convolution filters and Laplace's equation (Sheriff, 1994). The intervening yellow and blue regions of Figure 2 reflect the thicker and thinner crustal components, respectively.

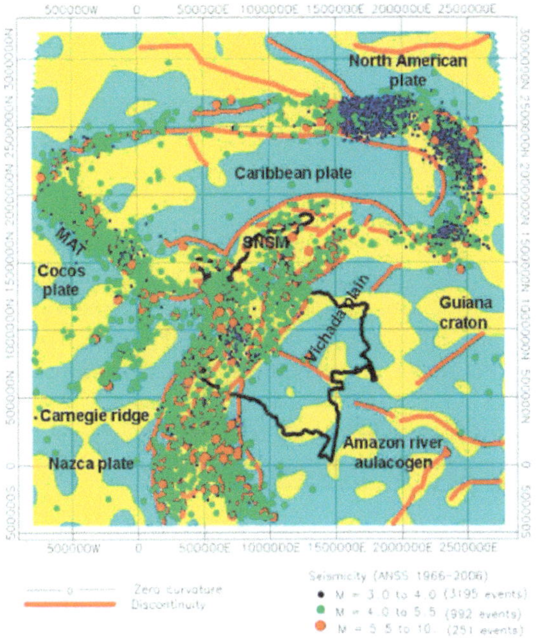

Figure 2. ANSS seismic catalog for 1966-2006 for three categories of earthquake magnitudes: M=3-4 (minor earthquakes), M= 4.1-5.5 (intermediate earthquakes), and M ¸ 5.5 to 10 (major earthquakes). Crustal discontinuities interpreted from gravity anomalies and zones of crustal thickness variations are also displayed. Yellow and blue zones correspond to thicker and thinner crust, respectively. This and the succeeding figures were created using the capabilities of Oasis Montaj (Geosoft, 2006).

SEISMIC DATA COMPILATION

For comparison with the crustal thickness estimates, the regional seismic data of northwestern South America were compiled from the Advanced National Seismic System catalog (ANSS; USGS, 2006). A more local study was also considered using the historic and instrumental seismic catalogs of the Red Sismológica Nacional of Colombia (RSNC; INGEOMINAS, 2006).

ANSS CATALOG

The ANSS catalog is a world-wide earthquake catalog created by merging the master earthquake catalogs from the contributing ANSS institutions and then removing duplicate solutions (USGS, 2006). The ANSS catalog currently consists of earthquake hypocenters, origin times, and magnitudes. The ANSS database was searched for data from 1966 to 2006 with Ritcher magnitudes M of 3-10.

The earthquake epicenters were converted from geographic to Cartesian coordinates in Figure 2 with the proportional symbols for Ritcher magnitudes given in the legend. Only those events with hypocenters from 0 to 60 km depths were considered in this study. The crustal discontinuities interpreted from the TCFAGA in Figure 1 are also superposed in Figure 2.

The seismic events in Figure 2 are concentrated along the plate boundary zones of the Nazca, Caribbean, North American and South American Plates, the North Andes, Panama and Costa Rica Microplates, the Cocos-Nazca spreading system, and the intraplate discontinuities. They are predominantly located in the thicker crustal sections and along the inferred edges of the thickness variations. However, the thicker oceanic sections of the Nazca, Malpelo and Cocos Ridges in the Pacific and the Beata Ridge in the Caribbean are aseismic.

Comparatively few earthquakes are concentrated in the thinner crustal regions with relatively negative TCFAGA. The Guiana Craton and the interior of Nazca and Caribbean oceanic Plates also lack significant seismic events and are seismically quiescent. To account for the intraplate earthquakes, gravity and magnetic anomalies that infer crustal discontinuities are useful to analyze deformation in the plate interior.

RSNC HISTORICAL AND INSTRUMENTAL SEISMIC CATALOG

The RSNC catalog was created by merging the master earthquake catalogs from the Observatorio Sismológico del Sur Occidente (OSSO), Centro Regional de Sismología Para America del Sur (CERESIS), Instituto Geofísico de los Andes, Red Sismológica Nacional, and regional institutions of northwestern South America and the Caribbean (INGEOMINAS, 2006). This catalog currently consists of earthquake epicenters, origin times, and magnitudes. The RSNC database compiled data from 1643 to 1991 for Ritcher magnitudes from M = 3.0 to M = 10.

The RSNC catalog contains an enormous number of events. Thus, the events were sorted by magnitude for the minor earthquakes (M = 3 to 4), moderate earthquakes (M= 4.1 to 5.0), and strong earthquakes (M > 5.0) as shown in Figures 3, 4, and 5 respectively.

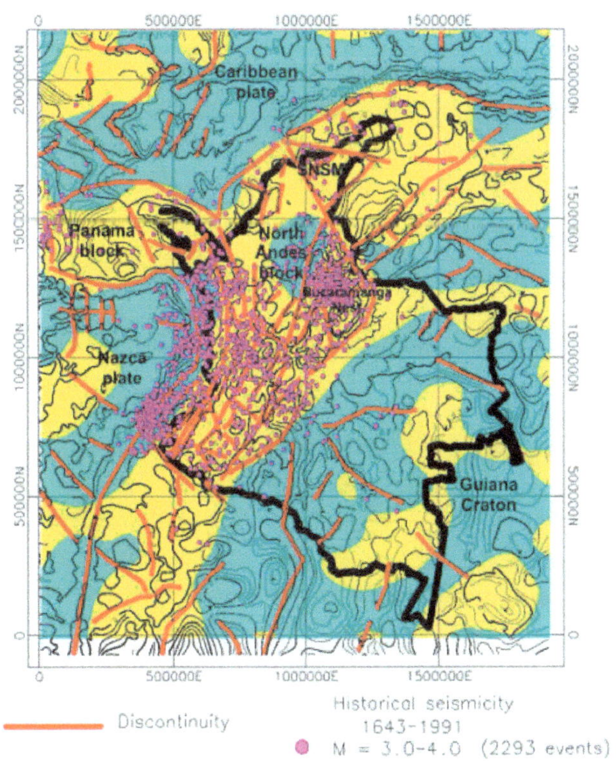

Figure 3. RSNC historical seismic catalog for 1643-1991 of earthquakes with magnitudes from M=3.0 to 4.0. Crustal discontinuities interpreted from gravity anomalies and the zero curvature contours of crustal thickness variations are also displayed. The yellow and blue zones correspond to thicker and thinner crust, respectively.

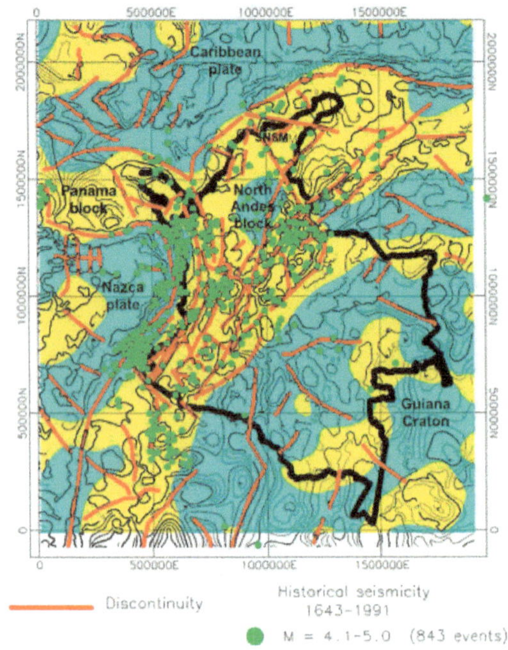

Figure 4. RSNC historical and instrumental seismic catalog for 1643-1991 of earthquakes with magnitudes from M=4.1 to 5.0. Crustal discontinuities interpreted from gravity anomalies and the zero curvature contours of crustal thickness variations are also displayed. The yellow and blue zones correspond to thicker and thinner crust, respectively.

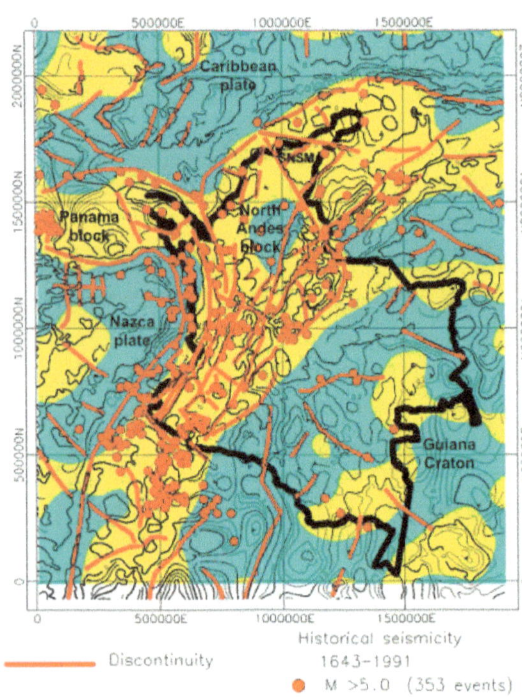

Figure 5. RSNC historical seismic catalog for 1643-1991 of earthquakes with magnitudes M > 5.0. Crustal discontinuities interpreted from gravity anomalies and the zero curvature contours of crustal thickness variations are also displayed. The yellow and blue zones correspond to thicker and thinner crust, respectively.

Minor and moderate seismic events in Figures 3 and 4, respectively, are concentrated along the Pacific subduction zone and intracrustal discontinuities of the North Andes Microplate. The concentration of events in the northern part of the eastern cordillera is known as the "Bucaramanga Nest." Few events are recorded for the Sierra Nevada of Santa Martha, Guiana Craton, and the northwestern flat lands of Colombia at the north of the Andes Mountains. At the continent, areas with thicker crust are more seismically active, while areas of thinner crust are less seismically active. An exception is the Sierra Nevada of Santa Martha with a thick crust and low seismic activity.

Major earthquakes in Figure 5 are located along the converging Nazca-North Andes and South America-North Andes continental boundaries. The major earthquakes are located further landward from the Pacific subduction zone relative to the minor and intermediate earthquakes that are concentrated closer to the coastline. The few seismic events of the Guiana Craton are mostly localized along local crustal discontinuities and are not associated with crustal thickness variations.

RECENT SEISMICITY IN THE RSNC CATALOG

The RSNC started operating since 1992 and has accumulated accurate instrumental information that consists of earthquake epicenters, hypocenters, origin times, and magnitudes. The epicenters were sorted by magnitude for minor events (M= 3 to 4) in Figure 6, moderate events (M= 4.1 to 5) in Figure 7, and major events (M > 5.0) in Figure 8.

The minor and moderate earthquakes in Figures 6 and 7, respectively, define seismic corridors oriented along the major crustal discontinuities of the North Andes Microplate. The Sierra Nevada of Santa Marta also shows more intensive seismic activity than previously recorded by the older data. Major earthquakes in Figure 8 are located along the major crustal discontinuities and plate boundary zones of the North Andes Microplate. In general, seismic events are associated within thicker crust, and therefore can be associated with tensional stress (Artyushkov, 1973).

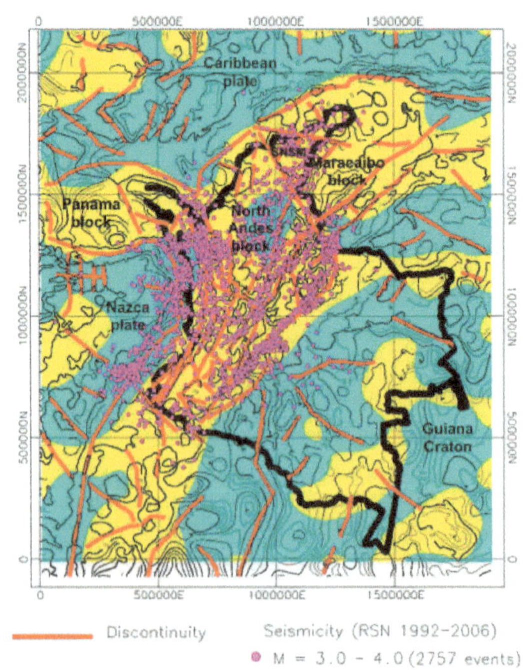

Figure 6. RSNC modern seismic catalog for 1992-2006 of earthquakes with magnitudes from M=3.0 to 4.0. Crustal discontinuities interpreted from gravity anomalies and the zero curvature contours of crustal thickness variations are also displayed. The yellow and blue

zones correspond to thicker and thinner crust, respectively.

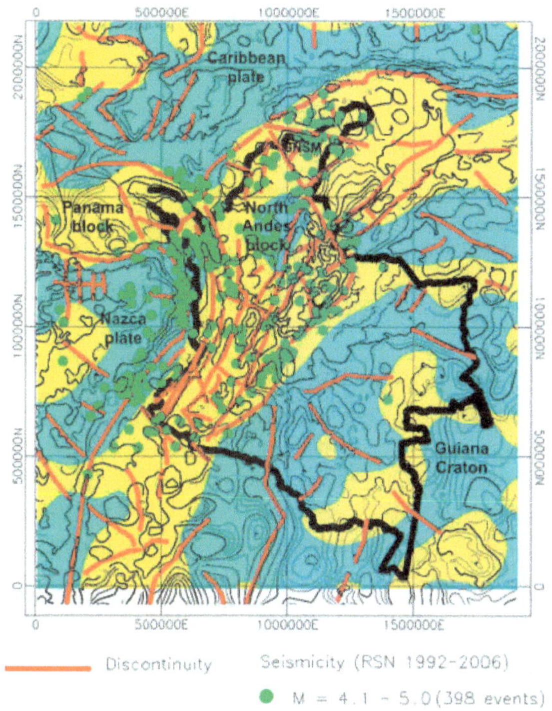

Figure 7. RSNC modern seismic catalog for 1992-2006 of earthquakes with magnitudes from M=4.1 to 5.0. Crustal discontinuities interpreted from gravity anomalies and the zero curvature contours of crustal thickness variations are also displayed. The yellow and blue zones correspond to thicker and thinner crust, respectively.

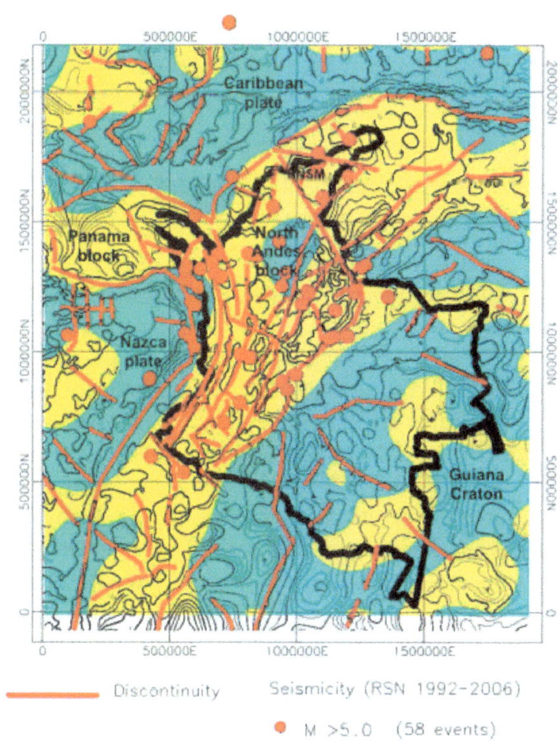

Figure 8. RSNC modern seismic catalog for 1992-2006 of earthquakes with magnitudes M > 5.0. Crustal discontinuities interpreted from gravity anomalies and the zero curvature contours of crustal thickness variations are also displayed. The yellow and blue zones correspond to thicker and thinner crust, respectively.

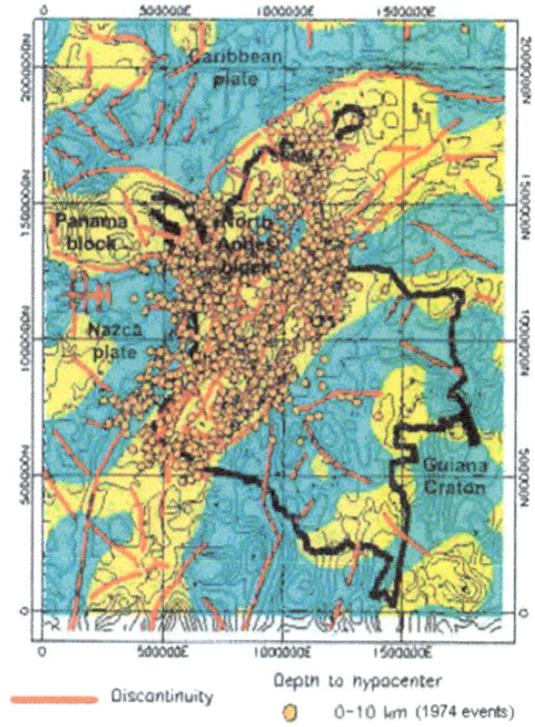

Figure 9. RSNC modern seismic catalog for 1992-2006 of earthquakes with hypocenter depths from 0 km to 10 km. Crustal discontinuities interpreted from gravity anomalies and the zero curvature contours of crustal thickness variations are also displayed. The yellow and blue zones correspond to thicker and thinner crust, respectively.

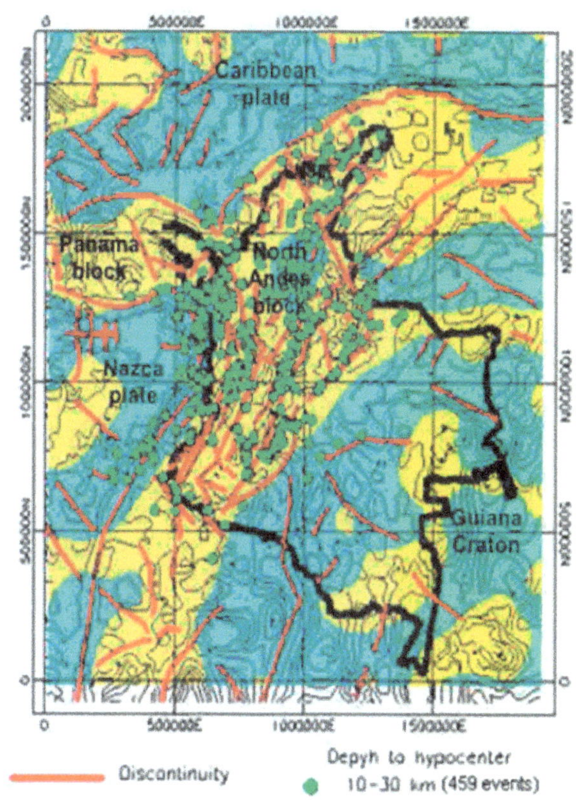

Figure 10. RSNC modern seismic catalog for 1992-2006 of earthquakes with hypocenter depths from 10 km to 30 km. Crustal discontinuities interpreted from gravity anomalies and the zero curvature contours of crustal thickness variations are also displayed. The yellow and blue zones correspond to thicker and thinner crust, respectively.

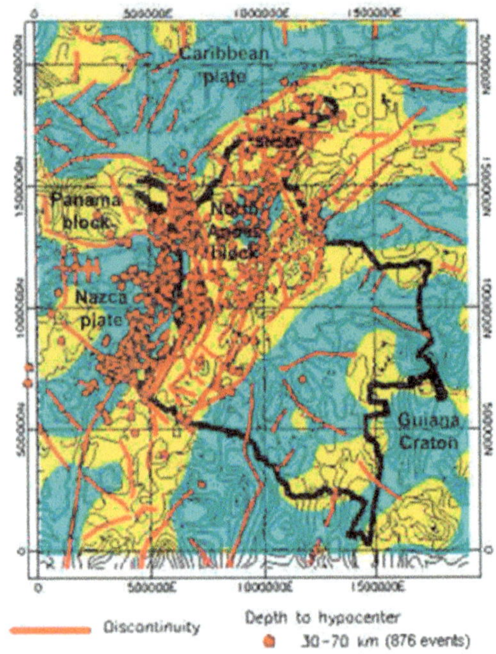

Figure 11. RSNC modern seismic catalog for 1992-2006 of earthquakes with hypocenter depths from 30 km to 70 km. Crustal discontinuities interpreted from gravity anomalies and the zero curvature contours of crustal thickness variations are also displayed. The yellow and blue zones correspond to thicker and thinner crust, respectively.

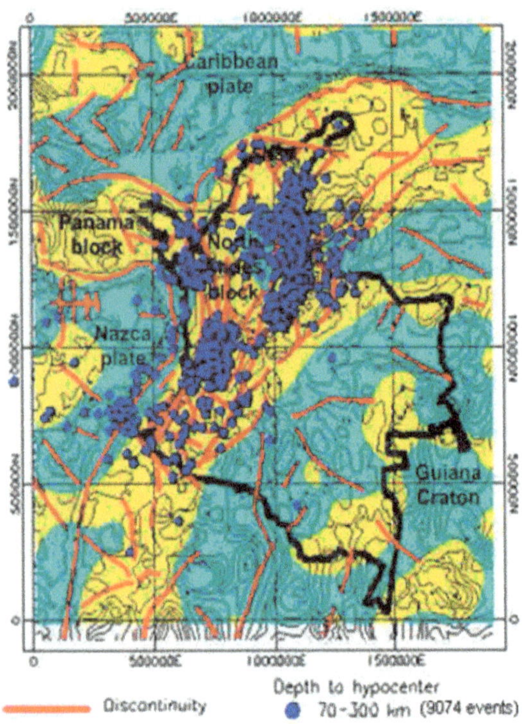

Figure 12. RSNC modern seismic catalog for 1992-2006 of earthquakes with hypocenter depths to the hypocenter from 70 km to 300 km. Crustal discontinuities interpreted from gravity anomalies and the zero curvature contours of crustal thickness variations are also displayed. The yellow and blue zones correspond to thicker and thinner crust, respectively.

Hypocenter depths from the modern RSNC were sorted into shallow earthquakes (from 0 to 10 km) in Figure 9, intermediate earthquakes (10 km to 30 km) in Figure 10, deep earthquakes (30 km to 70 km) in Figure 11, and very deep earthquakes (70 km to 300 km) in Figure 12. The tendency of the deepening hypocenters to migrate landward of the Pacific subduction zone is indicative of the oceanic plate subducting under the Andes at an angle that is less than vertical.

DISCUSSION

Gravity and topography are related to seismic activity by the frictional behavior of plate boundary zones and crustal discontinuities. When two plates push against each other, friction between the plates builds up causing the plates to lock. The continued build up of tectonic stress also causes the plates to deform, creating variations in topography and gravity. In addition to deforming the plates, the rocks of the sticking point can break when the stress build up exceeds the breaking strength of the rocks. The resulting strain released at the locking point produces the ground shaking of an earthquake. For northwestern South America, plate tectonic stress plays out notably in the ocean-continent, and ocean-ocean and continent-continent plate collision zones, and also along oceanic spreading ridges and transform faults. In the next sections, these plate boundaries and associated earthquakes zones areas are further analyzed.

OCEAN-CONTINENT COLLISION ZONES

In Figures 2 to 12 the subduction of the Nazca plate under the North Andes microplate is clearly parallel to the Andes Mountains. The collision caused the creation of the Andes Mountains by stacking thrust slices of the crust. The lateral movement of these thrust slices in the welded continent has continued after the collision generating minor, moderate and major earthquakes with shallow, intermediate and deep hypocenter depths defining seismic corridors that are oriented along the subduction zone and the Andes Mountains. The general pattern of the seismicity and focal mechanism solutions of earthquakes are in agreement with the easterly subduction of the Nazca Plate beneath the Andes.

The intensive seismicity along the Middle American Trench (MAT) has a broad pattern reflecting the low angle of inclination of the subducting Nazca and Cocos plates under the Central American Continent. The seismicity along the Caribbean-North Andes plate boundary is relatively low, with few minor (Figures 3, 6, and 9), moderate (Figures 4, 7, and 8), and major (Figures 10 and 11) historical seismic events. This seismicity reflects the predominantly dextral lateral movement of the Caribbean Plate with respect to the North Andes Microplate and South American Plate that may include minor subduction that is incipient when compared with the subduction of the Nazca Plate underneath the Andes Mountains.

OCEAN-OCEAN COLLISION ZONES

Figure 2 shows a broad arcuate seismic zone along the Lesser Antilles. A large number of events occur along the Caribbean-North American plate boundary, which dips to the west away from the underthrusting oceanic plate. Shallow, intermediate, and deep focus earthquakes occur at progressively greater distances from the site of underthrusting. The topographic expression of the trench is largely obscured by thick sediment fill derived from the Orinoco River of Venezuela, but its presence is indicated by the belt of negative TCFAGA in Figure 1. The seismic events are located in thicker oceanic crust associated with the doubling of the collisional oceanic crust. Focal mechanism solutions of the Lesser Antilles reflect regional compressional stress. However, the relatively thicker oceanic crust reflects local tensional stress.

CONTINENT-CONTINENT COLLISION ZONES

The continent-continent collision boundary zone between the North Andes Microplate and the South American plate inFigures 2 to 12 is characterized by intensive seismicity. Minor, moderate, and major historical and recent seismic events in Figures 3 to 8 have occurred along the Llanos Front Fault. Shallow events in Figure 9 can be associated with tensional stress of the sub-Andean thrust belt or fault zone, which is located on a relatively thicker continental crust that implies the local presence of additional tensional stress.

OCEANIC SPREADING RIDGES

Minor and moderate earthquakes occur along the Galapagos spreading ridge and the Panama fracture zone in Figure 2. Focal mechanism solutions indicate regional tensional events associated with plate accretion and strike-slip events where the ridges are offset by transform faults. However, the Panama fracture zone is located on relatively thinner oceanic crust that may be locally associated with compressional stress.

TRANSFORM FAULTS

The Southern Panama Fault Zone constitutes a conservative plate boundary between the Nazca Plate and Panama Microplate, where the plates are in tangential contact and are affected by limited or no subduction or accretion. Therefore, events along this fracture are rare as shown in Figure 2.

CRATONIC ZONE

The intraplate areas of the South American Plate are relatively aseismic. However, a few minor, moderate, major historical and recent seismic events in Figures 3 to 8 have occurred at shallow, intermediate, deep, and very deep hypocenter depths in Figures 9 to 12. These events are located along some of the crustal discontinuities interpreted from the surface FAGA in Figure 1. Their occurrences suggest the presence of continental sutures in the South American plate, with infrequently seismic activity. Although, these earthquakes are rare, they are important because they indicate the nature of directions of stress within the South American Plate. The instrumental seismic events in the above plate boundary zones are analyzed in the next section for their focal mechanisms.

FOCAL MECHANISM SOLUTIONS

Figure 13 gives the compressional stress regime for the study region from stress inversions of the focal mechanisms of the instrumental earthquakes of northwestern South America (Vargas and Duran, 2005). These solutions infer the type of faulting or focal mechanisms from the radiated seismic energy. The Costa Rica-Panama-North Andes converging margins show compressional stress oriented west-east, generating the relative movement of local tectonic blocks of the North Andes Microplate. Shallow earthquakes associated with the Pacific subduction zone show compressional stress oriented NW-SE. Vectors of displacement and compressional stress in southwestern Colombia are oriented west-east. The deep seismicity in the Bucaramanga nest reflects compressional stress oriented NNW – SSE associated with the subducting Caribbean Plate under the North Ande microplate. Complementary examples of global focal mechanisms for the study area can also be obtained from the Harvard Centroid Moment Tensor catalog (Harvard, 2006). These results reflect the inversions of long-period body and surface waves for the source moment tensors and best-fitting double-couple solutions.

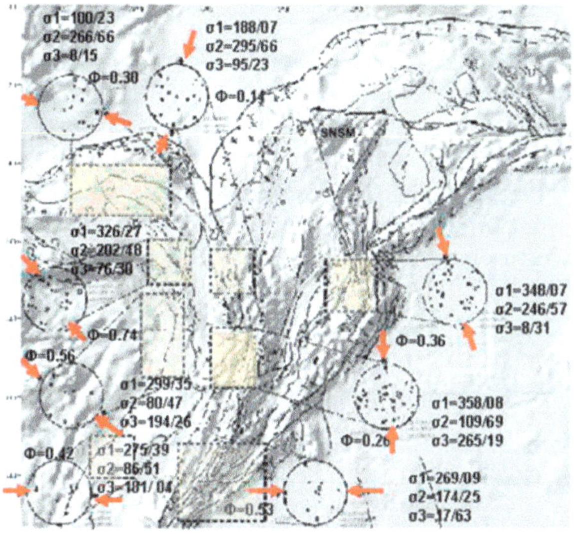

Figure 13. Main stress axes determined by the Reches Method (Reches, 1983; Reches, 1987) from 94 focal mechanism solutions of the Central Moment tensor catalog (CMT) reported by Harvard University (Harvard, 2006) for the period 1976-2000 and event magnitudes M = 5.0 (Vargas and Duran, 2005).

CONCLUSIONS

Seismicity in northwestern South America is mainly restricted to the plate boundary zones of the Caribbean, North America, and South America Plates and the Panama, Costa Rica and North Andes Microplates. Seismic corridors are oriented predominantly SW-NW, showing their association with compressional stress generated by the convergence of the Nazca and South American plates, and the shortening of the continental crust and mountain building of the Andes Mountains. The seismicity is extremely high along the northern margin of the Pacific subduction zone.

The hypocenter depths show the tendency of the earthquakes to migrate from the coastline eastwards as the hypocenter depths increase, confirming the eastward motion of the Nazca Plate under the North Andes Microplate. The spatial distribution of earthquakes is consistent with crustal discontinuities interpreted from the gravity anomalies. Seismic events are preferentially concentrated in zones of relatively thicker continental crust. This crust is inherently under tension by virtue of its enhanced thickness, but it is also being subjected to overwhelming compression from plate subduction forces. The analysis of the isostatic state of compensation from terrain and gravity anomalies of northwestern South America help to differentiate probable earthquake zones. Integrating the local stress of the crustal thickness variations with regional tectonic stresses may help to predict local ground motions. The focal mechanisms reflect regional stress and more local stress from the thickness variations of the crust.

REFERENCES

1. Artyushkov, E.V. (1973). Stresses in the lithosphere caused by crustal thickness inhomogeneities. Journal of Geophysical Research. 78. 7675-7708.
2. Bird, P. (2003). An Updated digital model of plate boundaries. Geochemistry, Geophysics Geosystems G3. 4 (3). 1-52.
3. Cediel, F., Shaw, R. P., and Caceres, C. (2003). Tectonic assembly of the Northern Andes Block. The Circum-Gulf of Mexico and Caribbean Hydrocarbon habitats, basin formation and plate tectonics: AAPG Memoir. 79. 815-848.
4. Geosoft (2006). Geosoft Oasis Montaj: Software for Earth Sciences Mapping and Processing.http://www.geosoft.com.
5. Harvard, (2006). Centroid Moment Tensor Catalog (CTM).http://www.seismology.harvard.edu/CMTsearch.html.
6. Hernandez, O. (2006). Tectonic analysis of northwestern South America from integrated satellite, airborne and surface potential field anomalies. PhD Dissertation, Department of Gelogical Sciences. The Ohio State University, Columbus, OH. USA. 177 pages.
7. INGEOMINAS, (2006). Red Sismológica Nacional de Colombia (RSNC) Catalogo de sismos.
8. Kim, J.W., von Frese, R.R.B., and Rae, K.H. (2000). Crustal modeling from spectrally correlated free-air and terrain gravity data A case study of Ohio. Geophysics. 65 (4). 1057-1069.
9. Shearer, P. M. (1999). Introduction to Seismology. Cambridge University Press, New York. 259 pages.
10. Song T. R. A. and Simons, M. (2003). Large trench parallel gravity variations predict seismogenic behavior in subduction zones. Science, Reports. 301 (5633). 630-633.
11. Turcotte D.L. and Shubert, G. (2002). Geodynamics, second edition. Cambridge University press, New York. 456 pages.
12. USGS, (2006). Advance National Seismic System (ANSS).http://earthquake.usgs.gov/research/monitoring/anss/
13. Vargas C. A. and Duran, J. P. (2005). State of stress of northwestern South America. Earth Sciences Res.J. 9 (1). 41-49.
14. von Frese, R. R. B. and Tan, L. (1999). Antarctic crustal modeling from spectral correlation of free-air gravity anomalies with the terrain. Journal of Geophysical research. 104, B11, 25- 275-25,296.

CHAPTER 8

SEISMOGENIC STRESS AND TECTONIC MOVEMENT

Jiren Xu and Zhixin Zhao

INTRODUCTION

The continental crust of the Indian plate collides with and subducts underneath Eurasia along great thrusts at the southern edge of the Himalayas. The compressive tectonic force due to the collision along the Himalayas between the Indian and Eurasian plates results in orogenic motions in the Himalayas, Tibetan plateau, and Altyn and Qilian mountains [1, 2]. Such violent crustal movements and frequent and strong earthquakes extend to the Tianshan Mountains [3, 4]. The compressive tectonic force affects crustal motions in the west of China and as far as Mongolia [5, 6]. Such motion and deformation make the Tibetan plateau and its surrounding areas one of the most active regions in terms of tectonics and seismicity in the world [7, 8]. The synchronous temporal variations of the seismic activities on the Tibetan plateau region and its vicinity are related to the compressional collision between the Indian and Eurasian plates [9, 10]. Many papers have reported significant findings in the fields of tectonics and seismicity in this region [11, 12]. Besides the compressive tectonic motions, some extensional tectonic motions have been reported on the Tibetan plateau. The normal faulting type motions on the Tibetan plateau is attributable to the gravitational collapse of the high plateau [13]. Extensional crevices and grabens have been detected on the Tibetan plateau [14]. LANDSAT image data shows that extensional activity appears in the Tibetan region [15]. A Late Cenozoic E-W extension occurred on the Tibetan p1ateau [16]. The southern part of the plateau is cut by N-S trending normal faults that are the surface expression of regional stretching in an ESE direction, approximately parallel to the Himalayas [17].

In this analysis, the focal mechanisms for studying the tectonic characteristics in and around the plateau are analyzed. The extensional lithosphere motion on the Tibetan plateau and tectonic implications are investigated in detail, based on the analyses of types of earthquake occurrences and the stress field, geothermal as well as GPS data.

Figure 1 shows the tectonic scheme of the Tibetan plateau and surrounding regions as a background map, including topographic information elevation and some faults. The elevation chart images the outline of the plateau.

Citation: Jiren Xu and Zhixin Zhao, "Seismogenic Stress and Tectonic Movement," International Journal of Geophysics, vol. 2009, Article ID 897424, 8 pages, 2009. doi:10.1155/2009/897424.

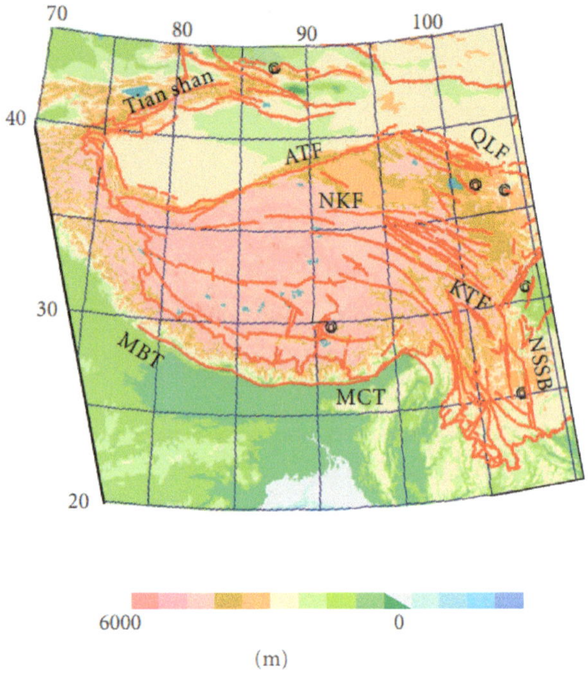

Figure 1: Topographic and tectonic map of the Tibetan Plateau and surrounding regions. Solid lines show major faults (ATF: Altyn Tagh Fault; NKF: North Kunlun Fault; KTF: Kang Ding Fault; MCT: Main Central Thrust; MBT: Main Boundary Thrust; QLF: Qilian fault). NSSB: southern segment of the North-South Seismic belt.

DATA OF FAULT PLANE SOLUTIONS

Focal mechanism solutions of 924 earthquakes with magnitude equal to or greater than M4.5 on the Tibetan plateau and its vicinities during the period between 1933 and 2003 were analyzed and used in this study. In which, the selected events greater than or equal to 5.0 (Mb) are more than 900. The depths of the events are less than 87 km in the lithosphere. The majority of events used in this analysis occurred mainly in the crust. The 217 fault plane solutions of events were determined form the distribution of the polarities of initial P-wave motion on the focal hemisphere by the authors referring to data from the Bulletin of the International Seismological Center (ISC). Another 253 solutions determined by P-wave first motions are selected from the other authors referring to data from the Chinese Seismic Net and the Bulletin of the International Seismological Center (ISC) using P-wave first motions. (e.g., [4]). The above 470 solutions were determined using the first polarities of P-waves of long- and short- period waveforms by the analog and digital records. The earthquake data from the Chinese Seismic Net have been collected from digital records since 2000.

All the solutions used in the analysis are well selected based on the contradictory rate of the plus to minus for the first polarity of P-wave motion. The focal mechanism solutions with symbolic contradictory ratio of P-wave onsets less than 10 percent are of about 89 percent of the total solutions. Solutions with symbolic contradictory ratios between 11 and 15 percent are of 11 percent of the total solutions.

The others 454 CMT solutions (determined by the Centroid Moment tensor solutions) were selected from the Harvard University and USGS. CMT solutions for events with magnitude greater than 5.0 from 1977 to 2003 were also used in this study (e.g., [18, 19]).

The characteristics of crust and lithosphere motions were analyzed based on the investigations of fault motions of events in and around the plateau. Earthquakes are divided into three types—normal, thrust, and strike-slip faulting events—depending on the plunge angles of the compressive stress P- and extensive stress T-axes. The statistical effect of characteristics of fault plane solutions were employed for analyses of the tectonics in and around plateau.

ACTIVITY OF NORMAL FAULTING EARTHQUAKES

Earthquakes can generally be understood as brittle rock ruptures in the crust and lithosphere. So the study of faulting types of earthquakes may help us to understand the characteristics of lithosphere motion. The form of occurrence of earthquakes reflects the stress regime in a region. It is necessary to analyze fault types of earthquakes for investigation of tectonic motion characteristics on the Tibetan plateau and its surroundings. Figure 2(a) shows the fault plane solutions of events. Figure 2(b) visualizes the distributions of normal, thrust, and strike-slip faulting events that are divided according to the plunge angles of their compressive P- and extensional T-axes. It can be seen from Figure 2 that many events with normal faulting are clearly concentrated in the region with altitudes greater than about 4000 m on the Tibetan plateau. Normal faulting type regions AB and CD encircled by dashed line in Figure 2 are zoned based on the azimuths of T-axes.

In the surroundings of the normal faulting event region, the region in Tibet characterized by abundant normal faults is bounded on the north by the discontinuous zone of strike-slip and thrust faults in the Altyn and Qilian mountains. A number of large strike-slip faulting events occurred along the lines of the Karakorum and Altyn mountains. In the Qilian Mountains many thrust faulting events occurred besides the strike-slip faulting events. Seismic activity with normal faulting events disappears on the northern boundary of the plateau, where the style of seismic activity differs from that in the high-altitude region of the plateau. The thrust faulting events are clearly predominant in the Tianshan Mountains. The thrust faulting events occurred mainly along the western part of the Himalayan front to the Kashmir region. A number of strike-slip faulting events occurred along the eastern Himalayan front, except for some thrust faulting type events. The focal mechanism solutions show mainly strike-slip faulting events in the southern segment of the NSSB (North-South Seismic Belt), the eastern boundary of the plateau. Many reverse and strike-slip faulting events occurred in the low surrounding regions of the Tibetan plateau. It may related with the strong compressive stress from the relative movement between the Indo-Australian and Eurasian plates acting on the surroundings of the plateau [20].

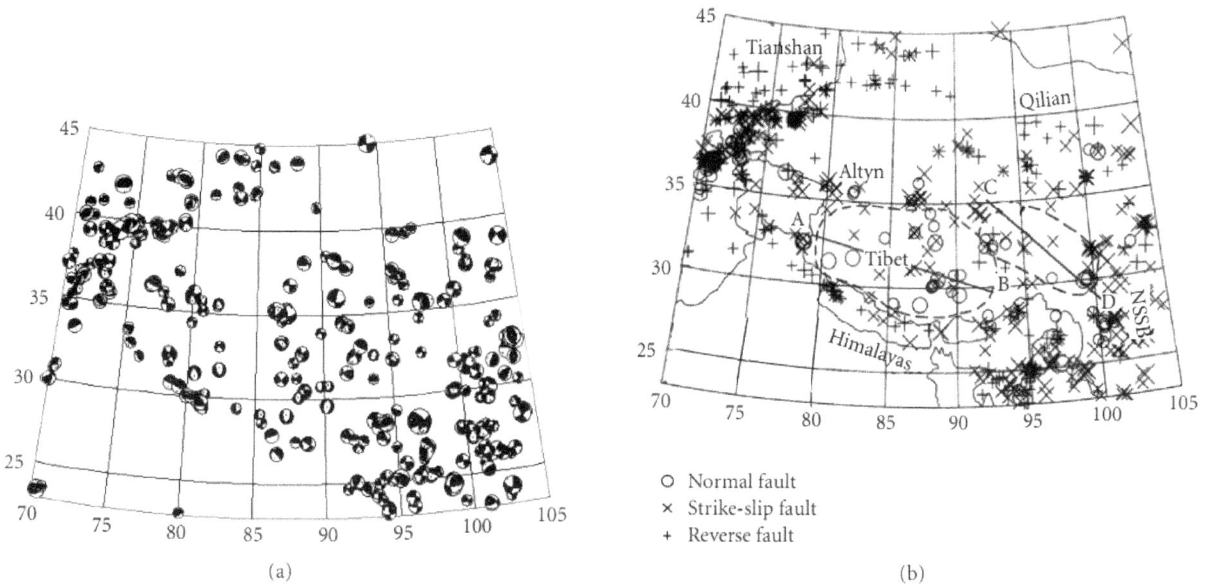

Figure 2: (a) Fault plane solutions of events. (b) Distribution of normal, thrust, and strike-slip faulting events. The Tibetan plateau is surrounded by the Himalayas, Altyn, Qilian, and NSSB. Open circle: normal faulting event; Cross: strike-slip faulting event; Plus: reverse faulting event. Regions AB (the western part of the normal faulting event region, Lhasa block) and CD (the eastern part of the normal faulting event region, Qiangtang-Changdu region) enclosed with dashed lines.

DISTRIBUTIONS OF THE HORIZONTAL COMPONENTS OF THE P- AND T-AXES

The stress field in and around the Tibetan plateau was investigated by analyzing the distributions of the P- and T-axes of focal mechanism solutions for analyses of style of earthquake occurrence. Figures 3 and 4 show the horizontal projections of the P- and T-axes, respectively, in the last 70 years or so. It can be seen from the figures that the P-axes of events lie mainly in N-S to NE-SW directions on the plateau and its surrounding regions, with a few exceptional events. On the Himalayan fronts, the P-axes are almost in an NNE SSW direction and are perpendicular to the Himalayan arc. The horizontal projections of some T-axes are very small there. The horizontal components of the P-axes are on average greater than those of the T-axes. This shows that the tectonic movements in the Himalayan front region are characterized by compressive stress in an NNE-SSW direction.

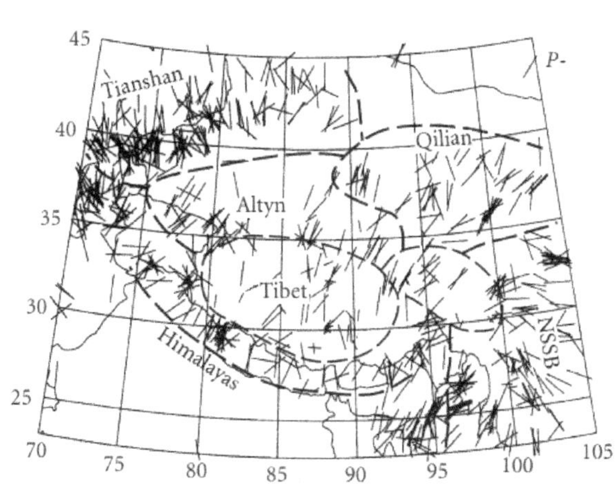

Figure 3: Distribution of compressive axes (P-) in and around the Tibetan plateau.

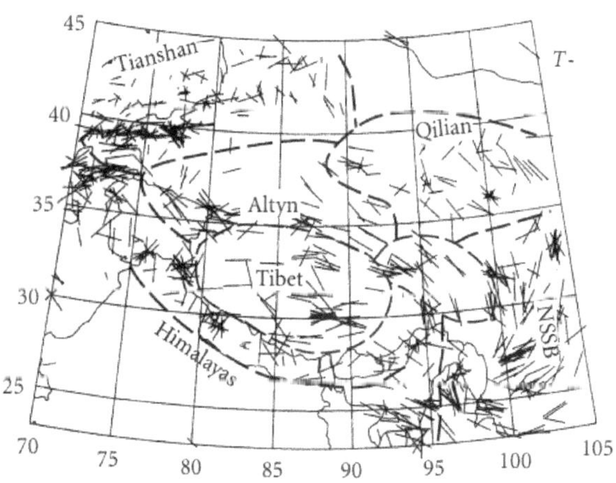

Figure 4: Distribution of extensional axes (T-) in and around the Tibetan plateau.

In the region north of the Himalayas and on the central Tibetan plateau (regions AB and CD enclosed with dashed lines in Figure 2) the horizontal components of the P-axes as a whole become less than those of the T-axes, although the P-axes basically do not change their orientations compared with those in the southern Himalayas. Most of T-axes lie nearly in an E-W direction in the AB part. T-axes are nearly in an NW-SE or NNW-SSE direction in the CD part. The horizontal projections of the T-axes of some events are markedly large, especially in the regions of high elevation on the Tibetan plateau.

Figure 5 shows the statistical charts of azimuths and dips of the compressive (P) and extensional (T) axes in the Lhasa block (AB part in Figure 2). The extensive stress in the nearly E-W direction dominates the seismic fault activities. The characteristics of the tectonic stress field in the regions of high elevation on the Tibetan plateau are clearly different from those in the southern margin of the Himalayan front.

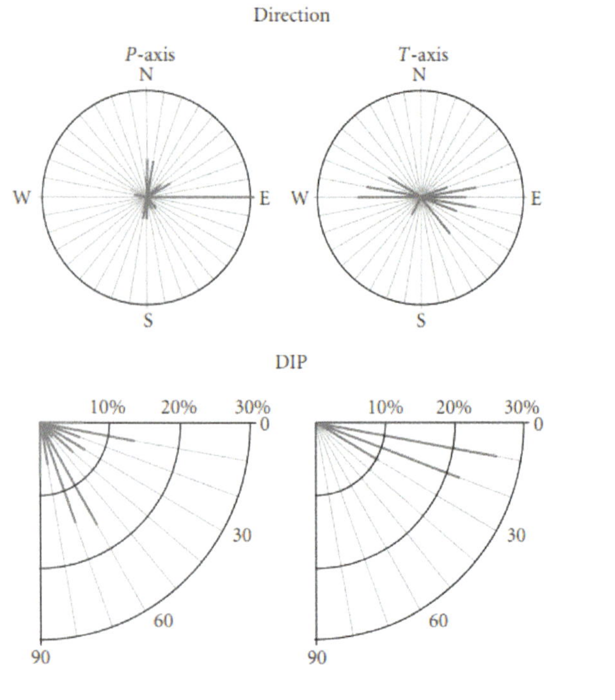

Figure 5: Statistical charts of azimuths and dips of compressive (P) and extensional (T) axes in the southern and central part of the Tibetan plateau (AB region in Figure 2).

The P-axes are mainly in an N-S to NE-SW direction on the northern borders of the Tibet plateau, Karakorum, Altyn, and Qilian Mountains. The orientations of the T-axes are in the general direction E-W to NNW-SSE. The horizontal projections of the P- and T-axes are great and have almost same size as the whole of the Altyn Mountains. The horizontal components of some T-axes are somewhat small in the Qilian Mountains region, while those of the P-axes are great. Strong compressive stress has been identified in the region north of the Qilian Mountains. The horizontal projections of the P-axes are large and lie in a general N-S direction in the Tianshan Mountains region. The horizontal components of the T-axes are quite small there.

In the western part of the southern segment (south of 33°N) of the NSSB (along about 96–104°E) east of the Tibetan plateau, the horizontal components of the P-axes lie almost in an NNE-SSW direction in the southern segment, changing to nearly NE-SW in the northern region of about latitude 33°N. The P-axes lie in a general E-W direction in the regions around 33°N in the NSSB. The horizontal components of the T-axes lie in and around the NW-SE direction in the southern segment of the NSSB, which forms the eastern boundary of the Tibetan plateau. The horizontal components of the P- and T-axes are almost the same in the eastern boundary.

PROJECTIONS FOR P- AND T-AXES ALONG PROFILES

Tectonic stress characteristics in the AB region of the central Tibetan plateau and its eastern vicinity, CD region, encircled with dashed lines as shown in Figure 2 are analyzed in detail for a further understanding of the tectonic implications of normal fault earthquakes on the plateau altitudes. Figures 6 and 7 show the projections for T-axes and stereonet diagrams of focal mechanism solutions on the vertical section beneath the lines AB and CD in the region in Figure 2. The maximum hypocentral depths of events are 78 km and 50 km in Figures 6 and 7, respectively. In the top profile of Figure 6, most of the T-axes lie in a nearly horizontal direction, with large components along the line AB, in a general E-W direction. These characteristics suggest that a horizontal extensional stress in a general E-W direction probably dominates the seismic fault motions in this region. Most of events occurring in this region are normal-faulting-type. The characteristics of normal-faulting events can also be identified in the bottom profile. This suggests that the horizontal motions of earthquake faults extend mainly in a general E-W direction.

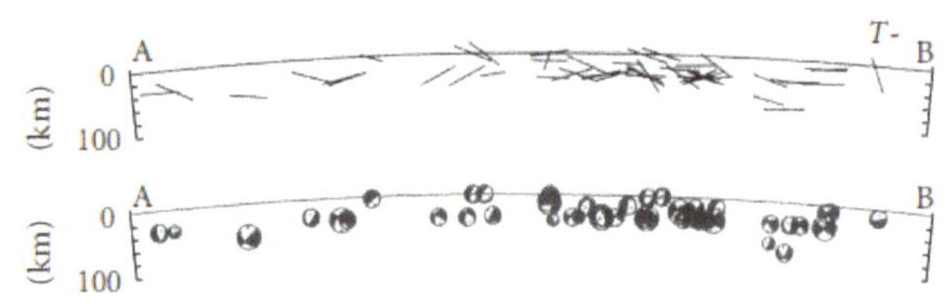

Figure 6: Projections of the vertical section along the line AB in a general E-W direction for the T-axes and equal area projection diagrams of focal mechanism solutions of events (1933–2003, M ≥ 4.5) in the region encircled by a dashed line in Figure 2.

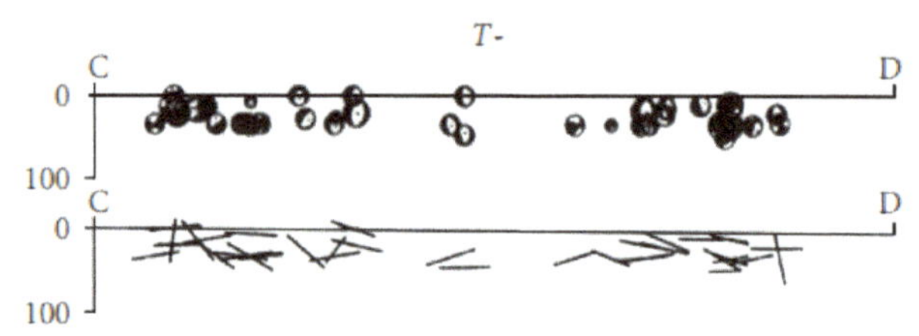

Figure 7: Projections of the vertical section along the line CD near the NW-SE direction for equal area projection diagrams and T-axes of focal mechanism solutions of events (1933–2003, M ≥ 4.5) in the region encircled by the dashed line in Figure 2.

With regard to the characteristics of the T-axes of focal mechanism solutions of events in profile CD in Figure7, most of the T-axes lie in a generally horizontal direction, with large components, along the line CD in a nearly NW-SE direction in Figure 7. These characteristics suggest that a horizontal extensional stress in the NW-NE direction probably dominates the seismic fault motions in the region CD in Figure 2. The characteristics of the stereonet diagrams of events can also be identified in the top profile.

NORMAL-FAULTING EARTHQUAKE REGION

The motional characteristics of normal-faulting-type events in Figure 2 are further analyzed by investigating the focal mechanism solutions. Figure 8 shows the stereonet diagrams of focal mechanism solutions only for normal faulting events with magnitudes greater than M5 in and around the Tibetan plateau. It can be seen that many normal-faulting-type events are concentrated on the Tibetan plateau. The strikes of many fault planes seem to be in a general N-S direction on high altitudes of the Tibetan plateau, corresponding to the region encircled by dashed line AB in Figure 2. While the strike directions of the fault planes of normal faulting events occurring in the Tianshan Mountains and southern segment of the BSSN are scattered in different directions.

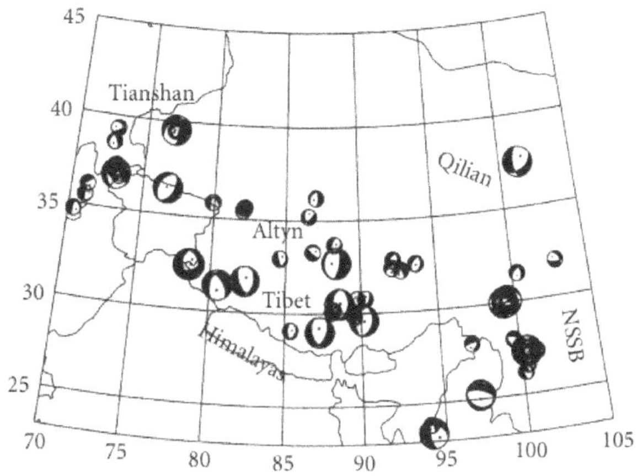

Figure 8: Distribution of the fault plane solutions for the normal fault-type events (1933-2003, M ≥ 5) in and around the Tibetan plateau.

The characteristics of tectonic motions due to earthquakes can be further determined by analyzing in detail the strike directions of the fault planes of the events. Figure 9 shows the equal area projection diagrams of focal mechanism solutions of events in the AB region in Figure 2. The total number of events is 54 in the region enclosed by a dashed line in Figure 9. Few of them are thrust-fault-type events; some belong to the category of strike-slip-faulting events. The majority of them are normal-fault-type earthquakes, and 25 of them are almost perfect normal faulting events in which the strike directions of the two node planes lie in a general N-S direction. Their P-axes are almost located at centers of the equal area projection diagrams. This implies that the P-axes are almost perpendicular to the surface, and the slip vectors of earthquake faults have major components in the E-W direction. Such events perhaps result in major horizontal extensions in the E-W direction whichever of the two node planes is taken as the source fault. Similarly, major horizontal extensions appeared in the NW-SE or NNW-SSE direction in the CD region in Figure 2.

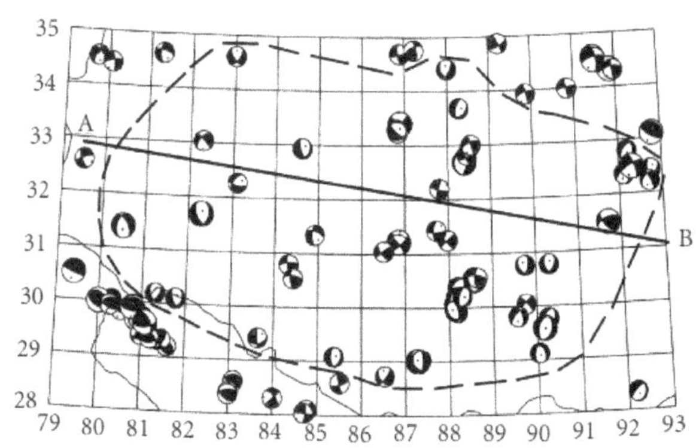

Figure 9: Distribution of the fault plane solutions of earthquakes in and around the AB region of the Tibetan plateau in Figure 2

Based on the above analyses, we can conclusively state that the crust extends in a general E-W direction in the AB region and NW-SE or NNW-SSE in the CD region at the altitudes of the Tibetan plateau due to the normal-faulting-type earthquakes.

CRUSTAL MOVEMENTS FROM GPS AND GEOTHERMAL DATA

The crustal east-west extensional motions are also confirmable by analyzing GPS and geothermal data. The rates of horizontal displacement lie in a general N-S direction on the western plateau and the western edge of the Kunlun Mountains. The eastward annually rates are between 15 and 20 mm/a in the central plateau [8]. The annually rates of displacements lie identically in a generally easterly direction on the eastern boundary of the Tibetan plateau, in the southern segment of the NSSB, almost same as those in the central plateau. The eastward horizontal displacement components increase on average from the western plateau to the eastern margin of the plateau, in the northern segment of the NSSB, and through the central plateau. The rapid clockwise flow occurs around the syntaxis of the Himalayas on the eastern margin of the plateau [22]. The above analyses based on GPS findings at the surface imply that there exists an eastward crustal movement on the central and east plateau. This coincides with the extensional motion in the E-W and NW-SE directions due to the normal faulting events in the present analysis. Deformation of a continuous medium at depth best describes the present-day tectonics of the Tibetan plateau.

Geological tectonic and geothermal activities in the Lhasa block of the plateau are shown in Figure 10. Large mountains and early major compressive or compressive-shear faults mainly lie in an E-W direction in the plateau. The north boundary of the Lhasa block is the Bangong-Nujiang suture zone, and the southern boundary is the Yarlung Zangbo Jiang suture zone. A number of faulted-depression zones with nearly N-S or NNE strike direction, however, occur in the plateau. The faulted-depression zones with middle or small scales were probably formed by extensions in a generally east-west direction. Many extensional faulting activities along a general NEE direction exist on the northern part of the plateau also [23]. The Tibetan plateau is also a geothermally active zone. Many hot springs and geysers are located on the plateau. It is interested that many hydrothermal activities are gathered along the above extensional faulted depression zones. According to a discussion of the lithospheric thickness and hypocentral depth the faulted-depression zones with hydrothermal activities probably indicate that cracks occur in the depth due to extensional motion in the plateau also. The hypocenters of some normal faulting events mentioned in this paper extend also as deep as about 87 km. The eastward extensional motions due to the normal faulting events probably extend to near the mantle.

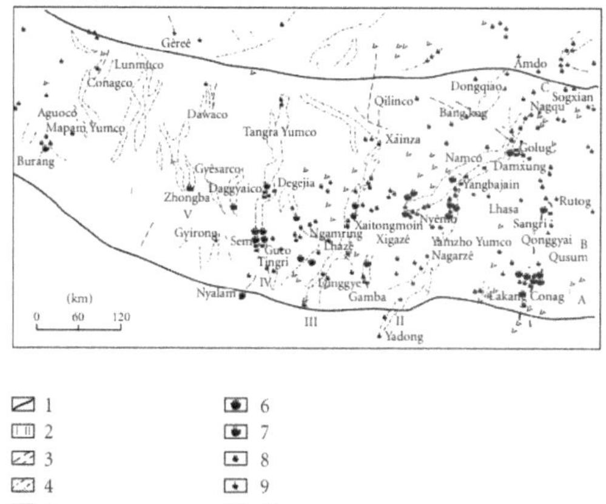

Figure 10: Distribution of hydrothermal manifestations in the Lhasa block of the Tibetan plateau (about 80–94°E, 27–33°N). 1: Early ma-

jor compressive or compressive-shear fault: the upper one is the Bangong-Nujiang suture zone, and the lower one is the Yarlung Zangbo Jiang suture zone; 2: the number of active tectonic zone; 3: boundary fault in active tectonic zone; 4: faulted-depression zone in active tectonic zone; 5: site of sampling; 6: boiling spring; 7: hydrothermal explosion, geyser and boiling fountain; 8: thermal spring; 9: hot spring; 10: untested hydrothermal area (according to [21]).

Figure 11 shows the eastward extension regions on the central Tibetan plateau as the result. The extensional-motion regions marked by open rhombs and the direction of the eastward extension motions marked by open arrow on the plateau altitudes were substantiated based on the present analyses. The other tectonic properties are referenced the study of Avouac and Tapponnier [24].

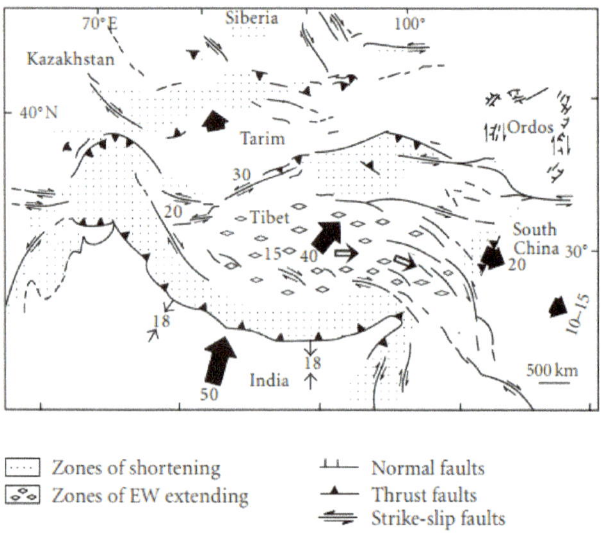

Figure 11: Eeastward extension regions on the central Tibetan plateau. Open rhomb area: extensional motion region obtained from the present analysis; open arrows: extensional direction of the plateau altitudes obtained in the present analysis (other tectonic properties are referenced the study of Avouac and Tapponnier [24]).

DISCUSSION

The characteristics and mechanism of extensional motions at the high altitudes of the Tibetan plateau have been analyzed [25–29]. The activities of normal faulting earthquakes concentrated at high altitudes were investigated based on distribution characteristics of stress field, GPS data, as well as geothermal data in this paper. The extensional-stress-predominating earthquake activity at these altitudes superposes on the compressive stress predominating seismic activities on the edges of the plateau. The eastward extensional motion at high altitudes becomes an obviously important tectonic activity regime, although shortening deformation and movement in a nearly N-S direction are revealed there too.

E-W extension shown by the normal faulting events behind the Himalayas may be related to tectonic stresses due to collision causing a wide distribution of P-axes in the general NNE-SSW or N-S directions in western China [13, 30]. The strong compressive stress due to the northward movement of the Indian plate also causes the Tibetan plateau rise gradually [31]. The weight of a high plateau and the buoyancy acting on the bottom of the plateau crust from the mantle may create an additional vertical compressive stress which can result in the relaxation of the topography and corresponding thinning of the crust together with the normal faulting activity. The altitude of the Tibetan plateau is so high with respect to its surroundings that active normal faulting may also be caused by gravitational collapse toward the plateau margins [25, 32, 33]

A large number of thrust and strike-slip fault events occur in the front of the Himalayan Mountains. Many strike-slip fault events occur in the Karakorum and Altyn mountains. Thrust-faulting events occur frequently in regions of the Qilian and Tianshan mountains [34]. These findings show that strong compressive stresses exist on the southern, western, and northern boundaries of the plateau. Moreover, the GPS data show that the northward annual slip rates on the northern margin of the plateau, for example, in the Altyn and Qilian mountains, are one tenth of those on the Himalayan front [22, 35–37]. These findings imply that the northward motion of the Tibetan plateau shortened in the N-S direction probably encounters strong obstructions at the western and northern margins [38].

A stress field in an NWW-SEE direction appears on the CD region in Figures 2 and 6. On the eastern margin of the Tibetan plateau, however, in the southern end of the NSSB, events are mainly of the strike-slip faulting type. The annually rates of crustal horizontal displacements are eastward and southward there. The extensional motion due to the relaxation of topography and/or gravitational collapse on the high plateau may easily become an eastward motion. The E-W stretching due to the normal fault events on the plateau is also consistent with eastward "escape-block" tectonics [24, 39].

Seismic velocity is rapid beneath the western Himalayas and western Tibet and is less rapid beneath north-central Tibet [40, 41]. The Tarim, Qaidam, Turfan, and Junggar basins on the western and northern sides of the plateau are underlain by large, rapid-velocity anomalies at depths of 100 km and 300 km [42]. The rapid-velocity anomalies are probably related to the ancient cold subduction materials beneath the Eurasian lithosphere of the Tethyan oceanic lithosphere [43]. It is a younger subduction history that dip angle of subduction is about $10 \pm 3°$[44]. The younger subduction motions encounter strong resistance from the front plate [45, 46]. The seismic velocity beneath the NSSB is less than that beneath the central plateau [47]. The image implies that the mantle materials under the plateau thrust beneath the Yangtze block. This may imply that hotter materials lie beneath the eastern part of the plateau. So the hotter materials possibly modulate continuously northward convergence movements from the Indian plate in the Tibetan plateau by the eastward extrusion of mantle material. The latter may drag the lithosphere eastwards.

ACKNOWLEDGMENTS

The authors are most grateful to the two reviewers for their helpful comments. This study was supported partly by the Nature Science Foundation of China (Grant no. 40674026) and Commonweal Special Science Foundation of China (Grant no. 200811037).

REFERENCES

1. P. Molnar and P. Tapponnier, "Cenozoic tectonics of Asia: effects of a continental collision," Science, vol.

189, no. 4201, pp. 419–426, 1975.

2. P. Molnar, P. England, and J. Martinod, "Mantle dynamics, uplift of the Tibetan Plateau, and the Indian monsoon," Reviews of Geophysics, vol. 31, no. 4, pp. 357–396, 1993. ·
3. C. Qin, C. Papazachos, and E. Papadimitriou, "Velocity field for crustal deformation in China derived from seismic moment tensor summation of earthquakes," Tectonophysics, vol. 359, no. 1-2, pp. 29–46, 2002. · ·
4. Z. Xu, "Modern tectonic stress in East Asia," Acta Seismologica Sinica, vol. 23, pp. 492–501, 2001.
5. R. Van der Voo, W. Spakman, and H. Bijwaard, "Tethyan subducted slabs under India," Earth and Planetary Science Letters, vol. 171, no. 1, pp. 7–20, 1999. · ·
6. E. T. Brown, R. Bendick, D. L. Bourlès, et al., "Slip rates of the Karakorum fault, Ladakh, India, determined using cosmic ray exposure dating of debris flows and moraines," Journal of Geophysical Research B, vol. 107, no. 9, p. 2192, 2002. ·
7. Y. Wang and H. Xu, "Convergence rate of the India plate to Eurasia subduction beneath Qing-Zang plateau—inversion result from the GPS observed data," Chinese Journal of Geophysics, vol. 46, pp. 185–190, 2003.
8. Q. Wang, P.-Z. Zhang, J. T. Freymueller, et al., "Present-day crustal deformation in China constrained by global positioning system measurements," Science, vol. 294, no. 5542, pp. 574–577, 2001. · ·
9. Z. Zhao, K. Oike, K. Matsumura, and Y. Ishikawa, "Stress field in the continental part of China derived from temporal variations of seismic activity," Tectonophysics, vol. 178, no. 2–4, pp. 357–372, 1990. ·
10. Z. Zhao, J. Xu, and S. Horiuchi, "Differentiation operation in the wave equation for the pseudospectral method with a staggered mesh," Earth, Planets and Space, vol. 53, no. 5, pp. 327–332, 2001. ·
11. Z. Zeng, Z. Ding, and Q. Wu, "Seismological evidence for the mulitiple incomplete crustal subductions in Himalaya and south Tibet," Chinese Journal of Geophysics, vol. 43, pp. 780–797, 2000.
12. Z. Fu, J. Xu, and J. Huang, "Numerical simulation of the compression uplift of the Qinghai-Xizang plateau," Chinese Journal of Geophysics, vol. 43, no. 3, pp. 346–355, 2000.
13. J. Xu, Z. Zhao, Y. Ishikawa, and K. Oike, "Properties of the stress field in and around west China derived from earthquake mechanism solutions," in Bulletin of the Disaster Prevention Research Institute, vol. 38, pp. 49–78, Kyoto University, Kyoto, Japan, 1988.
14. J. F. Deway and J. M. Bird, "Mountain belts and new global tectonics," Journal of Geophysical Research, vol. 75, no. 14, pp. 2625–2647, 1970. ·
15. P. Molnar, "A review of the seismicity and the rates of active underthrusting and deformation at the Himalaya," Journal of Himalayan Geology, vol. 1, pp. 131–154, 1990.
16. J. Ni and J. E. York, "Late Cenozoic tectonics of the Tibetan plateau," Journal of Geophysical Research, vol. 83, no. B11, pp. 5377–5384, 1978.
17. R. Armijao, P. Tapponnier, J. L. Merccier, and T.-L. Han, "Quaternary extension in southern Tibet: field observations and tectonic implications," Journal of Geophysical Research, vol. 91, pp. 13083–13872, 1986.
18. A. M. Dziewonski and J. H. Woodhouse, "An experiment in systematic study of global seismicity: centroid-moment tensor solutions for 201 moderate and large earthquakes in 1981," Journal of Geophysical Research, vol. 88, no. 4, pp. 3247–3271, 1983. ·
19. A. M. Dziewonski, A. Friedman, D. Giardini, and J. H. woodhouse, "A experiment in systematic study of Geobal seismicity of 1982: centroid-moment tensor solution fot 308 earthquake," Physics of the Earth and Planetary Interiors, vol. 33, pp. 76–90, 1983.
20. C. Froidevaux and Y. Ricard, "Tectonic evolution of high plateaus," Tectonophysics, vol. 134, no. 1–3, pp. 227–238, 1987. ·
21. H. Tonglin, "Relationship of the active structural system to geothermal activity in southern xizang (Tibet)," in Himalayan Geology, pp. 45 58, Geological Press, Beijing, China, 1981.
22. P.-Z. Zhang, Z. Shen, M. Wang, et al., "Continuous deformation of the Tibetan Plateau from global positioning system data," Geology, vol. 32, no. 9, pp. 809–812, 2004. ·
23. L. Yalin, W. Chengshan, Y. Haisheng, et al., "Characteristics of the shuanghu graben and cenozoic extension in northern Tibet," Science in China, Series D, vol. 44, supplement 1, pp. 285–291, 2001. ·
24. J.-P. Avouac and P. Tapponnier, "Kinematic model of active deformation in central Asia," Geophysical Research Letters, vol. 20, no. 10, pp. 895–898, 1993.
25. P. M. Blisniuk, B. R. Hacker, J. Glodny, et al., "Normal faulting in central Tibet since at least 13.5 Myr ago," Nature, vol. 412, no. 6847, pp. 628–632, 2001. ·
26. H. Z. Wu, P. Ye, and D. Hu, Evolvement for Crustal Deformation and Tectonic Physiognomy, Geological Press, Beijing, China, 2003.
27. J. F. Dewey, "Extensional collapse of orogens," Tectonics, vol. 7, no. 6, pp. 1123–1139, 1988. ·
28. P. Molnar and P. Tapponier, "Active tectonics of Tibet," Journal of Geophysical Research, vol. 83, no. B11, pp. 5361–5375, 1978.
29. S. L. Klemperer, "Crustal flow in Tibet: geophysical evidence for the physical state of Tibetan lithosphere, and inferred patterns of active flow," in Channel Flow, Ductile Extrusion and Exhumation in Continental Collision Zones, R. D. Law, M. P. Searle, and L. Godin, Eds., vol. 268, pp. 39–70, Geological Society, London, UK, 2006.
30. J. Xu and K. Oike, "Stress characteristics in the southern segment of the North-South Seismic Belt," Acta Seismologica Sinica, vol. 17, pp. 31–40, 1995 (Chinese).
31. R. S. Yeats and R. J. Lillie, "Contemporary tectonics of the Himalayan frontal fault system: folds, blind thrusts and the 1905 Kangra earthquake," Journal of Structural Geology, vol. 13, no. 2, pp. 215–225, 1991. ·
32. S. Robert, K. S. Yeats, and R. A. Clarence, The Geology

of Earthquakes, Oxford University Press, Oxford, UK, 1997.

33. J. Lavé, J. P. Avouac, R. Lacassin, P. Tapponnier, and J. P. Montagner, "Seismic anisotropy beneath Tibet: evidence for eastward extrusion of the Tibetan lithosphere?" Earth and Planetary Science Letters, vol. 140, no. 1–4, pp. 83–96, 1996. ·

34. T. M. Harrison, P. Copeland, W. S. F. Kidd, and A. Yin, "Raising tibet," Science, vol. 255, no. 5052, pp. 1663–1670, 1992. ·

35. R. Bendick, R. Bilham, J. Freymueller, K. Larson, and G. Yin, "Geodetic evidence for a low slip rate in the Altyn Tagh fault system," Nature, vol. 404, no. 6773, pp. 69–72, 2000. ·

36. R. Hetzel, S. Niedermann, M. Tao, et al., "Low slip rates and long-term preservation of geomorphic features in Central Asia," Nature, vol. 417, no. 6887, pp. 428–432, 2002. · ·

37. Z.-K. Shen, M. Wang, Y. Li, et al., "Crustal deformation along the Altyn Tagh fault system, western China, from GPS," Journal of Geophysical Research B, vol. 106, no. 12, pp. 30607–30621, 2001. ·

38. Z. Chen, B. C. Burchfiel, Y. Liu, et al., "Global positioning system measurements from eastern Tibet and their implications for India/Eurasia intercontinental deformation," Journal of Geophysical Research B, vol. 105, no. 7, pp. 16215–16227, 2000. ·

39. P. Tapponnier, "The Ailao Shan-Red River metamorphic belt: tertiary left-lateral shear between Indochina and South China," Nature, vol. 294, pp. 405–410, 1990.

40. L. Bourjot and B. Romanowicz, "Crust and upper mantle tomography in Tibet using surface waves,"Geophysical Research Letters, vol. 19, no. 9, pp. 881–884, 1992. ·

41. R. Zhou, S. P. Grand, F. Tajima, and X.-Y. Ding, "High velocity zone beneath the southern Tibetan plateau from P-wave differential travel-time data," Geophysical Research Letters, vol. 23, no. 1, pp. 25–28, 1996. ·

42. Y. Zhang and T. Tanimoto, "High-resolution global upper mantle structure and plate tectonics," Journal of Geophysical Research, vol. 98, pp. 9793–9823, 1993.

43. P. Molnar, "The geologic evolution of the Tibetan plateau," American Scientist, vol. 77, no. 4, pp. 350–360, 1989. ·

44. W. Zhao, K. D. Nelson, and Project INDEPTH Team, "Deep seismic reflection evidence for continental underthrusting beneath southern Tibet," Nature, vol. 366, no. 6455, pp. 557–559, 1993. · ·

45. J. Xu and K. Yoshiteru, "Geometry of slab, intraslab stress field and its tectonic implication in the Nankai trough, Japan," Earth, Planets and Space, vol. 54, no. 7, pp. 733–742, 2002. ·

46. J. Xu, Z. Zhao, Y. Kono, and H. Kinoshita, "Regional characteristics of stress field and its dynamics in and around the Nankai Trough, Japan," Acta Geophysica Sinica, vol. 46, no. 4, pp. 488–494, 2003. ·

47. Y. Zhang, "Three dimension upper mantle structure beneath East Asia and its tectonic implication," inMantle Dynamics and Plate Interaction in East Asia, M. E. J. Flower, S.-L. Chung, C.-H. Lo, and T.-Y. Lee, Eds., pp. 11–23, American Geophysical Union, Washington, DC, USA, 2000.

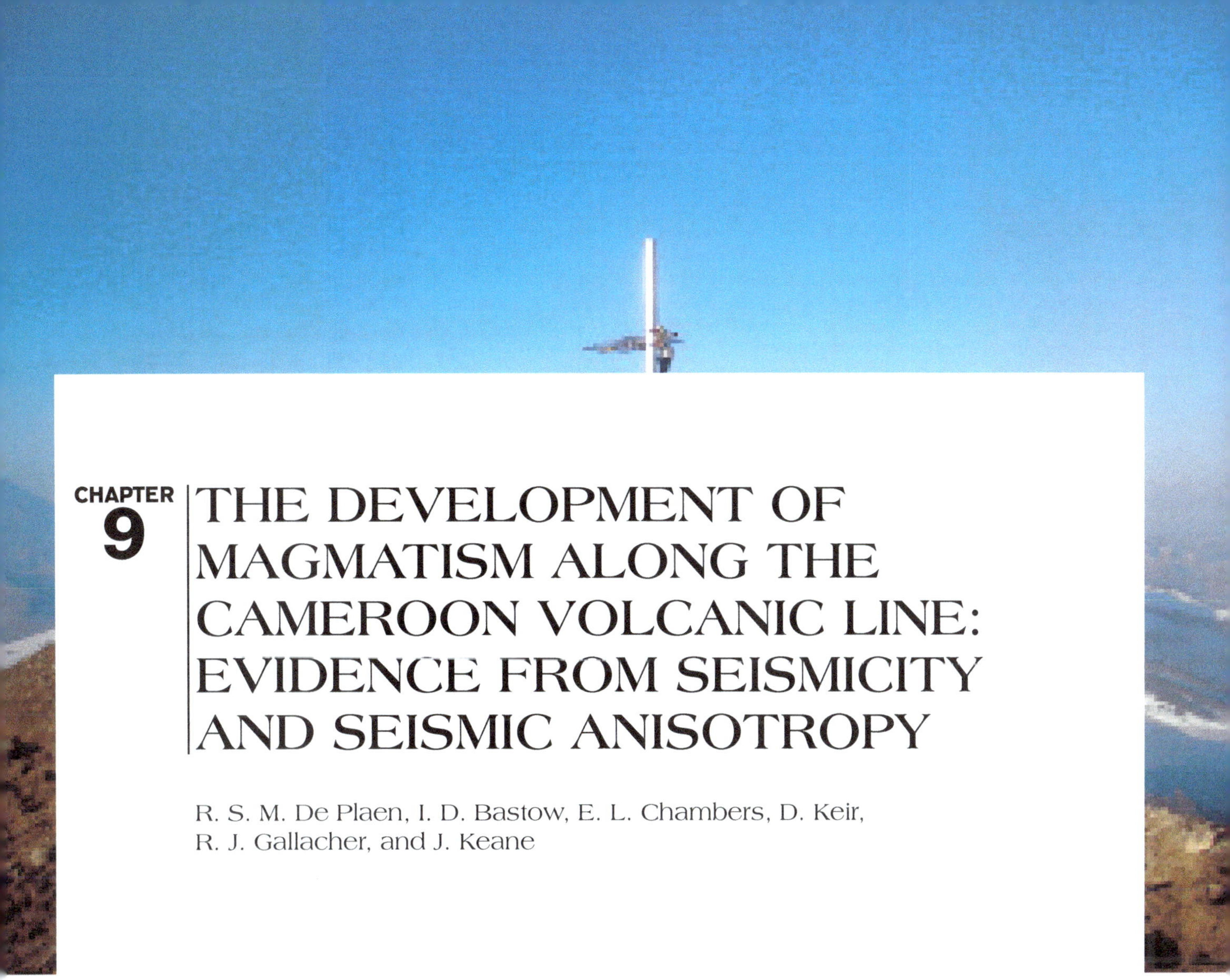

CHAPTER 9 THE DEVELOPMENT OF MAGMATISM ALONG THE CAMEROON VOLCANIC LINE: EVIDENCE FROM SEISMICITY AND SEISMIC ANISOTROPY

R. S. M. De Plaen, I. D. Bastow, E. L. Chambers, D. Keir, R. J. Gallacher, and J. Keane

OVERVIEW

The Cameroon Volcanic Line (CVL) lies to the north of the Congo Craton in West Africa and straddles the continent-ocean boundary (Figure 1). Unlike Hawaii, the archetypal example of a linear hotspot chain, the CVL displays no clear age progression along its length [e.g., Fitton and Dunlop, **1985**; Fitton, **1987**; Poudjom Djomani et al., **1997**], thus rendering it difficult to explain by the traditional idea of a plate moving over a relatively stationary hotspot [e.g., Morgan, **1983**]. Africa's very slow movement NW since ~30Ma [e.g., Steinberger and Torsvik, **2008**] may contribute to the longevity and lack of age progression along the CVL, but the paucity of detailed constraints on the upper mantle structure beneath the region means the causative mechanism(s) for the volcanic line remains uncertain. The nature of the magmatic plumbing system along the CVL also remains poorly constrained. Petrological studies indicate that the CVL's mostly alkaline basaltic volcanoes have erupted relatively low volumes of igneous rocks [e.g., Fitton, **1980**; Suh et al., **2003**]. These silica-poor lavas have not undergone appreciable fractionation at depth, which raises the question of whether or not the crust beneath the CVL is characterized by cooled mafic intrusions and/or present-day melt, as is often observed elsewhere at continental hotspots and rifts worldwide [e.g., Maguire et al., **2006**; Thybo and Nielsen, **2009**].

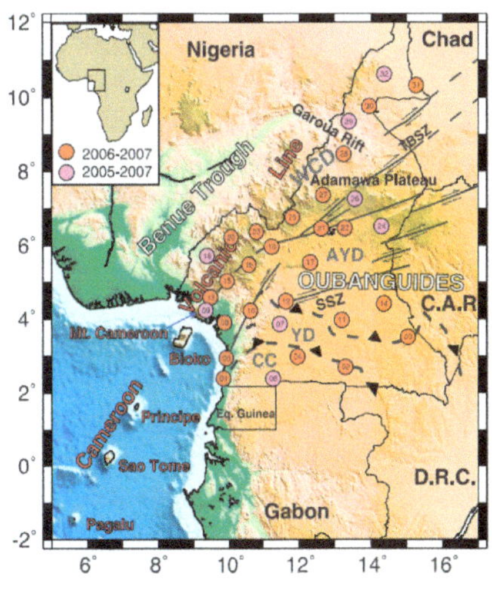

Figure 1. Topographic map of Cameroon showing the locations of Cameroon Broadband Seismic Experiment (CBSE) seismograph sta-

Citation: DePlaen,R.S.M.,I.D.Bastow, E. L. Chambers, D. Keir, R. J. Gallacher, and J. Keane (2014), The development of magmatism along the Cameroon Volcanic Line: Evidence from seismicity and seismic anisotropy, J. Geophys. Res. Solid Earth, 119, 4233–4252, doi:10.1002/2013JB010583.

tions. Numbers are station codes (Table 1). Also shown are major geologic features [e.g., Toteu et al., **2004**; Nzenti et al., **2006**; Reusch et al., **2010**] such as the CVL and the Benue Trough. CC: Congo Craton (Archean), the features north of which are parts of the Oubanguides belt. Heavy dashed lines are thrust faults. The domains are modified from Toteu et al. **[2004]** and Nzenti et al. **[2006]**. YD: Yaounde Domain (Neoproterozoic), AYD: Adamawa-Yade Domain (Paleoproterozoic), and WCD: Western Cameroon Domain (Neoproterozoic). Heavy black lines are major strike-slip faults of the Central African Shear Zone as defined by Ngako et al. **[2003]**, Toteu et al. **[2004]**, and Nzenti et al. **[2006]**. SSZ: Sanaga Shear Zone and TBSZ: Tchollir-Banyo Shear Zone.

Several models have been proposed to explain the development of the CVL. These can be grouped as plume and nonplume based. The plume-based models range from a single plume under a moving plate [e.g., Burke, **2001**; Lee et al., **1994**; Morgan, **1983**; Van Houten, **1983**] to multiple plumes [e.g., Ngako et al., **2006**] beneath the region. It has also been suggested that lateral flow of buoyant asthenosphere from the Afar hotspot in East Africa may now be contributing to the volcanism along the CVL [Ebinger and Sleep, **1998**]. Nonplume models include decompression melting beneath reactivated shear zones in Cameroon [Fairhead, **1988**; Fairhead and Binks, **1991**], or small-scale mantle convection [e.g., Meyers et al., **1998**; King and Anderson, **1998**; King, **2007**] in the zone between two deeply rooted cratons that was reactivated during the Mesozoic breakup of Gondwana. Milelli et al. **[2012]** and Fourel et al. **[2013]** use scaled laboratory models and analytical solutions to investigate lithospheric instabilities in the vicinity of large lateral variations in lithospheric thickness, as is observed along the CVL between the deeply rooted West African and Congo Cratons, and the continent-ocean boundary. These models suggest that lithospheric instabilities at the edge of a continent can develop over long timescales, leading to small rates of upwelling and decompression melting along linear zones, as is observed in Cameroon.

Each of the aforementioned scenarios for CVL development predicts different patterns of mantle flow, which in turn are expected to be manifested as seismic anisotropy, measurable via analysis of shear wave splitting [e.g., Silver and Chan, **1991**], as outlined below. Olivine is the most abundant mineral in the upper mantle and is highly anisotropic. Strain results in development of crystallographic preferred orientation of olivine, leading to alignment of aaxes in the flow direction [e.g., Zhang and Karato, **1995**; Bystricky et al., **2000**; Tommasi et al., **2000**]. If a shear wave travels through such a medium, the faster traveling component will travel parallel to the a axes of the crystal [e.g., Silver and Chan, **1991**]. Shear wave splitting parameters φ (the orientation of the fast shear wave) and δt(the lag time between the fast and slow shear wave) can therefore be related, for example, to asthenospheric flow [e.g., Vinnik et al., **1989**; Fouch et al., **2000**; Sleep et al., **2002**], the preferential alignment of fluid or melt [e.g., Blackman and Kendall, **1997**], preexisting anisotropy frozen in the lithosphere [e.g., Silver and Chan, **1988**; Vauchez and Nicolas, **1991**; Bastow et al., **2007**], or any combination thereof. In the case of Cameroon, vertically aligned olivine in a mantle plume or the upward/downward limb of a small-scale convection cell would be expected to be manifested as near-zero splitting in SKS phases. In contrast, horizontal asthenospheric flow from the Afar hotspot would be manifested as marked shear wave splitting in SKS phases, with the fast shear wave polarization direction parallel to the direction of flow. Lithospheric fossil anisotropy would be expected to parallel trends of major faults or rock fabrics, lithospheric thickness variations, or trends inferred from potential field data such as gravity and magnetics.

In this study, we perform shear wave splitting analysis of SKS phases recorded by the recent Cameroon Broadband Seismic Experiment (CBSE) [e.g., Reusch et al., **2010**; Tokam et al., **2010**] to investigate upper mantle anisotropy beneath the CVL and the edge of the neighboring Congo Craton. These constraints on mantle anisotropy allow us to explore the extent to which the assembly and subsequent breakup of Gondwana has imparted a fossil anisotropic fabric on the Cameroon lithosphere. By constraining the directionality of the flowing asthenosphere beneath the CVL, we are also able to discriminate between competing models for its development. If the lithospheric magmatic plumbing system of the CVL is characterized by large volumes of partial melt, aligned within SE-NW striking faults and fractures of the Central African Shear Zone (CASZ), this too will be manifested as strong seismic anisotropy. A study of seismicity along the CVL also provides an opportunity to examine the extent to which ongoing magmatism along the line is manifested as seismic activity: a common feature at hotspots such as Iceland [e.g., Key et al., **2011**] and Ethiopia [e.g., Ebinger et al., **2010**], and at hotspots in intraplate settings such as Galapagos and Hawaii [Tepp et al., **2014**], where fluid migration is ongoing.

GEOLOGY AND TECTONIC SETTING

Cameroon comprises several tectonic domains brought together during the Pan-African Cycle ~600Ma [Djomani et al., **1995**]. The Oubanguides Belt (Figure 1) is a part of the larger Neoproterozoic Pan-African-Brazilian Belt created during the formation of Gondwana with the collision of the São Francisco Craton, the Congo Craton, and the West African Craton [e.g.,Castaing et al., **1994**; Ngako et al., **2003**; Toteu et al., **2004**]. To the north of the Congo Craton, three main tectonic units are identified: the Neoproterozoic Yaounde Domain, the Paleoproterozoic Adamawa-Yade Domain, and the Neoproterozoic Western Cameroon Domain (Figure 1). The domain boundaries are defined by large shear zones such as the Sanaga Shear Zone and the Tchollir-Banyo Shear Zone (TBSZ), which form part of the larger Precambrian Central African Shear Zone (CASZ), which extends across the African continent from the Darfur region of Sudan, through Central Africa, to the Adamawa Plateau in Cameroon [e.g., Toteu et al., **2004**; Nzenti et al., **2006**; Castaing et al., **1994**]. The Pernambuco lineament in Brazil is probably the extension of the shear zone on the conjugate Atlantic-rifted margin [e.g. Dorbath et al., **1986**; Djomani et al., **1995**]. The opening of the South Atlantic during Cretaceous times (beginning ~130Ma) signaled the breakup of Gondwana with at least partial reactivation of strike-slip faults along the CASZ producing narrow subsiding rift basins infilled with Cretaceous-Tertiary sediments [e.g. Fairhead, **1988**]. Geological studies along the CASZ estimate ≥40km of dextral displacement [see, e.g., Daly et al., **1989**]. Mesozoic rifting also resulted in the formation of a series of major basins and grabens, such as the Benue Trough, to the west of the CVL (Figure 1) [e.g., Djomani et al., **1995**].

The Cenozoic CVL trends N30°E and extends from Pagalu Island to northernmost Cameroon (Figure 1). Volcanic activity along the CVL had begun on the continental and oceanic sectors by ~30Ma [e.g., Déruelle et al., **2007**]. Erosion has exposed more than 60 anorogenic ring complexes on the continental sector, ranging in age from 6 to 70Myr [e.g., Montigny et al., **2004**; Déruelle et al., **2007**]. By late Eocene/early Oligocene times volcanism had spread along the entire line, but without any clear age progression during its development [e.g., Fitton and Dunlop, **1985**; Fitton, **1987**; Poudjom Djomani et al., **1997**]. Most volcanic centers have been active in the past million years, though Mount Cameroon at the center of the line is the only one with a historically recorded eruption (in the year 2000) [e.g., Suh et al., **2003**; Déruelle et al., **2007**].

Igneous rocks along the CVL are mostly of alkaline basalt composition, and their isotopic and chemical composition is very similar in both the continental and the oceanic sectors [Fitton and Dunlop, **1985**; Fitton, **1987**]. This has been cited as evidence for a single source of volcanism beneath the CVL [e.g. Halliday et al., **1988**; Rankenburg et al., **2005**; Déruelle et al., **2007**]. Numerous petrological studies indicate that the CVL's volcanoes have mostly erupted low volumes of alkaline melt rapidly, with little-to-no fractionation within the crust during ascent to the surface. Thus, melt is being sourced almost directly from the mantle [e.g., Fitton and Dunlop, **1985**; Fitton, **1987**; Marzoli et al., **2000**; Suh et al., **2003**; Déruelle et al., **2007**]. The nature and extent of these lithospheric-scale conduits, however, remains elusive.

GEOPHYSICAL CONSTRAINTS

Mantle structure beneath Cameroon is shown in many global tomographic studies to be seismically slow [e.g., Montelli et al., **2006**; Ritsema et al., **2011**]. This is corroborated by observations from continent-scale surface wave studies [Pasyanos and Nyblade, **2007**; Priestley et al., **2008**; Fishwick, **2010**] that also indicate that the CVL mantle is characterized by slow seismic wave speeds. The lithosphere-asthenosphere boundary is <100km deep beneath the CVL, which in turn lies to the south of a failed continental rift zone (the Benue Trough). Beneath the neighboring Congo and West African Cratons, the lithosphere is ~250km thick [Fishwick, **2010**]. A recent study by Reusch et al. [**2011**] showed that the olivine phase transitions from olivine to wadsleyite (the "410") and from ringwoodite to perovskite+magnesiowüstite (the "660") are at normal depths beneath the CVL, suggesting that any thermal anomaly beneath the CVL is confined to the upper mantle. Dorbath et al. [**1986**] completed a teleseismic delay survey across the Adamawa Plateau, finding $\delta VP<2.5\%$ regional variations in seismic wave speed across the study area. More recently, Reusch et al. [**2010**] used body wave relative arrival time tomography to image the mantle seismic structure beneath the CVL. A continuous low-velocity zone ($\delta VS=-2$ to -3%) underlies the entire CVL to a depth of at least 300km and was attributed by Reusch et al. [**2010**] to a thermal anomaly of $\geq$280K. However, studies of absolute mantle wave speed beneath Cameroon are not as low as in other hotspots worldwide. In Ethiopia, for example, surface wave studies [Keranen et al., **2009**; Dugda et al., **2007**; Bastow et al., **2010**] reveal wave speeds of $\leq 4\text{km s}^{-1}$ for the Ethiopian rift: $\delta VS\geq 11\%$ slow. In contrast, similar constraints from Cameroon indicate S wave speeds consistently $\geq 4.1\text{km s}^{-1}$ beneath the CVL: $\delta VS\leq 8\%$ slow [e.g., Tokam et al., **2010**].

Flow models constrained by global tomography [e.g., Forte et al., **2010**] predict a strong lateral mantle flow field in the asthenosphere beneath Cameroon approximately parallel to the CVL. This would be expected to be manifest as CVL-parallel seismic anisotropy due to lattice preferred orientation of mantle minerals beneath our study area. The Forte et al. [**2010**] study also predicts a strong upwelling component to the mantle flow field from a region termed the West African Superplume.

Seismic anisotropy in Cameroon has been studied via analysis of SKS splitting by Koch et al. [**2012**]. They identify four regions of distinct anisotropy: (1) moderately strong NE-SW oriented fast polarization directions ($\delta t\approx 1.0$s) beneath the Congo Craton in the south, and the Garoua rift in the north (Figure 1); (2) weak anisotropy ($\delta t\approx 0.3$s) between the Congo Craton and the CVL; (3) N-S oriented fast polarization directions within the CVL, with $\delta t\approx 0.7$s. Koch et al. [**2012**] interpreted these observations as evidence for small-scale mantle convection beneath Cameroon.

Crustal structure in Cameroon has been investigated using both passive and active seismic sources [e.g., Stuart et al., **1985**; Dorbath et al., **1986**; Plomerova et al., **1993**; Meyers et al., **1998**; Tokam et al., **2010**; Gallacher and Bastow, **2012**] and potential field studies [e.g.,Fairhead and Okereke, **1987**; Toteu et al., **2004**; Tadjou et al., **2009**] that help to constrain the lithospheric subdivisions of the region. For example, thinned crust (~25km) is imaged beneath the Garoua Rift (Figure 1), while the transition from the CVL to the Congo Craton is characterized by a transition from ~36km thick crust to 43–48km thick crust [Tokam et al., **2010**; Gallacher and Bastow, **2012**]. Average crustal shear wave velocity also increases into the craton from the CVL: from $\sim 3.7\text{km s}^{-1}$ to $\sim 3.9\text{km s}^{-1}$ [Tokam et al., **2010**]. Bulk crustal Vp/Vsratios along the CVL are markedly low (~1.74) compared to other hotspots worldwide, an observation that Gallacher and Bastow [**2012**] cited as evidence for the lack of melt fractionation and intrusion during the ~30Ma development of the CVL.

Gravity studies have been conducted in the region to delineate tectonic boundaries, to explore crustal thinning, and to evaluate variations in lithospheric rigidity. Most of them highlight the various major geologic and tectonic features in the region, such as the Benue Trough with a significant positive Bouguer anomaly, the Adamawa uplift with a negative anomaly [e.g.,Poudjom Djomani et al., **1992**, **1997**; Fairhead and Okereke, **1987**] and the Congo Craton with a negative anomaly [e.g., Poudjom Djomani et al., **1997**; Tadjou et al., **2009**]. Evidence for continental collision was found at the northern margin of the Congo Craton, as well as a positive anomaly associated with the CASZ that bisects the Adamawa uplift [e.g., Poudjom Djomani et al., **1997**; Tadjou et al., **2009**]. For a recent review of broadband seismological constraints in Cameroon, see Fishwick and Bastow [**2011**].

Effective elastic plate thickness (Te) depends on geothermal gradient, crustal composition, and crust and mantle lithospheric thickness [e.g., Lowry and Pérez-Gussinyé, **2011**]. Pérez-Gussinyé et al. [**2009**] constrained lithospheric strength via study of Te across the African continent using coherence analysis of topography and Bouguer anomaly data. The weakest lithosphere was found in Ethiopia, a region of incipient continental breakup characterized by a markedly low wave speed [e.g., Bastow et al., **2010**], hot [Rooney et al., **2012**; Ferguson et al., **2013**] mantle. Channels of relatively weak lithosphere extend across the African continent from this region to Cameroon, where Te is also relatively low compared to the surrounding cratons. Pérez-Gussinyé et al. [**2009**] interpreted these observations as supportive of the model of Ebinger and Sleep [**1998**], who proposed that flow of buoyant upper mantle beneath continental lithosphere thinned extensively during Mesozoic rifting may now be contributing to volcanism along the CVL.

DATA AND ANALYSIS

Shear Wave Splitting

Data in this study come from the CBSE network of 32 portable broadband seismometers spaced 50–150km apart across the CVL and the western edge of the neighboring Congo Craton (Figure 1) [e.g., Reusch et al., **2010**; Tokam et al., **2010**]. The stations operated between January 2005 and February 2007. Waveforms for earthquakes of magnitude mb≥5.5 were visually inspected for high signal-to-noise ratio SKS phases, with earthquakes of epicentral distances ≤88°eliminated from the analysis in order to prevent contamination by other shear wave phases. Data were filtered prior to analysis with a zero-phase two-pole Butterworth filter with corner frequencies of 0.04–0.3Hz.

Traditional shear wave splitting analysis requires a single shear wave analysis window to be selected [Silver and Chan, **1991**]. The earlier study of the CVL by Koch et al. [**2012**] adopted such an approach. The choice of a final window is somewhat subjective, however, and in many cases different windows give very different results. To overcome this problem, we adopt here the semi-automated approach of Teanby et al. [**2004**], which is based on the method of Silver and Chan [**1991**].

For SKS waves, elliptical particle motion and energy on the tangential component seismogram are strong evidence for shear wave splitting. We rotated and time shifted the horizontal components to minimize the second eigenvalue of the covariance matrix for particle motion of a time window around the SKS phase. This procedure linearizes the particle motion and usually reduces the tangential component energy (assuming that the incoming SKS wave is radially polarized before entering the anisotropic medium). In the Teanby et al. [**2004**] approach, the splitting analysis is performed for 100 different windows, each encapsulating theSKS energy. Cluster analysis is then used to find pair of splitting parameters (φ, δt) that are stable over many different windows, which is not necessarily the pair with the lowest errors. An example of this procedure is shown in Figure 2. In the final analysis, a database of 78 splitting measurements was assembled from 23 earthquakes (Figure 3 and Table 1).

Table 1. Hypocentral Information for Earthquakes Used in the SKS Splitting Analysis

Date (dd/mm/yyyy)	Time (HH:MM)	Lat/Lon (°N/°E)	Depth (km)	Stations
02/03/2005	10:42	−6.53/129.93	202	CM06, CM18, CM26, and CM32
14/05/2005	05:05	0.59/98.46	34	CM29
21/05/2005	05:11	−3.29/−80.99	40	CM18 and CM24
15/06/2005	19:52	−44.87/−80.56	10	CM09
26/06/2005	14:11	−15.35/−72.96	111	CM26 and CM29
26/09/2005	01:55	−5.68/−76.40	115	CM24
15/10/2005	15:51	25.32/123.36	183	CM07
21/11/2005	15:36	31.02/130.00	145	CM06 and CM26
30/11/2005	16:53	6.27/124.03	13	CM07 and CM18
23/01/2006	20:50	6.86/−77.79	14	CM17, CM24, CM28, and CM30
27/01/2006	16:58	−5.47/128.13	397	CM01, CM02, CM05, CM06, CM13,
				CM15, CM18, CM19, CM20, CM21,

				CM22, CM23, CM24, CM25, CM26,
				CM27, and CM29
01/04/2006	10:02	22.87/121.28	9	CM28
30/04/2006	21:40	−27.21/−71.06	12	CM03, CM28, and CM32
16/07/2006	11:42	−28.72/−72.54	10	CM03, CM17, CM25, CM29, and CM30
24/08/2006	21:50	51.15/157.52	43	CM01, CM03, CM07, CM10, CM11,
				CM16, CM18, and CM31
16/09/2006	09:45	−3.08/129.44	17	CM02, CM03, and CM12
17/09/2006	09:34	−31.75/−67.18	142	CM31 and CM32
21/09/2006	18:54	−9.05/110.36	25	CM12 and CM32
30/09/2006	16:26	−15.59/−73.16	107	CM22
30/09/2006	17:50	46.35/153.17	11	CM02, CM07, CM10, CM16, CM18,
				CM22, and CM32
01/10/2006	09:06	46.47/153.24	19	CM02, CM10, CM11, CM16, CM18
				CM29, CM22, and CM32
15/11/2006	21:22	47.28/154.15	12	CM31
01/12/2006	03:58	3.40/99.09	206	CM01, CM05, CM10, CM15, and CM20
21/01/2007	11:27	1.07/126.28	22	CM12 and CM17

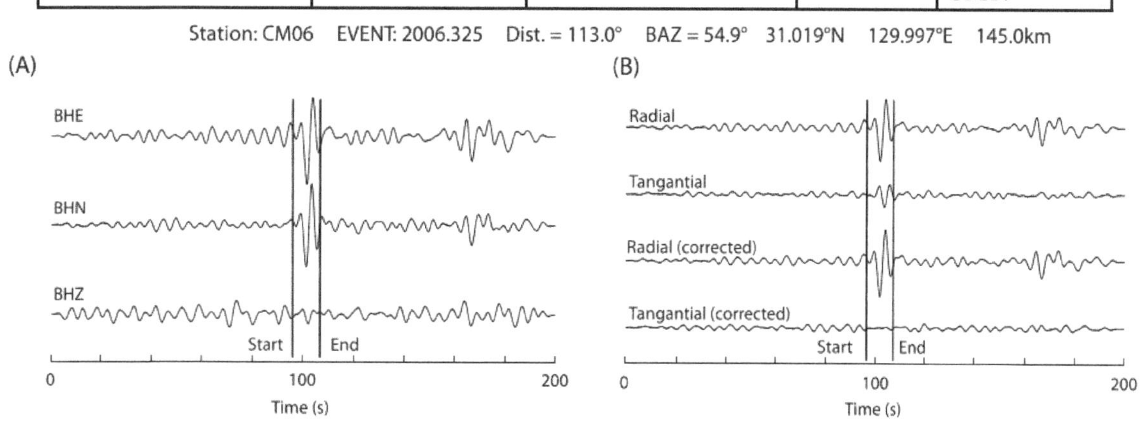

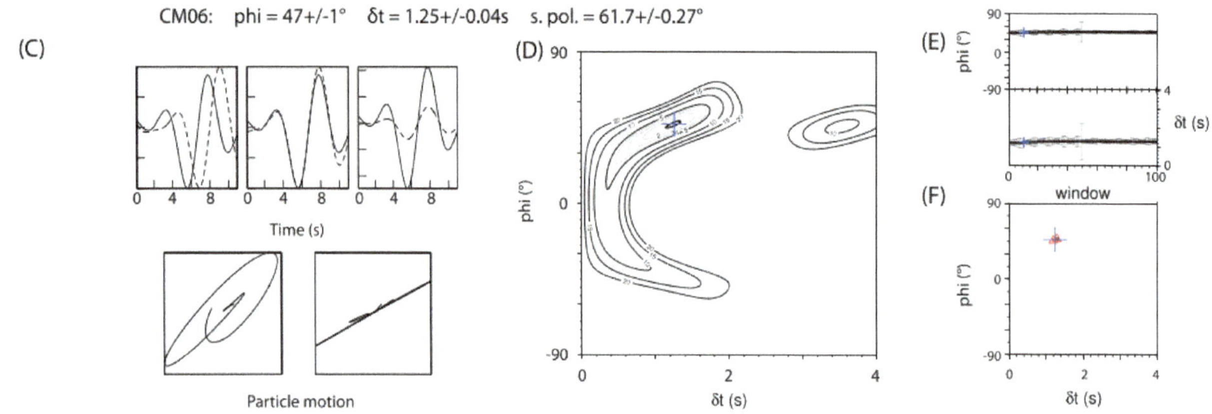

Figure 2. Example of a good shear wave splitting measurement for station CM06: (a) original three components (east, north, and vertical) of the SKS phase and the selection window used; (b) radial and tangential components before and after the analysis was performed; observe that energy on the tangential component has been removed; (c) match between the fast and slow waveform and the linearization of the particle motion; (d) error and uncertainty calculation, here a stable result and a well constrained 95% confidence contour (thick line) indicate a robust measurement; (e) measurements of φ and δt obtained from 100 different analysis windows plotted against window number; (f) cluster analysis of splitting parameters obtained from the 100 windows. Good results are stable over a large number of windows.

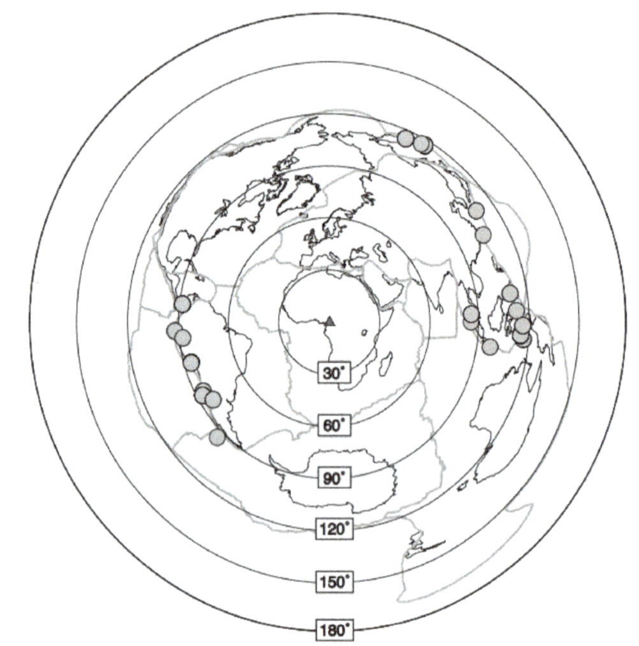

Figure 3. The global distribution of all the earthquakes used in the study plotted with an azimuthal equidistant map projection. The center of the CBSE network is marked by the triangle.

Seismicity

We analyzed seismicity in Cameroon during the period 28 January 2006 to 30 June 2006 to constrain patterns of earthquakes along the CVL. Continuous data from the Cameroon network is heavily contaminated by high-frequency cultural noise. We thus identified earthquakes on filtered seismograms, using a zero-phase Butterworth band-pass filter with corner frequencies of 0.5 and 4Hz. P and S wave arrivals are generally emergent, and their codas contain no significant energy at frequencies higher than 4Hz.

After waveform filtering, P and S wave arrival times were measured and used to compute earthquake locations using Hypoinverse-2000 [Klein, **2002**] with a 1-D velocity model previously employed in the region [e.g., Ubangoh et al., **1997**; Ateba and Ntepe, **1997**]. A Wadati diagram, plotted using the P and S wave local earthquake traveltimes, reveals a best fitVp/Vs ratio of 1.74 (Figure 6) and this value was used for the earthquake locations. Hypocenters were only obtained for earthquakes with four or more phase readings at three or more stations. The errors associated with earthquake location vary depending on the proximity of the earthquake to the seismic network. The depth distribution of the earthquakes is shown in Figures 7 and 8.

The majority of earthquakes are concentrated in the SW corner of the study region, near Mount Cameroon and extending offshore near the island of Bioko (Figure 8). Earthquakes near Mount Cameroon are proximal to the seismic network and have estimated horizontal errors of ±1km and vertical errors of ±2km. Those located offshore and outside the seismic network have horizontal errors of ±5km but vertical errors of up to ±10km. We therefore confidently interpret the distribution of hypocenters for earthquakes located onshore and only interpret epicenter distribution for earthquakes offshore. As far as we have been able to determine, the NE area of the CVL is aseismic, though we cannot preclude the possibility that this is due to the limited station density and relatively short period of network operation.

Local magnitudes (ML) of the earthquakes were computed, with maximum zero-to-peak amplitude measured on simulated horizontal-component Wood-Anderson displacement seismograms after removal of instrument response [e.g., Richter, **1935**; Keir et al., **2006**]. These measurements were used in conjunction with hypocentral distances to estimate MLusing the distance correction originally employed to estimate ML in California [Richter, **1935**]. Earthquake magnitudes are shown in Figure 8.

RESULTS AND COMPARISONS TO PREVIOUS STUDIES

Shear Wave Splitting

Individual shear wave splitting parameters are listed in Table 2, with associated hypocentral information in Table 1. Across the CBSE network δt varies from 0.6 to 1.95s, with several stations showing δt >1s (Figure 5). At stations where several shear wave splitting measurements were made, no significant back azimuthal variation in φor δt was observed (Table 2), with the implication that anisotropy in the mantle beneath can be approximated as a single, homogeneous, horizontal layer [e.g., Silver and Savage, **1994**]. We thus computed single-splitting parameters for each station using the stacking method of Restivo and Helffrich [**1999**]. The final splitting data set is presented in Figure 5.

Table 2. Shear Wave Splitting Parameters (Fast Polarization Direction, φ; Delay Time, δt) Are From the CBSE Networka

Station	**Station Lat/Lon**	***φ*(deg)**	***σφ*(deg)**	***δt*(s)**	***σδt*(s)**	**Event Lat/Lon**	**BAZ**
(*N*)	**(°N/°E)**					**(°N/°E)**	**(deg)**
CM02	2.70/13.29	47	8	1.25	0.10	−3.08/129.44	92.09
		46	3	1.08	0.03	−5.47/128.13	94.75
Stacked (4)		**36**	**1**	**1.15**	**0.03**	-	-
CM05	2.94/9.91	19	12	0.60	0.18	3.40/99.09	86.67
Stacked (2)		**22**	**9**	**0.60**	**0.14**	-	-
CM06	2.38/11.27	58	10	0.85	0.11	−6.53/129.93	96.10
		47	1	1.25	0.04	31.02/130.00	54.88
		70	5	1.08	0.09	−5.47/128.13	94.90
Stacked (3)		**42**	**1**	**0.98**	**0.04**	-	-
CM10	4.22/10.62	23	5	0.70	0.09	3.40/99.09	86.75
Stacked (4)		**28**	**4**	**0.65**	**0.04**	-	-
CM12	4.48/11.63	0	1	1.95	0.06	−3.08/129.44	91.11
Stacked (3)		**2**	**1**	**1.95**	**0.09**	-	-
CM15	5.03/9.93	41	17	0.47	0.17	3.40/99.09	86.71
Stacked (2)		**47**	**16**	**0.48**	**0.13**	-	-
CM17	5.55/12.31	19	12	1.50	0.21	−28.72/−72.54	241.08
		38	3	1.63	0.08	6.86/-77.79	276.80
Stacked (3)		**30**	**4**	**1.65**	**0.11**	-	-
CM21	6.47/12.62	72	4	1.48	0.08	−5.47/128.13	92.94
Stacked (1)		**73**	**4**	**1.50**	**0.08**	-	-
CM22	6.48/13.27	54	21	0.73	0.40	−5.47/128.13	92.99
		64	15	0.75	0.18	46.35/153.17	29.89
Stacked (3)		**62**	**4**	**0.80**	**0.07**	-	-
CM24	6.52/14.29	74	4	0.70	0.11	−5.47/128.13	93.06
Stacked (4)		**76**	**4**	**0.75**	**0.13**	-	-
CM26	7.26/13.55	30	10	0.93	0.14	−6.53/129.93	93.64
		27	6	0.95	0.14	−5.47/128.13	92.65
Stacked (4)		**57**	**4**	**0.63**	**0.06**	-	-
CM27	7.36/12.67	37	12	0.75	0.13	−5.47/128.13	92.52
Stacked (1)		**38**	**12**	**0.78**	**0.13**	-	-
CM29	9.35/13.39	65	2	1.23	0.08	−15.35/−72.96	254.37
		57	3	1.08	0.03	−5.47/128.13	91.66
Stacked (5)		**63**	**1**	**1.15**	**0.03**	-	-
CM31	10.33/15.26	66	7	0.82	0.06	51.15/157.52	24.32
Stacked (3)		**64**	**2**	**0.88**	**0.09**	-	-
CM32	10.62/14.37	44	15	0.88	0.34	−31.75/-67.18	237.46
		62	11	0.82	0.12	−6.53/129.93	92.08
		69	6	1.15	0.08	−9.05/110.36	97.81
Stacked (6)		**52**	**2**	**0.93**	**0.05**	-	-

[a]Errors associated with each measurement are the σφ and σδt; N is the number of individual measurements used to constrain

the stacked splitting parameters. BAZ: backazimuth. Bold text highlights the stacked results

For some station earthquake pairs in our analysis, no tangential component energy was noted, and the particle motion was linear prior to shear wave splitting analysis. This implies either that the mantle beneath the station is seismically isotropic or that the incoming shear wave energy is polarized parallel to or perpendicular to the true anisotropic fast direction that characterizes the subsurface. An example of a null splitting measurement is shown in Figure 4. Stations at which only null measurements were observed were predominantly in the southern to central portions of the CVL (Figure 5 and Table 3). Elsewhere, splitting parameters vary significantly in both φ and δt over relatively short (50–150km) length scales with φgenerally NNE-ENE (Figure 5 and Table 2).

Table 3. Null Shear Wave Splitting Observations

Station	**Station Lat/Lon (°N/°E)**	**BAZ (°)**	**Event Lat/Lon (°N/°E)**
CM01	2.39/9.83	86.66	3.40/99.09
		94.90	−5.47/128.13
		22.87	51.15/157.52
CM02	2.70/13.29	30.78	46.47/153.24
		30.92	46.35/153.17
CM03	3.52/15.03	91.77	−3.08/129.44
		241.36	−28.72/−72.54
		242.75	−27.21/−71.06
		25.46	51.15/157.52
CM05	2.94/9.91	94.60	−5.47/128.13
CM07	3.87/11.46	61.97	25.32/123.36
		81.71	6.27/124.03
		29.58	46.35/153.17
		23.47	51.15/157.52
CM09	4.23/9.33	225.40	−44.87/−80.56
CM10	4.22/10.62	28.87	46.47/153.24
		29.01	46.35/153.17
		22.93	51.15/157.52
CM11	3.98/13.19	30.36	46.47/153.24
		24.39	51.15/157.52
CM12	4.48/11.63	86.80	1.07/126.28
		98.40	−9.05/110.36
CM13	4.59/9.46	93.70	−5.47/128.13
CM15	5.03/9.93	93.48	−5.47/128.13
CM16	5.48/10.57	28.51	46.47/153.24
		28.65	46.35/153.17
CM16	5.48/10.57	22.65	51.15/157.52
CM17	5.55/12.31	86.41	1.07/126.28
CM18	5.72/9.35	266.79	−3.29/−80.99
		27.75	46.47/153.24
		94.17	−6.53/129.93
		80.63	6.27/124.03
		93.07	−5.47/128.13
		27.88	46.35/153.17
		21.91	51.15/157.52
CM19	5.97/11.23	93.07	−5.47/128.13
		22.92	51.15/157.52
CM20	6.22/10.05	86.75	3.40/99.09
		92.85	−5.47/128.13

CM22	6.48/13.27	254.20	−15.59/−73.16
CM23	6.37/10.79	92.83	−5.47/128.13
CM24	6.52/14.29	264.47	−5.68/−76.40
		267.34	−3.29/−80.99
		277.01	6.86/−77.79
CM25	6.76/11.81	240.99	−28.72/−72.54
		92.73	−5.47/128.13
CM26	7.26/13.55	254.43	−15.35/−72.96
		54.08	31.02/130.00
CM28	8.47/13.24	64.20	22.87/121.28
		276.90	6.86/−77.79
		242.42	−27.21/−71.06
CM29	9.35/13.39	90.22	0.59/98.46
		29.16	46.47/153.24
		241.19	−28.72/−72.54
CM30	9.76/13.95	241.29	−28.72/−72.54
CM31	10.33/15.26	237.64	−31.75/−67.18
		28.87	47.28/154.15
CM32	10.62/14.37	29.43	46.47/153.24
		29.56	46.35/153.17
		242.60	−27.21/−71.06

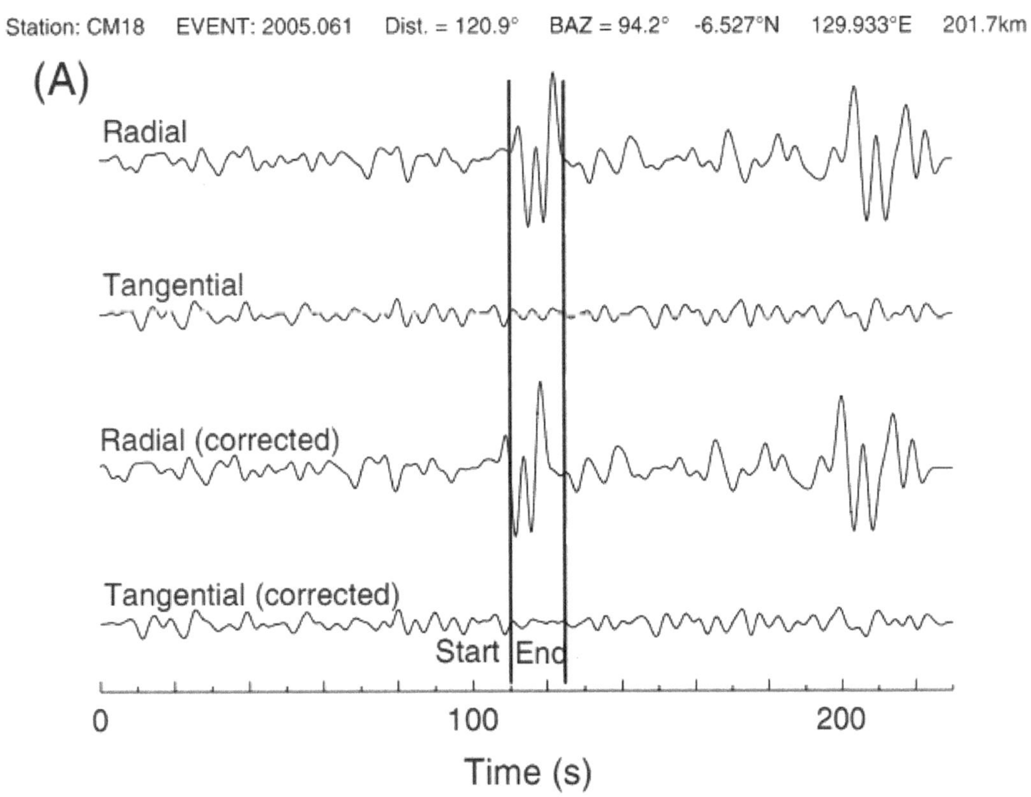

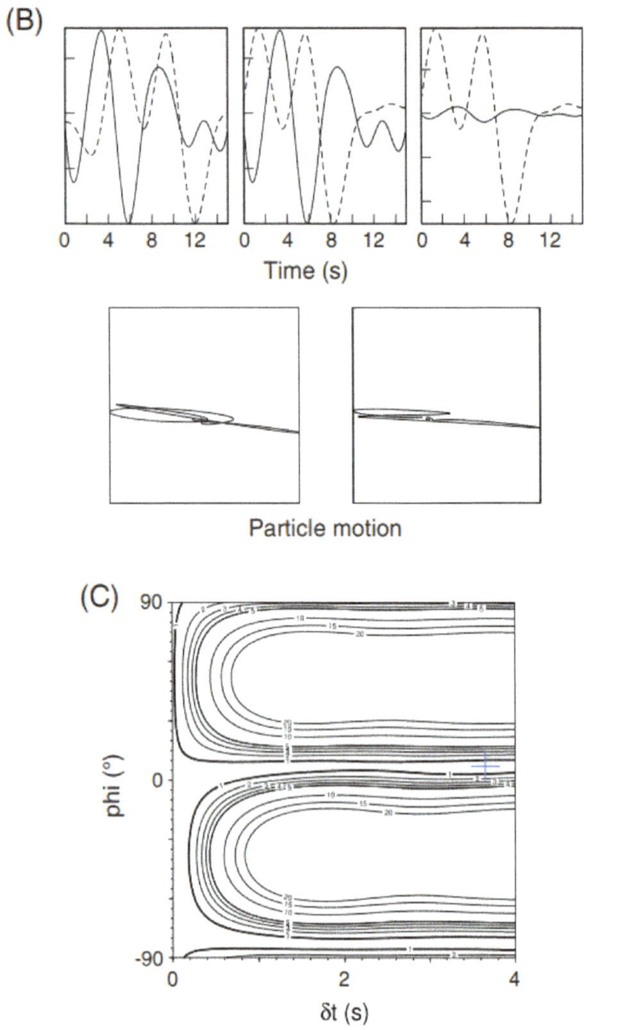

Figure 4. Example of a null measurement for station CM18. (a) There is no energy on the tangential component, and (b) the particle motion is already linear before the analysis is performed. These are evidence of the absence of splitting. The (c) stretched confidence contours are typical of a null measurement.

Table 4 shows a comparison of splitting parameters obtained in this study with that of Koch et al. [**2012**]. As mentioned earlier, the Koch et al. [**2012**] study identifies four regions of distinct anisotropy: (1) moderately strong NE-SW oriented fast directions ($\delta t \approx 1.0$ s) beneath the Congo Craton in the south, and the Garoua rift in the north (Figure 1); (2) weak anisotropy ($\delta t \approx 0.3$ s) between the Congo Craton and the CVL; (3) N-S oriented fast polarization directions within the CVL, with $\delta t \approx 0.7$ s. In line with the findings of our study, Koch et al. [**2012**] found no evidence in support of complex patterns of anisotropy (e.g., dipping or multiple anisotropic layers). The main discrepancies between the Koch et al. [**2012**] study and ours are at stations CM01, CM03, CM09, CMC11, CM13, CM16, CM19, CM20, CM23, CM25, CM28, and CM30. Here the earlier study presents splitting parameters (φ, δt) that contrast with the null measurements we present. With the exception of CM28 and CM30, the δt observations resulting from the stacking methodology adopted by Koch et al. [**2012**] are extremely small at these stations, ranging from 0.28 to 0.53s. Larger δt values at stations CM28 (1.05 ± 0.2s) and CM30 (0.88 ± 0.04s) contrast more significantly with our findings (Table 4).

Figure 5. Shear wave splitting parameters (blue arrows) from CBSE stations (white triangles) in Cameroon. Black crosses are null measurements (Table 3). APM: absolute plate motion from the HS3-Nuvel-1A model [Gripp and Gordon, **2002**] in the hotspot reference frame (white arrow). The domains are modified from Toteu et al. [**2004**] and Nzenti et al. [**2006**], YD: Yaounde Domain (Neoproterozoic); AYD: Adamawa-Yade Domain (Paleoproterozoic); and WCD: Western Cameroon Domain (Neoproterozoic). Heavy grey lines are major strike-slip faults associated with the CASZ as defined by Ngako et al. [**2003**], Toteu et al. [**2004**], and Nzenti et al. [**2006**]. SSZ: Sanaga Shear Zone and TBSZ: Tchollir-Banyo Shear Zone. Heavy dashed lines are thrust faults. The red line delineates the low-velocity zone at 200km depth from the P wave tomographic model of Reusch et al. [**2010**].

Table 4. A Comparison of Shear Wave Splitting Parameters From This Study With That of Koch et al. [**2012**]

	This study				***Koch et al.* [2012]**			
Station	***φ*(°)**	***σφ*(deg)**	***δt*(s)**	***σδt*(s)**	***φ*(deg)**	***σφ*(deg)**	***δt*(s)**	***σδt*(s)**
CM01	NULL				12	5	0.53	0.18
CM02	36	1	1.15	0.03	42	1	0.95	0.01
CM03	NULL				20	3	0.35	0.01
CM04	NO RESULT				47	3	0.60	0.05
CM05	19	9	0.60	0.14	22	3	0.70	0.06

CM06	42	1	0.98	0.04	56	1	0.78	0.01
CM07	NULL				NULL			
CM08	NO RESULT				25	3	1.10	0.04
CM09	NULL				26	9	0.43	0.10
CM10	28	4	0.65	0.04	42	3	0.75	0.05
CM11	NULL				21	3	0.25	0.04
CM12	2	1	1.95	0.09	7	3	0.78	0.08
CM13	NULL				11	12	0.45	0.16
CM14	NO RESULT				−32	3	0.35	0.05
CM15	47	16	0.48	0.13	9	1	1.00	0.09
CM16	NULL				26	4	0.28	0.04
CM17	30	4	1.65	0.11	26	2	1.33	0.04
CM18	NULL				NULL			
CM19	NULL				2	3	0.45	0.04
CM20	NULL				−1	3	0.48	0.07
CM21	73	4	1.50	0.08	56	1	0.70	0.05
CM22	62	4	0.80	0.07	72	1	1.10	0.01
CM23	NULL				−15	2	0.53	0.05
CM24	76	4	0.75	0.13	46	4	0.30	0.05
CM25	NULL				−15	2	0.53	0.05
CM26	57	4	0.63	0.06	74	3	0.63	0.01
CM27	38	12	0.78	0.13	44	8	0.43	0.09
CM28	NULL				15	7	1.05	0.20
CM29	63	1	1.15	0.03	57	1	0.93	0.04
CM30	NULL				57	4	0.88	0.04
CM31	64	2	0.88	0.09	44	4	0.95	0.06
CM32	52	2	0.93	0.05	41	4	0.83	0.08

Seismicity in Cameroon

During the 5 months analyzed, we located 203 local earthquakes with magnitudes ranging between 1.4 and 3 (Figure 8). The earthquakes are primarily concentrated near to Mount Cameroon (153 earthquakes), and the remainder are positioned offshore on the eastern side of Bioko (50 earthquakes). The rate of activity is variable with just 15 and 16 earthquakes reported for February and March 2006 respectively, increasing to 42 and 54 earthquakes in April and May 2006. Seventy-two earthquakes were reported for June 2006.

In contrast to previous studies that utilized seismic data only from stations located around Mount Cameroon, our study draws upon data from seismic stations located along the full length of the CVL. Most of the line is aseismic with activity confined to the region around Mount Cameroon and Bioko (Figure 6).

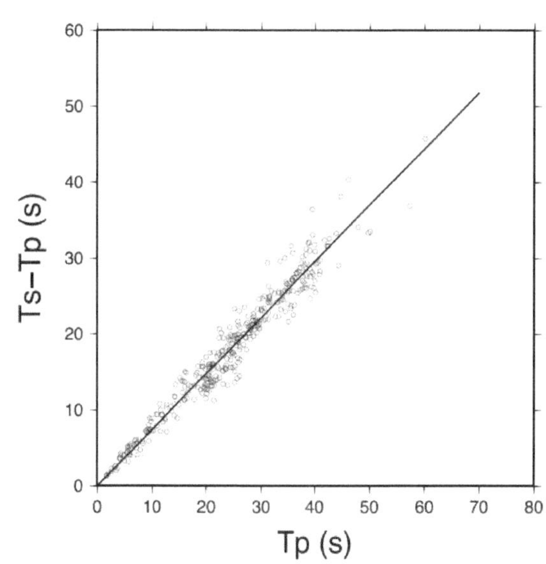

Figure 6. Wadati diagram showing the difference between S and P wave traveltimes versusP wave traveltime. The best fit line corresponds to a Vp/Vs ratio of 1.74.

Earthquakes near Mount Cameroon are distributed throughout the crust, with the majority ≤10km deep; eight lower crustal earthquakes occurred at 20–35km depth (Figures 7 and

8a). The shallow earthquakes form two main groups, with the majority forming a NNE-SSW striking, ~40km long cluster ~40km east of Mount Cameroon, close to an inferred major CASZ fault. A second cluster, ~15km long on the northern flank of Mount Cameroon is subparallel to faults that define the NW-SE striking Bokosso graben. Lower crustal earthquakes are located beneath the east and south flanks of Mount Cameroon within ~20km of the summit. Fifty earthquakes located in the Gulf of Guinea, mostly to the east of Bioko, define a ~20km wide, ~80km long cluster striking NNE-SSW (Figure 8).

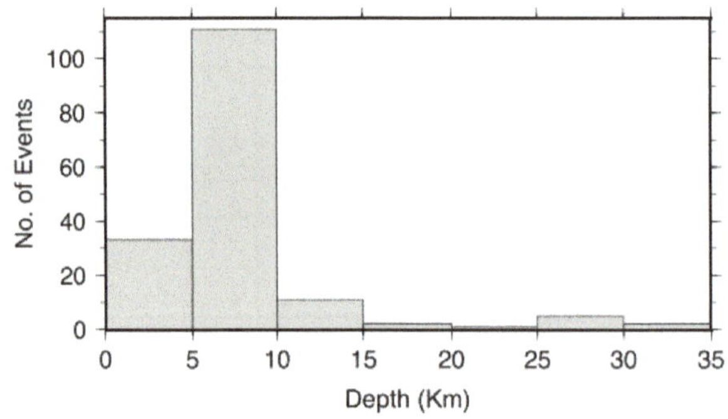

Figure 7. The depth distribution of earthquakes in Cameroon along the continental sector of the CVL. Earthquakes with depth uncertainties larger than ±2km are not plotted.

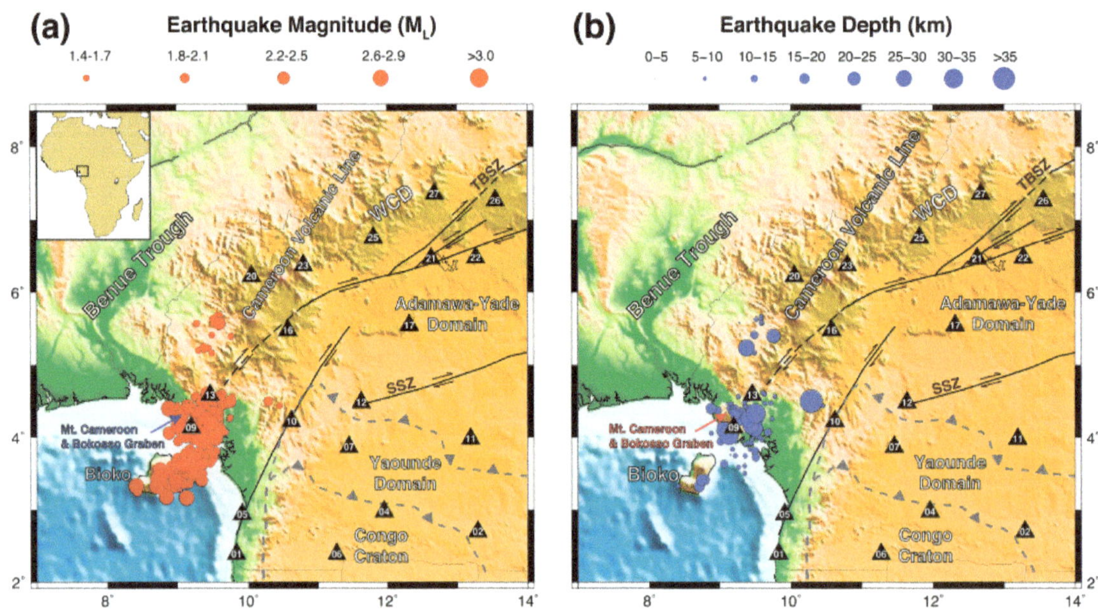

Figure 8. Seismicity along the CVL during February–May 2006. (a) Local earthquake magnitudes and (b) local earthquake depths in Cameroon. Heavy dashed lines are thrust faults. The domains are modified from Toteu et al. [**2004**] and Nzenti et al. [**2006**]. WCD: Western Cameroon Domain (Neoproterozoic). Heavy black lines are major strike-slip faults of the Central African Shear Zone. SSZ: Sanaga Shear Zone. Black numbered triangles are seismograph station locations.

The distribution of earthquakes in our study is broadly similar to that reported during continuous seismic monitoring in 1985–1992 following the 1982 eruption of Mount Cameroon. At that time, seismic activity was mainly in the upper crust beneath the Bokosso graben, and in distinct clusters 40km south, 40km east, and 10–20km southeast of Mount Cameroon [Ambeh and Fairhead, **1991**; Ubangoh et al., **1997**; Ateba and Ntepe, **1997**]. The latter cluster had earthquakes in the lower crust, as in our study, but also extending to 65km depth, within the mantle lithosphere [Ambeh and Fairhead, **1991**].

The rate of seismicity we report is toward the low end of that observed during 1985–1992. Our values are comparable to the 3–4 earthquakes per 10 days reported during 1991 and 1992, but lower than the ≤60 per 10 days reported during 1986, 1988, and in 1990 [Ubangoh et al., **1997**]. Importantly, however, our new results show that the first-order pattern of seismic activity that characterized Mount Cameroon during 1985–1992 persisted until 2006, except that we do not find significant seismicity beneath the summit of Mount Cameroon, as was observed during summit eruptions in 1999 and 2000 [Ateba et al., **2009**].

DISCUSSION

In this section, we use the results of our seismicity and SKS splitting studies to address outstanding questions concerning the reasons for the development of the CVL and the nature of the magmatic plumbing system beneath it.

Seismicity along the CVL

Our analysis of seismicity demonstrates that earthquake activity is focused on and around Mount Cameroon, the most recently active volcano of the CVL. This suggests a strong link between spatially localized magmatic processes and the generation of brittle failure in the crust. A notable feature of our results is the presence of lower crustal earthquakes at 20–35km depth beneath the south and east flanks of the volcano (Figures 7 and 8). Earthquakes of similar depth have been reported beneath Icelandic volcanoes such as Eyjafjallajokull and Askja [Tarasewicz et al., **2012**; Key et al., **2011**], beneath rift valley flank volcanoes in East Africa [e.g.,Keir et al., **2009**], and beneath intraplate volcanoes related to hotspots such as Kilauea volcano in Hawaii [e.g., Wolfe et al., **2003**]. Although we cannot preclude the possibility that this deformation is the result of localized deep crustal faulting beneath the CVL, the association between deep earthquakes and crustal magmatism in these studies suggests that brittle failure of the lower crust may be the result of elevated strain rates, induced by the flow of magma or the release of volatiles from the magma.

Upper crustal seismicity is identified on a fault system subparallel to the CASZ to the east of Mount Cameroon; it is also identified beneath the Bokosso graben, which is oriented orthogonal to the CASZ (Figure 8). Similar to the lower crustal seismicity, the shallow earthquakes are characterized by emergent phase arrivals and relatively low frequency content (<4Hz), again pointing toward the migration of magma or magma-derived volatiles through crustal fracture networks [e.g., Chouet and Matoza, **2013**]. Mount Cameroon last erupted 6 years before the time period of our study and the preexisting fault systems delineated by shallow earthquakes are not associated with magmatism. We thus favor the interpretation that the seismicity was induced by migration of pressurized volatiles, released during cooling of localized intrusions deeper in the lithosphere. This interpretation is supported by field measurements of ongoing gas release from the CASZ to the south and southeast of Mount Cameroon, coincident with the location of earthquakes [Ubangoh et al., **1997**; Turner et al., **2013**]. Gas release was also observed in 1907 on the NE flank of the volcano, as well as at the summit in 1861 [Ubangoh et al., **1997**].

Away from Mount Cameroon, the lack of seismic and volcanic activity (Figure 8) suggests that there is little evidence for active magma chambers and magma supply routes along the western sector of the CVL, consistent with the view that magmatism is dominated by low-volume alkaline basaltic volcanism, with associated short crustal residence times [e.g., Suh et al., **2003**]. Supporting evidence comes from constraints of bulk crustal Vp/Vs ratios along the CVL, which are remarkably low (~1.74)[Gallacher and Bastow, **2012**], with the implication that little mafic material is present at depth beneath the region, either as present-day melt, cooled gabbroic intrusions, or fractionated products of erupted lavas. The low Vp/Vs ratio observation from the Gallacher and Bastow [**2012**] receiver function study is corroborated by our seismicity data set (Figure 6). Oceanic intraplate shields such as Hawaii [e.g., Leahy and Park, **2005**] and Galapagos [e.g., Tepp et al., **2014**] have Vp/Vs~1.8, but the values we observe from Cameroon are markedly low when compared to other continental hotspots such as Ethiopia [e.g., Daly et al., **2008**], where large volumes of melt within the crust can push Vp/Vsratios to ≥1.80. Spatially localized magmatism along the CVL is thus likely delivered episodically from directly beneath individual volcanic centers during decadal to century timescale periods of intrusions and eruptions with associated degassing activity.

In the following sections, we explore the extent to which our SKS splitting study can inform about the deformation history of the Cameroon lithosphere, the CVL's magmatic plumbing system, the mantle flow field beneath it, and the implications of our results for the development of the chain.

Causes of Seismic Anisotropy in Cameroon

Patterns of seismic anisotropy can develop due to the preferential alignment of minerals in the crust and/or mantle, the preferential alignment of fluid or melt, or some combination thereof [e.g., Blackman and Kendall, **1997**]. A range of processes could lead to such anisotropy, including (1) asthenospheric flow parallel to absolute plate motion [e.g., Vinnik et al., **1989**;Bokelmann, **2002**; Assumpção et al., **2006**], (2) mantle flow around the continental keel beneath the Congo Craton [e.g., Fouch et al., **2000**], and (3) a preexisting fossil anisotropy frozen in the lithosphere [e.g., Silver and Chan, **1988**; Vauchez and Nicolas, **1991**; Bastow et al., **2011**]. Estimates of the amount of splitting that can be accrued in the crust vary from 0.1–0.3s [Silver, **1996**] to 0.1–0.5s [Barruol and Mainprice, **1993**]. Thus, a mantle contribution is required to explain the observations at a number of stations in Cameroon. At 12 of the 14 stations where we draw a null conclusion, Koch et al. [**2012**] either also present a null or$\delta t \leq 0.53$s (Table 4). These small δtvalues can likely be explained relatively easily with hypotheses concerning only the crust.

The direction of motion of the African Plate shows poor correlation with the φ values presented in Figure 5, thus ruling out basal drag as the principal cause of the observed anisotropy. Equally, there is little evidence in our results for lateral flow of asthenospheric material beneath the CVL. Our results therefore do not provide conclusive support for the hypothesis of Ebinger and Sleep [**1998**] that flow of buoyant asthenosphere beneath continental lithosphere thinned extensively during Mesozoic rifting may now be contributing to volcanism along the CVL. Laterally extensive horizontal asthenospheric flow patterns due to plate motions or mantle flow in the region would, using Fresnel zone arguments [e.g., Alsina and Snieder, **1995**], be expected to produce only gradual variations in φ and δt across the network [e.g., Forte et al., **2010**]. However, we cannot preclude the possibility that a narrow, focused asthenospheric flow channel exists beneath the region, to which our splitting study is insensitive.

If the CVL had experienced plate stretching during the opening of the South Atlantic, as was the case for the Benue Trough to the NW of our network, the lithosphere might be expected to have inherited an olivine lattice preferred orientation (LPO) fabric parallel to the approximately NW-SE oriented extension direction [e.g., Vinnik et al., **1992**; Gao et al., **1997**]. Alternatively, in hotspots such as Ethiopia, patterns of anisotropy are attributed to rift parallel-oriented melt pockets in the lithosphere [e.g., Kendall et al., **2006**; Bastow et al., **2010**]. The observations from Cameroon are not so easily explained by either of these mechanisms because much of the CVL is characterized by null measurements (Figure 5 and Table 3), and the larger values ofδt are found away from regions characterized by low mantle seismic wave speeds and Cenozoic volcanism (e.g., CM02 on the Congo Craton; Figure 5).

One potential reason for the apparent lack of melt in the CVL lithosphere compared to Ethiopia may be the considerably greater degree of focused tectonic faulting and fracturing in Ethiopia [e.g., Hayward and Ebinger, **1996**], which enhances focused magmatism along strike in the rift [Ebinger and Casey, **2001**; Rooney et al., **2011**]. Additionally, decompression melting in response to extension is expected to enhance melt volume production in Ethiopia compared to Cameroon, where there is no evidence for present-day extension. Thus, while we cannot preclude the possibility that poorly aligned, stationary melt exists in the lithosphere beneath the CVL, our study yields no evidence for significant magmatic extension beneath the CVL nor is there any evidence for strongly aligned melt along CASZ faults/fractures.

The null measurements we observe beneath much of the CVL (Figure 5 and Table 3) may be indicative of a widespread vertical LPO fabric beneath much of the region, with the aaxis of the mantle olivine crystals perpendicular to the particle motion of our near vertically propagatingSKS phases. If true, this latter flow direction would be consistent with hypotheses for the development of the CVL that include either a mantle plume or the upwelling limb of a small-scale convection cell beneath the region [e.g., Morgan, **1983**; Burke, **2001**; Reusch et al., **2010**]. Predictions of LPO mantle seismic anisotropy from tomography-based models of mantle convection [Forte et al., **2010**] also suggest a strong vertical component to the mantle flow field beneath Cameroon, consistent with our observations. The same study also predicts a horizontal flow component, however, for which our results do not provide compelling evidence. Alternatives to the isotropic or vertical flow field hypotheses exist though, not least the possibility that our back azimuthal event coverage is insufficient to characterize anisotropic fabrics beneath the stations we classify here as null. Plastic deformation of peridotites usually results in the alignment of the (100) axis of olivine parallel to the direction of flow [e.g., Zhang and Karato, **1995**; Bystricky et al., **2000**; Tommasi et al., **2000**]; however, the apparent isotropy inferred from our SKS splitting analysis could also result from a horizontal flow field in the event that olivine crystallographic orientations have axial b patterns [e.g., Vauchez et al., **2005**;Bascou et al., **2008**]. Null observations could also be the result of a two-layered model with orthogonal fast directions as has been suggested, for example, at station CAN in Australia [Barruol and Hoffmann, **1999**].

Splitting parameters, where observed in Cameroon, vary over length scales of ~100km (e.g., CM40–CM32; Figure 5), thus implying a relatively shallow anisotropic layer in some regions of the study area: a deep source of anisotropy would produce only subtle variations across the network. Lithospheric thickness is ~100km beneath the CVL, but as much as ~250km thick beneath the Congo Craton [e.g., Fishwick, **2010**]. Thus, in the following sections we explore the hypothesis that fossil lithospheric fabrics exist beneath some parts of Cameroon.

Fossil Lithospheric Anisotropy in Cameroon

Although there is some debate as to whether or not tectonic processes in the Archean were capable of imparting a seismic anisotropic fabric on the lithosphere [e.g., Fouch et al., **2004**], younger Precambrian fabrics have been identified in many areas worldwide. For example, the thick lithosphere of northern Canada retains a strong fossil lithospheric fabric of the ~1.8Ga Trans-Hudson Orogen [Bastow et al., **2011**], which marked the final stages of the assembly of Laurentia. Precambrian fossil lithospheric fabrics are also found, for example, in southern Africa [e.g., Silver et al., **2001**; Fouch et al., **2004**] and northern Europe (Fennoscandia) [e.g.,Plomerová and Babuska, **2010**]. Unlike Cameroon, these regions have since experienced no significant tectonic reworking.

Cameroon comprises several tectonic domains that span more than 3Ga of the geological record, including the Archean Ntem Complex in southernmost Cameroon [e.g., Tchameni et al., **2001**], the Neoproterozoic Yaounde Domain [Oliveira et al., **2006**], and the Oubanguides Belt, which were assembled during the Pan-African Orogeny ~600Ma (Figure 1). Our observations do not match the orientation of these terrane boundaries, however, thus indicating that a Precambrian fossil lithospheric fabric associated with Gondwana assembly is no longer present. At stations CM02, CM03, and CM06, for example, anisotropic fabrics are oblique to the trend of the suture between the Ntem Complex and Yaounde domains. Instead, φ values at these stations correspond more closely to the SW-NE trend of the CASZ. It is difficult to determine the extent of Mesozoic reworking in parts of Cameroon because surface geology is masked by dense rain forest. However, further north, where structural geological constraints are available [e.g., Toteu et al., **2004**; Nzenti et al., **2006**; Ngako et al., **2003**], anisotropic fast directions at stations CM17, CM21, CM22, CM24, CM26, and CM27 mirror the trends of the large-scale strike-slip faults that reactivated during the early stages of Gondwana breakup in Mesozoic times (Figure 5). In the northernmost part of the study area, anisotropic fast directions at stations CM29, CM31, and CM32 are also likely related to shear zones in the lithosphere, with apparently little influence from the obliquely (~N125°E) trending Garoua rift. Narrow pull-apart basins along the CASZ in Cameroon might have resulted in the development of anisotropic strain fabrics perpendicular to those seen in Figure 5 and have thus not resulted in the same lithospheric-scale deformation as the strike-slip faults.

A striking observation in Figure 5 is the abundance of null measurements along the CVL, below which mantle seismic wave speeds are known to be relatively slow [Reusch et al., **2010**]. Our study thus yields no evidence for a well-preserved fossil lithospheric fabric beneath much of the CVL. The Mesozoic reactivation of the CASZ is known to have affected this region in the form of strike-slip faulting, but we observe no evidence for the well-defined lithospheric anisotropic fabrics seen elsewhere in Cameroon. One possible explanation for this is that hotspot activity during both the breakup of Gondwana and the subsequent development of the CVL has destroyed the older fossil lithospheric fabrics. Hotspot tectonism associated with the Mesozoic opening of the South Atlantic ~130Ma may have thermomechanically eroded Precambrian age fossil lithospheric fabrics beneath the CVL, though its thermal influence would have long since dissipated and the lithosphere rethickened. To some extent, the lithosphere may have remained relatively thin on account of ongoing small-scale convection between the two-flanking cratons. The new, thin lithosphere would have been expected to inherit a new fossil anisotropic signature during cooling. However, anisotropic basal drag fabrics do not develop easily during slow absolute plate motion [e.g., Debayle and Ricard, **2013**] as is observed for present-day Cameroon. The young lower lithosphere of the CVL may thus not have inherited a new fossil anisotropic signature. On the other hand, studies of mantle seismic anisotropy from the Ronda Massif in Spain [Vauchez et al., **2005**] and from Polynesia [Tommasi et al., **2004**] conclude that fossil fabrics are inherited from older deformation and that partial melting and melt percolation under static conditions cannot erase preexisting olivine crystal preferred orientations. Thus, we cannot completely preclude the possibility that the apparent lack of fossil lithospheric fabrics beneath the western end of the continental CVL is simply the result of inadequate data coverage in our study. The null observations could also be the result of a two-layer model with orthogonal fast directions [e.g., Barruol and Hoffmann, **1999**]. Either way, the paucity of melt in the CVL lithosphere implied by both our seismicity and SKS splitting studies certainly does not support the view that metasomatism has greatly modified the lithosphere, at least during the past 30Ma.

Implications for the Origins of the CVL

While our study is not able to discriminate unambiguously between many of the competing hypotheses for the development of the CVL (e.g., plume versus small-scale convection), an actively upwelling component of mantle flow seems plausible beneath the CVL. The lack of evidence for aligned melt in the lithosphere, however—a common feature at other hotspots such as Ethiopia [e.g., Kendall et al., **2006**; Bastow et al., **2010**; Keir et al., **2011**]—is consistent with the findings of petrological studies that indicate only minimal fractionation of magmas within the shallow lithosphere prior to rapid eruption [Fitton and Dunlop, **1985**; Fitton, **1987**;Marzoli et al., **2000**; Déruelle et al., **2007**; Suh et al., **2003**]. Our local earthquake traveltime data set reveals the CVL crust has a low Vp/Vs ratio (1.74; Figure 8), an observation in agreement with teleseismic receiver function studies of bulk crustal structure [Gallacher and Bastow, **2012**] that also provides little evidence for large-scale modification of the crust beneath the CVL. The sporadic development of alkaline basaltic volcanism in Cameroon over the past ~30Ma may thus be explained better by lower melt volume hypotheses such as shear zone reactivation [e.g. Fairhead, **1988**] or lithospheric instabilities at the edge of continents [e.g., Milelli et al., **2012**].

Recent work by Milelli et al. [**2012**] and Fourel et al. [**2013**] used scaled laboratory models and analytical solutions to investigate the effects of large lateral variations in lithospheric thickness on lithospheric stability. Their work demonstrates that lithospheric instabilities can develop over long timescales with small rates of upwelling and decompression melting as is observed in Cameroon. Over time, removal of small volumes of lithosphere beneath the CVL would be expected to erode the CASZ-related fossil anisotropic fabrics observed elsewhere in Cameroon. In between delamination events, new lithosphere formed by conductive plate cooling might be expected to inherit a new LPO anisotropic fabric, but Africa's very slow movement NW since ~30Ma [e.g., Steinberger and Torsvik, **2008**] may mean that basal drag fabrics beneath the CVL are minimal and the new lithosphere thus assumes a near-isotropic seismic signature. Figure 9illustrates schematically how the CVL's magmatic plumbing system and underlying mantle flow field compares to those observed beneath the Ethiopian hotspot province, which is superposed by a mature rift zone, in contrast to the CVL.

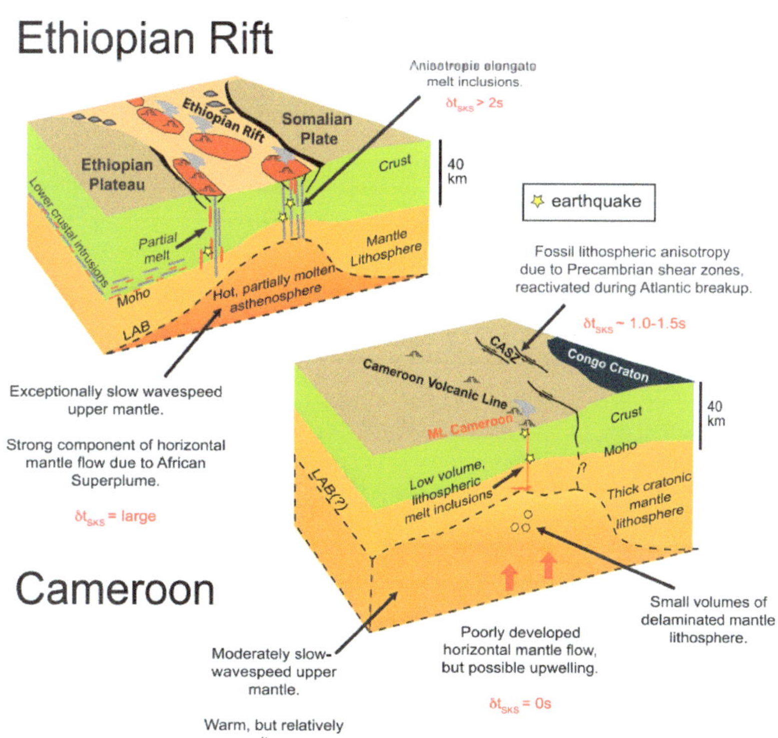

Figure 9. A summary conceptual cartoon for the Cameroon crust and mantle compared to that in the magmatically active Ethiopian rift. LAB: lithosphere-asthenosphere boundary; CASZ: Central African Shear Zone; and δtSKS: the splitting delay time observed in SKS phases. Modified after Gallacher and Bastow [**2012**].

CONCLUSIONS

We have performed seismicity and SKS splitting studies of the Cameroon Volcanic Line (CVL) with a view to understanding better the tectonic/geodynamic processes responsible for its development. The seismicity study indicates the region is markedly less seismically active than other continental intraplate hotspot worldwide, with little evidence for magmatism-related crustal deformation away from Mount Cameroon, which last erupted in 2000. The local earthquake traveltime data set reveals low crustal Vp/Vs ratios of ~1.74, which match earlier receiver function-derived values for the bulk crust in Cameroon [Gallacher and Bastow, **2012**].

The SKS splitting study yields no evidence for strongly aligned anisotropic melt inclusions beneath the CVL, as observed at some hotspots elsewhere. Instead, anisotropic fast polarization directions in Cameroon, where observed, parallel the Precambrian Central African Shear Zone (CASZ). However, along the western continental portion of the CVL, only null splitting measurements are found. We cannot completely preclude the possibility that this is the result of two anisotropic layers with orthogonal fast directions. Alternatively, the nulls may imply the CVL asthenosphere is isotropic or characterized by a vertically oriented olivine LPO fabric due, perhaps, to a mantle plume, or the upward limb of a small-scale convection cell between the Congo and West African Cratons. Precambrian CASZ fossil anisotropic fabrics along the CVL may have been thermomechanically eroded during the Mesozoic breakup of Gondwana (~130Ma), with an isotropic lower lithosphere subsequently reforming thermally during the slow movement of the African plate. Small-scale lithospheric delamination during the 30Ma recent development of the line may also have contributed to the erosion of the lithosphere beneath the CVL. This would have served both to erode CASZ fossil anisotropic fabrics and to generate the spatially and temporally sporadic low-volume alkaline basaltic volcanism observed along the line (Figure 9).

ACKNOWLEDGMENTS

The CBSE network was operated by Pennsylvania State University. Data from the CBSE network were all sourced freely from the IRIS Data Management Center (DMC)http://www.iris.edu/dms/nodes/dmc. We thank G. Helffrich, M. Kendall, J. Di Leo, K. Cashman, L. Caricchi, and S. Sparks for helpful discussions about Cameroon. Insightful reviews from C. Ebinger and A. Tommasi greatly improved the contribution. R.G. is supported by a NERC studentship. I.B. is funded by the Leverhulme Trust. This work was partially funded by Natural Environment Research Council grant number NE/K500926/1.

REFERENCES

1. Alsina, D., and R. Snieder (1995), Small-scale sublithospheric mantle deformation: Constraints from SKS splitting observations, Geophys. J. Int., 123, 431–448.
2. Ambeh, W., and J. Fairhead (1991), Regular, deep seismicity beneath Mt Cameroon volcano: Lack of evidence for tidal triggering, Geophys. J. Int., 106(1), 287–291.
3. Assumpção, M., M. Heintz, A. Vauchez, and M. Silva (2006), Upper mantle anisotropy in SE and central Brazil from SKS splitting: Evidence of asthenospheric flow around a cratonic keel, Earth Planet. Sci. Lett., 250(1–2), 224–240, doi:10.1016/j.epsl.2006.07.038.
4. Ateba, B., and N. Ntepe (1997), Post-eruptive seismic activity of Mount Cameroon (Cameroon), West Africa: A statistical analysis, J. Volcanol. Geotherm. Res., 79(1–2), 25–45.
5. Ateba, B., C. Dorbath, L. Dorbath, N. Ntepe, M. Frogneux, J. Delmond, and D. Manguelle (2009), Eruptive and earthquake activities related to the 2000 eruption of Mount Cameroon volcano (West Africa), J. Volcanol. Geotherm. Res., 179(3–4), 206–216, doi:10.1016/j.jvolgeores.2008.11.021.
6. Barruol, G., and R. Hoffmann (1999), Upper mantle anisotropy beneath the Geoscope stations, J. Geophys. Res., 104(B5), 10,757–10,773, doi:10.1029/1999JB900033.
7. Barruol, G., and D. Mainprice (1993), A quantitative evaluation of the contribution of crustal rocks to the shear-wave splitting of teleseismic SKS waves, Phys. Earth Planet. Int., 78, 281–300.
8. Bascou, J., G. Delpech, A. Vauchez, B. Moine, J. Cottin, and G. Barruol (2008), An integrated study of microstructural, geochemical, and seismic properties of the lithospheric mantle above the Kerguelen plume (Indian Ocean), Geochem. Geophys. Geosyst., 9(4), Q04036, doi:10.1029/2007GC001879.
9. Bastow, I., T. Owens, G. Helffrich, and J. Knapp (2007), Spatial and temporal constraints on sources of seismic anisotropy: Evidence from the Scottish highlands, Geophys. Res. Lett., 34(5), L05305, doi:10.1029/2006GL028911.
10. Bastow, I., S. Pilidou, J.-M. Kendall, and G. Stuart (2010), Melt-induced seismic anisotropy and magma assisted rifting in Ethiopia: Evidence from surface waves, Geochem. Geophys. Geosyst., 11, Q0AB05, doi:10.1029/2010GC003036.
11. Bastow, I., D. Thompson, J. Wookey, J. Kendall, G. Helffrich, D. Snyder, D. Eaton, and F. Darbyshire (2011), Precambrian plate tectonics: Seismic evidence from northern Hudson Bay, Canada, Geology, 39(1), 91–94, doi:10.1130/G31396.1.
12. Blackman, D., and J.-M. Kendall (1997), Sensitivity of teleseismic body waves to mineral texture and melt in the mantle beneath a mid-ocean ridge, Phil. Trans. R. Soc. Lond., 355, 217–231.
13. Bokelmann, G. (2002), Which forces drive North America?, Geology, 30(11), 1027–1030.
14. Burke, K. (2001), Origin of the Cameroon line of volcano-capped swells, J. Geol., 349–362.
15. Bystricky, M., K. Kunze, L. Burlini, and J.-P. Burg (2000), High shear strain of olivine aggregates: Rheological and seismic consequences, Science, 290(5496), 1564–1567.
16. Castaing, C., J. Feybesse, D. Thiéblemont, C. Triboulet, and P. Chèvremont (1994), Palaeogeographical reconstructions of the Pan-African/Brasiliano orogen: Closure of an oceanic domain or intracontinental

convergence between major blocks?, Precambrian Res., 69(1–4), 327–344.

17. Chouet, B., and R. Matoza (2013), A multi-decadal view of seismic methods for detecting precursors of magma movement and eruption, J. Volcanol. Geotherm. Res., 252, 108–175, doi:10.1016/j.jvolgeores.2012.11.013.
18. Daly, E., D. Keir, C. Ebinger, G. Stuart, I. Bastow, and A. Ayele (2008), Crustal tomographic imaging of a transitional continental rift: The Ethiopian rift, Geophys. J. Int., 172(3), 1033–1048, doi:10.1111/j.1365-246X.2007.03682.x.
19. Daly, M., J. Chorowicz, and J. Fairhead (1989), Rift basin evolution in Africa: The influence of reactivated steep basement shear zones, in Inversion Tectonics, vol.44, edited by M.A. Cooper and G.D. Williams, pp. 309–334, Geol. Soc. Spec. Pub., London, doi:10.1144/GSL.SP.1989.044.01.17.
20. Debayle, E., and Y. Ricard (2013), Seismic observations of large-scale deformation at the bottom of fast-moving plates, Earth Planet. Sci. Lett., 376, 165–177, doi:10.1016/j.epsl.2013.06.025.
21. Déruelle, B., I. Ngounouno, and D. Demaiffe (2007), The Cameroon Hot Line (CHL): A unique example of active alkaline intraplate structure in both oceanic and continental lithospheres, Comptes Rendus Geosci., 339(9), 589–600.
22. Djomani, Y., J. Nnange, M. Diament, C. Ebinger, and J. Fairhead (1995), Effective elastic thickness and crustal thickness variations in West Central Africa inferred from gravity data, J. Geophys. Res., 100(B11), 22,047–22,070, doi:10.1029/95JB01149.
23. Dorbath, C., L. Dorbath, J. Fairhead, and G. Stuart (1986), A teleseismic delay time study across the Central African shear zone in the Adamawa region of Cameroon, West Africa, Geophys. J. R. Astr. Soc., 86(3), 751–766.
24. Dugda, M., A. Nyblade, and J. Julià (2007), Thin lithosphere beneath the Ethiopian plateau revealed by a joint inversion of Rayleigh wave group velocities and receiver functions, J. Geophys. Res., 112, B08305, doi:10.1029/2006JB004918.
25. Ebinger, C., and M. Casey (2001), Continental breakup in magmatic provinces: An Ethiopian example, Geology, 29, 527–530.
26. Ebinger, C., and N. Sleep (1998), Cenozoic magmatism throughout East Africa resulting from impact of a single plume, Nature, 395, 788–791.
27. Ebinger, C., A. Ayele, D. Keir, J. Rowland, G. Yirgu, T. Wright, M. Belachew, and I. Hamling (2010), Length and timescales of rift faulting and magma intrusion: The Afar rifting cycle from 2005 to present, Ann. Rev. Earth Planet. Sci., 38, 439–466.
28. Fairhead, J. (1988), Mesozoic plate tectonic reconstructions of the Central South Atlantic Ocean: The role of the West and Central African rift system, Tectonophysics, 155(1–4), 181–191.
29. Fairhead, J., and R. Binks (1991), Differential opening of the Central and South Atlantic oceans and the opening of the West African rift system, Tectonophysics, 187(1–3), 191–203.
30. Fairhead, J., and C. Okereke (1987), A regional gravity study of the West African rift system in Nigeria and Cameroon and its tectonic interpretation, Tectonophysics, 143(1–3), 141–159.
31. Ferguson, D., J. Maclennan, I. Bastow, D. Pyle, S. Jones, D. Keir, J. Blundy, T. Plank, and G. Yirgu (2013), Melting during late-stage rifting in Afar is hot and deep, Nature, 499(7456), 70–73, doi:10.1038/nature12292.
32. Fishwick, S. (2010), Surface wave tomography: Imaging of the lithosphere-asthenosphere boundary beneath Central and Southern Africa?, Lithos, 120(1–2), 63–73.
33. Fishwick, S., and I. Bastow (2011), Towards a better understanding of African Topography, a review of passive-source seismic studies of the African crust and upper mantle, in The Formation and Evolution of Africa: A Synopsis of 3.8 Ga of Earth History, vol.357, edited by D. Van Hinsbergen et al., pp. 343–371, Geol. Soc. Lond. Spec. Pub., London, doi:10.1144/SP357.19.
34. Fitton, J. (1980), The Benue Trough and Cameroon Line—A migrating rift system in West Africa, Earth Planet. Sci. Lett., 51(1), 132–138.
35. Fitton, J. (1987), The Cameroon line, West Africa: A comparison between oceanic and continental alkaline volcanism, in Alkaline Igneous Rocks, vol.30, edited by J.G. Fitton and B.G.J. Upton, pp. 273–291, Geol. Soc. Lond. Spec. Pub., London, U. K.
36. Fitton, J., and H. Dunlop (1985), The Cameroon Line, West Africa, and its bearing on the origin of oceanic and continental Alkali Basalt, Earth Planet. Sci. Lett., 72(1), 23–38.
37. Forte, A., R. Moucha, N. Simmons, S. Grand, and J. Mitrovica (2010), Deep-mantle contributions to the surface dynamics of the North American continent, Tectonophysics, 481, 3–15, doi:10.1016/j.tecto.2009.06.010.
38. Fouch, M., A. Fischer, E. Parmentier, M. Wysession, and T. Clarke (2000), Shear wave splitting, continental keels, and patterns of mantle flow, J. Geophys. Res., 105, 6255–6276, doi:10.1029/1999JB900372.
39. Fouch, M., P. Silver, D. Bell, and J. Lee (2004), Small-scale variations in seismic anisotropy near Kimberley, South Africa, Geophys. J. Int., 157, 764–774, doi:10.1111/j.1365-246X.2004.02234.x.
40. Fourel, L., L. Milelli, C. Jaupart, and A. Limare (2013), Generation of continental rifts, basins, and swells by lithosphere instabilities, J. Geophys. Res. Solid Earth, 118, 3080–3100, doi:10.1002/jgrb.50218.
41. Gallacher, R., and I. Bastow (2012), The development of magmatism along the Cameroon Volcanic Line: Evidence from teleseismic receiver functions, Tectonics, 31, TC3018, doi:10.1029/2011TC003028.
42. Gao, S., P. Davis, H. Liu, P. Slack, A. Rigor, Y. Zorin, V. Mordvinova, V. Kozhevnikov, and N. Logatchev (1997), SKS splitting beneath the continental rift zones, J. Geophys. Res., 102, 22,781–22,797.
43. Gripp, A., and R. Gordon (2002), Young tracks of hotspots and current plate velocities, Geophys. J. Int., 150(2),

321–361, doi:10.1046/j.1365-246X.2002.01627.x.

44. Halliday, A., A. Dickin, A. Fallick, and J. Fitton (1988), Mantle Dynamics: A Nd, Sr, Pb and O Isotopic Study of the Cameroon Line Volcanic Chain, J. Petrol., 29(1), 181–211, doi:10.1093/petrology/29.1.181.

45. Hayward, N., and C. Ebinger (1996), Variations in the along–axis segmentation of the Afar rift system, Tectonics, 15, 244–257.

46. Keir, D., G. Stuart, A. Jackson, and A. Ayele (2006), Local earthquake magnitude scale and seismicity rate for the Ethiopian Rift, Bull. Seismol. Soc. Am., 96(6), 2221–2230.

47. Keir, D., I. Bastow, K. Whaler, E. Daly, D. Cornwell, and S. Hautot (2009), Lower crustal earthquakes near the Ethiopian rift induced by magmatic processes, Geochem. Geophys. Geosyst., 10, Q0AB02, doi:10.1029/2009GC002382.

48. Keir, D., C. Pagli, I. Bastow, A. Ayele, and E. Ababa (2011), The magma-assisted removal of Arabia in Afar: Evidence from dike injection in the Ethiopian rift captured using InSAR and seismicity, Tectonics, 30(2), TC2008, doi:10.1029/2010TC002785.

49. Kendall, J.-M., S. Pilidou, D. Keir, I. Bastow, G. Stuart, and A. Ayele (2006), Mantle upwellings, melt migration and the rifting of Africa: Insights from seismic anisotropy, in The Afar Volcanic Province within the East African Rift System, vol. 259, edited by G. Yirgu, C. J. Ebinger, and P. K. H. Maguire, pp. 271–293, Geol. Soc. Lond. Spec. Pub., London, U. K.

50. Keranen, K., S. Klemperer, J. Julia, J. Lawrence, and A. Nyblade (2009), Low lower crustal velocity across Ethiopia: Is the main Ethiopian Rift a narrow rift in a hot craton?, Geochem. Geophys. Geosyst., 10, Q0AB01, doi:10.1029/2008GC002293.

51. Key, J., R. White, H. Soosalu, and S. Jakobsdóttir (2011), Multiple melt injection along a spreading segment at Askja, Iceland, Geophys. Res. Lett., 38, L05301, doi:10.1029/2010GL046264.

52. King, S. (2007), Hotspots and edge-driven convection, Geology, 35(3), 223, doi:10.1130/G23291A.1.

53. King, S., and D. Anderson (1998), Edge–driven convection, Earth Planet. Sci. Lett., 160, 289–296.

54. Klein, F. (2002), User's Guide to HYPOINVERSE-2000: A Fortran Program to Solve for Earthquake Locations and Magnitudes, pp. 1–123, Open-File Report 2002-171,U.S. Geological Survey, Menlo Park, Calif.

55. Koch, F., D. Wiens, A. Nyblade, P. Shore, R. Tibi, B. Ateba, C. Tabod, and J. Nnange (2012), Upper mantle anisotropy beneath the Cameroon Volcanic Line and Congo Craton from shear wave splitting measurements, Geophys. J. Int., 190, 75–86, doi:10.1111/j.1365-246X.2012.05497.x.

56. Leahy, G., and J. Park (2005), Hunting for oceanic island Moho, Geophys. J. Int., 160(3), 1020–1026.

57. Lee, D., A. Halliday, J. Fitton, and G. Poli (1994), Isotopic variations with distance and time in the volcanic islands of the Cameroon Line: Evidence for a mantle plume origin, Earth Planet. Sci. Lett., 123(1–3), 119–138.

58. Lowry, A., and M. Pérez-Gussinyé (2011), The role of crustal quartz in controlling Cordilleran deformation, Nature, 471(7338), 353–357.

59. Maguire, P., G. R. Keller, S. L. Klemperer, G. D. Mackenzie, K. Keranen, S. Harder, B. O'Reilly, H. Thybo, L. Asfaw, M. A. Khan, and M. Amha (2006), Crustal structure of the northern Main Ethiopian Rift from the EAGLE controlled-source survey; a snapshot of incipient lithospheric break-up, in The Afar Volcanic Province Within the East African Rift System, vol. 259, edited by G. Yirgu, C. J. Ebinger, and P. K. H. Maguire, pp. 269–291, Geol. Soc. Spec. Pub., London, U. K.

60. Marzoli, A., E. Piccirillo, P. Renne, G. Bellieni, M. Iacumin, J. Nyobe, and A. Tongwa (2000), The Cameroon Volcanic Line revisited: Petrogenesis of continental basaltic magmas from lithospheric and asthenospheric mantle sources, J. Petrol., 41(1), 87–109.

61. Meyers, J., B. Rosendahl, G. Harrison, and Z.-D. Ding (1998), Deep–imaging seismic and gravity results from the offshore Cameroon Volcanic Line, and speculation of African hotlines, Tectonophysics, 284, 31–63.

62. Milelli, L., L. Fourel, and C. Jaupart (2012), A lithospheric instability origin for the Cameroon Volcanic Line, Earth Planet. Sci. Lett., 335-336, 80–87, doi:10.1016/j.epsl.2012.04.028.

63. Montelli, R., G. Nolet, F. Dahlen, and G. Masters (2006), A catalogue of deep mantle plumes: New results from finite-frequency tomography, Geochem. Geophys. Geosyst., 7, Q11007, doi:10.1029/2006GC001248.

64. Montigny, R., I. Ngounouno, and B. Deruelle (2004), Âges KAr des roches magmatiques du fossé de Garoua (Cameroun): Leur place dans le cadre de la Ligne du Cameroun, Comptes Rendus Geosci., 336(16), 1463–1471.

65. Morgan, J. (1983), Hotspot tracks and the early rifting of the Atlantic, Tectonophysics, 94(1–4), 123–139.

66. Ngako, V., P. Affaton, J. Nnange, and T. Njanko (2003), Pan-African tectonic evolution in Central and Southern Cameroon: Transpression and transtension during sinistral shear movements, J. Afr. Earth Sci., 36(3), 207–214.

67. Ngako, V., E. Njonfang, F. Aka Tongwa, P. Affaton, and J. Nnange Metuk (2006), The North-South Paleozoic to quaternary trend of alkaline magmatism from Niger-Nigeria to Cameroon: Complex interaction between hotspots and precambrian faults, J. Afr. Earth Sci., 45(3), 241–256.

68. Nzenti, J., B. Kapajika, G. Wörner, and T. Lubala (2006), Synkinematic emplacement of granitoids in a Pan-African shear zone in Central Cameroon, J. Afr. Earth Sci., 45(1), 74–86.

69. Oliveira, E., S. Toteu, M. Araújo, M. Carvalho, R. Nascimento, J. Bueno, N. McNaughton, and G. Basilici (2006), Geologic correlation between the Neoproterozoic Sergipano Belt (NE Brazil) and the Yaoundé Belt (Cameroon, Africa), J. Afr. Earth Sci., 44(4–5), 470–478.

70. Pasyanos, M., and A. Nyblade (2007), A top to bottom lithospheric study of Africa and Arabia, Tectonophysics, 444(1–4), 27–44.

71. Pérez-Gussinyé, M., M. Metois, M. Fernández, J. Vergés, J. Fullea, and A. Lowry (2009), Effective elastic thickness of Africa and its relationship to other proxies for lithospheric structure and surface tectonics, Earth Planet. Sci. Lett., 287(1–2), 152–167.
72. Plomerová, J., and V. Babuska (2010), Long memory of mantle lithosphere fabric—European LAB constrained from seismic anisotropy, Lithos, 120(1), 131–143, doi:10.1016/j.lithos.2010.01.008.
73. Plomerova, J., V. Babuška, C. Dorbath, L. Dorbath, and R. Lillie (1993), Deep lithospheric structure across the Central African shear zone in Cameroon, Geophys. J. Int., 115(2), 381–390.
74. Poudjom Djomani, Y., M. Diament, and Y. Albouy (1992), Mechanical behaviour of the lithosphere beneath the Adamawa Uplift (Cameroon, West Africa) based on gravity data, J. Afr. Earth Sci., 15(1), 81–90.
75. Poudjom Djomani, Y., M. Diament, and M. Wilson (1997), Lithospheric structure across the Adamawa Plateau (Cameroon) from gravity studies, Tectonophysics, 273(3–4), 317–327.
76. Priestley, K., D. McKenzie, E. Debayle, and S. Pilidou (2008), The African upper mantle and its relationship to tectonics and surface geology, Geophys. J. Int., 175(3), 1108–1126.
77. Rankenburg, K., J. Lassiter, and G. Brey (2005), The role of continental crust and lithospheric mantle in the Genesis of Cameroon Volcanic Line Lavas: Constraints from isotopic variations in Lavas and Megacrysts from the Biu and Jos Plateaux, J. Petrol., 46(1), 169–190.
78. Restivo, A., and G. Helffrich (1999), Teleseismic shear wave splitting measurements in noisy environments, Geophys. J. Int., 137, 821–830.
79. Reusch, A., A. Nyblade, D. Wiens, P. Shore, B. Ateba, C. Tabod, and J. Nnange (2010), Upper mantle structure beneath Cameroon from body wave tomography and the origin of the Cameroon Volcanic Line, Geochem. Geophys. Geosyst., 11, Q10W07, doi:10.1029/2010GC003200.
80. Reusch, A., A. Nyblade, R. Tibi, D. Wiens, P. Shore, A. Bekoa, C. Tabod, and J. Nnange (2011), Mantle transition zone thickness beneath Cameroon: Evidence for an upper mantle origin for the Cameroon Volcanic Line, Geophys. J. Int., 187(3), 1146–1150, doi:10.1111/j.1365-246X.2011.05239.x.
81. Richter, C. (1935), An instrumental earthquake magnitude scale, Bull. Seismol. Soc. Am., 25(1), 1–32.
82. Ritsema, J., A. Deuss, H. van Heijst, and J. Woodhouse (2011), S40RTS: A degree-40 shear-velocity model for the mantle from new Rayleigh wave dispersion, teleseismic traveltime and normal-mode splitting function measurements, Geophys. J. Int., 184(3), 1223–1236, doi:10.1111/j.1365-246X.2010.04884.x.
83. Rooney, T., I. Bastow, and D. Keir (2011), Insights into extensional processes during magma assisted rifting: Evidence from aligned scoria cones, J. Volcanol. Geotherm. Res., 83–96, doi:10.1016/j.jvolgeores.2010.07.019.
84. Rooney, T., C. Herzberg, and I. Bastow (2012), Elevated mantle temperature beneath East Africa, Geology, 40, 27–30, doi:10.1130/G32382.1.
85. Silver, P. (1996), Seismic anisotropy beneath the continents, Annu. Rev. Earth Planet. Sci., 24, 385–432.
86. Silver, P., and G. Chan (1991), Shear wave splitting and subcontinental mantle deformation, J. Geophys. Res., 96(B10), 16,429–16,454.
87. Silver, P., and W. Chan (1988), Implications for continental structure and evolution from seismic anisotropy, Nature, 335(6185), 34–39.
88. Silver, P., and M. Savage (1994), The interpretation of shear wave splitting parameters in the presence of two anisotropic layers, Geophys. J. Int., 119, 949–963.
89. Silver, P., S. Gao, and K. Liu (2001), Mantle deformation beneath Southern Africa, Geophys. Res. Lett., 28, 2493–2496.
90. Sleep, N., C. Ebinger, and J.-M. Kendall (2002), Deflection of mantle plume material by cratonic keels, J. Geol. Soc. Lond., 199, 135–150.
91. Steinberger, B., and T. Torsvik (2008), Absolute plate motions and true polar wander in the absence of hotspot tracks, Nature, 452(7187), 620–623, doi:10.1038/nature06824.
92. Stuart, G., J. Fairhead, L. Dorbath, and C. Dorbath (1985), Crustal structure of the Adamawa Plateau Cameroon, Sci. Tech. Rev., 1(1-2), 25–35.
93. Suh, C., S. Ayonghe, R. Sparks, C. Annen, J. Fitton, R. Nana, and A. Luckman (2003), The 1999 and 2000 eruptions of Mount Cameroon: Eruption behaviour and petrochemistry of lava, Bull. Volcanol., 65(4), 267–281.
94. Tadjou, J., R. Nouayou, J. Kamguia, H. Kande, and E. Manguelle-Dicoum (2009), Gravity analysis of the boundary between the Congo Craton and the Pan-African Belt of Cameroon, Austrian J. Earth Sci., 102(1), 71–79.
95. Tarasewicz, J., R. White, A. Woods, B. Brandsdóttir, and M. Gudmundsson (2012), Magma mobilization by downward-propagating decompression of the Eyjafjallajökull volcanic plumbing system, Geophys. Res. Lett., 39, L19309, doi:10.1029/2012GL053518.
96. Tchameni, R., K. Mezger, N. Nsifa, and A. Pouclet (2001), Crustal origin of early proterozoic syenites in the Congo Craton (Ntem Complex), South Cameroon, Lithos, 57(1), 23–42.
97. Teanby, N., J.-M. Kendall, and M. Van der Baan (2004), Automation of shear-wave splitting measurements using cluster analysis, Bull. Seismol. Soc. Am., 94(2), 453–463.
98. Tepp, G., C. Ebinger, M. Ruiz, and M. Belachew (2014), Imaging rapidly deforming ocean island volcanoes in the western Galápagos archipelago, Ecuador, J. Geophys. Res. Solid Earth, 119, 442–463, doi:10.1002/2013JB010227.
99. Thybo, H., and C. Nielsen (2009), Magma-compensated crustal thinning in continental rift zones, Nature, 457, 873–876, doi:10.1038/nature07688.
100. Tokam, A., C. Tabod, A. Nyblade, J. Julià, D. Wiens, and M. Pasyanos (2010), Structure of the crust beneath

Cameroon, West Africa, from the joint inversion of Rayleigh wave group velocities and receiver functions, Geophys. J. Int., 183(2), 1061–1076.

101. Tommasi, A., D. Mainprice, G. Canova, and Y. Chastel (2000), Viscoplastic self-consistent and equilibrium-based modeling of olivine lattice preferred orientations: Implications for the upper mantle seismic anisotropy, J. Geophys. Res., 105(B4), 7893–7908, doi:10.1029/1999JB900411.

102. Tommasi, A., M. Godard, G. Coromina, J. Dautria, and H. Barsczus (2004), Seismic anisotropy and compositionally induced velocity anomalies in the lithosphere above mantle plumes: A petrological and microstructural study of mantle xenoliths from French Polynesia, Earth Planet. Sci. Lett., 227(3), 539–556.

103. Toteu, S., J. Penaye, and Y. Djomani (2004), Geodynamic evolution of the Pan-African Belt in Central Africa with special reference to Cameroon, Can. J. Earth Sci., 41(1), 73–85.

104. Turner, M., M. Reagan, S. Turner, R. Sparks, H. Handley, G. Girard, and C. Suh (2013), Timescales of magma degassing-insights from U-series disequilibria, Mount Cameroon, West Africa, J. Volcanol. Geotherm. Res., 262, 38–46, doi:10.1016/j.jvolgeores.2013.06.003.

105. Ubangoh, R., B. Ateba, S. Ayonghe, and G. Ekodeck (1997), Earthquake swarms of Mt Cameroon, West Africa, J. Afr. Earth Sci., 24(4), 413–424.

106. Van Houten, F. (1983), Sirte Basin, north-central Libya: Cretaceous rifting above a fixed mantle hotspot?, Geology, 11(2), 115–118.

107. Vauchez, A., and A. Nicolas (1991), Mountain building: Strike-parallel motion and mantle anisotropy, Tectonophysics, 185(3), 183–201.

108. Vauchez, A., F. Dineur, and R. Rudnick (2005), Microstructure, texture and seismic anisotropy of the lithospheric mantle: Insights from the Labait volcano xenoliths (Tanzania), Earth Planet. Sci. Lett., 232, 295–314.

109. Vinnik, L., V. Farra, and B. Romanowicz (1989), Azimuthal anisotropy in the Earth from observations of SKS at Geoscope and NARS broadband stations, Bull. Seismol. Soc. Am., 79(5), 1542–1558.

110. Vinnik, L., L. Makeyeva, A. Milev, and A. Usenko (1992), Global patterns of azimuthal anisotropy and deformation in the continental mantle, Geophys. J. Int., 111, 433–447.

111. Wolfe, C., P. Okubo, and P. Shearer (2003), Mantle fault zone beneath Kilauea volcano, Hawaii, Science, 300(5618), 478–480, doi:10.1126/science.1082205.

112. Zhang, S., and S.-I. Karato (1995), Lattice preferred orientation of olivine aggregates deformed in simple shear, Nature, 375, 774–777.

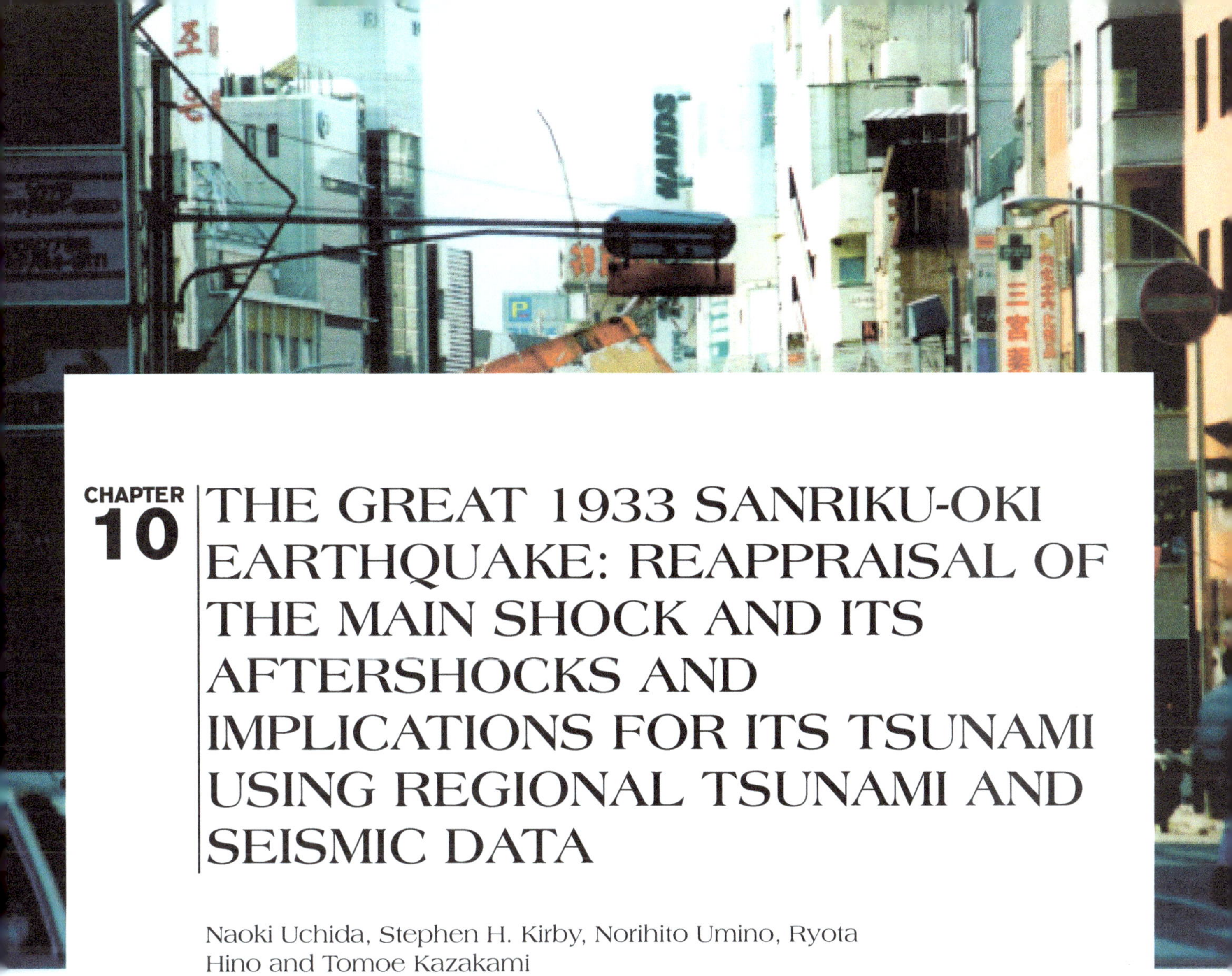

CHAPTER 10 THE GREAT 1933 SANRIKU-OKI EARTHQUAKE: REAPPRAISAL OF THE MAIN SHOCK AND ITS AFTERSHOCKS AND IMPLICATIONS FOR ITS TSUNAMI USING REGIONAL TSUNAMI AND SEISMIC DATA

Naoki Uchida, Stephen H. Kirby, Norihito Umino, Ryota Hino and Tomoe Kazakami

Keywords: Earthquake dynamics, Earthquake source observations, Seismicity and tectonics, Subduction zone processes, Dynamics: seismotectonics, Pacific Ocean

INTRODUCTION

The 1933 Sanriku-oki earthquake offshore northern Honshu, Japan (M_w8.4) is the largest earthquake that has recognized to date in the outer-rise/outer-trench-slope regions of the Earth. The ground shaking at Sanriku coast about 300 km to the west of this source was not so large, but the rupture caused a huge tsunami that led to about 3000 deaths (The Central Meteorological Observatory 1933). The maximum tsunami inundation height was 23 m at Ryouri-Shirahama town along the Sanriku coast (Kunitomi 1933; Fig. 1a). The focal mechanism is considered to be a high-angle normal fault based on the first-motion polarity of seismic waveforms and the hypocentral depth is shallow based on its earthquake location (Kanamori 1971; Okal *et al.* 2016).

(a)

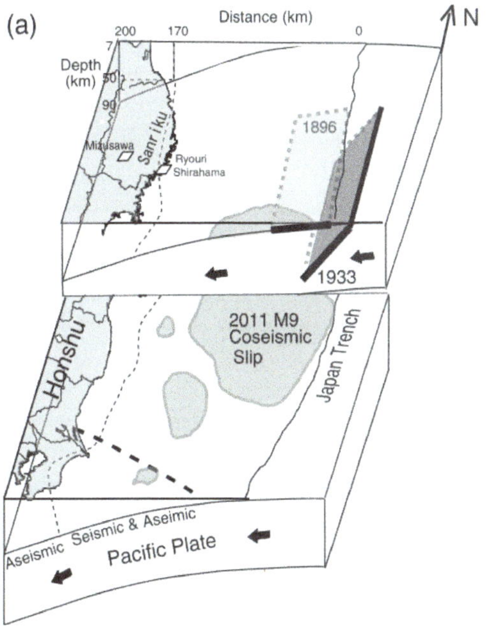

Citation: Naoki Uchida, Stephen H. Kirby, Norihito Umino, Ryota Hino and Tomoe Kazakami, "The great 1933 Sanriku-oki earthquake: reappraisal of the main shock and its aftershocks and implications for its tsunami using regional tsunami and seismic data," doi: 10.1093/gji/ggw234.

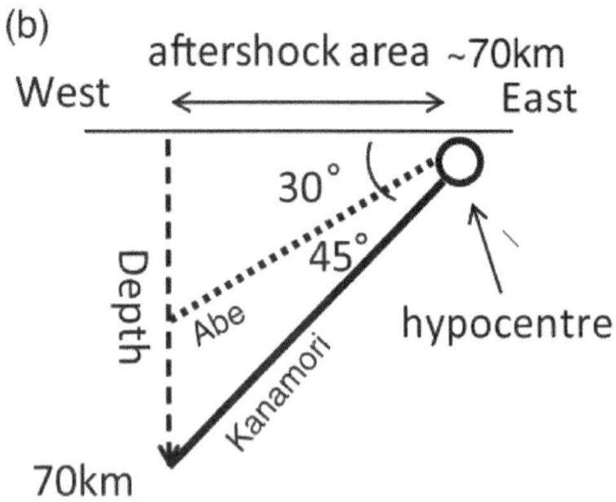

Figure 1. (a) Schematic diagram of the Japan trench subduction system showing the approximate location of the 1933 fault model (Kanamori1971), the 1896 earthquake fault model (Tanioka & Satake 1996) and the 2011 Tohoku-oki earthquake fault model (Iinuma *et al.* 2012). The location of Ryori-Shirahama town and Mizusawa latitude observatory are also shown. (b) East–west cross-section showing the Kanamori (1971), Abe (1978) and Okal *et al.* (2016) models for the 1933 rupture.

Outer-rise/outer-trench-slope faulting is considered a pathway for the water supply and hydration into the deep Earth, because such brittle faults provide a means by which seawater may infiltrate into the mantle part of incoming plate and hydrate it (Peacock 2001; Ranero *et al.* 2003; Lefeldt*et al.* 2009). This hydration process is also important for the source process of deep intraplate earthquakes where subsequent dehydration is a plausible cause of the intermediate-depth earthquake occurrence (Kirby*et al.* 1996; Kita *et al.* 2006; Kawakatsu & Watada 2007; Tsuji *et al.* 2008; Nakajima *et al.* 2009) and a water flux that leads to arc magmatism (Iwamori 2000; Hasegawa & Nakajima 2004). Seismic wave speed reduction and inferred weakening of the incoming plate near the trench is evident in tomographic joint inversions of seismic refraction and wide-angle reflection data (Ivandic *et al.* 2008).

The occurrence of outer-rise/outer-trench-slope earthquakes following large interplate earthquakes is well documented for many such earthquakes (Christensen & Ruff 1988; Wartman *et al.* 2009). The M_w 9.0 2011 Tohoku-oki earthquake that occurred in the same subduction zone as the 1933 event was also followed by M_w 7.6 and many other outer-trench-slope earthquakes (Asano *et al.* 2011; Lay *et al.* 2011; Obana *et al.*2012). Lay *et al.* (2011) argue that the 1933 earthquake followed the 1896 tsunami earthquake that occurred in the adjacent inner trench slope region (Fig. 1a; Tanioka & Satake 1996). However, the reverse case of triggering of interplate earthquakes by outer rise/outer trench slope earthquake is less common. Lay *et al.* (2010) suggest that the 2009 Samoa-Tonga event (M_w 8.1) is also a rare example of an outer trench slope earthquake triggering an interplate faulting subevent (M_w 7.8) that occurred 2 min after the first normal-faulting subevent. However, Beavan*et al.* (2010) suggest for the same earthquake that a slow interplate earthquake occurred first and triggered the outer-trench-slope earthquake. Todd & Lay (2013) showed an example of a large 2011 Kermadic outer-trench-slope intraplate earthquake that triggered interplate earthquakes.

The 3-month aftershock map of the 1933 earthquake from the Japan Meteorological Agency›s (JMA)›s standard earthquake location catalogue shows events that are distributed over a wide area, not only under the outer-trench-slope region but also under the inner-trench-slope region. Thus, it is important to investigate whether these inner-trench-slope aftershocks are due to location errors of their hypocentres or if these events represent seismicity in the megathrust zone that was activated following the 1933 outer-trench-slope normal-faulting earthquake. The recent slip history and mode in this area are also important for understanding the coupling state and stress state of the shallow plate boundary, especially since large interplate thrust slip occurred near the trench during the 2011 Tohoku-oki earthquake (Fujiwara *et al.* 2011; Iinuma *et al.* 2011). Such comparisons are also useful for understanding the relationship of the 1933 events to the tsunami earthquake that occurred in 1896 (Tanioka & Satake 1996) in the adjacent forearc.

THE SOURCE MODELS OF THE GREAT 1933 SANRIKU-OKI EARTHQUAKE

To model observed long-period surface waves and near-field displacement waveforms, Kanamori (1971) constructed a seismic rupture model as a 185 × 100 km seismogenic fault dipping 45° towards N90°W (Fig. 1b) that explains the 1933 main-shock waveforms with a uniform slip of 3.3 m. The fault length (185 km) was based on the one-day aftershock distribution and the dip angle was based on a first motion fault plane solution (45°). The 100-km fault width was based on a 70-km horizontal width of the aftershocks and the fault dip angle (70/cos45° ≈ 100 km) suggesting the fault rupture extended to a maximum depth of about 70 km (Fig. 1b). From an estimated thickness of the oceanic lithosphere of 70 km (Kanamori & Press 1970), he concluded that rupture cut through the entire thickness of the Pacific-Plate lithosphere. Although, some recent studies suggest thicker lithosphere for the study region (80–100 km; Zhao *et al.* 1994; Tonegawa *et al.* 2006; Kawakatsu *et al.* 2009), the depth extent of the rupture is still large fraction of the lithosphere thickness. However, observations of focal mechanisms of recent events (2000–2006) near the 1933 earthquake suggest a stress-state reversal from a shallower level (normal faulting at 10–20 km depth) to deeper level (reverse faulting at 25–40 km depth) (Gamage *et al.* 2009). Therefore, a full lithosphere rupture during the 1933 main shock is unlikely to have occurred given the present-day the stress state.

From the point of view of the initial motion of tsunami waves, Abe (1978) noted that the Kanamori›s (1971) fault model could not explain the tsunami first motions at the Sanriku coast, to the west of the main-shock epicentre (Fig. 1a)—Kanamori›s model predicted a downward first motion, but all of the four Sanriku-coast tide-gauges in the Abe›s paper show an upward tsunami first motions. To fit observed tsunami waveforms, Abe proposed the same west-dipping fault but with a smaller dip angle (30°) than the Kanamori›s model (45°) (Fig. 1b). Aida (1977) suggests from tsunami wave simulation that a model with half the width of Kanamori (1971) fault model better explains the tsunami waveforms. These fault models do not suggest full-lithosphere rupture and, hence there is a disparity between the Kanamori›s rupture model and the tsunami observations. Although outer-trench-slope seismicity in the source area is dominantly of the normal faulting type (Koyama *et al.* 1973) as expected in a plate bending environment, the existence of deeper reverse fault earthquakes (41 km) was found by (Seno & Gonzalez 1987). Gamage *et al.* (2009) found a shallow double-planed seismic zone with a depth separation of 28–32 km in the forearc region including the region beneath the Japan trench. They showed the upper-plane events are characterized by normal faulting and lower-plane events are characterized by reverse faulting. The 2011 March 11 M_w 9.0 Tohoku-oki interplate earthquake triggered a pulse of outer-rise seismicity after the earthquake (Lay *et al.* 2011; Obana *et al.*2012). Obana *et al.* (2012) indicate that tensional stresses extended to depths of as much as 40 km after the Tohoku-oki earthquake, significantly deeper than the depth limit of normal faulting earthquake before the earthquake (≈20 km). The faulting mechanism of the 1933 event is relevant to the rheological properties of the incoming Pacific plate and the stress state near the plate boundary.

In this study, we relocated the aftershocks of the 1933 earthquake using modern methods, including both a regional velocity structure and a double-difference method (Zhang & Thurber 2003). In addition, we employed new observations of the regional waveform characteristics of aftershocks that occurred in the near-trench region to test the distribution of earthquakes. We interpret these hypocentres as well as the main-shock focal mechanism and tsunami waveforms to discuss the depth extent of the main shock and possibility of the activation of multiple faults during and/or after the 1933 earthquake. We also show that triggering of interplate aftershocks by stress transfer from the 1933 main shock best satisfies the aftershocks under the shallow inner trench slope.

DATA AND TEST RELOCATION

We use JMA's phase data for the hypocentre relocations of the 1933 earthquake and its aftershocks. It is important to know accuracy of the data and the stability of the hypocentre location when discussing results from this dataset that has relatively large uncertainties. To estimate the accuracy of the phase data in 1933, we re-picked the earthquake phases by using smoked-paper records obtained at the Mizusawa station where timing was established using on-site astronomical observations at the Mizusawa Astronomical observatory (39.13°N, 141.13°E, about 300 km west of the main shock, Fig. 1). We successfully obtained 27 *S–P* times. The *S–P* time versus P-origin time plot (called Wadati diagram; (Umino *et al.* 2006)) is shown in Fig. 2 together with that for JMA data and the catalogue of Mizusawa Latitude Observatory (The International Latitude Observatory of Mizusawa 1984). Here, we used JMA origin times from the JMA catalogue. In this figure, the data are supposed to be distributed along a line whose slope is determined from the ratio of *P*- to *S*-wave velocity but there are large departures from this line. From visual inspection of the waveforms and the Wadati diagram, we consider there is about a 5 s uncertainty stemming from station clock inaccuracy and the resolution of time on the seismogram as well as drum-speed drift. Next, we performed test hypocentre relocations using several methods to compare the resulting scatter in the locations over characteristics for 3 months after the main shock using a 1-D velocity structure. The JMA catalogue shows an earthquake distribution both to the east and west of the Japan trench and the main-shock hypocentre is located slightly east of the off-trench aftershock distribution (Fig. 3a). When the actual velocity structure varies both with latitude and longitude as well as with depth, and cannot be modelled well by 1-D velocity structure, the use of stations at different ranges from the hypocentre can cause a systematic shift of apparent hypocentre locations. Since the 1933 earthquake is large, there are phase data from distant stations (more than 1500 km away) compared with relatively small aftershocks recorded only at regional stations. Therefore we relocated the earthquakes by restricting the phase data to regional distances (squares in Fig. 3b) to obtain better-resolved relative locations for the aftershocks and main shock. Relocation by using the routine Tohoku university's procedure (Hasegawa *et al.* 1978) using regional stations shows that the main-shock hypocentre is located inside the off-trench aftershock cloud (Fig. 3b). A 1-D velocity structure is also that used for the routine hypocentre determination of Tohoku University. The phase data are from JMA for all test relocations other than global ones that used those from the International Seismological Summary (ISS; Fig. 3e). The *S–P* time relocation method does not rely on the absolute timing of the *P* wave which may suffer from inaccurate timing of the stations. Here we used a grid search method to find the best-fit event locations to the observed *S–P* times. The results show that the main-shock hypocentres shifted slightly south and toward the trench. Aftershocks still ranged both to the east and west of the trench (Fig. 3c). The *P*-only relocation does not suffer from relatively inaccurate *S*-wave picking. However, this relocation result (Fig. 3d) is rather scattered because of poor constraint on the distance from the network on land. Relocation using global teleseismic station data with relatively good station coverage usually does not suffer as much from regional structure heterogeneity (Hino *et al.* 2009) because the seismic rays plunge downward where the subduction-related complexity is thought to be small. The relocation using global data was performed by the interactive iterative method of (Wysession *et al.* 1991). The teleseismic main-shock location is closer to the trench compared with the JMA catalogue; aftershocks to the west of the trench are relocated relatively north (Fig. 3e). The detailed feature and implications are discussed in our companion paper (Okal *et al.* 2016). The relocation by the double difference method (Waldhauser & Ellsworth 2000) improves the relative location between events. The results show the main shock is located within the earthquake cloud in the off trench locations and the earthquake distribution is relatively tight.

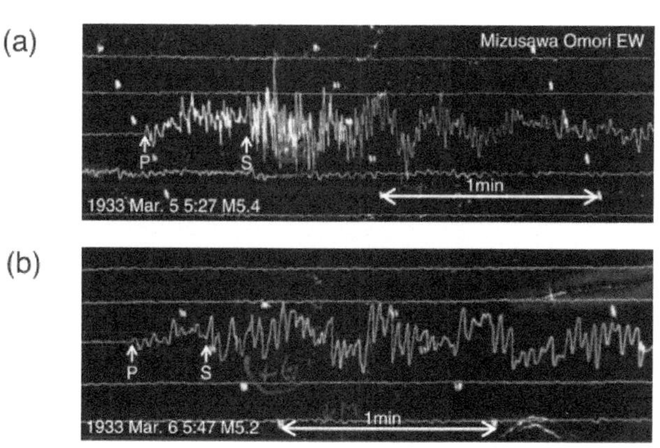

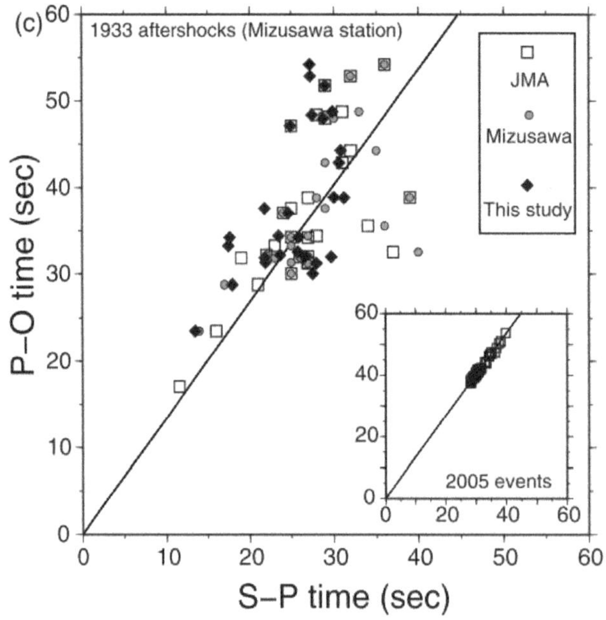

Figure 2. (a,b) Examples of two smoked-paper 1933 aftershock waveforms recorded at the Mizusawa station. The record was written on a 16-s-period E–W Omori type seismometer (45 kg mass weight). The JMA origin times are shown in the left bottom of each trace. (c) The *S*-wave/*P*-wave arrival time difference versus *P*-wave arrival time – origin time at Mizusawa station for the 1933 aftershocks (1933 March 3 to May 24). The origin times are from the JMA catalogue. Squares, circles and diamonds are JMA phase data, the Mizusawa catalogue and our re-picked Mizusawa phase data, respectively. Right bottom inset is the same for the large figure but for the aftershocks of the 2005 M_w 7.0 outer-trench-slope earthquake (2007 May 5–31, Hino *et al.* 2009) recorded at several land stations. The black lines for both plots show a least-squares fit that corresponds to Vp/Vs = 1.75 for the 2005 aftershocks.

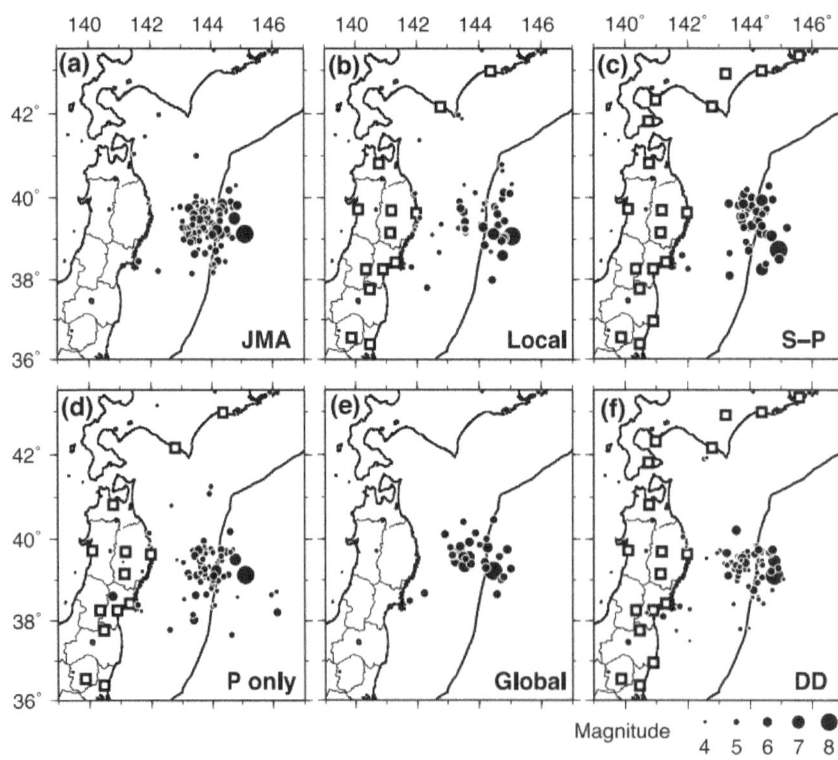

Figure 3. Epicentre distribution of the 1933 main shock and 3-month aftershocks. (a) JMA catalogue. (b) Relocation using phase data only from local stations and Tohoku-University's location routine that uses a 1-D velocity structure (Hasegawa *et al.*1978). Epicentres with 0.5° or less error are selected here. (c) Relocation using a grid search of *S–P* times. (d) Relocation using local stations and Tohoku-University's routine 1-D velocity structure and only *P*waves. (e) Relocation *P*-wave phase data from global stations (Okal *et al.* 2016). (f) Relocation result using the double-difference method. In panels (b)–(d) and (f), the stations used for the relocation are shown by white squares. For all panels, the earthquake magnitudes are from the JMA catalogue.

Using these relocation methods we found that the main-shock epicentre is located inside the aftershock cloud off the trench except for the *P*-wave-only relocations that have relatively weak constraint on the distance from the seismic network (east-west direction). The aftershocks are located both east and west of the trench regardless the relocation method and which body-wave phase data were used (*P*, *P* and *S*, *P–S*).

SYSTEMATIC HYPOCENTRE MISLOCATION NEAR THE TRENCH AND CONSTRUCTION OF A 3-D VELOCITY STRUCTURE OFF SANRIKU

It is generally difficult to determine hypocentre depth accurately for shallow earthquakes occurring in the vicinity of a trench. Recent earthquake relocations by using OBS (ocean bottom seismometer) data near the Japan trench indicate large differences (~20 km) between well constrained OBS hypocentres and the hypocentres in the JMA or the Tohoku University's earthquake catalogue that are determined only from land seismic network data (Hino *et al.* 2000, 2009; Obana *et al.* 2012). These OBS locations provide accurate hypocentre locations in the near-trench region because the stations used for the relocations are located above the earthquake hypocentres and appropriate local velocity structures are used. Fig. 4(a) compares the epicentres from JMA (black dots) and OBS (collared dots) locations of events since 1992 (Hino *et al.*2000, 2009; Obana *et al.* 2012) and Fig. 4(c) shows time residuals for each P phase at land stations by assuming the OBS hypocentre and origin time are correct (see the figure caption for the detail). The OBS locations tend to be west (landward) of apparent JMA›s locations for events to the west of the trench (inner-trench-slope region) and tend to be east (seaward) of apparent JMA›s locations for the events to the east of the trench (outer trench region). The positive observed minus calculated *P*-wave arrival times ($O - C$) for many events to the west of the trench suggest the actual averaged velocity along the path from the events to land station is slower than the 1-D velocity structure and the negative $O - C$ for events to the east of the trench suggest that the averaged velocity along the path from these events to the land stations is faster than the 1-D velocity structure. Therefore, we consider velocity heterogeneity in the offshore region is causing systematic shifts of the earthquake hypocentres determined by the land network using a commonly adopted 1-D velocity structure.

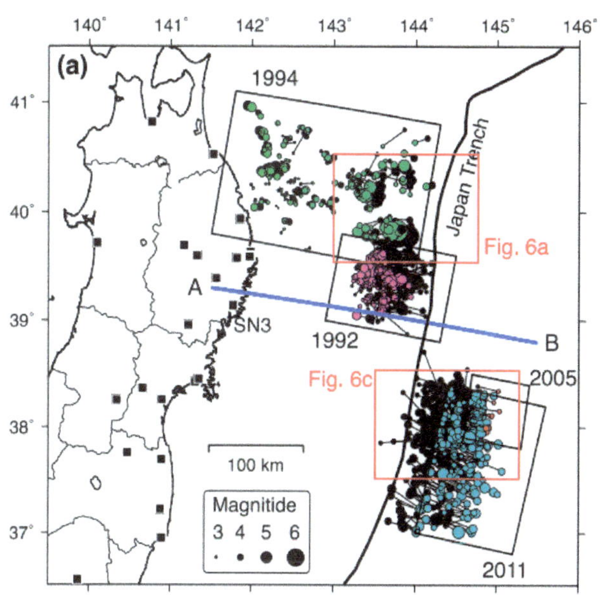

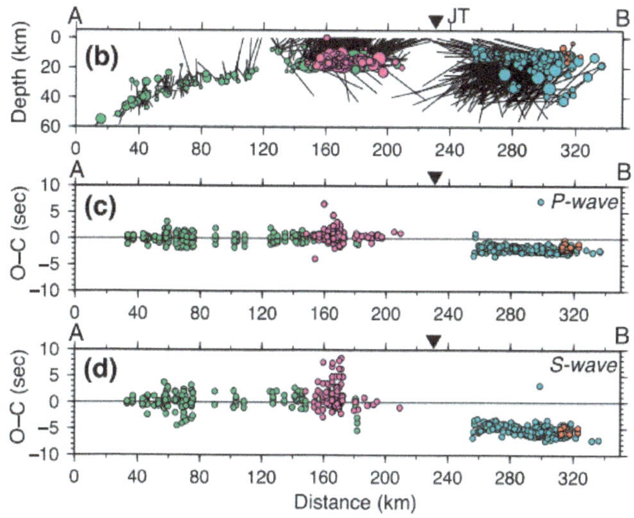

Figure 4. (a) Comparisons of the hypocentres from the JMA catalogue (black circles) and OBS relocations for the aftershocks of the 1992 M6.9 earthquake (Hino *et al.*1996; pink), the 1994 M7.6 Sanriku-oki earthquake (Hino *et al.* 2000; green), the 2005 M7.0 earthquake (Hino*et al.* 2009; red) and the off-trench aftershocks of the 2011 M9.0 Tohoku earthquake and M7.6 outer-trench-slope earthquake (Obana*et al.* 2012). Rectangles show the selected regions where OBS stations were distributed in which the hypocentres are located with good accuracy. (b) Cross-section of hypocentres projected on trench-perpendicular line A–B shown in Fig. 4(a). The colours are the same as in panel (a). Thin lines join the OBS-relocated hypocentres (circle) with the hypocentres from the JMA catalogue. Inverted triangles show the position of the trench along this section. (c,d) Observed (*O*) minus Calculated (theoretical) (*C*) traveltimes as a function of the distance along the line A–B (a) for *P*-wave (c) and *S*-wave (d) readings. We plotted $O - C$ for each phase data set obtained at the on-land stations (squares). For the calculation of (*O*), we used the JMA phase data and the OBS-relocated origin time. For the calculation of (*C*), we used 1-D routine velocity structure of Tohoku University (Hasegawa *et al.* 1978). Colours are the same as in panel (a).

To incorporate this velocity heterogeneity, we construct a 3-D velocity structure and relocated earthquake by the structure. Previous seismic tomography and receiver function studies suggest, for the subducting plate, a faster subducting plate mantle (Tonegawa *et al.* 2006; Hino *et al.*2009; Kawakatsu *et al.* 2009) and a slower offshore crust (Matsuzawa*et al.* 1987; Kawakatsu & Watada 2007; Tsuji *et al.* 2008; Yamamoto *et al.*2008; Hino *et al.* 2009) compared to the average 1-D velocity structure beneath the Japanese island. Structure constraints from refraction and reflection experiments suggest a relatively shallow Moho and a faster mantle for the offshore forearc region (Ito *et al.* 2004; Ito *et al.* 2005). Using these constraints, we constructed a 3-D velocity structure as shown in Fig. 5. In this structure, we used the 3-D depth of the plate boundary (Nakajima & Hasegawa 2006; Kita *et al.* 2010) as a reference. The structure model consists of layers with seismic velocity that depends only the distance from the surface. Note that for the velocity within the subducting plate the distance from the subducting plate surface is used instead of the depth. For the crustal part of the subducting plate (0–15 km from the interplate surface), the velocity determined from the distance from the surface was increased with depth according to observed *P*-wave speed inferred from S-to-P converted waves (Shiina *et al.* 2013).

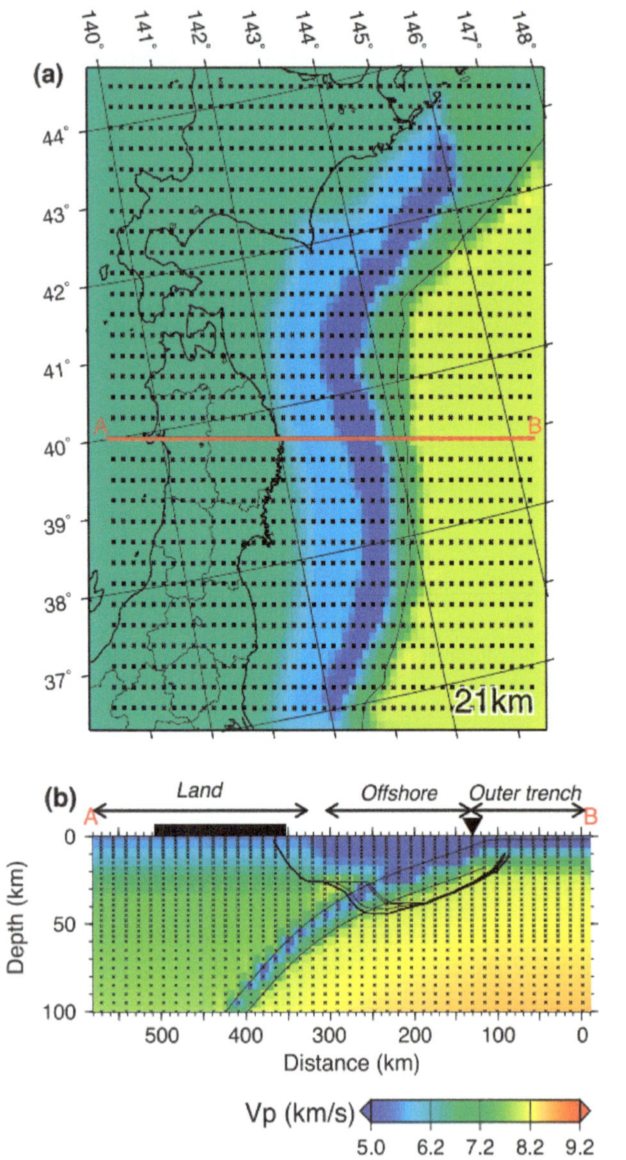

Figure 5. 3-D *P*-wave velocity structure used in this study based on marine and land velocity structure from Tohoku University investigation (Hasegawa*et al.* 1978; Hino *et al.* 2000,2009). Small crosses show grid points. (a) The velocity along a depth section at 21 km. (b) Cross-section along the line A–B shown in (a). Land, offshore and outer trench slope show the areas where each 1-D velocity structure was used. Black curved line indicates the slab top surface. Thick black line and inverted triangle show land area and the Japan Trench, respectively.

To test the relocation procedure using this 3-D velocity structure, we compared our locations of earthquakes with the locations based on OBS observations. Our relocation was performed by using land data only and the tomoFDD hypocentre relocation method (Zhang & Thurber 2003). Both earthquakes in the outer trench/outer rise region and inner trench region relocated closer to the OBS hypocentres (Fig. 6). The depths of the relocated earthquakes were greatly improved greatly by this procedure (Figs 6b and d). These results suggest that the use of a 3-D velocity structure greatly improves the absolute locations of these offshore earthquakes.

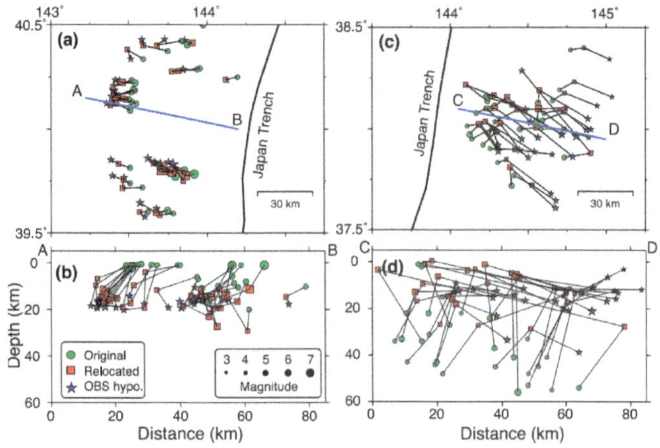

Figure 6. Comparisons of the relocated hypocentres using the 3-D velocity structure (red squares), OBS data (blue stars) and data from the original JMA catalogue (green circles). The hypocentres for the same event are connected by thin lines. Map view (a) and (b) cross-section along the line A–B of selected events under the inner trench slope. Earthquakes with M3.5 or larger that occurred after the 1994 M7.6 interplate earthquake are shown here. Map view (c) and (d) cross-section along the line C–D of selected off-trench earthquakes. Earthquakes that occurred after the 2005 M7.0 and 2011 M9.0 earthquakes are shown here. These two areas are also shown in Fig. 4(a).

1933 HYPOCENTRE RELOCATIONS USING THE 3-D VELOCITY STRUCTURE

Fig. 7 show the 1933 main-shock and aftershock distribution before and after relocation using the foregoing procedure. Before relocation, the hypocentres are diffusely scattered, especially in depth and the main-shock hypocentre is located outside of the dense earthquake cluster. The relocation result shows that the earthquake distribution for both in the inner- and outer-trench-slope regions is largely shallow (depth ≤60 km below the sea level). Note that we adopted the hypocentre of main shock from global station data due to insufficient number of local phase data. The cross-section in the outer trench region shows that the aftershocks did not occur on a single plane. Red symbols show the events in the outer-trench area, which does not show simple fault plane. These earthquakes appear to be separated from the earthquakes in the offshore forearc region (black) by a relatively sparse seismicity region (Fig. 7c). To test this hypocentre distribution, we checked the waveforms for these events at Mizusawa station (Fig. 8a). The colour in Fig. 8 is coded according to the dominant frequency in the first 40 s window of the waveform. We determined the dominant frequency from the smoked-paper records by counting the number of peaks in the event time window. The distribution shows that high-frequency events are mostly located east of the trench and low frequency events tend to be located west of the trench region. Gamage *et al.* (2009) showed that the waveform characteristics are different for the hypocentres of recent events east and west of the trench. High frequency waveforms characterize earthquakes east of the Japan Trench and low-frequency waveforms characterize those earthquakes that locate west of the trench. The relationship applies very clearly to the dominant frequency of the 1933 aftershock and suggests that our locations relative to the trench position are accurate. Many aftershocks do indeed occur west of the trench. The activity of earthquakes show that the earthquakes to the west of the trench started to occur within 2 hours after the main shock (Figs 9aand b), suggesting the triggered seismicity started very early stage of the aftershock activities.

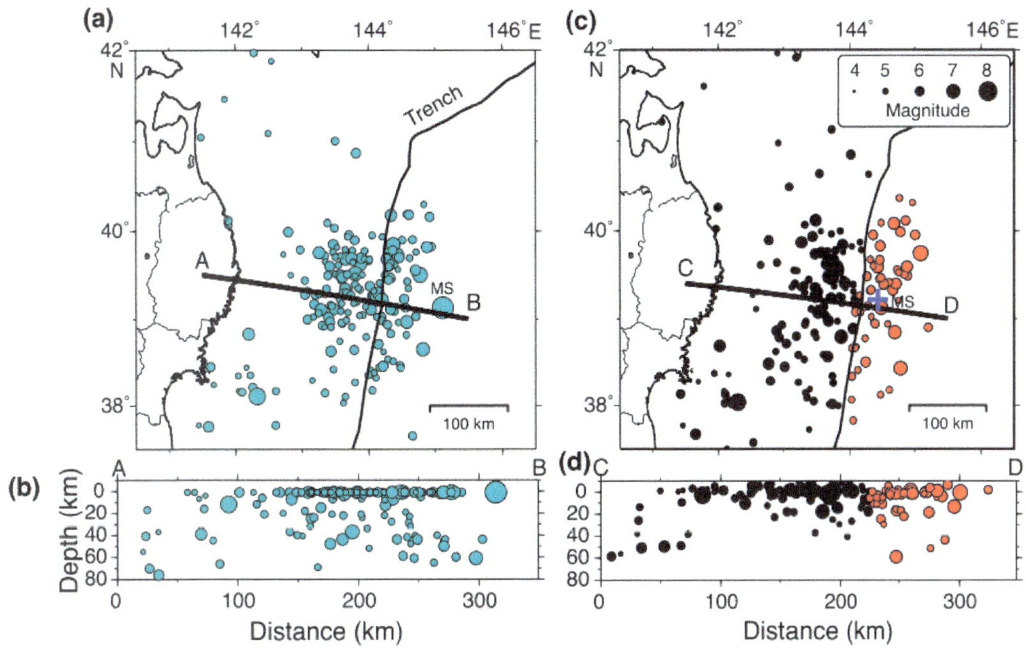

Figure 7. Hypocentre distribution of the 1933 main shock and its aftershocks. (a,b) JMA catalogue; (c,d) the results from our relocation by the 3-D velocity structure. Relocated epicentre of the main shock (cross) is from global stations. For panels (c) and (d), the red and black circles show hypocentres of outer trench and inner trench earthquakes, respectively.

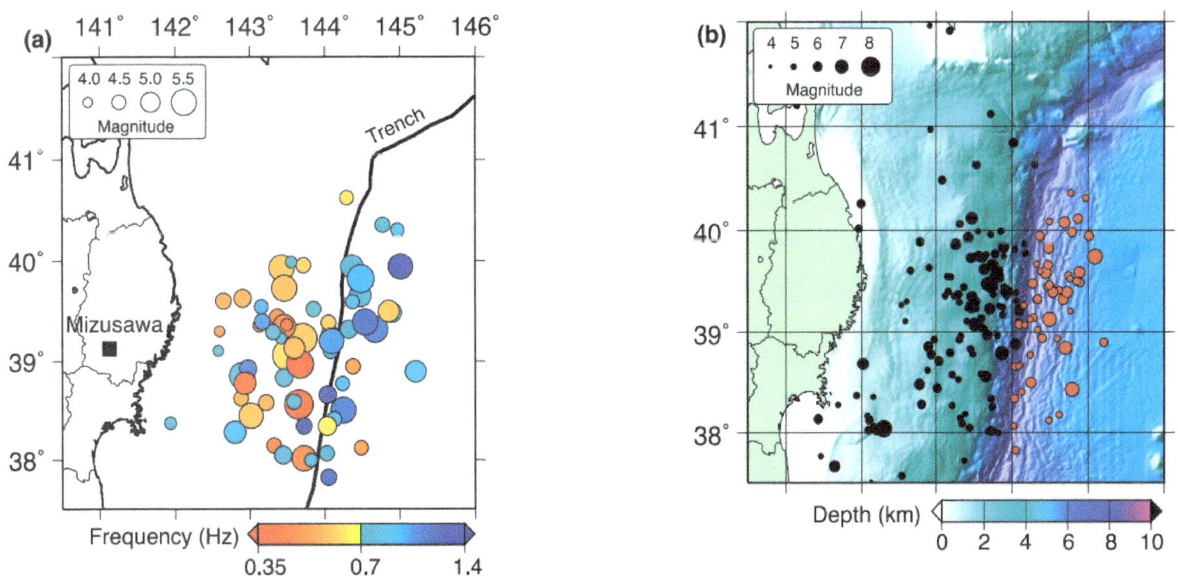

Figure 8. (a) Dominant frequency (colour) plotted at the location of the hypocentres. The black square shows the location of the seismic station (Mizusawa) where the dominant frequency was estimated from the waveforms recorded there. (b) Three months relocated aftershocks plotted on the bathymetry. Colours for the hypocentres are the same as in Fig. 7(c).

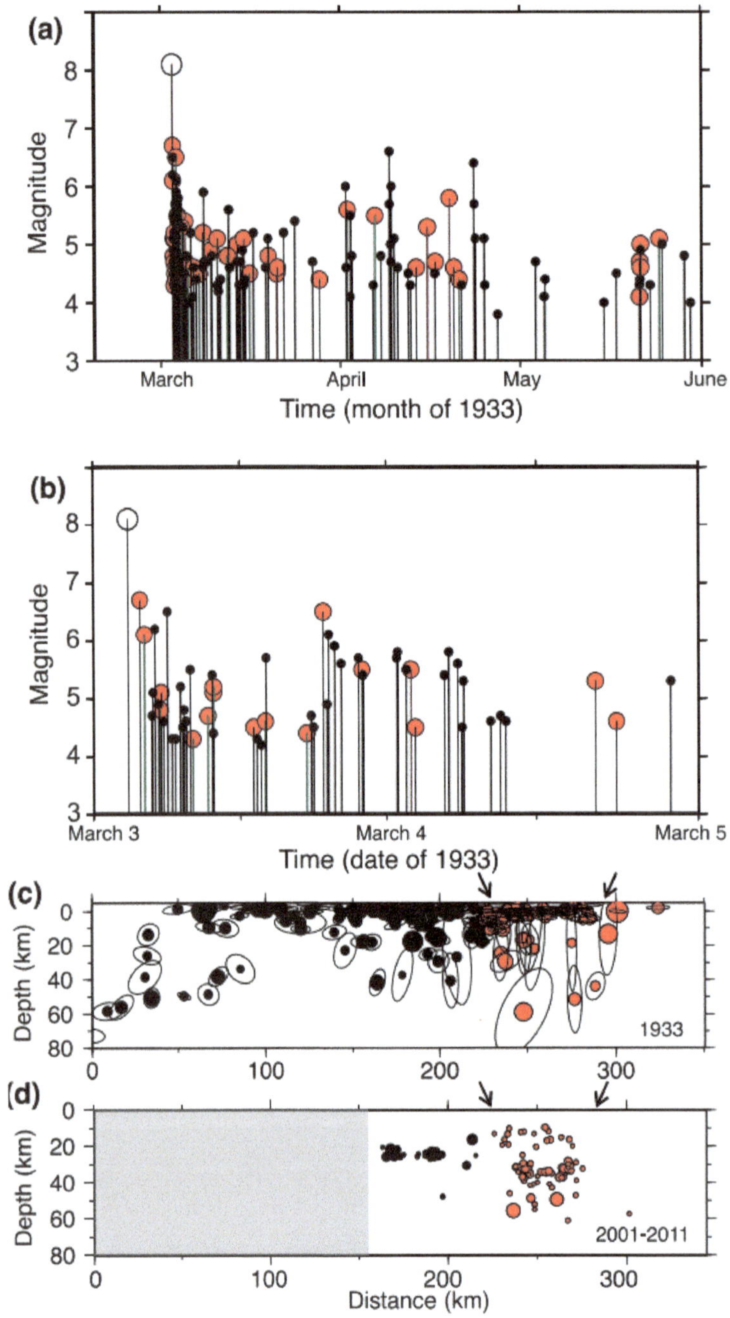

Figure 9. (a,b) Magnitude–time plot of relocated earthquakes. Colours are the same as in Figs 7(c,d) and 9(c). (c) Relocated earthquake locations with error ellipsoid in the same cross-section in Fig. 7(d). The error ellipsoid was estimated from 98 times relocations with bootstrap resampled residuals. Relocated earthquake locations for recent earthquakes from October 2001 to 2011 March 10. The location of the cross-section is the same as Fig. 9(c) and the band width is 20 km.

DISCUSSION

Triggering of the Inner Trench Earthquakes

We found many earthquakes occurred beneath the inner-trench-slope region after the 1933 earthquake. To investigate the cause of these earthquakes in the megathrust zone, we estimated Coulomb stress change for interplate thrust earthquakes due to a giant off-trench normal faulting earthquake. Although the earthquake alignment in the cross-section is ambiguous due to small number of earthquakes and their scatter nearest the Japan trench, we take high angle (60° dip) rupture planes as working hypotheses. The bathymetry of the outer trench slope in this region shows both east-dipping and west-dipping high-angle fault scarps (Kobayashi*et al.* 1998) and Okal *et al.* (2016) estimated normal faulting earthquake 61° dip towards the trench. Low-angle normal fault plane (30° dip) that was suggested by Abe (1978) is mechanically very unfavourable and inconsistent with *P*-wave first motions for the 1933 earthquake. Considering wide distribution of trench-normal seismicity, single high dip-angle fault within shallow depth (depth ≤60 km) cannot cover all the epicentre in the outer-trench area. Therefore, we consider compound rupture of two normal faults facing with each other in addition to single west-dipping normal. The compound rupture still satisfies the first motion observation of (Kanamori 1971) but with a steeper land-facing fault that is suggested by Okal *et al.* (2016). The compound rupture model have a 60 degree dip, a 58 km down-dip width and a 278 km fault plane length whose depth range extends from 10 to 60 km from sea level or 4 to 54 km below the seafloor, and conventional single fault model have the same dimensions (Fig. 10, rectangles). The average slip amounts are 4.5 and 9 m for the compound model and single model, respectively. The total moment is 4.3×10^{21} Nm for both models which is the same as Kanamori›s model (Kanamori 1971) but smaller than more recent result [7×10^{21} Nm (Okal *et al.* 2016)]. Although there are difference in the moment, they does not affect to the spatial pattern and polarity of the Coulomb stress change. This calculation was made using an elastic half-space following Okada (1992), assuming a shear modulus of 3.0×10^{10}N m^{-2}, a Poisson's ratio of 0.25, and an apparent coefficient of friction of 0.3. The Coulomb stress changes for thrust earthquakes on the plate boundary (Nakajima & Hasegawa 2006; Kita *et al.* 2010) show that events under the inner trench slope within 27 d after the 1933 earthquakes are mostly located in the area having positive stress changes from the compound model (Fig. 10a). There are large areas with Coulomb stress greater than the ~0.01 MPa, values which is the band of stress change that promotes seismicity in other tectonic settings (Stein *et al.* 1992; Harris *et al.* 1995; Toda *et al.* 1998). The single fault model also show positive Coulomb stress change near the trench where high aftershock seismicity occurred, but the contours with positive stress change greater than ~0.01 MPa do not contain all of the relocated aftershocks. Therefore, the spatial extent of aftershock is more consistent with compound rupture and suggests that the seismicity in the megathrust zone was probably triggered by the compound outer-trench-slope normal faulting. We now discuss in the next section these fault models in the context of the tsunami records.

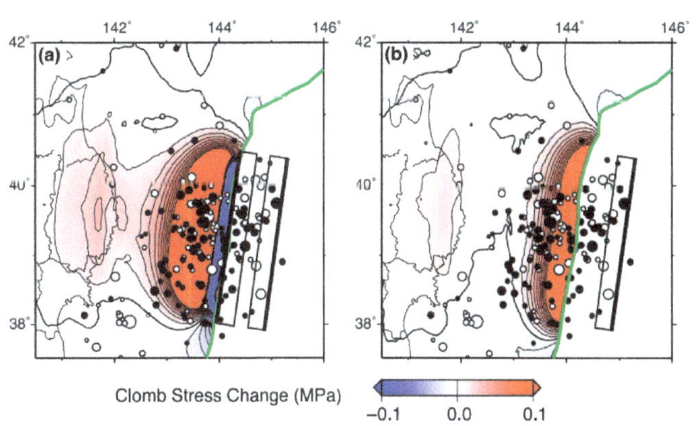

Figure 10. Changes in Coulomb failure stress on the plate boundary (colour-scale and contours) caused by two 60° normal faults shown by rectangles. The thick line is shallower edge of the faults. Panel (a) is for compound rupture of two normal faults dipping toward each other and panel (b) is for single west-dipping fault. The contour interval is 0.01 MPa; contours larger than 0.1 MPa are not drawn. For the calculation of the Coulomb failure stress change, the slip direction of the target earthquake is assumed to be parallel to the direction of relative plate motion. Black and white circles are relocated earthquakes that occurred from 1933 March 3 to 31 and 1933 April 1 to August 28, respectively.

Source Faults and the Tsunami Paradox

To test the possibility of multiple faults in the shallow outer-trench-slope area, we calculated tsunami waveforms by solving the finite difference approximation of the linear long-wave equations (Satake 1995) for the compound-rupture model and single-fault model defined in the previous section.

Fig. 11 shows the bathymetry, fault models and tide-gauge observation points where we could obtain the near-field tide gauge records and its locations in 1933. The compound-rupture model and single-fault model corresponds to the rupture of faults A + B and fault B, respectively. The tide gauge records at Kushiro (KSR), Hachinohe (HCH) and Kesen-numa (KSN) (Sekiguchi & Nakano 1933) are digitized and corrected for curvature and inclination due to finite arm length. Then the tidal components are estimated by a moving window analysis and the components are removed from the tide gauge record to compare with the Tsunami modelling results. The observed tsunami waveform show small but clear upward first motions for HCH and KSN, located to the west of the source faults (Figs 12b and c). The first motion polarity at Kushiro, located to the north of the source faults is unclear due to relatively large noise to signal (Fig. 12a).

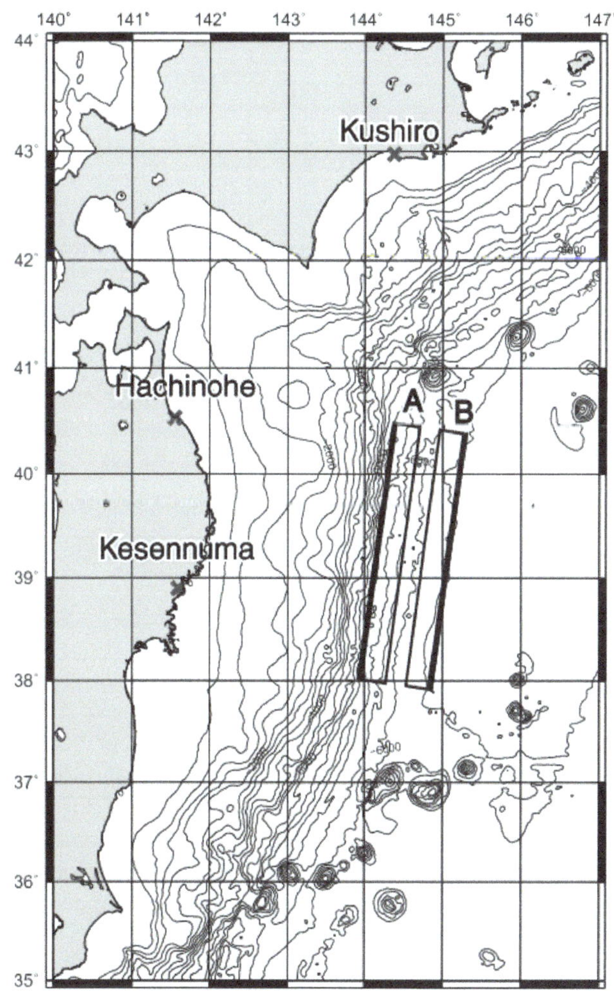

Figure 11. Distribution of tide gauge stations (cross), fault models (rectangles A and B) and bathymetry used for the theoretical tsunami calculations (contours).

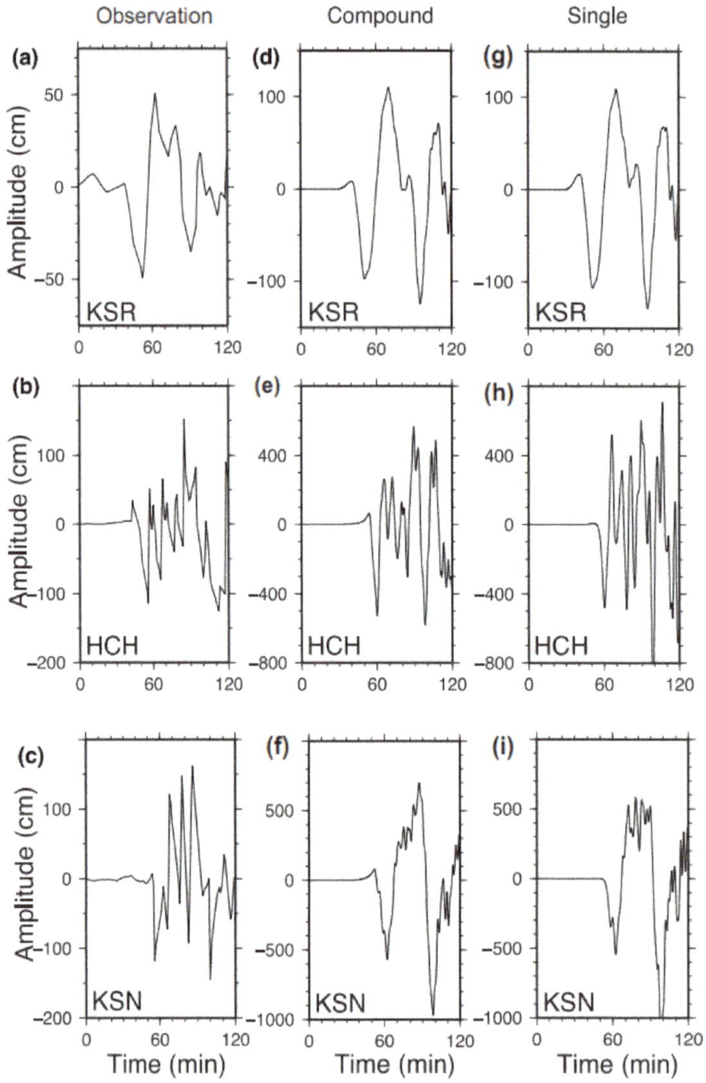

Figure 12. Observed and calculated tsunami waveforms for Kushiro (KSR), Hachinohe (HCH) and Kesen-numa (KSN) tide gauge stations. (a–c) Observed waveforms after the correction for curvature and inclination due to the finite arm length and subtracting the tidal components. (d–f) Theoretical waveform from a compound rupture model. (g–i) Theoretical waveform for a single fault model.

The tsunami calculation results for these stations show that the compound-rupture model can explain the upward first-motion polarity for HCH and KSN (Figs 12e and f). The overall features of waveforms at about 1 hour from the onset of Tsunami waves are also very similar except for their amplitudes for the compound-rupture model. On the other hand, the single-fault model cannot explain the tsunami first motions at HCH and KSN. The east dipping fault (Fault A) in the compound model closer to the Sanriku coast contributes an upward first motion of the tsunami waves that is a robust observation in the tide-gauge waveforms (Abe 1978). The tsunami source model of (Hatori 1974) from back propagation of the tsunami suggest that western limit of the tsunami source is located not only in outer-trench-slope region but extends about 50 km west of the Japan trench. Thus, this also suggests that a single west-dipping fault (Fault b) cannot explain the tsunami waveforms well unless the fault is very close to the Japan trench or the fault dip is small.

Despite the similarity of the shape of the modelled Tsunami waveform with observation, the calculated amplitudes for the compound model are about 3–6 times larger than observations. The possibilities for the causing this discrepancy includes an assumption of too much slip (moment) for the earthquake, limitations of the frequency-responses of the tide gauges for short wavelength waves, or inadequate modelling of local bathymetry. Because this earthquake occurred early in the instrumental record, precise information on these limitations are not available. However we consider these cases do not affect the polarity of the initial tsunami wave that is a reliable feature for both observation and modelling by compound rupture model.

In Tohoku area, the occurrence of possible future off-trench earthquake triggered by M9 Tohoku-oki earthquake of 2011 is of concern. The automatic detection of compound rupture or doublet is still difficult (Beavan *et al.* 2010; Lay *et al.* 2010). If a subevent of a great earthquake closer to the land-area occurred simultaneously with the initial subevent further from the trench, it would cause earlier tsunami arrivals and in some cases larger tsunami waves on the coast than a single off-trench earthquake, an important scenario that we should keep in mind.

Depth Limit of the Rupture

Although the phase data for our relocations has an approximate 5 s standard error, the estimation of hypocentre uncertainty from a bootstrap method suggests that this shallow earthquake distribution is robust (Fig. 9c). 90 per cent of the aftershocks are located shallower than 18 km The stress distribution in the oceanic lithosphere entering the deep sea trench is a superposition of bending forces, gravitational pull of cold sinking slab, and interplate locking along the megathrust boundary (Christensen & Ruff 1988). When considering the gravitational pull of cold sinking slab, a slab at the 50–200 km depth range in this subduction zone is not under deviatoric tension (Hasegawa *et al.* 1978; Yoshii 1979; Kawakatsu & Seno 1983; Kita *et al.* 2010). Bending produces a stress state with shallow extension and deeper compression. It seems unlikely that interplate locking acts as the cause of near-trench deviatoric tension because a locked interplate fault would tend to shelter the up-dip region from the slab-pull (Lay *et al.* 2010). In our study we found few reliable aftershocks (depth ≧50 km). Although rupturing without producing aftershocks may possible, the aftershock distribution does not support full-plate offshore Pacific plate rupture. The aftershock distribution that we document suggests the whole lithosphere was not under deviatoric tension at the time of the 1933 earthquake.

Comparison with Recent Seismicity and Bathymetry

We have compared the aftershock distribution of the 1933 earthquake with recent seismicity from October 2001 to 2011 March 10 that was relocated using the same procedure as we used for the aftershocks of the 1933 earthquake (Figs 13a and b). Here we selected the time period when the High-Sensitivity Seismograph Network (Hi-net) phase data are available and the seismicity is not affected by the 2011 Tohoku-oki earthquake to obtain accurate and stable seismicity.

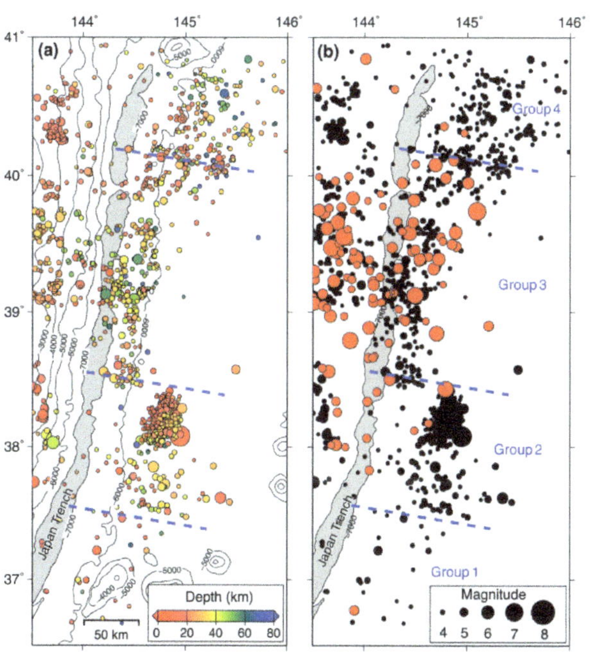

Figure 13. Comparison of the relocated 1933 aftershock epicentres and recent seismicity. (a) Epicentre distribution of recent earthquakes from October 2001 to 2011 March 10 (circle). The circles are colour coded according to hypocentral depth. (b) The comparison of the epicentre distribution of recent earthquakes (black circles) and the 6-month aftershocks of the 1933 earthquake (red circles). Grey area shows the area with seafloor depths ≥7000 m defining the position of the Japan Trench. Contours in panel (a) represent seafloor depths in 1000 m intervals. Note that the recent Group 2 off-trench events are located in the same band of seismicity as the 1933 aftershocks. Note the sharp boundaries between Group 3 and Groups 2 and 4.

We subdivided and classified the off-trench region of the Japan Trench into numbered Group Sectors based on the morphologies and distributions of fault scarps in relation to the local orientation of the Japan Trench and orientations of magnetic anomalies that indicate the seafloor spreading fabric. Such structural fabric can be reactivated in the zone of deformation near trenches (Masson 1991; Kobayashi *et al.* 1998). To do this, we relied upon the paper by Masao Nakanishi (2011) who mapped the off-trench normal fault scarps seaward of the Japan and Kuril Trenches, their spacings, the directions that these scarps dip, and their fault throws along selected profiles. Nakanishi›s work was based on swath mapping data collected from various sources. Kobayashi *et al.* (1998) also noted the abrupt change in fault-scarp orientations, spacing, and average scarp throws between Group Sectors 3 and 4 with the Group Sector 4 scarps being attributed to reactivation of Pacific-Plate brittle weakness parallel to the seafloor spreading fabric. We point out that the Group 3 Sector is where most of the off-trench 1933 aftershocks occurred and where narrow near-trench bands of present-day earthquakes have also occurred near the trench. The ~trench-parallel horst and graben structure in this Sector 3 indicates that both ocean-ward and trench-ward dipping faulting occur in this sector and that maximum fault throws approach 500 m near the Japan Trench.

Recent seismicity is largely shallow (depth ≤60 km below the sea level) and also consistent these four sector groupings of earthquakes along the trench based on fault scarp morphology (Fig. 13a). The 6-month aftershocks of the 1933 earthquake (red circles in Fig. 13b) are mainly distributed in Group 3 where the recent earthquakes are relatively concentrated near the trench (Fig. 13a, black circles in Fig. 13b). In this region there are several trench-parallel alignments of seismicity. This linear trend at shallow depths was also reported using OBS locations in 1981 by Kasahara *et al.* (1982). As noted above the bathymetry in this region shows largely trench parallel horst-and-graben fault scarps (Fig. 8b) a structure that is in consistent with our 1933 earthquake›s fault models. Kobayashi *et al.* (1998) suggests that the crust is weakest along the inherited spreading fabric, second weakest probably along the non-transform offset direction and strongest in directions very oblique to these orientations. The fault scarps and trench-parallel bands of earthquakes crosscut the spreading fabric at angles as high as 58° in the latitude range for the 1933 aftershocks indicating that the 1933 rupture and subsequent seismogenesis has occurred on faults that were newly created in the off-trench region with total normal fault throws approaching 500 m near the trench axis (Nakanishi 2011). We have investigated the fine structure of recent earthquakes in the Group 3 Sector and show a cross section of the hypocentres of these events that includes the hypocentre of the 1933 main shock (Fig. 9d). This cross section resembles the cross section of our 1933 main shock (Fig. 9c), but with many more events. We recognize multiple fault-like alignments evident in both the epicentres of off-trench earthquakes in trench parallel bands (Fig. 13a) and in cross sections of hypocentres that were relocated using the same procedure that we used in relocating the 1933 aftershocks. In particular we point out multiple alignments of modern hypocentres into steeply dipping fault-like structures, including one such alignment is closest to the trench that dips seaward. Obana *et al.* (2016 and personal communications in May 2016) also reported two such trench-parallel bands of seismicity and a seaward-dipping structure in preliminary locations of off-trench earthquakes detected by a two-month OBS deployment in late summer of 2015. He and his colleagues interpreted these fault-like alignments, as we do, are possibly continuing aftershocks of the great 1933 Sanriku-oki earthquake more than 80 years after the earthquake and disastrous tsunami (Obana*et al.* 2016). The foregoing interpretation of a two-fault compound source with ruptures dipping toward each other for the 1933 main shock and aftershocks and present-day off-trench earthquakes is obviously of a simplification of a more complex distributed extensional regime in the outer rise and inner trench slope of the 1933 source region. Until we have long-term deep-water OBS deployments in the off-trench region of the Group 3 Sector and detailed seismic reflection and coring in that area to test this model, it should be considered provisional.

That such aftershocks could still be occurring is remarkable and is consistent with long-duration aftershocks of large intraplate earthquakes in the distributed deformation in continental interiors where average loading rates are very low (Stein & Liu 2009). Their paper gives insights into long-duration aftershocks using rate-and-state friction theory by equating low stressing rates with low average slip rates in such regions. Kirby (2016, unpublished data) has estimated average slip rates of about 0.5 mm yr^{-1} in the Sanriku source region based on the evolution of scarp heights during the time interval from the Pacific Plate moving over the outer rise to the trench at the latitude of the 1933 main shock at 39.2°. The Stein and Liu model indicated aftershock durations of centuries for great earthquakes for the low average slip rates in our setting. A slow rate of decay of seismicity rate of 1933 aftershocks is also consistent with nearly constant seismicity rates evident after just a couple of months in 1933 (Figs 13a and b). In our view, the occurrence of fault-like alignments of modern off-trench earthquake hypocentres re-enforces our interpretation of very sparse 1933 aftershock hypocentres that both seaward-dipping and oceanward-dipping faulting to depths exceeding 50 km occurred during the 1933 main-shock rupture. The relocated epicentre and focal mechanism based on first motions and waveform modelling (Okal *et al.* 2016) indicate that the initial main-shock rupture was probably trench-ward dipping at about 60° and located in the band of off-trench earthquakes most seaward of the trench, but that its moment tensor was also consistent with the total moment release from a compound rupture. As we discussed earlier in this paper, the upward first motion of the tide-gage records of the tsunami waves also point to an ocean-ward-dipping rupture nearest the trench and Japan coastline as a part of this compound earthquake and tsunami source.

CONCLUSIONS

Land-station based hypocentre determinations by using 3D velocity structure was applied in this paper to the off-Sanriku near-trench earthquakes where systematic hypocentre mislocations are recognized in the previous studies. The improvement of hypocentre locations were confirmed by an examination of recent earthquakes that are more accurately located based on OBS data above the study area. Aftershocks that occurred just after the 1933 Sanriku-oki main shock are located in a roughly 280-km-long zone in the off-trench region that is separated from the aftershock seismicity under the marine forearc trench slope. The off-trench earthquakes are shallow (depth ≦50 km) and at a maximum depth below the seafloor of about 43 km. There are west dipping and east dipping earthquake alignments in the trench-normal cross-section. Although the aftershock depth distribution is not closely constrained, the shallow aftershock distribution suggests a shallow rupture depth range and possibly a compound rupture for the 1933 main shock. We found the compound rupture model explains better the polarity of tsunami waves at the Sanriku coast than a single west-dipping normal fault and better fits the Coulomb stress change in relation to aftershock locations the west of the trench. The occurrence of aftershocks both east and west of the trench was confirmed by an investigation of waveform frequency differences between these two classes of shallow earthquakes which is also seen in more recent earthquakes with better location accuracies. The relocated earthquakes under the offshore forearc are shallow (depth ≦30 km) and occur where recent activity of interplate thrust earthquakes is high. This suggests the deformation of the 1933 outer-rise earthquake triggered interplate earthquakes by a stress transfer process. Recent (2001–2012) seismicity in the 1933 of-trench source area relocated by the same method shows that the present seismicity in the off-trench region of northern Honshu can be divided into several groups of earthquakes along the trench; one group roughly corresponds to the locations of the off-trench aftershocks of the 1933 earthquake. Comparison of the 1933 rupture dimensions based on our relocations with the morphologies of fault scarps in the outer trench slope suggest that the rupture was largely limited to the region where fault scarps are trench parallel and cross cut seafloor spreading fabric. These suggest bending and along-strike structural segmentation largely controls the horizontal and vertical extent of the fault. The triggering of interplate seismicity suggests that the great outer-rise/outer-trench-slope earthquake environment is closely linked to the earthquake cycle of interplate earthquake. The depth extent for this largest outer rise/outer-trench-slope earthquake also provides a constraint on the stress state of the subducting plate and the water transported to deep Earth into the subducting slab.

ACKNOWLEDGMENTS

We thank the International Latitude Observatory of Mizusawa, Yoshiaki Tamura and Toshio Kono for smoked-paper seismogram images of the 1933 aftershocks, and the Japan Meteorological Agency for earthquake phase data. We also thank E. Okal for the discussion on the hypocentre relocation and providing hypocentre relocation results by global station data, K. Obana for providing us the hypocentres of off-trench earthquakes based on OBS observations, S. Kita for providing programs for constructing offshore velocity structure and earthquake relocations, M. Okada for the locations of tide gauge stations in 1933, Y. Tanioka for the assistance in digitizing the tsunami waves, and W. Ellsworth, T. Matsuzawa,

A. Hasegawa and H. Kanamori for discussion. Some of the figures were drawn using Generic Mapping Tools software (Wessel & Smith1995).

REFERENCES

1. Abe K. A dislocation model of the 1933 Sanriku earthquake consistent with the tsunami waves. J. Phys. Earth 1978;26:381-396.
2. Aida I. Simulations of large tsunamis occurring in the past off the coast of the Sanriku district. Bull. Earthq. Res. Inst. Univ. Tokyo 1977;52:71-101.
3. Asano Y. Spatial distribution and focal mechanisms of aftershocks of the 2011 off the Pacific coast of Tohoku Earthquake. Earth Planets Space 2011;63:669-673.
4. Beavan J., Wang X., Holden C., Wilson K., Power W., Prasetya G., Bevis M., Kautoke R. Near-simultaneous great earthquakes at Tongan megathrust and outer rise in September 2009. Nature 2010;466:959-963.
5. Christensen D.H., Ruff L.J. Seismic coupling and outer rise earthquakes. J. geophys. Res. 1988;93:13 421–13 444.
6. Fujiwara T., Kodaira S., No T., Kaiho Y., Takahashi N., Kaneda Y. The 2011 Tohoku-Oki earthquake: displacement reaching the trench axis. Science 2011;334:1240.
7. Gamage S.S.N., Umino N., Hasegawa A., Kirby S.H. Offshore double-planed shallow seismic zone in the NE Japan forearc region revealed by sP depth phases recorded by regional networks. Geophys. J. Int. 2009;178:195-214.
8. Harris R.A., Simpson R.W., Reasenberg P.A. Influence of static stress changes on earthquake locations in southern California. Nature 1995;375:221-224.
9. Hasegawa A., Nakajima J. Geophysical constraints on slab subduction and arc magmatism. In: Sparks R.S.J., Hawkesworth C.J., editors. The State of the Planet: Frontiers and Challenges in Geophysics. AGU; 2004. p. 81-93.
10. Hasegawa A., Umino N., Takagi A. Double-planed structure of the deep seismic zone in the northeastern Japan arc. Tectonophysics 1978;47:43-58.
11. Hatori T. Tsunami sources on the Pacific side in norsteast Japan. Zisin(2) 1974;27:321-337.
12. Hino R., Kanazawa T., Hasegawa A. Interplate seismic activity near the northern Japan Trench deduced from ocean bottom and land-based seismic observations. Phys. Earth planet. Inter. 1996;93:37-52.
13. Hino R., Ito S., Shiobara H., Shimamura H., Sato T., Kanazawa T., Kasahara J., Hasegawa A. Aftershock distribution of the 1994 Sanriku-oki earthquake (Mw 7.7) revealed by ocean bottom seismographic observation. J. geophys. Res. 2000;105:21 697–21 710.
14. Hino R. Insight into complex rupturing of the immature bending normal fault in the outer slope of the Japan Trench from aftershocks of the 2005 Sanriku earthquake (Mw = 7.0) located by ocean bottom seismometry. Geochem. Geophys. Geosyst. 2009;10:Q07O18. doi:10.1029/2009GC002415.
15. Iinuma T., Ohzono M., Ohta Y., Miura S. Coseismic slip distribution of the 2011 off the Pacific coast of Tohoku Earthquake (M 9.0) estimated based on GPS data—Was the asperity in Miyagi-oki ruptured? Earth Planets Space 2011;63:643-648.
16. Iinuma T. Coseismic slip distribution of the 2011 off the Pacific Coast of Tohoku Earthquake (M 9.0) refined by means of seafloor geodetic data. J. geophys. Res. 2012;117. doi:10.1029/2012JB009186.
17. Ito A., Fujie G., Tsuru T., Kodaira S., Nakanishi A., Kaneda Y. Fault plane geometry in the source region of the 1994 Sanriku-oki earthquake. Earth planet. Sci. Lett. 2004;223:163-175.
18. Ito A., Fujie G., Miura S., Kodaira S., Kaneda Y., Hino R. Bending of the subducting oceanic plate and its implication for rupture propagation of large interplate earthquakes off Miyagi, Japan, in the Japan Trench subduction zone. Geophys. Res. Lett. 2005;32:L05310. doi:10.1029/2004GL022307.
19. Ivandic M., Grevemeyer I., Berhorst A., Flueh E.R., McIntosh K. Impact of bending related faulting on the seismic properties of the incoming oceanic plate offshore of Nicaragua. J. geophys. Res. 2008;113:B05410. doi:10.1029/2007JB005291.
20. Iwamori H. Deep subduction of H2O and deflection of volcanic chain towards backarc near triple junction due to lower temperature. Earth planet. Sci. Lett. 2000;181:41-46.
21. Kanamori H. Seismological evidence for a lithospheric normal faulting — the Sanriku earthquake of 1933. Phys. Earth planet. Inter. 1971;4:289-300.
22. Kanamori H., Press F. How thick is the lithosphere? Nature 1970;226:330-331.
23. Kasahara J., Nagumo S., Koresawa S., Nishi Y., Sugimoto H. A linear trend of hypocenter distribution in the outer slope region of the Japan Trench revealed by OBS array: preliminary report. Bull. Earthq. Res. Inst. Univ. Tokyo 1982;57:83-104.
24. Kawakatsu H., Seno T. Triple seismic zone and the regional variation of seismicity along the Northern Honshu Arc. J. geophys. Res.: Solid Earth 1983;88:4215-4230.
25. Kawakatsu H., Watada S. Seismic evidence for deep-water transportation in the mantle. Science 2007;316:1468-1471.
26. Kawakatsu H., Kumar P., Takei Y., Shinohara M., Kanazawa T., Araki E., Suyehiro K. Seismic evidence for sharp lithosphere-asthenosphere boundaries of oceanic plates. Science 2009;324:499-502.
27. Kirby S., Engdahl R.E., Denlinger R. Intermediate-depth intraslab earthquakes and arc volcanism as physical expressions of crustal and uppermost mantle metamorphism in subducting slabs. In: Bebout G.E., Scholl D.W., Kirby S.H., Platt J.P., editors. Subduction Top to Bottom. AGU; 1996. p. 195-214.
28. Kita S., Okada T., Nakajima J., Matsuzawa T., Hasegawa A. Existence of a seismic belt in the upper plane of the double seismic zone extending in the along-arc direction

at depths of 70–100 km beneath NE Japan. Geophys. Res. Lett. 2006;33. doi:10.1029/2006GL028239.

29. Kita S., Okada T., Hasegawa A., Nakajima J., Matsuzawa T. Existence of interplane earthquakes and neutral stress boundary between the upper and lower planes of the double seismic zone beneath Tohoku and Hokkaido, northeastern Japan. Tectonophysics 2010;496:68-82.

30. Kobayashi K., Nakanishi M., Tamaki K., Ogawa Y. Outer slope faulting associated with the western Kuril and Japan trenches. Geophys. J. Int. 1998;134:356-372.

31. Koyama J., Horiuchi S., Hirasawa T. Earthquake generating stress in north-east Japan arc inferred from superposition of the initial motions of P-waves. Zisin(2) 1973;26:241-253.

32. Kunitomi S. On the Sanriku-oki strong motion and Tsunami. Kenshin-jihou 1933;7:1-43.

33. Lay T., Ammon C.J., Kanamori H., Rivera L., Koper K.D., Hutko A.R. The 2009 Samoa-Tonga great earthquake triggered doublet. Nature 2010;466:964-968.

34. Lay T., Ammon C.J., Kanamori H., Kim M.J., Xue L. Outer trench-slope faulting and the 2011 Mw 9.0 off the pacific coast of Tohoku earthquake. Earth Planets Space 2011;63:713-718.

35. Lefeldt M., Grevemeyer I., Goßler J., Bialas J. Intraplate seismicity and related mantle hydration at the Nicaraguan trench outer rise. Geophys. J. Int. 2009;178:742-752.

36. Masson D.G. Fault patterns at outer trench walls. Mar. Geophys. Res. 1991;13:209-225.

37. Matsuzawa T., Umino N., Hasegawa A., Takagi A. Estimation of thickness of a low-velocity layer at the surface of the descending oceanic plate beneath the northeastern Japan arc by using synthesized PS-wave. Tohoku Geophys. J. 1987;31:19-28.

38. Nakajima J., Hasegawa A. Anomalous low-velocity zone and linear alignment of seismicity along it in the subducted Pacific slab beneath Kanto, Japan: reactivation of subducted fracture zone? Geophys. Res. Lett. 2006;33. doi:10.1029/2006GL026773.

39. Nakajima J., Tsuji Y., Hasegawa A. Seismic evidence for thermally-controlled dehydration reaction in subducting oceanic crust. Geophys. Res. Lett. 2009;36:L03303. doi:10.1029/2008GL036865.

40. Nakanishi M. Bending-related topographic structures of the subducting plate in the Northwestern Pacific Ocean. In: Ogawa Y., Anma R., Dilek Y., editors. Accretionary Prisms and Convergent Margin Tectonics in the Northwest Pacific Basin. Springer; 2011. p. 1-38.

41. Obana K. Normal-faulting earthquakes beneath the outer slope of the Japan Trench after the 2011 Tohoku earthquake: implications for the stress regime in the incoming Pacific plate. Geophys. Res. Lett. 2012;39:L00G24. doi:10.1029/2011GL050399.

42. Obana K., Nakamura Y., Kaiho Y., Yamamoto Y., Kodaira S., Fujie G. JpGU 2016 Meeting. 2016. Seismicity observations in the source region of the 1896 Meiji-Sanriku and 1933 Showa-Sanriku Earthquakes. SSS02-14.

43. Okada Y. Internal deformation due to shear and tensile faults in a half-space. Bull. seism. Soc. Am. 1992;82:1018-1040.

44. Okal E.A., Kirby S.H., Kalligeris N. The Showa Sanriku earthquake of 02 March 1933: a global seismological reassessment. Geophys. J. Int. 2016. in press, doi:10.1093/gji/ggw206.

45. Peacock S.M. Are the lower planes of double seismic zones caused by serpentine dehydration in subducting oceanic mantle? Geology 2001;29:299-302.

46. Ranero C.R., Phipps Morgan J., McIntosh K., Reichert C. Bending-related faulting and mantle serpentinization at the Middle America trench. Nature 2003;425:367-373.

47. Satake K. Linear and nonlinear computations of the 1992 Nicaragua earthquake tsunami. Pure appl. Geophys. 1995;144:455-470.

48. Sekiguchi R., Nakano M. Report on Sanriku Tsunami based on tide gauge. Kenshin-jihou 1933;7:71-79.

49. Seno T., Gonzalez D.G. Faulting caused by earthquakes beneath the outer slope of the Japan Trench. J. Phys. Earth 1987;35:381-407.

50. Shiina T., Nakajima J., Matsuzawa T. Seismic evidence for high pore pressures in the oceanic crust: implications for fluid-related embrittlement. Geophys. Res. Lett. 2013;40:2006-2010.

51. Stein R.S., King G.C.P., Lin J. Change in failure stress on the Southern San Andreas fault system caused by the 1992 magnitude = 7.4 Landers earthquake. Science 1992;258:1328-1332.

52. Stein S., Liu M. Long aftershock sequences within continents and implications for earthquake hazard assessment. Nature 2009;462:87-89.

53. Tanioka Y., Satake K. Fault parameters of the 1896 Sanriku Tsunami Earthquake estimated from tsunami numerical modeling. Geophys. Res. Lett. 1996;23:1549-1552.

54. The Central Meteorological Observatory. Damage investigation of the Sanriku Tsunami. Kenshin-jihou 1933;7:105-159.

55. The International Latitude Observatory of Mizusawa. Seismological Observations at Mizusawa for the Period Between 1902–1967. Mizusawa-insatsu; 1984.

56. Toda S., Stein R.S., Reasenberg P.A., Dieterich J.H., Yoshida A. Stress transferred by the 1995 Mw = 6.9 Kobe, Japan, shock: effect on aftershocks and future earthquake probabilities. J. geophys. Res. 1998;103:24 543–24 565.

57. Todd E.K., Lay T. The 2011 Northern Kermadec earthquake doublet and subduction zone faulting interactions. J. geophys. Res. 2013;118:249-261.

58. Tonegawa T., Hirahara K., Shibutani T., Fujii N. Lower slab boundary in the Japan subduction zone. Earth planet. Sci. Lett. 2006;247:101-107.

59. Tsuji Y., Nakajima J., Hasegawa A. Tomographic evidence for hydrated oceanic crust of the Pacific slab beneath northeastern Japan: implications for water transportation in subduction zones. Geophys. Res. Lett. 2008;35:L14308. doi:10.1029/2008GL034461.

60. Umino N. Revisiting the three M ~ 7 Miyagi-oki earthquakes in the 1930s: possible seismogenic slipon asperities that were re-ruptured during the 1978 M7.4 Miyagi-oki earthquake. Earth Planets Space 2006;58:1587-1592.

61. Waldhauser F., Ellsworth W.L. A double-difference earthquake location algorithm: method and application to the Northern Hayward Fault, California. Bull. seism. Soc. Am. 2000;90:1353-1368.

62. Wartman J.M., Kita S., Kirby S.H., Choy G.L. American Geophysical Union, Fall Meeting 2009 abstract. 2009. A global outer-rise/outer-trench-slope (OR/OTS) earthquake study. #S53A-1472.

63. Wessel P., Smith W.H.F. New version of generic mapping tools released. EOS, Trans. Am. geophys. Un. 1995;79:329.

64. Wysession M., Okal E., Miller K. Intraplate seismicity of the Pacific Basin, 1913–1988. Pure appl. Geophys. 1991;135:261-359.

65. Yamamoto Y. Spatial heterogeneity of the mantle wedge structure and interplate coupling in the NE Japan forearc region. Geophys. Res. Lett. 2008;35:L23304. doi:10.1029/2008GL036100.

66. Yoshii T. A detailed cross-section of the deep seismic zone beneath northeastern Honshu, Japan. Tectonophysics 1979;55:349-360.

67. Zhang H., Thurber C.H. Double-difference tomography: the method and its application to the Hayward Fault, California. Bull. seism. Soc. Am. 2003;93:1875-1889.

68. Zhao D., Hasegawa A., Kanamori H. Deep structure of Japan subduction zone as derived from local, regional and teleseismic events. J. geophys. Res. 1994;99:22 313–22 329.

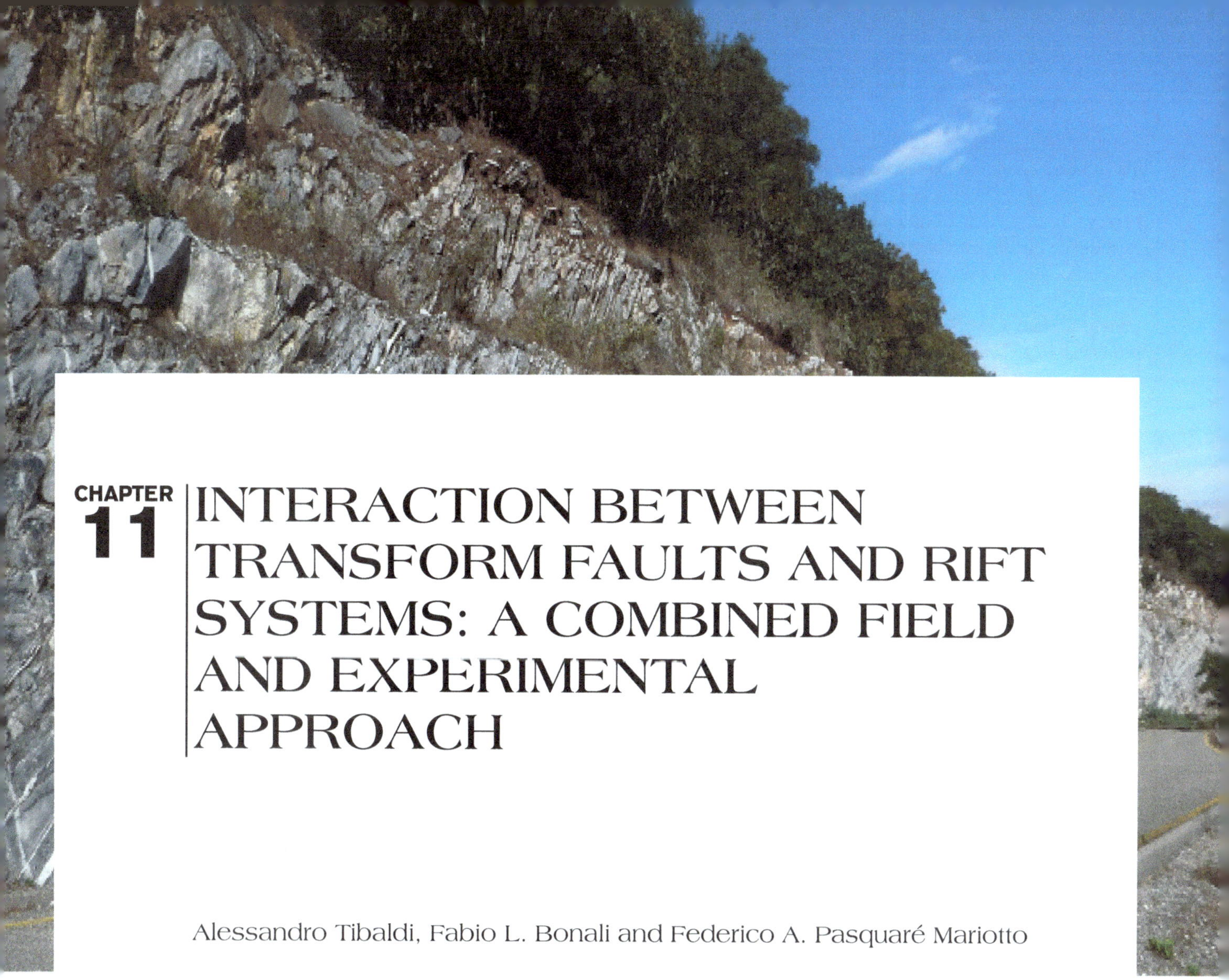

CHAPTER 11

INTERACTION BETWEEN TRANSFORM FAULTS AND RIFT SYSTEMS: A COMBINED FIELD AND EXPERIMENTAL APPROACH

Alessandro Tibaldi, Fabio L. Bonali and Federico A. Pasquaré Mariotto

We present a detailed field structural survey of the area of interaction between the active NW-striking transform Husavik-Flatey Fault (HFF) and the N–S Theystareykir Fissure Swarm (TFS), in North Iceland, integrated by analog scaled models. Field data contribute to a better understanding of how transform faults work, at a much higher detail than classical marine geophysical studies. Analog experiments are conducted to analyse the fracture patterns resulting from different possible cases where transform faulting accompanies or postpones the rift motions. Different tectonic block configurations are also considered and results are compared with field data in order to study as thoroughly as possible the interaction between the HFF and the TFS as well as the possible prolongation of the HFF into the TFS. West of the intersection between the transform fault (HFF) and the rift zone (TFS), the former splays with a gradual bending giving rise to a leading extensional imbricate fan. The westernmost structure of the rift, the N–S Gudfinnugja Fault (GF), is divided into two segments: the southern segment makes a junction with the HFF and is part of the imbricate fan; north of the junction instead, the northern GF appears right-laterally offset by about 20 m. Southeast of the junction, along the possible prolongation of the HFF across the TFS, the strike of the rift faults rotates in an anticlockwise direction, attaining a NNW–SSE orientation. Moreover, the TFS faults north of the HFF prolongation are fewer and have smaller offsets than those located to the south. Through the comparison between the structural data collected in the field at the HFF–TFS connection zone and a set of scaled experiments, we confirm a prolongation of the HFF through the rift, although here the transform fault has a much lower slip-rate than west of the junction. Our data suggest that transform fault terminations may be more complex than previously known, and propagate across a rift through a modification of the rift pattern.

Citation: Tibaldi A, Bonali FL and Pasquaré Mariotto FA (2016) Interaction between Transform Faults and Rift Systems: A Combined Field and Experimental Approach. Front. Earth Sci. 4:33. doi: 10.3389/feart.2016.00033.

INTRODUCTION

A transform fault is characterized by strike-slip motions that end abruptly and are turned into deformation at a companion structure. A classic example is given by a transform fault that merges into an expanding rift (Wilson, 1965). This type of intersection zone has been studied mainly by geophysical surveys in marine environments, using bathymetric maps and earthquake data. As a consequence, hardly ever available data are detailed enough to allow for a precise evaluation of the geometry, kinematics, and deformation at the termination of transform faults. Moreover, several mechanisms have been proposed to describe how such strike-slip faults terminate, but they are poorly resolved in several cases. More detailed data on transform fault terminations have been collected and studied on-land, especially in regard to the transition to reverse fault/trench systems (e.g., Davis and Burchfiel, 1973; Bellier and Sebrier, 1995; Quebral et al., 1996; Tsutsumi and Okada, 1996; Freymueller et al., 1999; Norris and Cooper, 2000), while fewer are related to on-land rift zones (e.g., Joffe and Garfunkel, 1987; Mouslopoulou et al., 2007a).

Terminations of transform faults can be of four main types: (i) the fault can end at the tip without any apparent connection with any other structure, with some secondary faults that can depart for a short distance, but the rock volume surrounding the fault tip appears basically not faulted at a regional scale of observation (Mouslopoulou et al., 2007b; Figure 1A); (ii) the transform fault ends via a branching of the main fault plane into a splay (Figure 1B). In this case, the main fault bends into a series of secondary faults with different orientations, decreasing the displacement on the main transform fault and redistributing the slip into the rock volume located on one side of the fault plane. Based on the geometry of the splay and the kinematics of the main fault, the resulting configuration can be an extensional or a contractional imbricate fan (Woodcock and Fischer, 1986); (iii) the transform fault tip coincides with the tip of a normal or a reverse fault (Figure 1C). An example can be observed at the intersection between the Sumatra Fault in Indonesia and the Sunda Strait Rift (Lelgemann et al., 2000), and at the tip coinciding with the Median Tectonic Line and the Yatsushiro graben in Japan (Kamata and Kodama, 1994); (iv) the transform fault terminates in correspondence of a through-going reverse or normal fault system (Figure 1D). This configuration can be observed, for example, where the Dead Sea fault intersects the Red Sea Rift (Joffe and Garfunkel, 1987).

Several examples of transform-rift interaction may be related to case (ii), such as at the intersection of the East Pacific Rise with the Tamayo Transform Fault (Gallo et al., 1984), the Siqueiros Transform fault system with the East Pacific Rise (Fornari et al., 1989), the Vema Fracture Zone (Mamaloukas-Frangoulis et al., 1991), the Oceanographer Transform fault (Karson et al., 1984) with the Mid-Atlantic Ridge, and several other sites (review in Şengör, 2014). This type of configuration is stable with the slip vector of the rift coinciding with the slip vector of the transform fault, especially if simplified as the scheme of Figure 1C. This scheme works also on a regional scale, like in the case of the Tjörnes transform fault zone (TFZ) that connects the Kolbeinsey ridge, a mid-ocean ridge off the north coast of Iceland, with the rift zone in North Iceland (Gudmundsson, 2007). However, if a more local scale is considered, the connection between a transform fault and a rift can be much more complex, especially if this is studied onshore with a detail that cannot be obtained in offshore settings. The TFZ, for example, when reaching the North Iceland Rift, exhibits a complex splay of oblique faults and tension fractures that gradually link the two main fault systems (Gudmundsson et al., 1993). Moreover, there is no connection point between the TFZ and a main fault of the North Iceland Rift, but instead a diffuse zone of fracturing that resembles more the type displayed in Figure 1D. The correct evaluation of the mode of interconnection between a transform fault and a rift zone requires also the assessment of fault kinematics, opening direction of the tension fractures, as well as the timing of their relative development.

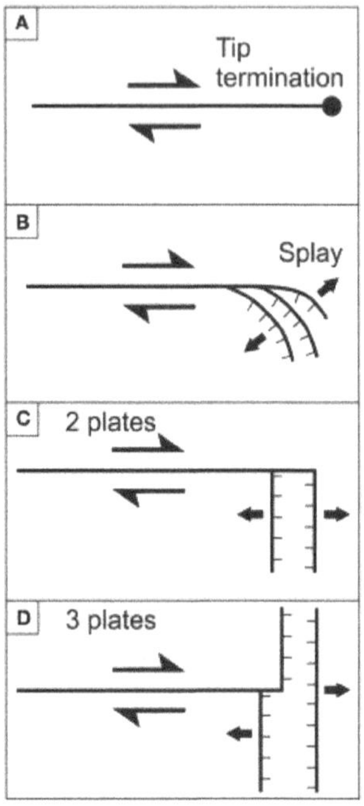

Figure 1. Schematic diagram illustrating the four basic types of strike-slip fault terminations. (A) The strike-slip fault ends at a tip surrounded by undeformed rocks; **(B)**the strike-slip fault terminates with a splay, which in this example creates a transtensional zone; **(C)** the strike-slip fault terminates in correspondence of a normal or reverse fault; **(D)**the strike-slip fault terminates against another through-going fault system. (**A**, **C**, and **D**modified after Mouslopoulou et al., 2007a).

The emerged oceanic rift zone of Iceland and its transform faults represent the ideal settings for the needed detailed observations and provide fundamental insights into the mechanisms of connection and propagation of strike-slip faults with normal faults. In North Iceland (Figure 2A), the Theistareykir Fissure Swarm (TFS; Figure 2B), the westernmost section of the Northern Volcanic Zone, is affected by active surface tectonic activity as testified to by the presence of both Holocene faults and tension fractures (Gudmundsson et al., 1993; Magnúsdóttir and Brandsdóttir, 2011;Hjartardóttir et al., 2015; Pasquarè Mariotto et al., 2015; Figure 2C). Most of these structures strike N00 to N20°. The western part of the TFS is connected to the major, active NW–SE transcurrent Husavik-Flatey Fault (HFF) belonging to the TFZ (Figure 2). According to previous research, the surface expression of the HFF ends at the junction with the westernmost normal fault of the TFS, known as Gudfinnugja Fault (GF; Gudmundsson et al., 1993; Magnúsdóttir and Brandsdóttir, 2011). More recently, a possible subsurface prolongation of the HFF toward the southeast has been proposed, on the basis of earthquake and fracture distribution near the Krafla fissure swarm

(Hjartardóttir et al., 2012) and of the geometric and kinematic characteristics of the fractures within the TFS (Tibaldi et al., 2016). The structural complexity of the TFS–HFF junction area is compensated by ideal geological and morphological settings: the area is completely flat, hence possible disturbances such as topographic effects on the geometry and kinematics of the structures can be ruled out. Moreover, the study area is essentially composed of a horizontal sequence of basaltic lavas dated at 14.5 and 2.4 ka BP west of the TFS and at 14.5 ka BP within the TFS (Saemundsson et al., 2012). Hence, also possible effects due to lithological heterogeneities at the surface can be disregarded; furthermore, offsets along the different structures can be directly compared.

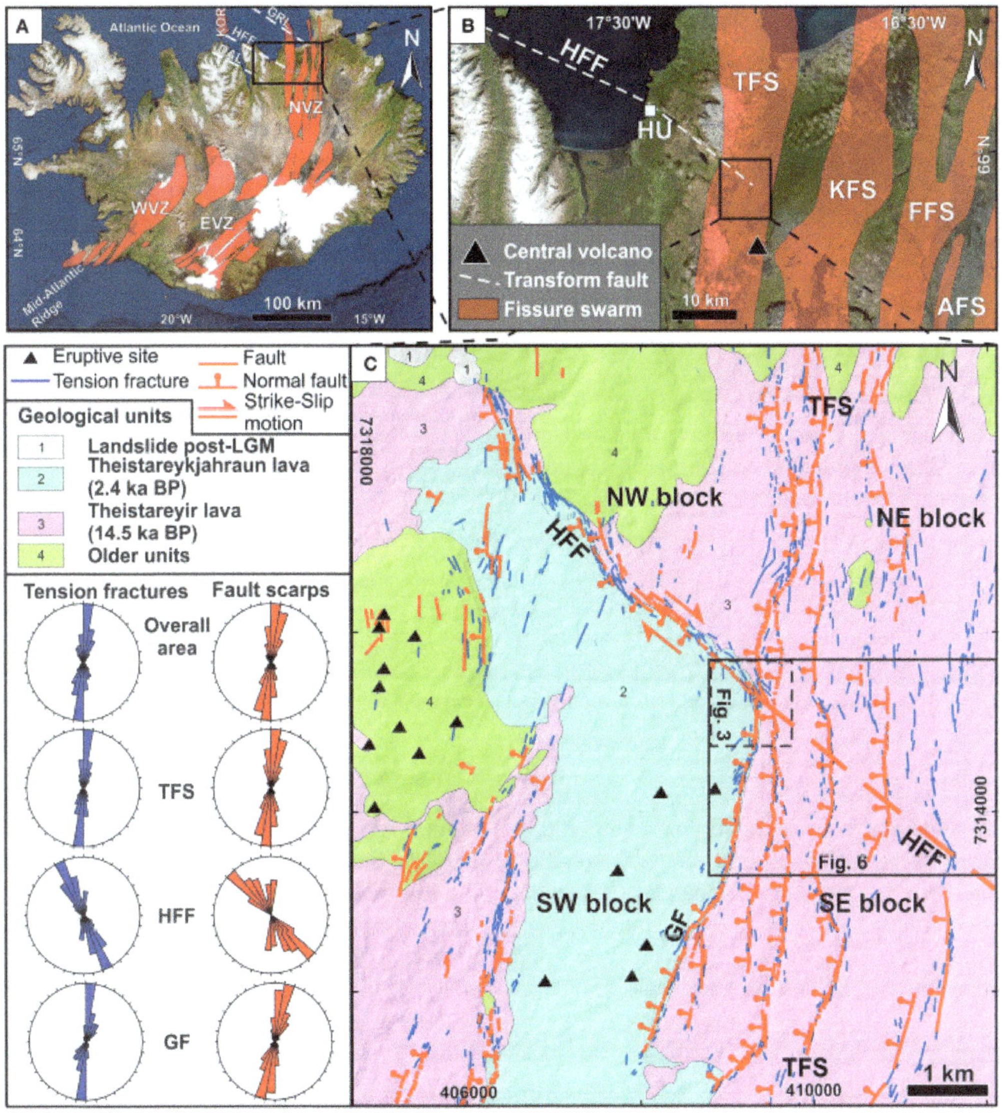

Figure 2. **(A)** Iceland with its volcanic zones, among which the North Volcanic Zone (NVZ), the West Volcanic Zone (WVZ), and the East Volcanic Zone (EVZ). The Tjörnes transform fault zone is composed of the Dalvik lineament (DAL), the Husavik-Flatey Fault (HFF), and the Grimsey lineament (GRL). KOR, Kolbeinsey ridge. **(B)** Tectonic setting of northeastern Iceland; orange stripes represent the main volcanic systems that compose the NVZ, among which the Theystareykir Fissure Swarm (TFS), one of the main objects of the present study. Black triangle shows location of Theistareykjabunga shield volcano. **(C)** Simplified geologic map of the study area (based on personal data and Saemundsson et al., 2012), integrated with a structural sketch map of the eastern portion of the HFF, and a major section of the TFS. Rose diagrams of faults and tension fractures are also provided.

We identified four sub-areas delimited by the intersection between the two major faults, the HFF and the GF (Tibaldi et al., 2016). As illustrated in Figure 2, the sub-areas are, respectively, located: (i) north of the HFF and west of the GF (NW block); (ii) south of the HFF and west of the GF (SW block); (iii) north of the prolongation of the HFF and east of the GF (NE block); (iv) south of the prolongation of the HFF and east of the GF (SE block).

In order to analyse at the highest level of detail the various aspects of the interconnection between a transform fault and a rift zone, we compared the structural data collected in the field at the HFF–TFS connection zone with a set of scaled experiments designed to reproduce the various possible motions under different block configurations. We used different experimental settings that involve: (i) a different number of moving blocks; (ii) the presence or absence of a ductile zone along the rift; (iii) the position of the tip point of the transform fault with respect to the rift, and (iv) the relative time distribution of the motions for the involved blocks. The goals of our study are: (i) to summarize the main geometric and kinematic characteristics of the area of connection between the HFF and the TFS; (ii) to show the different structures that may result from analog experiments focused on transcurrent-rift interactions in different settings; (iii) to compare the results in order to better understand how such interactions works and contribute to evaluating the possible prolongation of the HFF eastward beyond the TFS. The recognition of active transform faults propagating across rift zones may help to calculate the real length of these faults and related possible earthquake magnitude.

GEOLOGIC-TECTONIC FRAMEWORK

The North Iceland Rift coincides with the Northern Volcanic Zone (NVZ, Figure 2A) that comprises, in addition to the TFS, another four, about N–S-striking volcanic rift zones (Figure 2B), namely the Krafla (KFS), Fremrinámar (FFS), Askja (AFS), and Kverkfjöll volcanic systems (Hjartardóttir et al., 2015). Each of these systems consists of 5–20 km-wide and 60–100 km-long swarms of normal faults, eruptive and tension fractures (Figure 2C), and a main volcano (Saemundsson, 1974). Most research efforts focused on the Krafla system (e.g., Angelier et al., 1997;Acocella et al., 2000; Hjartardóttir et al., 2015), whereas the structures of the TFS were studied only recently in detail by Pasquarè Mariotto et al. (2015), Tibaldi et al. (2016), and regionally byHjartardóttir et al. (2015).

The transform fault (HFF), which runs 25 km onshore, was first studied by Einarsson (1958) and mapped by Saemundsson (1974), and later by Opheim and Gudmundsson (1989), Gudmundsson et al. (1993), Garcia et al. (2002), Garcia and Dhont (2005), and Magnúsdóttir and Brandsdóttir (2011). It is made of en-échelon, right-stepping, and left-stepping, NW–SE dextral strike-slip fault segments (Gudmundsson et al., 1993; Gudmundsson, 2007), which separate Tertiary rocks to the north from Upper Pleistocene rocks to the south (Saemundsson, 1974; Garcia et al., 2002). This fault is marked by land surface depressions occupied by lakes (sag ponds) with an estimated component of vertical offset at the surface that amounts to 200 m (Gudmundsson et al., 1993). In addition, Tryggvason (1973) suggested that such fault can reach an overall vertical offset of as much as 1400 m, although the location of this onshore measurement remains undefined. Saemundsson (1974), based on local geology, proposed a cumulated total component of right-lateral displacement ranging from 5 to 10 km near the town of Husavik whereas, according to Young et al. (1985) it may amount to 20 km.

The HFF joins with the GF, the westernmost N–S-striking normal fault of the TFS (Gudmundsson et al., 1993) making a 54–56° angle (Figure 2C). Near the junction area, along the HFF there are several transpressional and transtensional zones localized at NW–SE fault strands with left-stepping and right-stepping geometries, respectively (Gudmundsson et al., 1993; Pasquarè Mariotto et al., 2015). Approaching the junction with the GF, the HFF main fault plane rotates gradually and splays southwards into a set of minor faults that distribute the deformation.

The study area hosts the Theistareykjabunga shield volcano (Figure 2) and its Skildingahraun lava flows, whose age has been estimated at about 14.5 ka BP (Saemundsson et al., 2012). This age is based on tephrochronology obtained from a soil covering the lava, which contains the Vedde ash tephra layer (12.0 ka BP, M.Á. Sigurgeirsson, pers. comm.). Between the Vedde ash and the Skildingahraun lava there is another soil with a basaltic tephra layer that may be correlated with the tephra (GS-2/BAS-1) found in various cores and with an age close to 14.5 ka BP (M.Á. Sigurgeirsson, pers. comm.). About 2.4 ka BP, the latest products of this eruptive center were emplaced, forming the "Theistareykjahraun" lava flows, located between the shield and the HFF (Saemundsson et al., 2012). The latest major event that occurred nearby the studied area was the rifting episode at the Krafla volcanic system from 1975 to 1984 (Bjornsson, 1985; Buck et al., 2006), resulting in several surface meter-large horizontal and vertical offsets along nearby fractures (Tryggvason, 1980, 1984, 1986). Such rifting episode was capable of inducing stress changes in the nearby fault systems within the TFZ; in particular, the 1975 dyke intrusion triggered the 1976 Köpasker earthquake as a consequence of induced Coulomb stress on the Grímsey Oblique Rift (Maccaferri et al., 2013). The rift-induced effect on the HFF resulted in: (i) an increase in the seismicity rate along the central part of the HFF, with the seismicity progressively recovering from west to east; (ii) a decrease in seismic rate along the eastern part of HFF; (iii) a modification of the way in which the system releases seismic energy (Maccaferri et al., 2013).

FIELD DATA

The Husavik-Flatey Fault

The eastern portion of the HFF is composed of several fault segments mostly striking N125–135° (Figures 2C, 3A). These segments show right-lateral strike-slip motions with a local component of normal motions (Gudmundsson et al., 1993; Garcia and Dhont, 2005). Fault scarps face southward, suggesting transtensional kinematics with downthrow of the southern block. In two main zones, N125–135°-striking fault strands with a right-stepping geometry are linked by fault swarms striking from N340 to N00°, showing dip-slip normal kinematics (Pasquarè Mariotto et al., 2015). Several small, meter-sized pull-apart depressions are located at right-stepping overlap zones, whereas push ridges occur at left-stepping overlap zones. The surface trace of the HFF terminates against the GF through a gradual bending and a splay of the fault plane that causes the transition from an original NW–SE orientation to a N–S strike (Figures 3A,C; Gudmundsson et al., 1993).

The easternmost part of the HFF, west of the junction with the GF, offsets the lava flows of the Theistareykjabunga shield volcano dated at about 14.5 ka BP in the study area (Saemundsson et al., 2012). The amount of their surface strike-slip component of dislocation is difficult to quantify here, whereas the dip-slip component ranges from a minimum of 5 m to a maximum of 17 m. The dip-slip component increases at the easternmost termination of the HFF, in correspondence of the bending and splay of the main fault plane. The dip-slip displacements completely replace transcurrent ones when the splay fault strands attain the about N–S strike of the GF. We noticed that the easternmost fault strand shows the larger offset; hence, we can define this zone as an extensional leading imbricate fan.

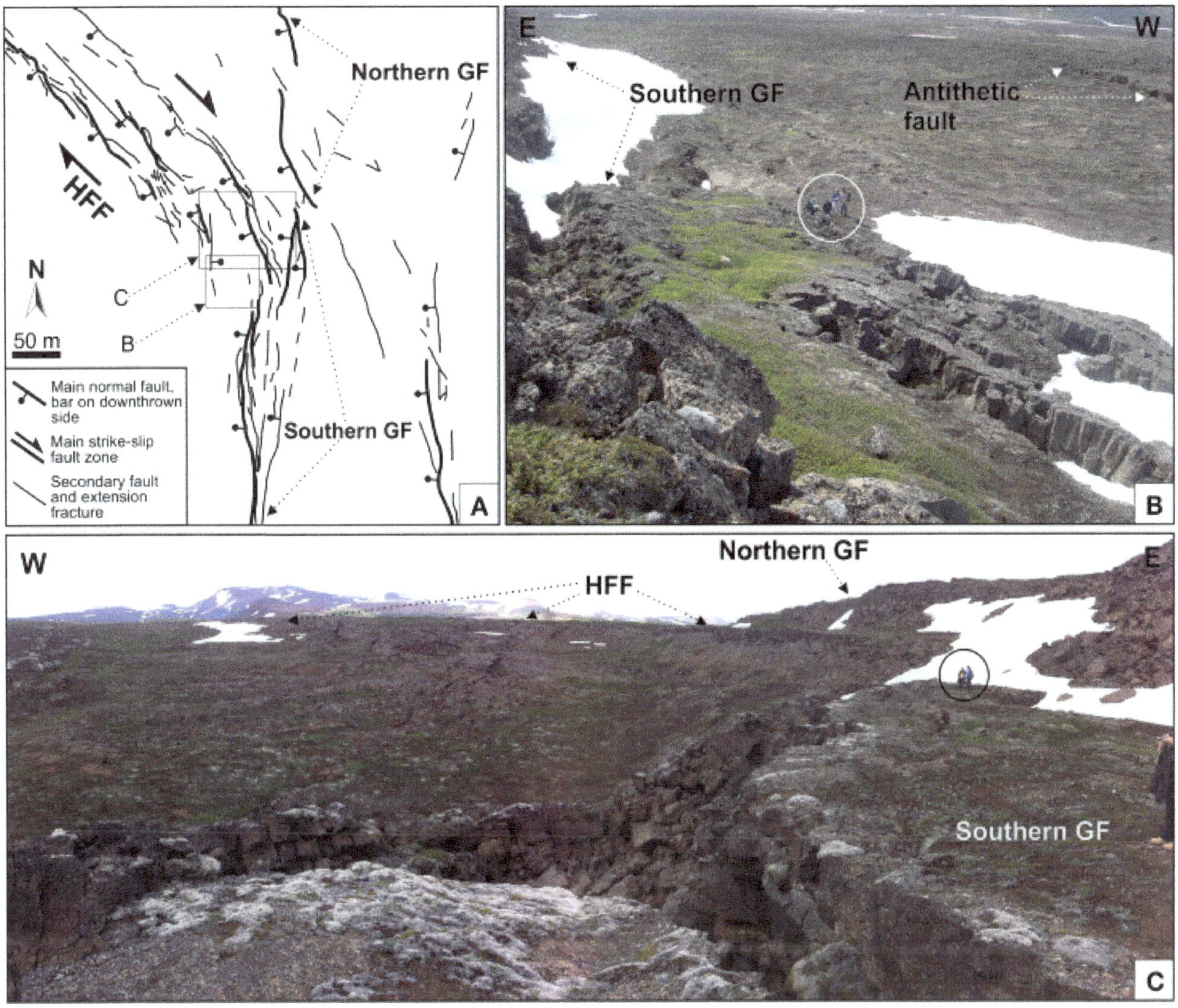

Figure 3. **(A)** Detailed structural sketch map of the triple junction between the Husavik-Flatey Fault (HFF) and the Theystareykir Fissure Swarm (TFS). The HFF terminates against the westernmost fault of the TFS, known as Gudfinnugja Fault (GF) through a gradual bending and a splay of the fault plane that from an original NW–SE orientation attains a N–S strike. **(B)** Photo of the southern GF and its antithetic normal fault, looking south. Note men in circle for scale. Location in Figure 3A. **(C)** Photo of the junction zone between the HFF and the southern GF, looking north. Note men in circle for scale. Location in Figure 3A.

The Theystareykir Fissure Swarm (TFS)

The TFS has a total length of 34 km; it is 8-km-wide in the study area south of the junction with the HFF, and narrows to < 3 km north of the junction. In the study area, normal faults and tension fractures strike N350–N20°, with an average of N00°. Several individual faults that characterize the TFS are spaced from 0.2 to 1 km apart from each other. Most individual fault segments gradually fade out and are replaced by tension fractures with the same orientation as the faults (Figure 2C). Other tension fractures are located far away from the faults.

Besides this general pattern, very detailed field surveys showed that, in correspondence of the possible prolongation of the HFF into the TFS, rift fractures tend to rotate (Figures 2C, 4). Here, faults strike N00 to N22° and N125 to N180°, as far as about 0.5 km from the theoretical prolongation of the HFF (Figure 4A). This means that along this strip coinciding with the possible prolongation of the HFF, there are structures with the same orientation as the HFF. At greater distances from the HFF prolongation, fault strikes tend to be oriented about N–S and, more exactly, they strike mostly N160–180° and N00–40°. In other words, structures with the same orientation as the HFF are missing in the TFS at distances >0.5 km both north and south of the prolongation of the HFF, whereas along the HFF prolongation, a generalized anticlockwise rotation of the TFS structures can be clearly observed. An identical situation can be observed also in regard to the strike of tension fractures, as shown in Figure 4B. Also the opening directions of tension fractures show a homogenous distribution in the range N65–120° as far as about 0.5 km from the HFF prolongation, whereas at shorter distances the range is wider, from N25 to N125° (Figure 4C). It is worth mentioning that about 50% of the tension fractures show opening directions that are oblique with respect to the fracture strike (Figure 4D), indicating a right-lateral component.

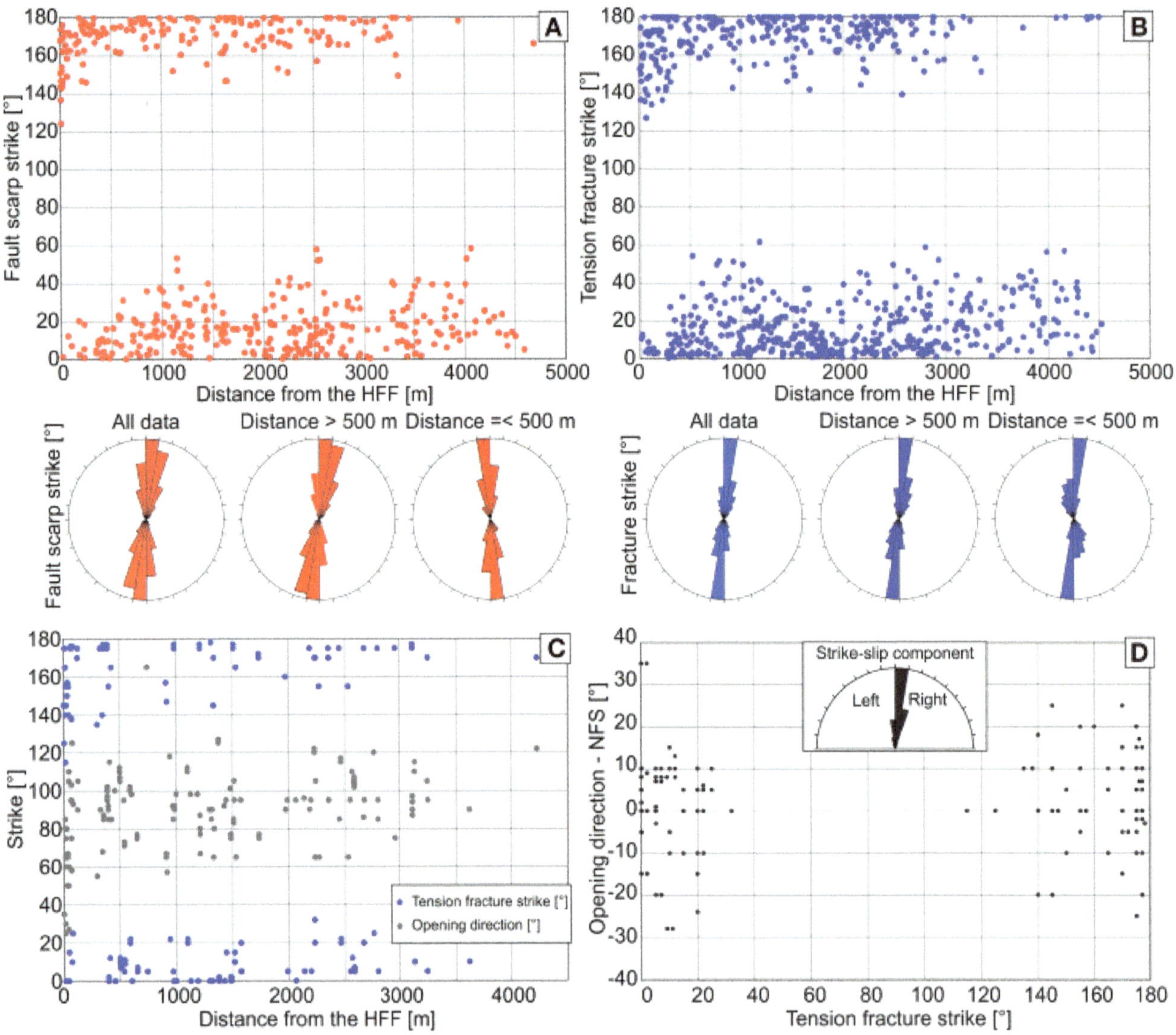

Figure 4. Structural data measured along the TFS. (A) Plot of the distribution of fault strikes measured in the field and on high-resolution satellite images (total = 434; *Y*axis), vs. the distance from the HFF prolongation in meters (*X* axis). Such distance is defined as the minimum distance from the surface projection of the fault (e.g., Bindi et al., 2011). Below the graph, rose diagrams with fault strike distribution are provided. **(B)** Plot of distribution of tension fracture strike measured in the field and on high-resolution satellite images (total = 722) (*Y* axis), vs. distance from the HFF prolongation in meters (*X* axis). Below the graph, rose diagrams with tension fracture strike distribution are provided. **(C)**Plot of distribution of opening directions measured in the field (total = 141) along tension fractures, vs. distance from the HFF prolongation in meters (*X* axis). **(D)** Graph of tension fracture strikes measured in the field (total = 141; *Y* axis), vs. the angle between the opening directions and the normal to the fracture strike (NFS). Positive values indicate a right-lateral component of the opening direction, negative values show a left-lateral component.

As far as faults are concerned, Tibaldi et al. (2016) recognized 10 fault zones in the study area (Figure 2). The westernmost structure is the GF, which is 5.7 km long and whose strike is mainly N00–10° but in detail it ranges between N350 and N20°. Although in previous works the GF was regarded as a single fault continuously running from south to north of the HFF tip, we document that it is actually composed of a series of discrete fault strands. Starting from the north, there is a NNE-striking tension fracture that gradually transitions into a fault scarp. The scarp faces to the west and is initially 1–3 m high, gradually increasing to 25 ± 2 m and then decreasing to 15 ± 2 m near the junction with the HFF (Figure 5). At a distance of 21 m from the junction, the fault scarp dies out and a fault bridge separates it from the GF scarp located south of the junction. At the junction with the HFF, the southern GF has a N160° strike and shows a dip-slip component of about 25 m. Farther south, the offset increases to the maximum value of 33 ± 2 m and locally there are antithetic faults creating a narrow asymmetric graben (e.g., Figure 3B). North of the junction, there are 14.5 ka old lavas at both fault sides. South of the junction, 14.5 ka old lavas crop out on the eastern block and 2.4 ka old lavas are observed on the western block (hanging wall). This indicates that the measured offsets south of the junction are minimum because part of the real total offset has been hidden by the younger lavas onlapping the fault scarp. This also means that the difference in offset along the GF is even larger between north and south of the junction with the HFF.

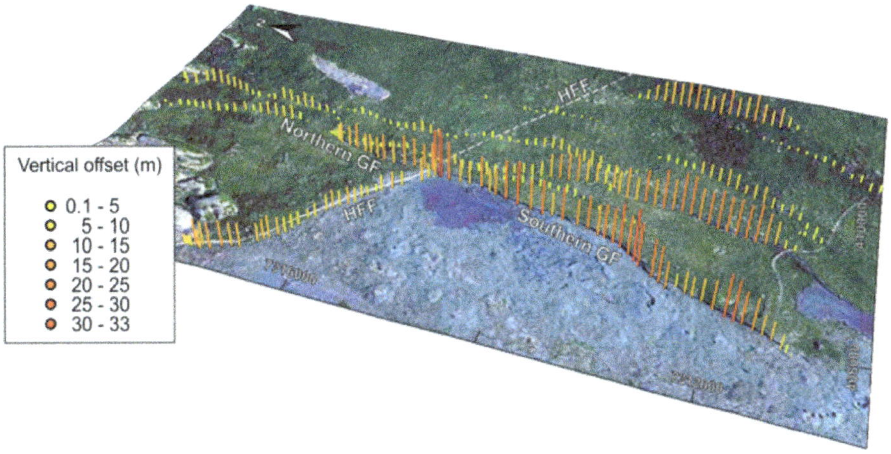

Figure 5. Bar graphs summarizing the offsets cumulated along all the faults located north of the possible prolongation of the HFF with respect to those located south. The graph clearly indicates that those located to the south have much larger offsets.

All the other single fault segments of the TFS range in length from 10 to 1339 m. The scarps always face to the west, indicating gradual downthrowing in the same direction. These faults affect the same lavas dated at 14.5 ka BP, which have the same horizontal attitude, and thus the offsets and geometry of the faults can be directly compared with each other. The offsets measured systematically by Tibaldi et al. (2016) indicate that the faults located north of the possible prolongation of the HFF have smaller offsets than those located to the south (Figure 5). The same authors also discovered that the faults of the TFS change their geometry along a strip corresponding to the possible prolongation of the HFF toward the south-east (Figure 6). To highlight this, they plotted only the segments of tension fractures and faults of the TFS that do not have the typical N00 to N20° orientation. These segments depict a corridor where the fractures strike exactly as the easternmost strands of the HFF west of the junction with the GF (N330 N340°), with an overall pattern that is consistent with the earlier stages of development of a transtensional fault system (Harding et al., 1985; Clifton et al., 2000; Agostini et al., 2009).

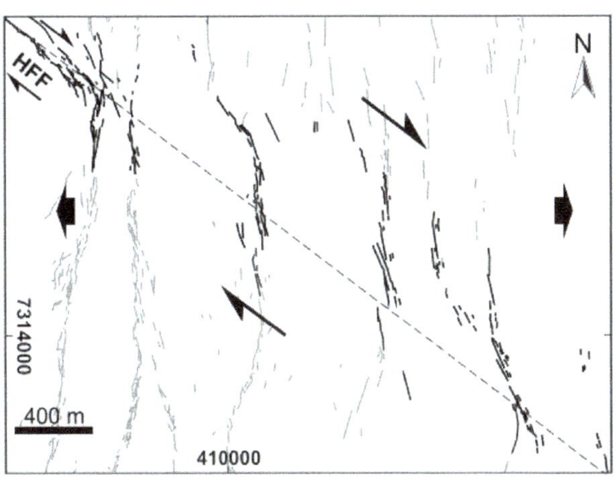

Figure 6. Plot of the segments of tension fractures and faults of the TFS that do not have the typical N00 to N20° orientation. These segments depict a corridor where the fractures strike exactly as the easternmost strands of the HFF (N330–N340°), with an overall pattern that is consistent with the earlier stages of development of a transtensional right-lateral fault system (modified after Tibaldi et al., 2016).

EXPERIMENTAL METHODOLOGY

Experimental Apparatus

The experimental apparatus consists of a rectangular box (70 × 65 × 10 cm) confined by three moveable and one fixed walls at the shorter sides, and two fixed walls at the longer sides (Figure7A). The three moveable walls are fixed to the underlying metal plates that mimic the tectonic blocks; three of such plates are moveable (moving metal plate NW, SW, SE), and one is fixed (fixed metal plate NE) to the basal wood table (Figure 7A).

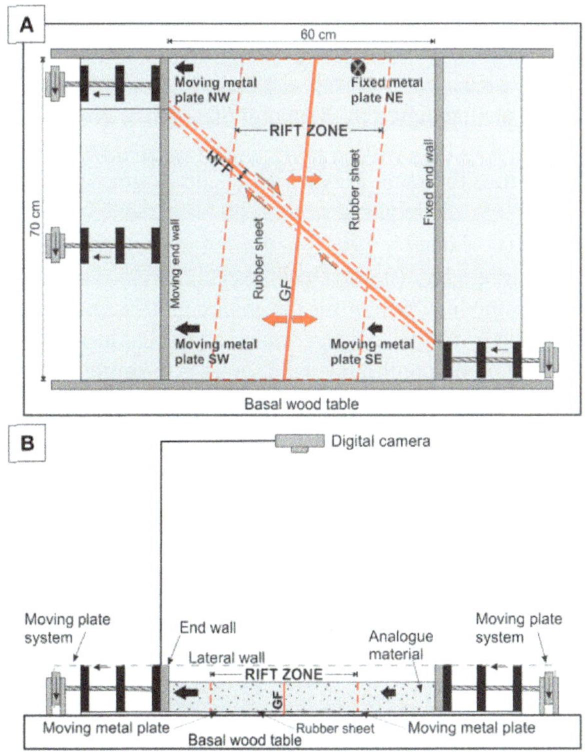

Figure 7. Sketch of the experiment apparatus: above in plan view and below in section view. (A) Three out of four plates can move together or singularly. **(B)** The angle α indicates the amount of transtension. The arrows indicate the vs. of plate movement and their dimension is proportional to the relative slip rate.

Three of them are anchored to their "moving plate system," capable of moving the plates very slowly (reaching high precision) and orthogonally to the sidewalls (Figures 7A,B), simulating the E–W opening direction (extensional stresses) in the studied area, that has been acting for the last 14.5 ka (Pasquarè Mariotto et al., 2015); such opening direction is also supported by GPS data from Jouanne et al. (2006).

The edge effect due to the sharp termination of the plate has been minimized by choosing the thinnest possible rigid aluminum plate, with a thickness of 0.24 mm. The geometry of the metal plates has been tailored in order to mimic as much as possible the fault geometry studied in the field. Based on recently collected field data (e.g., Pasquarè Mariotto et al., 2015), the TFS is assumed as trending N06° based on its overall geometry, as well as on the first westward normal fault of the TFS (the GF); the eastern part of the HFF, close to the triple junction, is assumed as N130°-striking (Figure 2C). A shown in Figure 7A, the GF and the HFF are highlighted with a red line and represent the limits of the four plates.

Scaling Procedure

Analog experiments were geometrically and dynamically scaled (e.g., Holohan et al., 2005; Norini et al., 2008 Ruch et al., 2012). Material properties and forces in the experiments should comply with the principle of geometric and dynamic similarity (Hubbert, 1951) and operate according to a consistent length ratio (h*) determined by $h^* = h_{Model}/h_{Nature}$ (Merle, 2015). For the geometric scaling of our models we chose a scaling ratio given by $l^* = l_{Model}/l_{Nature}$, $= 1 \times 10^{-4}$, such that 1 cm in the model scales to 100 m in nature. In agreement with this scaling, the length of the GF in the model is 60 cm, corresponding to its actual length (about 6 km) in nature (Pasquarè Mariotto et al., 2015). The dynamic scaling factor (stress ratio, $\sigma^* = \sigma_{Model}/\sigma_{Nature}$) is calculated from the equation $\sigma^* = \rho^* g^* l^*$. As the density of analog materials is smaller than that of natural rocks by a factor of ~2 (i.e., density ratio—$\rho^* = 0.5$), and gravitational acceleration is the same in both model and nature (i.e., gravity ratio—$g^* = 1$), the required stress ratio for a geometric scaling factor of 10^{-4} is $\sigma^* = 5 \times 10^{-5}$. As the cohesion of natural rocks lies in the range 10^5–10^7 Pa (Hoshino et al., 1972 Glicken et al., 1980), a suitable cohesion for the analog material needs to be in the range 5–500 Pa. Higher values of 10^7 Pa, in nature, are representative of small laboratory samples (Schultz, 1996) or intact rock (Handin, 1966; Jaeger and Cook, 1971).

In our experiments we used a mixture of two end-member analog materials, i.e., sifted sand and wheat flour, with the purpose of simulating brittle deformation in the upper crust (e.g., Walter and Troll, 2001; Holohan et al., 2005; Tibaldi et al., 2008; Acocella et al., 2013; Tortini et al., 2014). The density range for natural rocks is approximately 2000–3300 kg/m^3, with most rocks (granites, basalts, hard limestones, etc.) lying between 2600 and 3000 kg/m^3 (Goodman, 1989). Therefore, we used a mixture of wheat flour and sifted sand, both 50% in volume: (i) the dry, sifted sand is characterized by a grain size of 200÷ 350 μm, a bulk volume of 1470 kg/m^3 and a low cohesion (e.g., 30–40 Pa—Walter and Troll, 2001); (ii) the wheat flour has a bulk density of 700 kg/m^3 and a cohesion of 330 Pa (Krantz, 1991); (iii) our mixture has a density of 1215 kg/ m^3, a measured ϕ of 41° and a cohesion surely between the range 5–500 Pa, in order to reach a scaled value close to natural rocks (10^6 Pa; e.g., density ratio and cohesion range).

According to the Bukingam-∏theorem, or pi Theorem (Tibaldi, 1995; Galland et al., 2014; Merle, 2015, and references therein), seven variables minus three dimensions, equal to four independent dimensionless numbers (Table 1) are needed to maintain a value as close as possible between the field and the experimental system, in order to warrant similarity. Time was not scaled in our experiments, because the brittle behavior of sand and other Mohr–Coulomb materials is generally regarded as independent of the strain rate (Holohan et al., 2005; Merle, 2015).

Table 1. List of parameters and dimensionless numbers used in this work.

Parameters	Symbol	Dimensions		
Fault scarp height	H	L		
Fault scarp length	L	L		
Substrate thickness	T	L		
Rock density	ρ	ML^{-3}		
Internal friction	Φ	–		
Rock cohesion	c	$ML^{-1}T^{-2}$		
Gravity	g	LT^{-2}		
Dimensionless numbers	**Real values**	**Real values**	**Real values**	**Experimental values**
$\prod_1 = FS_h/FS_l$	30 m/6000 m	=0.005	3/600 mm	=0.005
$\prod_2 = L_{scarp}/H_{sub}$	6000 m/500 m	=12	600/50 mm	=12
$\prod_3 = \Phi$		30–45		40
$\prod_4 = \rho gh/C$		1.25–122.5		1.1907–119.07

The first and second dimensionless numbers account for the geometric ratio of the system (e.g.,Merle, 2015):

$$\prod_1 = \textit{vertical offset of GF/GF length}$$

In nature, the GF is 6-km-long and has a maximum offset of about 30 m; therefore, at the scale of the experiment, it has a length of 60 cm and a vertical offset of 3 mm.

$$\prod_2 = \textit{GF length/thickness of the substratum}$$

We consider a crust portion of about 500 m; hence, at the scale of the experiment, the GF has a length of 60 cm and the substratum thickness is about 5 cm.

The third dimensionless number accounts for the brittle behavior of the analog material and the substratum. It is the angle of internal friction:

$$\prod_3 = \Phi$$

Angles of internal friction of the sand (Φ_{sand} = ~33°) and the flour ($\phi_{wheatflour}$ = ~43°) closely resemble those in natural rocks (30–40°, Byerlee, 1978) as well as in volcanic rocks (Φ_{rocks} = 30–45°, Walter and Troll, 2001). Our analog material has an Φ = ~41°.

The fourth and last dimensionless number is the ratio between the gravity stress and the cohesion (Merle, 2015):

$\prod_4 = \rho gh/c$

In our experiments, h corresponds to the thickness of the substratum (500 m scaled to 5 cm in the model) and $\prod_4$ ranges from 1.19 to 119, with a cohesion for the analog material in the range 5–500 Pa; we used an analog material with a cohesion of about 150 Pa. An average density for natural rocks is about 2500 kg/m^3, but the range for sedimentary and magmatic rocks is from 2300 to 2900 kg/m^3 and their typical cohesion is between 10^5 and 10^7 Pa (Hoshino et al., 1972; Glicken et al., 1980). Thus, overall values of $\prod_4$ in nature range from 1.22 to 122, with a fixed density of 2500 kg/m^3 and varying the cohesion (10^5 and 10^7 Pa). Hence, the values of $\prod_4$ in the experiments correspond to those in natural systems. So the model has been geometrically and dynamically correctly scaled.

Quantification of Fault Slip and Experiment Set-Up

We conducted a series of analog modeling experiments by simulating the movement of the NW, SW, and SE blocks (Figure 2) with different combinations, always imparting an amount of westing movement based on GPS data, for the E–W opening direction (or west spreading), derived fromJouanne et al. (2006). Based on the published GPS measurements, we considered a westing movement of about 10 mm/yr for the NW block, 15 mm/yr for the SW block, and 6 mm/yr for the SE block. Since faults and tension fractures (Figure 2) affect the 14.5 ka-old lavas, a maximum westing movement of 145 m for the NW block, 217 for the SW block, and 87 m for the SE block has been reconstructed. At the scale of the model (10^{-4}), the above values correspond to about 14.5, 21.7, and 8.7 mm, respectively. Consequently, the NW block moves about 1.5 times faster than the SE one, and the SW block moves about 2.5 times faster than the SE one. Such configuration has been reproduced in the models, firstly by moving only the SW block, then the NW and the SW blocks, and then by moving contemporaneously all three blocks at the scaled westing velocity. Finally, a series of experiments has been conducted by creating first the rift zone and then the transform fault across it. At each step, the SE block has moved to the east of 0.5 mm, the NW one has moved of 0.75 mm and the SW one has moved of 1.25 mm, corresponding to 5, 7.5, and 12.5 m in nature.

By taking the NE block as fixed: *(i)* we reproduced extension in the northern part of the TFS by moving the NW block toward the west, *(ii)* as well as we reproduced the possible prolongation of the HFF across the TFS by moving also the SE block toward the west (Figure 7). The extension in the southern part of the TFS as well as the movement along the HFF was reproduced by moving the SW block at a rate higher than the NE and SE blocks, consistently with field and GPS data. Such data also suggest that the extension amount across the southern part of the TFS is greater than that going on in the northern TFS (roughly estimated in about 8–10 and 3–5 mm/yr, respectively); such setting has been mimicked in our analog experiments by choosing this block configuration.

The geological-structural evolution of the study area has been characterized by different episodes of lava emplacement and faulting events. In order to develop experiments which are as much as possible similar to reality, we simulated the evolution of the study area by creating a deposit of the granular mixture above the metal plates, then inducing fractures along the transform fault zone and the rift zone by deforming the granular mixture, then filling structural depressions and the adjoining areas with new granular material and then deforming it again.

Furthermore, we ran two different sets of experiments, the first with no ductile connection among the plates (the rubber sheet), and the second with a ductile connection as large as the TFS. During the experiments without the ductile connection we focused on the surface effect of propagation of a pre-existing deeper fracture. In the second series of experiments we added a 30-cm-wide rubber strip in order to: (i) locate the rift zone area (e.g., Morley, 1999); (ii) simulate a broad zone of extension (rift zone) between the two plates (McClay, 1990; Tron and Brun, 1991; McClay and White, 1995). In our model, the rubber sheet has a trend of N06°, similar to the TFS overall orientation. The TFS width (e.g., Figure 2) has been scaled to the model (Figure 7A). The rubber sheet has been centered over the edge of the metal plate and has been securely attached (e.g.,Clifton et al., 2000), so as to restrict extension to the central part of the model (e.g., McClay, 1990). Consequently, the angle between the zone of rifting and the extension direction was controlled by the orientation of the rubber sheet (e.g., McClay and White, 1995). Furthermore, we added a 4-cm-wide rubber strip along the HFF.

Assumptions and Limitations

Our box hosting the granular mixture was made with dimensions larger than those necessary for highlighting the actual area of interest near the transform fault—rift junction. This option was chosen in order to avoid side effects that might perturb the geometry and kinematics of the experiments. This correct procedure and scaling is testified to by the observation that, in all experiments, the measured surface deformation amounts, as well as the geometry and kinematics, do not vary at considerable distance form the area of interest, and perturbations do occur only at few-cm wide strips near the boundary walls. This suggests that boundary effects due to the confinement of the granular mixture within the box may be neglected.

In our models we simulated the pre-existence of fractures along the transform fault zone and the rift zone by deforming the 4 cm-thick granular mixture, then filling structural depressions and the adjoining areas with a 1 cm-thick layer of new granular material and then deforming it again. This was aimed at mimicking the evolution of the studied area in the most realistic way, keeping into account the existence of new lava flows covering the studied area over time (e.g., Saemundsson et al., 2012). However, we need to acknowledge that the interior of the resulting blocks is characterized by isotropic conditions: any pre-existing fracture and layering within the blocks that could affect the resulting deformation pattern is not included and is beyond the scope of the study.

During the experiments based on the set-up without the ductile connection between the plates (the rubber strip), as already introduced before, the edge effect due to the sharp termination of the plate has been minimized by choosing the thinnest possible aluminum plate (0.24 mm).

In regard to the influence of igneous intrusions on the formation of the structures in the studied area, we are aware that tension fracture opening can occur along a rift also as a consequence of magmatic pressure (Gudmundsson, 1987a,b, 1995; Opheim and Gudmundsson, 1989; Trippanera et al., 2014, 2015; Sigmundsson et al., 2015). In fact, it is well-known that, at the tip of an advancing dyke, tensile stresses can increase due to magma injection (Rubin and Pollard, 1988;Rubin, 1992) and exceed the Coulomb failure threshold, triggering earthquakes in the brittle crust (Brandsdóttir and Einarsson, 1979; Einarsson and Brandsdóttir, 1980; Feigl et al., 2000; Cattin et al., 2005). The effects of magmatic pressure at an advancing dyke have already been explored in several papers by field data, numerical modeling, and analog experiments (review in Rivalta et al., 2015, and reference therein). The present paper, instead, is mainly devoted to understanding the overall geometry and kinematics of the structures, and assessing the relative development of the transform and rift zones. Nevertheless, a mention of the implications of the possible role played by dykes in the study area is provided in the discussion section. It is worth mentioning that the strain rate may vary considerably in time, because intrusions can produce meter-scale deformation in few days (e.g., Metzger et al., 2011).

Finally, thermal effects are not included in our experiments. Although this is a common limitation of most analog experiments, we stress also the fact that possible differences in the thermal flux below the HFF respect to the TFS are not known in the study area and it would have been difficult to quantify them anyway. An anomalous heat flux should be present further south, where the Þeistareykjavirkjun geothermal plant is under construction.

EXPERIMENT RESULTS

Here, we describe only a selection of the total of 20 experiments we have conducted, and will dedicate another paper to the full description of all models. The examples reported below refer to the models that are more relevant to the discussion of the development of different structures around the HFF-rift junction. We will present experiments with a more and more complex set-up in order to assess the role played by the relative movement of each tectonic block in creating the final general structural pattern observed in the field (experiments 1, 2, and 3). This was done also in view of the previously suggested different models of transform fault termination. Experiment 4 has been carried out to simulate a broader zone of extension, and finally experiment 5 was conceived to verify the relative timing of tectonic events.

Under the set-up of the first series of experiments (no ductile connection among the plates, Table2), experiment 1 was run by moving only the plate south of the HFF (SW plate in Figure 7) toward the west with respect to the other fixed plates, mimicking a two-plate configuration (Figure 8A). After gradual increments in motion, we added another 1 cm-thick layer of granular material and then applied the successive deformation in gradual steps. In Figure 8A we show the step with an extensional cumulated deformation of 7.5 mm that corresponds to the slip cumulated by the "Theistareykjahraun" lavas at a rate of 6 mm/yr during the last 12.5 ka (Table 2). The movement created a series of faults all around the plate boundaries (Figure 8B); most notably, a series of en-échelon, strike-slip right-lateral faults with a left-stepping geometry formed, which show also a subordinate normal component. With reference to the north coinciding with the moving end-wall of the experiment box (Figure 7), these faults strike mostly N130–140° in the central and western part of this fault zone. Toward the east, the faults gradually bend attaining a N–S orientation. The transcurrent kinematics is gradually replaced by normal displacements with the formation of scarps facing both eastward and westward. The normal faults are accompanied by the development of parallel extension fractures.

Table 2. List of experiments done in the present work.

Nr.	Figure	Westing movement (mm)				Note
		NW	SW	NE	SE	
FIRST SERIES OF EXPERIMENTS—NO DUCTILE CONNECTION						
1	9	0	7.5	0	0	two-plate configuration
2	10	6	10	0	0	three-plate configuration
3	11	5.75	12.5	0	4	four-plate configuration
SECOND SERIES OF EXPERIMENTS—DUCTILE CONNECTION						
4	12	11	20	0	7.5	four-plate configuration
5	13A-B	7.5	25	0	9	four-plate configuration

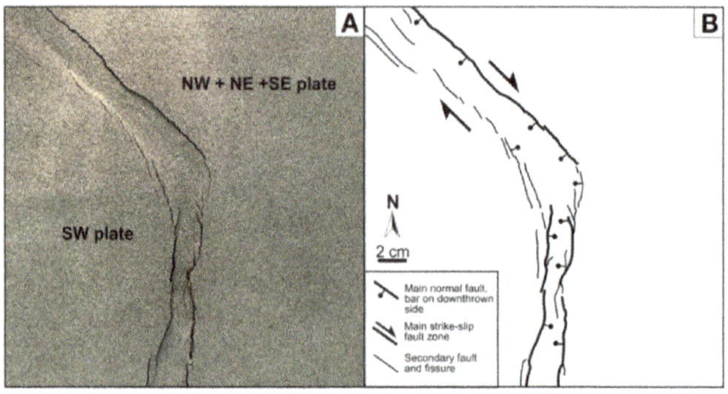

Figure 8. (A) Photo of the experiment 1, made with only the SW plate moving westward, and the other three plates fixed. **(B)** Sketch of the photo with indication of the main and secondary faults and extensional fractures. Note the development of the transform fault (HFF) with a right-lateral transtensional kinematics and of the southern GF with normal motions. This one-plate kinematics does not account for the other structures found in the field.

Experiment 2 (Table 2) uses a three-plate configuration (Figure 9A) obtained by moving contemporaneously the plates south and north of the HFF (SW and NW plates in Figure 7) toward the west at different velocities, and keeping the two plates east of the TFS locked. Also in this case, after the various gradual increments in motion, we added another layer of granular material; then we gradually imposed a successive westward motion totalling 10 mm at the SW plate, and a total westward motion of 6 mm to the NW plate. The plate movements resulted in a zone of en-échelon, normal right-lateral faults with a left-stepping geometry like in the previous set of experiments. With the set-up of experiment 2 (Figure 9), anyway, we can see the development of the rift zone north and south of the junction with the transcurrent fault zone. It is worth observing the greater number of normal faults and higher normal offsets south of the junction, consistent with the larger westward displacement of the plate located south of the HFF. All faults and extensional fractures developed along the rift have the same dominant orientation, although locally they show a gentle zig-zag pattern. In correspondence to the junction between the transcurrent fault and the rift, there is no offset of the rift faults but rather a complex zone with the formation of rhomboidal structures in plan view. There is no gradual bending of the transcurrent fault near the junction, but instead this fault maintains a rectilinear geometry until it ends in correspondence of the eastern-most fault of the rift.

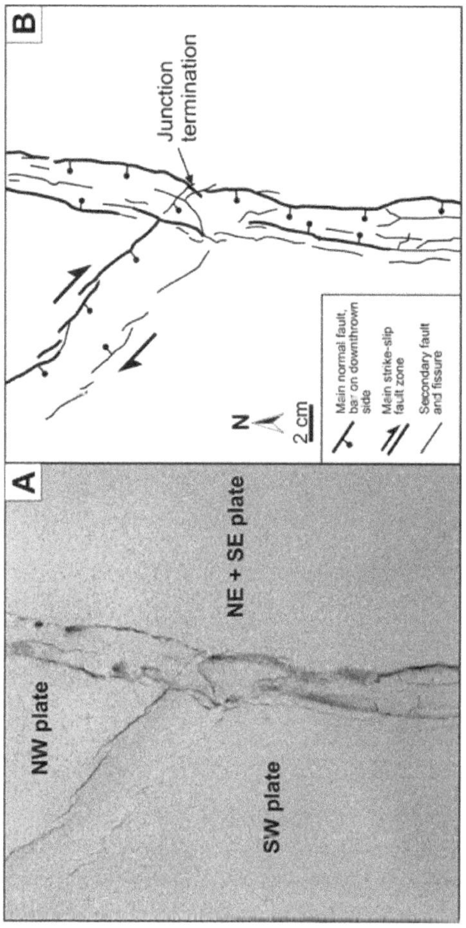

Figure 9. (A) Photo of the experiment 2, made with only the NW and SW plates moving westward, and the other two plates fixed. **(B)** Sketch of the photo with indication of the main and secondary faults and extensional fractures. Note the development of the transform fault (HFF) with a right-lateral transtensional kinematics and of the complete GF both north and south of the junction (indicated by the arrow). This two-plate kinematics does not account for the interruption of the GF north of the junction and its bending, as well as of the other field structures located further to the east.

Experiment 3 (Table 2 and Figure 10) is based on a four-plate configuration obtained by moving contemporaneously, with differential increments, the plates south, and north of the HFF toward the west, by moving at a lower rate toward the west the plate south-east of the TFS (SE plate) and keeping the plate north-east of the TFS (NE plate) locked. The procedure followed the same steps described for the previous experiments. The cumulated westward motion of the SW plate is 12.5 mm, for the NW plate it is 5.75 mm, and the total cumulated offset of the SE plate toward the west is 4 mm (Table 2). The plate movements induced the formation of the normal right-lateral faults between the NW and SW plates, which show a N130–140° strike in the central and western part of the transtension zone. At the eastern part instead, there is a gradual anticlockwise bending of these faults that attain a N–S strike in correspondence of the junction with the rift. It is also worth highlighting the formation of three main fault strands at the eastern termination of the right-lateral transtension zone, all affected by a larger component of normal motions with respect to the more western segments. This fault zone is connected with the southern segment of the rift zone, whereas there is no connection with the northern segment. The rift is characterized by a discontinuous development of faults and extension fractures, and it is particularly worth highlighting the gradual anticlockwise bending of the southern termination of the northern segment of the rift faults (arrow in Figure 10B). Southeast of the junction, a series of en-échelon left-stepping fractures developed. These structures lye above the boundary between the NE and the SE plate, whose differential movements mimicked a dextral transtensional zone.

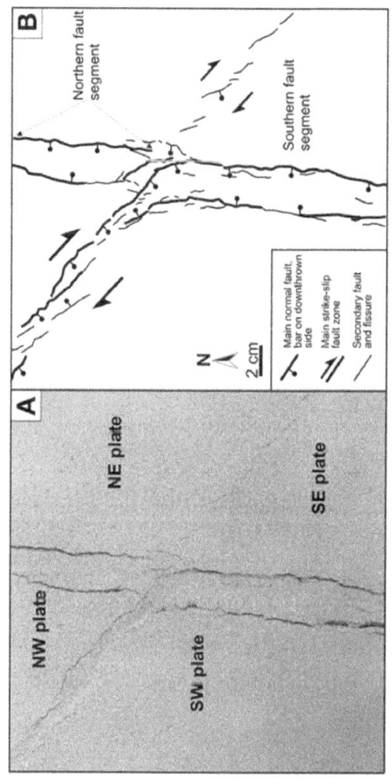

Figure 10. (A) Photo of the experiment 3, conducted by moving westwards at differential velocities and amount the NW, SW, and SE plates, and keeping the NE plate fixed. **(B)**Sketch of the photo with indication of the main and secondary faults and extensional fractures. Note the strong similarity with the real field data of the same area, as shown in Figure 3. The HFF gradually bends toward the south near the junction with the GF. North of it, the northern fault segment bends toward the SE. Southeast of the junction, a series of en-échelon, left-stepping fractures developed, as observed in the field.

A second set of experiments was run by adding a ductile connection among the plates, as described above. This 30-cm-wide rubber strip was used to locate the rift zone area (e.g., Morley, 1999) and to simulate a broad zone of extension (rift zone) between the two plates (McClay, 1990; Tron and Brun, 1991; McClay and White, 1995). Experiment 4 was conducted under the same set-up described for experiment 3—i.e., a four-plate configuration, and the metal plates were moved up to a similar cumulated total displacement as in the previous set-up (Table 2). It is possible to notice that the four-plate configuration with the ductile connection results in a diffuse zone of strain in correspondence of the rift zone

(Figure 11), giving rise to a wider rift than that in the previous experiments. The rift zone is characterized by a series of anticlockwise-rotated fractures in correspondence of the boundary between the NE and SE plate. West of the junction, the normal right-lateral fault zone developed.

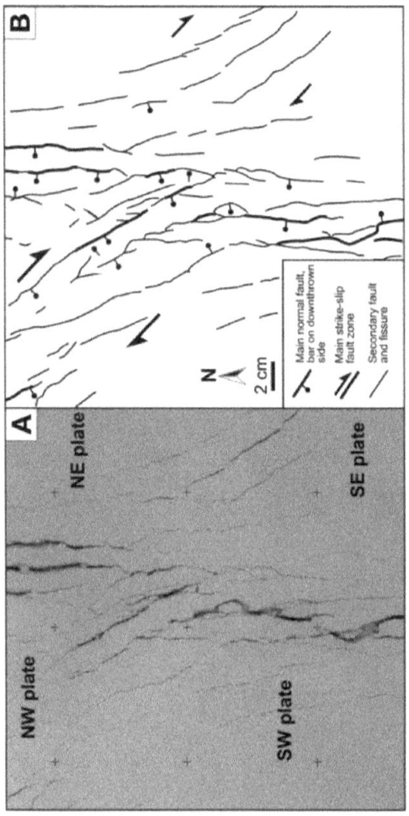

Figure 11. (A) Photo of the experiment 4, made by moving westwards at differential velocities and amount the NW, SW, and SE plates, and keeping the NE plate fixed, plus a diffuse zone of strain across the whole rift (TFS). **(B)** Sketch of the photo with indication of the main and secondary faults and extensional fractures. In this case several other faults and extensional fractures developed east of the GF, mimicking the overall structure of the real TFS. The change in strike of the TFS structures southeast of the HFF–GF junction is related to the presence of a NW–SE right-lateral shear zone below the rift, and supports our interpretation of the HFF prolongation.

Experiment 5 was conducted again with a four-plate configuration and the ductile lower zone but, in this case, we first moved the NW and SW plates, creating the rift, and later on we moved only the SW and SE plates inducing the HFF also across the rift (Figures 12A,B). This experiment, thus, mimics the recent propagation of the HFF toward the southeast, across an already developed rift zone (TFS). The metal plates were moved up to a total slip of 7.5 mm (NW plate), 25 mm (SW plate), and 9 mm (SE plate; Table 2). We can observe that this set-up produces a clockwise rotation and some offset of the rift faults. Most fault strands along the prolongation of the shear zone through the rift attain a N10 to N30° strike, which does not correspond to the field geometry of the actual TFS here. As a comparison, Figures 12C,D show another experiment done as the set-up of experiment 4 with the simultaneous development of both the rift and the shear zone. Note that, in this case, like in Figure 11 of experiment 4, the fractures along the prosecution of the shear zone across the rift strike more NNW–SSE, and thus they have a different geometry from the structures developed in experiment 5.

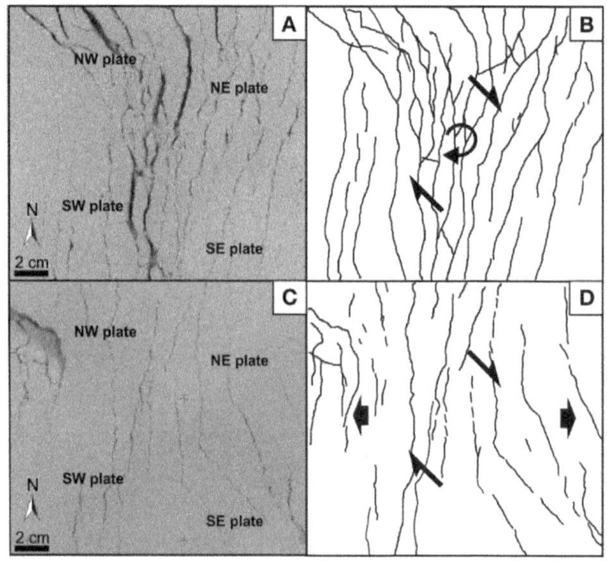

Figure 12. Photos of the analog experiments and sketch of the resulting fracture field under different time relationships between the development of the rift zone (TFS) and the SE-ward prolongation of the transform fault (HFF). In **(A)** and **(B;** experiment 5) we observe the geometry of the tension fractures and faults obtained first by forming the rift zone and then by inducing the HFF right-lateral shear zone across the rift. **(C)** and **(D)**. represent, instead, the geometry when the rift zone develops contemporaneously to the right-lateral shear zone across it (i.e., the HFF prolongation). Note that case **(C, D)**. is in better agreement with the field data.

DISCUSSION

The field data here presented allow to better define the geometric and kinematic characteristics of the studied faults, whereas the integration with analog models enables to constrain the space and time relations between the transform HFF and the TFS rift zone. The outcomes resulting from this combined approach will be discussed in the following sections.

General Features Based on Field Data

Starting with the HFF, this major fault shows a complex geometry and kinematics toward the junction with the TFS. With respect to the classical models of terminations of transform faults (Figure 1), it appears that the HFF has an intermediate character between several end-members; the fault ends via a branching, toward the south, of the main fault plane into various splays, which gradually bend (Figure 13). They are characterized by a normal component of movement that increases toward the east, and thus the overall geometry and kinematics resemble the model of Figure 1B in the form of a leading extensional imbricate fan. The leading (i.e., easternmost) splay fault merges with the GF, thus recalling also the model of Figure 1C. However, the possible prolongation of the HFF toward the southeast, across the TFS, is more similar to the model of Figure 1D, with the transform fault that terminates gradually in correspondence of a through-going normal fault system. These data, apart from improving the knowledge about this Icelandic zone, also indicate that the detail that can be obtained in the field enables to define the geometric setting of a transform-rift junction, which is indeed much more complex than those that can be reconstructed by means of offshore studies.

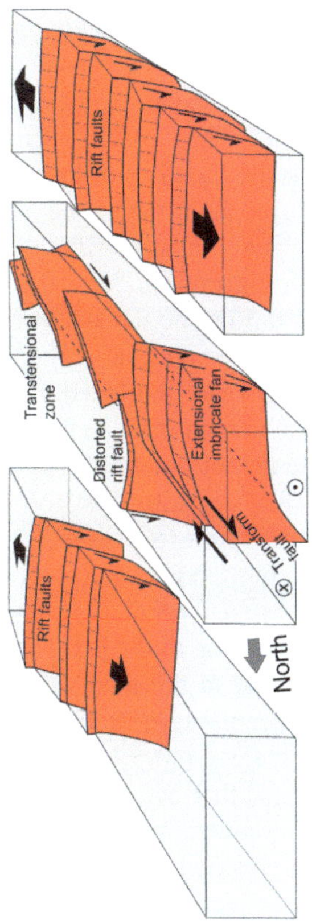

Figure 13. The model of transform fault termination suggested by our data: The main transcurrent fault plane gradually bends and splays giving rise to a releasing imbricate fan. Further east, the fault interacts within the rift zone producing en-échelon fissures and normal faults rotated anticlockwise with respect to the N–S rift structures. The size of the half arrows (along the transform fault) is proportional to the amount of strike-slip motions; the size of the full arrows is proportional to the extension amount in the rift.

The configuration of model of Figure 1D is associated with a three-plate configuration, which corresponds to: (i) the tectonic bock located south of the outcropping HFF, (ii) the block located north of it, and (iii) the block located east of the TFS. The latter corresponds to a zone of diffuse deformation as indicated by the presence of several extensional fractures and 10 major, normal faults, all of which are striking mostly around a N–S direction. The extension directions measured in the field at the TFS are mostly from N90 to N100°, and in the SW block (located south of the outcropping HFF) the dominant extension axes are N100–110° (Tibaldi et al., 2016). These axes are consistent with the component of extension recognized in the field along the HFF, which thus can be classified as a transtensional fault, at least along its eastern termination. This range of values is consistent with the GPS vectors measured in the same area by Jouanne et al. (2006) and Metzger et al. (2013).

We wish to underscore the lack of a direct connection between the GF north and south of the junction with the HFF. The detailed field data of Figure 6 shows that the segment of the GF north of the junction bends toward the SE and terminates without connecting with the GF south of the junction. The apparent offset between these two segments of the GF is about 20 m in a strike-slip right-lateral sense (see Figure 3).

All these observations represent also a robust dataset for planning the analog experiments that we conducted to better define the structural evolution of the interaction zone between the HFF and the TFS. From the field data alone, in fact, it is difficult to fully understand how the system works. Several different hypotheses can be proposed, with particular regard to the phases of development of the different faults. For our analog experiments we used different set-ups to reproduce the development of the HFF with respect to the TFS, as discussed below.

Prolongation of the HFF by Field Evidence and Analog Modeling

Previous research suggested that the HFF ends at the junction with the westernmost fault of the TFS, represented by the GF (Gudmundsson et al., 1993; Magnúsdóttir and Brandsdóttir, 2011). This observation is convincing if one looks at the surface trace of the already known HFF: this fault zone, in fact, is composed of a series of closely spaced fault scarps that mostly strike N130–140° and than bend into the GF, without any evident prolongation east of the junction. More recently, a possible subsurface prolongation of the HFF toward the southeast has been proposed, on the basis of earthquake and fracture distribution near the Krafla fissure swarm (Hjartardóttir et al., 2012). In correspondence of the TFS, Tibaldi et al. (2016) suggested the possible prolongation of the HFF southeast of the junction based on substantial field evidence, including changes in the TFS structures north of the prolongation of the HFF, with a decrease in: (i) the width of the TFS; (ii) the number of faults; (iii) the fault offset values. Moreover, iv) along the prolongation of the HFF there is a generalized anticlockwise rotation of the TFS structures (Figures 4, 6). We hypothesize that this evidence can be reconciled in terms a four-plate configuration given by the tectonic blocks located north and south of the already known trace of the HFF (NW and SW blocks in Figure 2), plus a third tectonic block in correspondence of the southeastern TFS (SE block in Figure 2) and a fourth tectonic block in the northeastern sector of the TFS (NE block in Figure 2). In order to test whether this block configuration might produce the observed structures, we have run a series of analog experiments with different set-ups and compared our experimental outcomes with field data.

Experiment 1 (Figure 8) created a series of major normal faults with angles of 50–60° with respect to the right-lateral strike-slip fault. This range matches the 50–60° angles measured in the field between the N130° and 140° segments of the HFF with respect to the GF. Anyway, this configuration, in which we moved only the plate south of the HFF, does not reproduce all the features of the studied area, and especially the normal fault segments present north and east of the junction.

The set-up of experiment 2 (Figure 9) created a series of major normal faults with an angle of 50–60° with respect to the right-lateral strike-slip fault, both north and south of the junction. This configuration reproduces correctly the HFF, the GF and its connection with the HFF, as well as the other, N–S-striking normal faults and extensional fractures, located north of the junction. In particular, the larger westward displacement of the plate located south of the HFF with respect to the plate located north of it, produced a greater amount of normal faults and larger normal offsets south of the junction. This is exactly what we found along the western sector of the TFS. By increasing the total motion of these two plates, we obtained an enlargement of the offsets of the faults along the rift zone. Anyway, the faults and extensional fractures developed along the rift have all the same orientation, and thus this set-up of the experiments is not able to mimic the variation in the geometry of the fractures of the TFS near the HFF–GF junction, which is characterized by the bending of the northern segment of the GF, and does not reproduce the fractures located along the possible prolongation of the HFF.

Experiment 3 (Figure 10) induced: right-lateral transtension along the HFF, the junction with the GF, larger normal motions south of the junction, the bending of the northern segment of the GF, and en-échelon fractures south-east of the junction. We wish to stress that only this plate configuration created an anticlockwise rotation of the fractures of the GF both north and south of the HFF–GF junction. If we compare the structural pattern around the junction in experiment 3 (Figure 10) with the field data for the same area shown in Figure 6, we can observe the strong similarity. It is also possible to notice, in the experiment, the development of a series of left-stepping, en-échelon structures southeast of the junction. These structures formed above the boundary between the NE (fixed) plate and the SE plate, which was moved very slowly westwards, hence mimicking a dextral, transtensional zone.

In experiment 4 (Figure 11), a wide rift zone fully develops and a more complex geometry occurs along an area that corresponds to the boundary between the underlying NE and SE plates, represented by an anticlockwise rotation of the TFS structures. This re-orientation of the rift zone fractures has been induced by the local orientation of the σ_3 linked to the right-lateral shear zone, and mimics the geometries found in the field at the TFS along the prolongation of the HFF, as shown in Figure 6 and resumed in Figure 13. Based on experiments 3 and 4 (Figures 10, 11), we argue that the four-plate configuration is the one that fits best with field data, and thus our analog experiments confirm the hypothesis proposed by Tibaldi et al. (2016) regarding the prolongation of the HFF across the TFS southeast of the junction with the GF. Our work indicates that such prolongation may amount to, at least, 3 km into the TFS.

In regard to the relative time of propagation of the HFF across the TFS, experiment 5 (Figures12A,B) indicates that, in the case of the HFF propagation toward the southeast after the development of the TFS, the rift fractures tend to rotate in a clockwise direction, and locally they are right-laterally offset, instead of assuming a NNW–SSE strike (anticlockwise respect to the dominant N–S orientation of the rift fractures) as observed in the experimental set-up of Figures10, 11, 12C,D and, especially, in the field. In our experiments, only in the case of contemporaneous motions between the rift and the right-lateral shear zone across it, it has been possible to reproduce all the features seen in the field. As a consequence, the faults and tension fractures of the TFS must have developed at the same time as the right-lateral motions along the southeastern prolongation of the HFF.

The relative motions among the various moving plates in the analog experiments are such that the amount of the right-lateral strike-slip movement between the NE and SE plate is much smaller than the strike-slip movement between the NW and SW plate. This implies that, in the experiments, the segment of the transform fault east of the junction is much less developed at the surface than the transform fault segment west of the junction. This is consistent with the geometry of the structures that developed during the experiments. West of the junction, in fact, where the cumulated offset is larger, the transform fault zone is expressed by a series of long en-échelon fracture segments oblique to the main fault zone, and a few segments parallel to the fault zone, most of which are interconnected. This fault architecture corresponds to the Riedel (R) and P structures that form during the development of a shear zone, following the classical nomenclature of Tchalenko (1970). East of the junction instead, the transform fault geometry is given by a series of shorter en-échelon fractures at higher angle respect to the general deformation zone. This architecture recalls the Riedel' (R') structures that form in the very early stage of a transtensional zone (Sylvester, 1988; Agostini et al., 2009). Moreover, we need to keep into account the role played by the combination of transform faulting and rifting, evidenced by both field data and analog modeling. The same two patterns of structures are present in the field at the two sides of the junction HFF-rift. Since the lava deposits east and west of the junction have the same age (14.5 ka BP, Saemundsson et al., 2012), we can conclude that two fracture patterns correspond to two stages of development of the HFF; the segment east of the junction is less developed because here the HFF moves at a much lower rate than west of the junction, at least from the perspective of a Holocene time window.

From a more general perspective, our results contribute to improving previous models on transform fault termination. The field and experimental data suggest that, over a given time window, a transform fault might end through a decrease in the slip rate across the rift, partitioning its motion in correspondence of more than one companion structure (Figure 13). But our model also implies that a transform fault tip may be unstable with time and that the fault may propagate across the rift; therefore, fault length can increase with time. Another possibility is that the horizontal position of the transform fault tip is stable with time, but its location cannot be recognized simply because it is covered by lava deposits which are younger than the latest fault motion. In this case, the next fault motion will occur via an upward propagation of the fault plane close to the tip.

Role of Magma Intrusions

In the previous sections we have contributed to explaining the geometry and kinematics of the structures at the zone of intersection between the HFF and the TFS, with a very good fit between field and analog data under a four-plate configuration. Considering the forces responsible for this deformation field, it is necessary to take into account also the role of igneous intrusions.

As classically accepted, the magmatic pressure from intruding dykes may induce tension fracture opening at depth and at the surface (Gudmundsson, 1987a,b, 1995; Opheim and Gudmundsson, 1989; Sigmundsson et al., 2015). An overpressured dyke exerts tensile stresses along its tip line, which are capable of producing an advancing fracture; once the fracture has formed, the magma intrudes along it and the dyke propagates (review in Tibaldi (2015), and reference therein). If a fracture already exists and has a suitable orientation with respect to the horizontal regional least principal stress (σ_3), dyke intrusion is facilitated. Dyke intrusion can release the strain cumulated under an extensional tectonic regime, such as the one in Iceland. Also in this case magma tends to propagate preferably along a plane that is normal to the direction of the maximum extensional strain. Iceland is subject to regional forces that pull apart the western Atlantic plate from the eastern Atlantic plate (Oskarsson et al., 1985; Sigmundsson, 2006). The observed velocity of spreading across the northern rift zone is 18.1 ± 1.2 mm/yr in direction N109.3° ± 4.2° consistent with plate models (Sella et al., 2002; Geirsson et al., 2006). The elastic strain accumulates in the crust until it is released by coseismic faulting and/or by dyke intrusion.

Based on the geometric and kinematic compatibility between the observed faults and tension fractures in the study area, and the data coming from regional tectonics and GPS-recorded movements, we argue that magma may intrude the TFS rift zone guided by the regional deformation field. Dykes propagate following the regional σ_3 that here is oriented about N100–110°. As dykes intrude, they release the accumulated regional strain and their emplacement is accompanied by rigid deformation of the host rocks. At the scale of the general studied area, this deformation is accommodated by the pattern of structures of the TFS described so far, whereas locally syn-dyke deformation may include bulging above the dyke. The syn-dyke deformation may be accompanied by the development of tension fractures and normal slip at faults dipping inward above the dyke intrusion zone (Mastin and Pollard, 1988 Trippanera et al., 2015) and possible inversion of fault kinematics with reverse slip at shallow level (Gudmundsson and Loetveit, 2005;Gudmundsson et al., 2008).

The fact that there are several Holocene lava eruptive sites south of the HFF and none north of it, suggests that most dykes should propagate with a limitation boundary represented by the HFF. This hypothesis may be explained in two ways: (i) the HFF might act, at least in part, as a barrier against the horizontal propagation of dykes, represented by a mechanical discontinuity where local stresses may re-orientate; this phenomenon has been proposed elsewhere for the vertical arrest of dykes by Gudmundsson (2002, 2011); (ii) the role of the HFF as an arrest boundary may also be more related to its oblique orientation with respect to the dyke intrusion direction; dykes that start intruding south of the HFF exert a lateral force on the SW and SE tectonic blocks (with reference to Figure 7). Moreover, another point to keep into consideration is the fact that the SW block is subject to extension (rift opening plus favorable kinematics from the HFF), whereas the NW block is subject to a lesser degree of extension (slower rift opening). Hence, potential intrusions north of the HFF and west of the GF would be substantially hindered. This may lead to preferential intrusion in the SW block. Once the process of strain release starts during dyking, these blocks move along the normal faults and along the HFF that acts as a weakness zone with respect to the more stable NW and NE blocks. Through this process the magma force exerted on the dyke walls is transformed into deformation of the SW and SE blocks, and the remaining magma overpressure might not be sufficient to allow intrusion further north across the HFF. Anyway, the possibility that (with a lower frequency) the magma driving overpressure may be large enough to sustain dyke propagation also through the HFF cannot be ruled out. Finally, it is worth mentioning the considerations of Maccaferri et al. (2013), who documented that large dyking events or rifting episodes in the NVZ have the potential to change the state of stress in the Tjörnes Fracture Zone (TFZ), hence including the HFF. According to these authors, major magmatic events in the NVZ (including the Theistareykir, Krafla, or Askja volcanoes) might transfer major positive Coulomb stresses to the currently locked portion of HFF, thus accumulating tectonic strain.

CONCLUSIONS

We integrated detailed field data with analog scaled experiments to better understand the interaction between an active transform fault and an active rift zone. The field observations come from recently-published data by Pasquarè Mariotto et al. (2015) and Tibaldi et al. (2016) on the transform HFF and the Theistareykir Fissure Swarm (TFS), integrated by new data on their junction zone.

The field data show that the western-most fault of the TFS, known as GF is not a continuous rectilinear fault running across the junction with the HFF; in fact, our data indicate that this N–S fault is broken into two main distinct segments that bend in an anticlockwise direction approaching the junction. Moreover, their tips are separated by about 20 m of apparent strike-slip right-lateral offset at the junction. Only the southern GF segment, which shows the larger normal offset, merges directly with the HFF. Southeast of the junction, there is substantial field evidence suggesting the prolongation of the HFF across the TFS: (i) an anticlockwise re-orientation of the faults and tension fractures of the TFS; (ii) the variation, north of this prolongation zone, of the TFS width; (iii) the decrease in the number of faults, and (iv) the decrease in offset amounts toward the north.

The analog experiments simulated different possible plate configurations, leading to a similarity with field truth only in the case of a four-plate configuration, represented by tectonic blocks north and south of the already known trace of the HFF that move westward at different velocities, plus a third tectonic block SE of the junction that also moves westward at a lower slip-rate, and a relatively fixed fourth tectonic block NE of the junction. This configuration and the field data imply a prolongation of the HFF southeast of the junction with the GF, and thus the presence of a right-lateral shear zone across the TFS that moves at very low Holocene slip-rate.

These findings support the hypothesis that the HFF is longer than previously known, implying a re-evaluation of the seismic hazard posed by this structure to the town of Husavik, located directly on the fault trace, and to the surrounding region. Moreover, the structural architecture described here demonstrates that a transform fault—rift junction may be much more complex than the commonly accepted models, and that detailed analyses of the geometry, kinematics, and deformation amounts along the rift zone are necessary to recognize the possible through-going nature of the zone of interaction. Our model implies that a transform fault may propagate horizontally across the rift and the fault tip changes position with time. The real tip position might also be hidden below younger lava deposits and thus the fault plane will propagate upward to the surface during the next earthquake. With particular regard to the latter point, our analog experiments show that if the transform fault crosses part of the rift below younger deposits, this may be recognized by a distortion of the rift pattern with its normal faults that re-orientate perpendicularly to the local σ_3 of the shear zone and/or assume a transtensional kinematics.

Author Contributions

AT coordinated the research, collected the field data, wrote most of the sections of the paper; FB prepared and carried out the experiments and contributed to the related data sections and discussion in the paper, he contributed also to field data collection; FP collected the field data and contributed to their description and discussion in the text.

Conflict of Interest Statement

The authors declare that the research was conducted in the absence of any commercial or financial relationships that could be construed as a potential conflict of interest.

ACKNOWLEDGMENTS

This work is a contribution to the International Lithosphere Program—Task Force II. Funding for field work came from ILP and from the University of Milan Bicocca. M. V. Bonali is fully acknowledged for his help in building the experimental apparatus. We wish to acknowledge the Editor, V. Acocella, one anonymous reviewer and Dr. Joel Ruch, for their very helpful suggestions and comments that helped us to substantially improve the quality of our work.

REFERENCES

1. Acocella, V., Gudmundsson, A., and Funiciello, R. (2000). Interaction and linkage of extension fractures and normal faults: examples from the rift zone of Iceland. *J. Struct. Geol.* 22, 1233–1246. doi: 10.1016/S0191-8141(00)00031-6
2. Acocella, V., Neri, M., and Norini, G. (2013). An overview of experimental models to understand a complex volcanic instability: application to Mount Etna, Italy. *J. volcanol. Geotherm. Res.* 251, 98–111. doi: 10.1016/j.jvolgeores.2012.06.003
3. Agostini, A., Corti, G., Zeoli, A., and Mulugeta, G. (2009). Evolution, pattern, and partitioning of deformation during oblique continental rifting: inferences from lithospheric-scale centrifuge models. *Geochem. Geophys. Geosyst.* 10:Q11015. doi: 10.1029/2009GC002676
4. Angelier, J., Bergerat, F., Dauteuil, O., and Villemin, T. (1997). Effective tension-shear relationships in tension fissure swarms, axial rift zone of northeastern Iceland. *J. Struct. Geol.* 19, 673–685. doi: 10.1016/S0191-8141(96)00106-X
5. Bellier, O., and Sebrier, M. (1995). Is the slip rate variation on the Great Sumatran Fault accommodated by fore-arc stretching?*Geophys. Res. Lett.* 22, 1969–1972. doi: 10.1029/95GL01793
6. Bindi, D., Parolai, S., Oth, A., Abdrakhmatov, K., Muraliev, A., and Zschau, J. (2011). Intensity prediction equations for Central Asia. *Geophys. J. Int.* 187, 327–337. doi: 10.1111/j.1365-246X.2011.05142.x
7. Bjornsson, A. (1985). Dynamics of crustal rifting in Iceland. *J. Geophys. Res.* 90, 151–162. doi: 10.1029/JB090iB12p10151
8. Brandsdóttir, B., and Einarsson, P. (1979). Seismic activity associated with the September 1977 deflation of the Krafla central volcano in NE-Iceland. *J. Volc. Geotherm. Res.* 6, 197–212. doi: 10.1016/0377-0273(79)90001-5
9. Buck, W. R., Einarsson, P., and Brandsdóttir, B. (2006). Tectonic stress and magma chamber size as controls on dike propagation: constraints from the 1975–1984 Krafla rifting episode. *J. Geophys. Res.* 111:B12404. doi: 10.1029/2005JB003879
10. Byerlee, J. (1978). Friction of rocks. *Pageoph* 116, 615–626. doi: 10.1007/BF00876528
11. Cattin, R., Doubre, C., De Chabalier, J.-B., King, G., Vigny, C., Avouac, J.-P., et al. (2005). Numerical modelling of Quaternary deformation and post-seismic displacement in the Asal–Ghoubbet rift (Djibouti, Africa). *Earth Planet. Sci. Lett.* 239, 352–367. doi: 10.1016/j.epsl.2005.07.028
12. Clifton, A. E., Schlische, R. W., Withjack, M. O., and Ackermann, R. V. (2000). Influence of rift obliquity on fault-population systematics: results of experimental clay models. *J. Struct. Geol.* 22, 1491–1509. doi: 10.1016/S0191-8141(00)00043-2
13. Davis, G. A., and Burchfiel, C. B. (1973). Garlock fault: an intracontinental transform structure, southern California. *Geol. Soc. Am. Bull.* 84, 1407–1422.
14. Einarsson, P., and Brandsdóttir, B. (1980). Seismological evidence for lateral magma intrusion during the July 1978 deflation of the Krafla volcano in NE-Iceland. *J. Geophys.* 47, 160–165.
15. Einarsson, T. (1958). A survey of the geology of the area Tjörnes-Bárðardalur in northern Iceland including paleomagnetic studies. *Soc. Sci. Islandica* 32, 1–79.
16. Feigl, K., Gasperi, J., Sigmundsson, F., and Rigo, A. (2000). Crustal deformation near Hengill volcano Iceland 1993–1998: coupling between magmatic activity and faulting inferred from elastic modelling of satellite radar interferometry. *J. Geophys. Res.* 105, 25655–25670. doi: 10.1029/2000JB900209
17. Fornari, D. J., Gallo, D. G., Edwards, M. H., Madsen, J. A., Perfit, M. R., and Shor, A. N. (1989). Structure and topography of the Siqueiros Transform Fault system: evidence for the development of intra-transform spreading centers. *Mar. Geophys. Res.* 11, 263–299. doi: 10.1007/BF00282579
18. Freymueller, J. T., Murray, M. H., Segall, P., and Castillo, D. (1999). Kinematics of the Pacific–North America plate boundary zone, Northern California. *J. Geophys. Res. B Solid Earth Planets* 104, 7419–7441. doi: 10.1029/1998JB900118
19. Galland, O., Burchardt, S., Hallot, E., Mourgues, R., and Bulois, C. (2014). Dynamics of dikes versus cone sheets in volcanic systems. *J. Geophys. Res.* 119, 6178–6192. doi: 10.1002/2014jb011059
20. Gallo, D. G., Kidd, W. S. F., Fox, P. J., Karson, J. A., MacDonald, K., Crane, K., et al. (1984). Tectonics at the intersection of the East Pacific Rise with the Tamayo Transform Fault. *Mar. Geophys. Res.* 6, 159–185. doi: 10.1007/BF00285958

21. Garcia, S., Angelier, J., Bergerat, F., and Homberg, C. (2002). Tectonic analysis of an oceanic transform fault zone revealed by fault–slip data and earthquake focal mechanisms: the Husavik–Flatey Fault, Iceland. *Tectonophysics* 344, 157–174. doi: 10.1016/S0040-1951(01)00282-7

22. Garcia, S., and Dhont, D. (2005). Structural analysis of the Húsavík–Flatey Transform Fault and its relationships with the rift system in Northern Iceland. *Geodin. Acta* 18, 31–41. doi: 10.3166/ga.18.31-41

23. Geirsson, H., Árnadóttir, T., Völksen, C., Jiang, W., Sturkell, E., Villemin, T., et al. (2006). Current plate movements across the Mid-Atlantic Ridge determined from 5 years of continuous GPS measurements in Iceland. *J. Geophys. Res.* 111, B09407. doi: 10.1029/2005JB003717

24. Goodman, R. E. (1989). *Introduction to Rock Mechanics, 2nd Edn*. New York, NY: John Wiley and Sons.

25. Gudmundsson, A. (1987a). Geometry, formation and development of tectonic fractures on the Reykjanes Peninsula, Southwest Iceland. *Tectonophysics* 139, 295–308.

26. Gudmundsson, A. (1987b). Tectonics of the thingvellir fissure swarm, SW Iceland. *J. Struct. Geol.* 9, 61–69. doi: 10.1016/0191-8141(87)90044-7

27. Gudmundsson, A. (1995). Infrastructure and mechanics of volcanic systems in Iceland. *J. Volcanol. Geotherm. Res.* 67, 1–22. doi: 10.1016/0377-0273(95)92782-Q

28. Gudmundsson, A. (2002). Emplacement and arrest of sheets and dykes in central volcanoes. *J. Volcanol. Geotherm. Res.* 116, 279–298. doi: 10.1016/S0377-0273(02)00226-3

29. Gudmundsson, A. (2007). Infrastructure and evolution of ocean-ridge discontinuities in Iceland. *J. Geodyn.* 43, 6–29. doi: 10.1016/j.jog.2006.09.002

30. Gudmundsson, A. (2011). Deflection of dykes into sills at discontinuities and magma-chamber formation. *Tectonophysics* 500, 50–64. doi: 10.1016/j.tecto.2009.10.015

31. Gudmundsson, A., Brynjólfsson, S., and Jónsson, M. T. (1993). Structural analysis of a transform fault-rift zone junction in North Iceland. *Tectonophysics* 220, 205–221. doi: 10.1016/0040-1951(93)90232-9

32. Gudmundsson, A., Friese, N., Galindo, I., and Philipp, S. L. (2008). Dike-induced reverse faulting in a graben. *Geology* 36, 123–126. doi: 10.1130/G24185A.1

33. Gudmundsson, A., and Loetveit, I. F. (2005). Dyke emplacement in a layered and faulted rift zone. *J. Volcanol. Geotherm. Res.*144, 311–327. doi: 10.1016/j.jvolgeores.2004.11.027

34. Glicken, H., Janda, R. J., and Voight,. B. (1980). Catastrophic landslide/debris avalanche of May 18, 1980, Mount St. Helens volcano. *EOS Trans. Am. Geophys. Union* 61:1135.

35. Handin, J. (1966). "Strength and ductility," in *Handbook of Physical Constants*, Vol. 97, ed S. P. Clark (Boulder, CO: Geol. Soc. Am. *Memoir)*, 223.

36. Harding, T. P., Vierbuchen, R. C., and Christie-Blick, N. (1985). "Structural styles, plate tectonic settings, and hydrocarbon potential of divergent (transtensional) wrench faults," in *Strike-Slip Deformation, Basin Formation, and Sedimentation*, Vol. 37, eds K. T. Biddle, N. Christie-Blick (New York, NY: Columbia University; Society of Economic Paleontologists and Mineralogists Special Publication), 51–77.

37. Hjartardóttir, Á. R., Einarsson, P., Bramham, E., and Wright, T. J. (2012). The Krafla fissure swarm, Iceland, and its formation by rifting events. *Bull. Volcanol.* 74, 2139–2153. doi: 10.1007/s00445-012-0659-0

38. Hjartardóttir, Á. R., Einarsson, P., Magnúsdóttir, S., Björnsdóttir, Þ., and Brandsdóttir, B. (2015). "Fracture systems of the Northern Volcanic Rift Zone, Iceland – An onshore part of the Mid-Atlantic plate boundary," in *Magmatic Rifting and Active Volcanism*, eds T. J. Wright, A. Ayele, D. J. Ferguson, T. Kidane, and C. Vye-Brown (London: Geological Society; Special Publications), 420.

39. Holohan, E. P., Troll, V. R., Walter, T. R., Münn, S., McDonnell, S., and Shipton, Z. K. (2005). Elliptical calderas in active tectonic settings: an experimental approach. *J. Volcanol. Geotherm. Res.* 144, 119–136. doi: 10.1016/j.jvolgeores.2004.11.020

40. Hoshino, K., Koide, H., Inami, K., Iwamura, S., and Mitsui, S. (1972). Mechanical properties of Japanese Tertiary sedimentary rocks under high confining pressures. *Geol. Surv. Jap.* 244, 1–200.

41. Hubbert, M. K. (1951). Mechanical basis for certain familiar geologic structures. *Geol. Soc. Am. Bull.* 62, 355–372. doi: 10.1130/0016-7606(1951)62[355:MBFCFG]2.0.CO;2

42. Jaeger, J. C., and Cook, N. G. W. (1971). *Fundamental of Rocks Mechanics*. New York, NY: Chapman and Hall.

43. Joffe, S., and Garfunkel, Z. (1987). Plate kinematics of the circum Red Sea – a re-evaluation. *Tectonophysics* 141, 5–22. doi: 10.1016/0040-1951(87)90171-5

44. Jouanne, F., Villemin, T., Berger, A., and Henroit, O. (2006). Rift-transform junction in North Iceland: rigid blocks and narrow accommodation zones revealed by GPS 1997-1999-2002. *Geophys. J. Int.* 167, 1439–1446. doi: 10.1111/j.1365-246X.2006.03107.x

45. Kamata, H., and Kodama, K. (1994). Tectonics of an arc–arc junction: an example from Kyushu Island at the junction of the Southwest Japan Arc and the Ryukyu Arc. *Tectonophysics* 233, 69–81. doi: 10.1016/0040-1951(94)90220-8

46. Karson, J. A., Fox, P. J., Sloan, H., Crane, K. T., Kidd, W. S. F., Bonatti, E., et al. (1984). The geology of the Oceanographer Transform: the ridge-transform intersection. *Mar. Geophys. Res.* 6, 109–141. doi: 10.1007/BF00285956

47. Krantz, R. W. (1991). Measurements of friction coefficients and cohesion for faulting and fault reactivation in laboratory models using sand and sand mixtures. *Tectonophysics* 188, 203–207. doi: 10.1016/0040-1951(91)90323-K

48. Lelgemann, H., Gutscher, M.-A., Bialas, J., Flueh, E., Weinrebe, W., and Reichert, C. (2000). Transtensional

basins in the western Sunda Strait. *Geophys. Res. Lett.* 27, 3545–3548. doi: 10.1029/2000GL011635

49. Maccaferri, F., Rivalta, E., Passarelli, L., and Jónsson, S. (2013). The stress shadow induced by the 1975–1984 Krafla rifting episode. *J. Geophys. Res.* 118, 1109–1121. doi: 10.1002/jgrb.50134
50. Mastin, L. G., and Pollard, D. D. (1988). Surface deformation and shallow dike intrusion processes at Inyo Craters, Long Valley, California. *J. Geophys. Res.* 93, 13221–13235. doi: 10.1029/JB093iB11p13221
51. Magnúsdóttir, S., and Brandsdóttir, B. (2011). Tectonics of the Þeistareykir fissure swarm. *Jökull* 61, 65–79.
52. Mamaloukas-Frangoulis, V., Auzende, J. M., Bideau, D., Bonatti, E., Cannat, M., Honnorez, J., et al. (1991). *In-situ* study of the eastern ridge-transform intersection of the Vema Fracture Zone. *Tectonophysics* 190, 55–71. doi: 10.1016/0040-1951(91)90354-U
53. McClay, K. R. (1990). Deformation mechanics in analogue models of extensional fault systems. *Geol. Soc. Lond. Spec. Publ.* 54, 445–453. doi: 10.1144/GSL.SP.1990.054.01.40
54. McClay, K. R., and White, M. J. (1995). Analogue modelling of orthogonal and oblique rifting. *Mar. Petrol. Geol.* 12, 137–151. doi: 10.1016/0264-8172(95)92835-K
55. Merle, O. (2015). The scaling of experiments on volcanic systems. *Front. Earth Sci.* 3:26. doi: 10.3389/feart.2015.00026
56. Metzger, S., Jónsson, S., and Geirsson, H. (2011). Locking depth and slip-rate of the Húsavík Flatey fault, North Iceland, derived from continuous GPS data 2006-2010. *Geophys. J. Int.* 187, 564–576. doi: 10.1111/j.1365-246X.2011.05176.x
57. Metzger, S., Jonsson, S., Danielsen, G., Hreinsdottir, S., Jouanne, F., Giardini, D., et al. (2013). Present kinematics of the Tjörnes Fracture Zone, North Iceland, from campaign and continuous GPS measurements. *Geophys. J. Int.* 192, 441–455. doi: 10.1093/gji/ggs032
58. Morley, C. K. (1999). How successful are analogue models in addressing the influence of pre-existing fabrics on rift structure?. *J. Struct. Geol.* 21, 1267–1274. doi: 10.1016/S0191-8141(99)00075-9
59. Mouslopoulou, V., Nicol, A., Little, T. A., and Walsh, J. J. (2007a). Displacement transfer between intersecting strike-slip and extensional fault systems. *J. Struct. Geol.* 29, 100–116. doi: 10.1016/j.jsg.2006.08.002
60. Mouslopoulou, V., Nicol, A., Little, T. A., and Walsh, J. J. (2007b). "Terminations of large strike-slip faults: an alternative model from New Zealand," *in Tectonics of Strike-Slip Restraining and Releasing Bends, Vol. 290, eds W. D. Cunningham and P. Mann (London: Geological Society; Special Publications*), 387–415.
61. Norini, G., Capra, L., Groppelli, G., and Lagmay, A. M. F. (2008). Quaternary sector collapses of Nevado de Toluca volcano (Mexico) governed by regional tectonics and volcanic evolution. *Geosphere* 4, 854–871. doi: 10.1130/GES00165.1
62. Norris, R. J., and Cooper, A. F. (2000). Late Quaternary slip rates and their significance for slip partitioning on the Alpine Fault, New Zealand. *J. Struct. Geol.* 23, 507–520. doi: 10.1016/S0191-8141(00)00122-X
63. Opheim, J., and Gudmundsson, A. (1989). Formation and geometry of fractures, and related volcanism, of the Krafla fissure swarm, northeast Iceland. *Bull. Geol. Soc. Am.* 101, 1608–1622.
64. Oskarsson, N., Steinthorsson, S., and Sigvaldason, G. E. (1985). Iceland geochemical anomaly: origin, volcanotectonics, chemical fractionation and isotope evolution of the crust. *J. Geophys. Res. 90*, 10011–10025. doi: 10.1029/JB090iB12p10011
65. Pasquarè Mariotto, F., Bonali, F. L., Tibaldi, A., Rust, D., Oppizzi, P., and Cavallo, A. (2015). Holocene displacement field at an emerged oceanic transform-ridge junction: the Husavik-Flatey Fault - Gudfinnugja Fault system, North Iceland. *J. Struct. Geol.*75, 118–134. doi: 10.1016/j.jsg.2015.03.011
66. Quebral, R. D., Pubellier, M., and Rangin, C. (1996). The onset of movement on the Philippine Fault in eastern Mindano: a transition from a collision to a strike-slip environment. *Tectonics* 15, 713–726. doi: 10.1029/95TC00480
67. Rivalta, E. Taisne, B., Bunger, A. P., and Katz, R. F. (2015). A review of mechanical models of dike propagation: schools of thought, results and future directions. *Tectonophysics* 638, 1–42. doi: 10.1016/j.tecto.2014.10.003
68. Rubin, A. (1992). Dike-induced faulting and graben subsidence in volcanic rift zones. *J. Geophys. Res.* 97, 1839–1858. doi: 10.1029/91JB02170
69. Rubin, A. M., and Pollard, D. D. (1988). Dike induced faulting in rift zones in Iceland and Afar. *Geology* 16, 413–317.
70. Ruch, J., Acocella, V., Geshi, N., Nobile, A., and Corbi, F. (2012). Kinematic analysis of vertical collapse on volcanoes using experimental models time series. *J. Geophys. Res.* 117, B7. doi: 10.1029/2012JB009229
71. Saemundsson, K. (1974). Evolution of the axial rifting zone in northern Iceland and the Tjornes fracture zone. *Geol. Soc. Am. Bull.* 85, 495–504.
72. Saemundsson, K., Hjartarson, A., Kaldal, I., Sigurgeirsson, M. A., Kristinsson, S. G., and Vikingsson, S. (2012). *Geological Map of the Northern Volcanic Zone, Iceland. Northern Part 1: 100.000.* Reykjavik: Iceland GeoSurvey and Landsvirkjun.
73. Schultz, R. A. (1996). Relative scale and the strength and deformability of rock masses. *J. Struct. Geol.* 18, 1139–1149. doi: 10.1016/0191-8141(96)00045-4
74. Sella, G. F., Dixon, T. H., and Mao, A. (2002). REVEL: a model for recent plate velocities from space geodesy. *J. Geophys. Res.*107, 2081. doi: 10.1029/2000JB000033
75. Şengör, A. C., (2014). *Transform Fault. Earth Sciences Series, Encyclopedia of Marine Geosciences*. Berlin: Springer.
76. Sigmundsson, F. (2006). *Iceland Geodynamics: Crustal Deformation and Divergent Plate Tectonics.* Chichester: Springer Science & Business Media; Praxis Publisher

Ltd.

77. Sigmundsson, F. A., Hooper, S., Hreinsdóttir, K. S., Vogfjörd, B. G., Ófeigsson, E., Rafn Heimisson, S., et al. (2015). Segmented lateral dyke growth in a rifting event at Bárðarbunga volcanic system, Iceland. *Nature* 517, 191–195. doi: 10.1038/nature14111

78. Sylvester, A. G. (1988). Strike-slip faults. *Geol. Soc. Am. Bull.* 100, 1666–1703.

79. Tchalenko, J. S. (1970). Similarities between shear zones of different magnitudes. *Geol. Soc. Am. Bull.* 81, 1625–1640. doi: 10.1130/0016-7606(1970)81[1625:SBSZOD]2.0.CO;2

80. Tibaldi, A. (1995). Morphology of pyroclastic cones and tectonics. *J. Geophys. Res.* 100, 24521–24535. doi: 10.1029/95JB02250

81. Tibaldi, A. (2015). Structure of volcano plumbing systems: a review of multi-parametric effects. *J. Volcanol. Geoth. Res.* 298, 85–135. doi: 10.1016/j.jvolgeores.2015.03.023

82. Tibaldi, A., Bonali, F. L., Einarsson, P., Hjartardóttir, Á. R., and Pasquarè Mariotto, F. (2016). Partitioning of Holocene kinematics and interaction between the Theistareykir Fissure Swarm and the Husavik-Flatey Fault, North Iceland. *J. Struct. Geol.* 83, 134–155. doi: 10.1016/j.jsg.2016.01.003

83. Tibaldi, A., Pasquarè, F. A., Papanikolaou, D., and Nomikou, P. (2008). Tectonics of Nisyros Island, Greece, by field and offshore data, and analogue modelling. *J. struct. Geol.* 30, 1489–1506. doi: 10.1016/j.jsg.2008.08.003

84. Tortini, R., Bonali, F. L., Corazzato, C., Carn, S. A., and Tibaldi, A. (2014). An innovative application of the Kinect in Earth sciences: quantifying deformation in analogue modelling of volcanoes. *Terra Nova* 26, 273–281. doi: 10.1111/ter.12096

85. Trippanera, D., Porreca, M., Ruch, J., Pimentel, A., Acocella, V., Pacheco, J., et al. (2014). Relationships between tectonics and magmatism in a transtensive/transform setting: an example from Faial Island (Azores, Portugal). *Geol. Soc. Am. Bull.* 126, 164–181. doi: 10.1130/B30758.1

86. Trippanera, D., Ruch, J., Acocella, V., and Rivalta, E. (2015). Experiments of dike-induced deformation: Insights on the long-term evolution of divergent plate boundaries. *J. Geophys. Res.* 120, 6913–6942. doi: 10.1002/2014JB011850

87. Tron, V., and Brun, J. P. (1991). Experiments on oblique rifting in brittle-ductile systems. *Tectonophysics* 188, 71–84. doi: 10.1016/0040-1951(91)90315-J

88. Tryggvason, E. (1973). Seismicity, earthquake swarms, and plate boundaries in the Iceland region. *Bull. Seismol. Soc. Am.* 63, 1327–1348.

89. Tryggvason, E. (1980). Subsidence events in the Krafla area, North-Iceland, 1975-1979. *J. Geophys. Zeitschrift Geophys.* 47, 141–153.

90. Tryggvason, E. (1984). Widening of the Krafla fissure swarm during the 1975–1981 volcano-tectonic episode. *Bull. Volcanol.* 47, 47–69. doi: 10.1007/BF01960540

91. Tryggvason, E. (1986). Multiple magma reservoirs in a rift zone volcano: ground deformation and magma transport during the September 1984 eruption of Krafla, Iceland. *J. Volcanol. Geoth. Res.* 28, 1–44. doi: 10.1016/0377-0273(86)90003-X

92. Tsutsumi, H., and Okada, A. (1996). Segmentation and Holocene surface rupture faulting on the Median Tectonic Line, southwest Japan. *J. Geophys. Res.* 101, 5855–5871. doi: 10.1029/95JB01913

93. Walter, T. R., and Troll, V. R. (2001). Formation of caldera periphery faults: an experimental study. *Bull. Volcanol.* 63, 191–203. doi: 10.1007/s004450100135

94. Wilson, J. T. (1965). A new class of faults and their bearing on continental drift. *Nature* 207, 343–347. doi: 10.1038/207343a0

95. Woodcock, N. H., and Fischer, M. (1986). Strike-slip duplexes. *J. struct. Geol.* 8, 725–735. doi: 10.1016/0191-8141(86)90021-0

96. Young, K. D., Jancin, B., and Orkan, N. I. (1985). Transform deformation of tertiary rocks along the Tjornes Fracture Zone, North Central Iceland. *J. Geophys. Res.* 90, 9986–10010. doi: 10.1029/JB090iB12p09986

CHAPTER 12

EVOLUTION OF THE EARTH: WATER, PLATE TECTONICS, AND DELAMINATION OF ANTI-CONTINENT

Yoshiyuki Tatsumi, Takeshi Sato and Shuichi Kodaira

Keywords: Earth's crust Continent Anti-continent Delamination Water Plate tectonics

INTRODUCTION

Known terrestrial planets possess a common internal structure with a metallic core, a rocky mantle, and an evolved crust, although the presence of two layers in the core, i.e., liquid and solid, is uncertain in terrestrial planets other than the Earth (Fig. 1). Thus, it would seem appropriate to ascribe a fundamentally common mechanism to the formation and evolution of these inner planets. However, current surface features of terrestrial planets are highly variable, suggesting that different tectonic regimes have operated during the evolution of these planets. One such example is topography. The Earth exhibits a bimodal height distribution at the surface (Fig. 1) in contrast to the rather even surface of other terrestrial planets. This unique "lowland vs. highland" topography of the Earth is a consequence of the distribution of continental and oceanic crust on the fluidal and denser mantle, each with distinct densities and thicknesses that are on average 2700 vs. 2900 kg/m^3 and 40 vs. 6 km, respectively. The contrast in density is due to a difference in their average compositions; the continental crust is intermediate or andesitic, containing ~60 wt% SiO_2, whereas the oceanic crust is mafic or basaltic, containing ~50 wt% SiO_2 (e.g., Christensen and Mooney 1995; Kelemen 1995; Taylor and McLennan 1995; Rudnick 1995). In contrast, other terrestrial planets are covered with a uniform crust that is broadly mafic in composition (e.g., Taylor and McLennan,2009). Therefore, it may be appropriate to regard the Earth as unique in our solar system, being the only "water planet," "shore planet," and "andesite planet." In this study, we examined the mechanism of continental crust formation and emphasized the role of the delamination of mafic components in the creation of the evolved, andesitic crust found on the Earth, by analyzing the seismic and petrologic crust-mantle structure of the juvenile Izu-Bonin-Mariana arc.

Citation: Yoshiyuki Tatsumi, Takeshi Sato and Shuichi Kodaira, Evolution of the Earth: water, plate tectonics, and delamination of anti-continent, DOI: 10.1186/s40623-015-0267-2

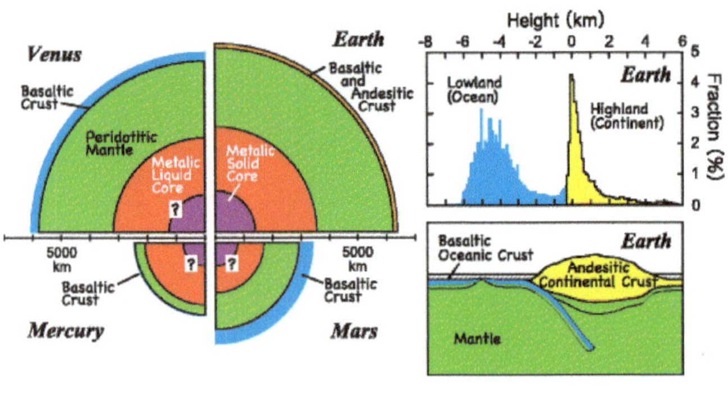

Figure 1: Inner structures of the terrestrial planets (*left*) and characteristics of the Earth (*right*). Although the terrestrial planets show a common layered structure, the Earth is distinct in its bimodal height distribution at its surface, caused by magmatism associated with plate tectonics.

Two Types of Earth's Crusts: a Consequence of Plate Tectonics

The presence of a high-temperature metallic core, common to all terrestrial planets (Fig. 1) (Aitta, 2012; Smith et al.,2012; Stevenson, 2001), causes convection in the mantle with a high Rayleigh number. Since mantle viscosity strongly depends on temperature (e.g., Karato and Wu, 1993), the top thermal boundary layer, the lithosphere, is very viscous and stiff. Stagnant lid convection, involving heat transfer by conduction in the upper lid underlain by a convecting mantle, should be the most likely mode of mantle convection and is typical of mantles in other terrestrial planets such as Venus, Mercury, and Mars (e.g., Solomatov, 1995; Schubert et al., 2001), where the entire planetary surface consists of one continuous plate.

The Earth, on the other hand, is unique among the terrestrial planets of the solar system because the lithosphere or lid is broken into multiple tectonic plates that move and founder into the mantle, i.e., the operation of plate tectonics. For such a distinctive regime to exist, stresses associated with mantle convection have to exceed the strength of the rigid lithosphere, which may happen with weakening of the lithosphere. The presence of water is one possible mechanism that reduces the yield strength of the lithosphere and friction on faults (e.g., Moresi and Solomatov,1998; Valencia et al., 2007; Korenaga 2009). The timing of the onset of plate tectonics is still debated. For example, Stern (2005) suggests the initiation of modern-style plate tectonics at 1 Gya based on the absence of ophiolites and ultra-high-pressure metamorphism before that era. Hopkins et al. (2010) considered that zircon dating back to >4.2 Gya may indicate the operation of plate tectonics in the Hadean. On the other hand, there is mounting evidence that both ocean formation and plate tectonics operation took place in the early Archean (3.6–3.9 Ga) (Nutman et al.,2002; Komiya et al., 1999; Shirey et al., 2008). Once plate tectonics started operating, magmatism occurred in two different tectonic settings: at divergent plate boundaries such as mid-ocean ridges and continental rifts, and at convergent plate boundaries such as arc-trench systems and collision zones. The two distinct types of crust that characterize the Earth are believed to have been created at these two different plate boundaries.

The basaltic composition of Earth's oceanic crust has been inferred from the analyses of rocks sampled from both oceanic crust and their analogs—ophiolites that once formed the ocean floor before being obducted onto the continent (e.g., Dewey and Bird, 1971; Coleman, 1977; Dilek and Furnes 2011). It has been established that the oceanic crust forms the top part of the oceanic plate and is created at mid-ocean ridges.

The continental crust is too thick to sample directly in its entirety. Therefore, its composition has been estimated indirectly by comparing the seismic velocity structure of the crust and the measured velocities of various rocks under crustal P-T conditions, in addition to studies of crustal xenoliths. The average P-wave velocity (V_P) of the continental crust is ~6.5 km/s, suggesting that the bulk continental crust is andesitic in composition (e.g., Christensen and Mooney 1995). Andesite, the name of which is derived from the Andes Mountains, and its plutonic equivalents such as tonalite and granodiorite represent magmatism in subduction zones where active arcs form because of plate subduction. As a result, the "andesite model" was put forward by Taylor (1967) in which continental crust is created at subduction zones by magmatism. However, it is also now well known that while andesites are predominant in continental arcs with mature continental crust, basalts are the major volcanic products at intra-oceanic arcs with a juvenile crust (e.g., Gill, 1981; Ewart, 1982; Tatsumi and Eggins, 1995). This leads to a fundamental problem with the andesite model; it cannot explain the creation of the initial continental crust but only explains the generation of the continental crust under conditions of pre-existing continental crust.

Izu-Bonin-Mariana (IBM) arc: a Site of Original Continental Crust Formation?

The IBM arc, which extends 2800 km southward from the Izu Peninsula (Fig. 2), has evolved due to the subduction of the Pacific Plate since 50 Ma, accompanied by back-arc rifting from 25 to 15 Ma that separated the IBM arc from the remnant Kyushu-Palau ridge, creating the Shikoku Basin and Parece Vela Basin (Fig. 2). Back-arc rifting was also initiated immediately behind the volcanic front in the central part of the Mariana arc after 2.8 Ma (Taylor 1992; Ishizuka et al. 2002).

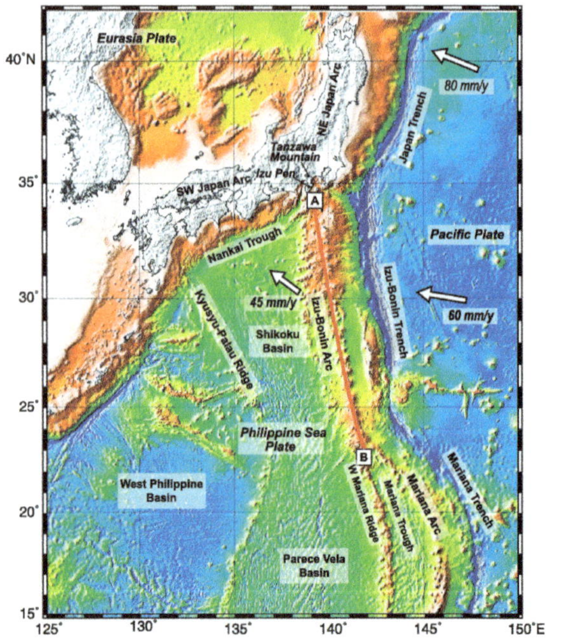

Figure 2: Bathymetry of the IBM intra-oceanic arc. *A–B* is the along-arc cross section shown in Fig. 6.

One marked structural feature of the IBM arc is the presence of a layer in the middle of the crust with a V_p of 6.0–6.8 km/s, equivalent to that of the average continental crust and typical of intermediate-composition plutonic rocks (Christensen and Mooney, 1995). This middle crust layer is underlain by a lower crust, a low-V uppermost mantle layer, and a normal upper mantle that exhibits a V_p of 6.9–7.2, 7.2–7.6, and >7.8 km/s, respectively (Fig. 3) (Suyehiro et al., 1996; Takahashi et al., 2007, 2008; Sato et al., 2009). This characteristic layered structure has been confirmed to exist throughout the entire IBM arc (Kodaira et al., 2007a,b; Takahashi et al., 2008; Sato et al.,2009).

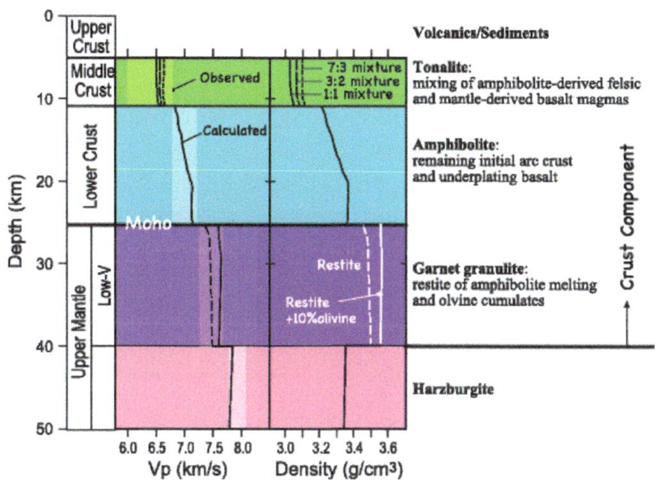

Figure 3: V_p structure of the IBM arc and its petrological interpretation. V_p and density calculated for the inferred lithologies in the sub-IBM arc crust and mantle are also shown. The calculated V_p values (*lines*) are consistent with the observed V_p structure (*light area*). Note that a density inversion is predicted for the low-V uppermost mantle layer

The presence of plutonic rocks with andesitic compositions in the IBM middle crust is supported by the recovery of intermediate plutonic rock xenoliths in arc lavas, exposures along the IBM and Kyushu-Palau arc systems (e.g., Sakamoto et al., 1999), the exposure of tonalitic plutons in the Tanzawa Mountains where the IBM arc collides with the Japanese islands (Kawate and Arima, 1998; Tani et al., 2010), and the presence of a ~150-km-long felsic melt layer in the Mariana arc (Stern et al., 2013). Furthermore, a middle crust layer with a V_p of 6.0–6.8 km/s is not unique to the IBM arc; it has also been documented in other intra-oceanic arcs such as the Tonga arc (Crawford et al., 2003) and the Kuriles (Nakanishi et al., 2009). These lines of evidence provide compelling support for a challenging hypothesis; the ingredients for continental crust are created at modern intra-oceanic arcs. Thus, the continents are born in the oceans.

Tatsumi et al. (2008) examined whether petrologic models, including those in which andesitic magma is generated by crustal melting and those in which it is generated by mantle-melting-derived basaltic magma mixing with crust-melting-derived felsic magma, could lead to the genesis of the IBM crust and mantle. They found that the lithologies predicted by these models exhibited a V_p consistent with the observed values. However, this modeling was based on a melting regime obtained by compiling the results of melting experiments for a variety of basalts with and without H_2O. Since this study, new constraints on crustal melting and andesitic magma formation in the IBM arc have been provided by Tatsumi and Suzuki (2009), who thoroughly investigated the liquid line of descent for a representative IBM basalt at crustal pressures. Thus, it is important to examine the implications that these latest results have for the lithology and the physical properties of the IBM crust and mantle.

Seismic vs. petrologic structure of the IBM crust and mantle

Arc evolution Model and Lithology of Crust and Mantle

The seismic structure of the sub-IBM arc crust and mantle (Fig. 3) has been petrologically modeled by Tatsumi et al. (2008, 2014) and is schematically shown in Fig. 4. The compositions of the inferred crust/mantle components used here are listed in Table 1. A representative IBM basalt (basalt in Table 1), which was the initial material used in Tatsumi and Suzuki (2009), was assumed to form the initial arc crust. Although boninite is known to be the lithology for an initial IBM arc (e.g., Ishizuka et al., 2011), it may not be distributed widely, particularly as the intermediate middle crust lithology along the current IBM arc, since this hydrous magma forms amphibole-rich pluton with a V_p greater than observed values (Tatsumi et al., 2008). Successive magmatic underplating caused the partial melting of the initial crust to produce a rhyolitic melt after a 5–10 % partial melting, leaving a restite or a melting residue (the compositions of these components inferred from experiments are listed in Table 1). Mixing of this rhyolite magma with an IBM basalt magma produced andesite magmas (andesites 1–3 in Table 1 with different mixing ratios), which ascended and solidified to form a middle crust layer. A noteworthy difference in chemistry between the inferred IBM andesite and the average continental crust was a strong depletion in potassium and other incompatible elements in IBM andesites (Table 1). This was due to the production of IBM magmas from depleted peridotite or harzburgite (Tatsumi, 2000), whereas continental crusts may have been created above more fertile peridotites or even enriched mantle/crust (Albarède, 1998; Clift and Vannucchi, 2004).

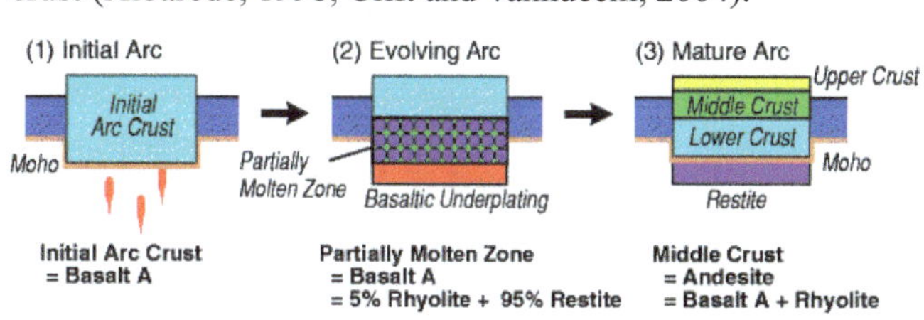

Figure 4: Schematic diagram showing the model of arc crust evolution from Tatsumi et al. (2008).

Table 1: Compositions of inferred basaltic and rhyolitic end-member magma and restite

	This study								Tatsumi et al. (2008)			
	Basalt	Rhyolite	Andesite 1 (1:1 mixture)	Andesite 2 (3:2 mixture)	Andesite 3 (7:3 mixture)	Restite	Restite + 10 % olivine	Harzburgite	Basalt	Rhyolite	Restite	Mixed
SiO_2	49.40	75.00	62.20	59.64	57.08	47.76	47.00	45.40	50.00	75.00	47.22	60.00
TiO_2	0.70	0.40	0.55	0.58	0.61	0.72	0.65		0.80	0.30	0.86	0.60
Al_2O_3	18.34	13.52	15.93	16.41	16.89	18.64	16.78		19.10	14.00	19.67	17.06
FeO[a]	11.02	2.23	6.63	7.51	8.39	11.58	11.76	5.70	10.20	2.00	11.11	6.92
MgO	6.41	0.68	3.55	4.12	4.69	6.78	10.75	48.90	6.00	0.20	6.64	3.68
CaO	12.42	5.12	8.77	9.50	10.23	12.89	11.60		12.10	3.00	13.11	8.46
Na_2O	1.60	1.78	1.69	1.68	1.66	1.59	1.43		1.60	4.50	1.28	2.76
K_2O	0.10	1.26	0.68	0.56	0.45	0.03	0.02		0.20	1.00	0.11	0.52
Total	100.00	100.00	100.00	100.00	100.00	100.00	100.00	100.00	100.00	100.00	100.00	100.00
H_2O	0.50		0.50	0.50	0.50				0.20			0.50

[a]FeO, total Fe as FeO

Mixing ratio represents basalt/rhyolite

Since the basalt used in this study as a representative IBM basalt was not a primary magma that was in equilibrium with upper mantle peridotites, olivine cumulates may have formed during magmatic differentiation and would have likely mixed with the restite. The restite with and without 10 % olivine cumulate were then considered to be a possible crust/mantle component. The sub-IBM arc mantle was assumed to consist of harzburgite (Table 1) as it is commonly found as xenoliths in serpentine mud volcanoes in the IBM forearc (Ishii et al., 2000).

In summary, the new modeling suggested that the plutonic equivalent of the mixed andesitic magma was the lithology of the middle crust, the representative IBM basalt was the lithology of the lower crust, the restite with and without olivine cumulates was the lithology of the low-V uppermost mantle, and harzburgite was the lithology of the normal upper mantle (Table 1).

Physical Properties of Crust and Mantle

Once the chemical compositions of each layer of the sub-IBM arc crust and mantle were defined, the mineral assemblages and physical properties of these layers were obtained and compared to the observed layered structure.

The amount of H_2O in magma governs the amount of hydrous phases in a solidified rock and may be a critical control on its physical properties. It was possible to estimate the H_2O content in magmas by mixing the basalt magma, assumed to contain 0.5 wt% H_2O (Tatsumi and Suzuki, 2009), with a rhyolitic magma that was produced by partial melting of the hydrous basalt. Assuming that ~10 % partial melting was required to produce rhyolitic magma, the rhyolite would contain ~5 wt% H_2O. An intermediate magma with ~60 % SiO_2 produced by a ~3:2 mixture of basaltic and rhyolitic magmas would thus contain ~3.2 wt% H_2O. This value was much higher than the 0.3–0.5 wt% H_2O inferred from the amount of hydrous phases contained in the tonalitic plutonic rocks from the IBM (Tatsumi et al., 2008). With the current data available, however, it was difficult to further constrain the H_2O content in IBM magmas. Therefore, we tentatively accepted 0.5 wt% as the H_2O content of both mixed intermediate and original basaltic magmas.

Assuming the distribution of crust and mantle components outlined above, subsolidus phase equilibria at pressures and temperatures relevant to crust-to-mantle conditions were obtained using the free energy minimization algorithm, Perple_X (Connolly, 1990; 2005), along inferred geothermal gradients. The temperature of the mantle wedge at 70 km depth was fixed at 1400 °C, the temperature required for arc basalt magma generation (Tatsumi et al., 1983). The surface temperature was fixed at 0 °C and the temperature at the Moho was set at 750 °C, the solidus temperature of a basaltic rock with 0.5 wt% H_2O. The subsolidus temperature at the Moho was consistent with the observation that the Moho discontinuity was clearly defined beneath the IBM arc (Kodaira et al., 2007a; Takahashi et al., 2008). On the other hand, the lower crust should be partially molten beneath the region where the arc crust was being created and evolved by the localized heat supply from mantle-derived basalt magmas. This temperature was ~120 °C lower than the Moho temperature inferred by Tatsumi et al. (2008). This may be due to the greater amount of H_2O in the basalt composition used in this study (0.5 compared to 0.2 wt% in the earlier study). Oxygen fugacity (f_{O2}) was fixed at an approximate QFM buffer as a stable phase condition, and hence, the physical properties of the lower crust were relatively insensitive to f_{O2} (Behn & Kelemen, 2006). The results of the calculations are shown in Fig. 5 and listed in Table 2.

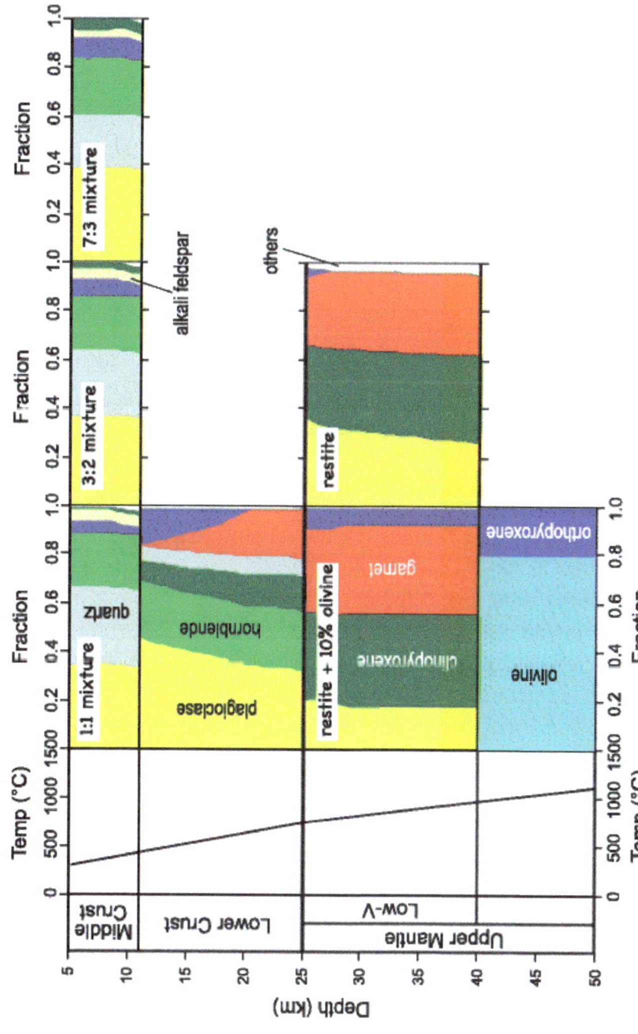

Figure 5: Mineral assemblages in the sub-IBM arc crust and mantle along a possible geotherm for the inferred crust and mantle components listed in Table 1 for different basalt-rhyolite mixtures. Compared to the model in Tatsumi et al. (2008), this model uses a lower temperature to maintain subsolidus temperatures at the Moho.

Table 2: Mineral assemblages and physical properties of inferred lithologies for sub-IBM crust and mantle

Layer	Middle crust									Lower crust				Upper mantle								
Composition	1:1 mixture			3:2 mixture			7:3 mixture			Basalt				Restite			Restite + 10 % olivine			Harzburgite		
Depth (km)	5	8.5	11	5	9	11	5	10	11	11	12	20.5	25	25	27	40	25	30	40	40	50	70
T (°C)	300	380	430	300	390	430	300	410	430	430	450	650	750	750	780	970	750	820	970	970	1113	1400
Quartz	31.7	31.6	31.9	26.9	26.8	27.0	22.0	21.8	22.0	6.2	6.3	7.2	7.3	1.5	2.6	4.1						
Plagioclase	34.8	34.8	33.6	36.8	37.0	35.9	38.8	38.8	38.1	45.3	44.5	34.4	32.2	36.6	32.2	26.0	20.4	17.1	17.1			
Alkali feldspar	4.5	4.5	4.5	3.8	3.7	3.8	3.1	3.0	3.1	0.7	0.7	0.7	0.7	0.2	0.2	0.2	0.2	0.2	0.2			
Hornblende	22.0	22.0	22.1	22.2	22.2	22.3	22.5	22.5	22.6	23.4	23.4	24.2	24.4									
Garnet		0.4	2.3		0.2	1.9		0.4	1.4		1.4	19.3	20.5	27.1	32.1	33.5	34.6	36.8	37.0			
Olivine																				80.0	80.0	80.0
Orthopyroxene	5.1	4.8	3.5	6.9	6.9	5.7	8.8	8.6	7.9	14.9	14.0	0.0	0.0	4.3			8.4	7.2	6.7	20.0	20.0	20.0
Clinopyroxene	1.3	1.3	1.4	2.6	2.6	2.7	4.1	4.1	4.2	8.7	8.9	13.3	14.4	29.8	32.3	35.6	35.6	37.7	38.5			
Ilmenite	0.6	0.6	0.6	0.7	0.7	0.7	0.7	0.7	0.7	0.9	0.9	0.9										
Rutile													0.5	0.6	0.6	0.6		0.5	0.5			
ρ (g/cm^3)	2.92	2.93	2.95	2.96	2.96	2.98	3.00	3.00	3.01	3.11	3.12	3.27	3.26	3.35	3.38	3.40	3.46	3.46	3.46	3.25	3.24	3.20
V_P (km/s)	6.50	6.51	6.54	6.57	6.56	6.59	6.63	6.63	6.65	6.82	6.85	7.13	7.14	7.33	7.45	7.48	7.60	7.64	7.62	7.86	7.81	7.51
V_S (km/s)	3.82	3.81	3.82	3.82	3.80	3.82	3.82	3.80	3.81	3.80	3.82	3.98	3.99	4.06	4.12	4.14	4.23	4.25	4.23	4.49	4.44	4.22
Poisson's ratio	0.24	0.24	0.24	0.24	0.25	0.25	0.25	0.25	0.25	0.27	0.27	0.27	0.27	0.28	0.28	0.28	0.27	0.28	0.28	0.26	0.26	0.27

The resulting mineral assemblages were then used to calculate the physical properties of each layer of the model IBM crust and mantle following the method of Hacker et al. (2003). The calculated seismic velocity and density, together with the observed V_p for each layer, are presented in Fig. 3 and Table 2.

One major difference between the calculation results presented here and in earlier efforts was that garnet was ubiquitous both in the lower crust and the uppermost mantle layers in the new results (Fig. 5 and Table 2), but appeared only in the lower portion of the low-V layer in the uppermost mantle in previous results (Tatsumi et al.,2008). This was due to the lower temperature used in this study to keep the lower crust beneath the non-volcanic region at subsolidus temperatures. The presence of garnet in the IBM crust was apparently inconsistent with the geochemical characteristics of crust-derived felsic volcanic rocks that may have been in this arc, suggesting that there was no garnet residue during crustal melting (Tamura et al., 2009). It should be emphasized, however, that the crust-mantle lithology obtained in this study formed under stationary, non-magmatic conditions instead of in higher temperature conditions associated with localized magmatism. Garnet may be replaced by pyroxene and plagioclase at higher temperatures (Tatsumi et al. 2008) when crustal melting takes place, producing felsic magmas with no garnet signatures.

Arc crust Evolution via Formation and Delamination of Anti-continent

Figure 3 shows that the seismic velocities inferred from the petrology of the likely lithologies of the sub-IBM arc crust and mantle provided a V_p structure consistent with that observed. This suggested that the sub-IBM arc crust and uppermost mantle may be composed of these lithologies, which have evolved through the processes shown in Fig. 4. In this case, the Moho discontinuity, which defines the boundary between the crust and mantle, may be the boundary between the remaining basaltic, initial arc crust and the restite of crustal melting following the extraction of rhyolitic melts. However, this interpretation requires the crustal component, which is a restite of crustal melting and thus originally forms within the crust (above the Moho), to be transferred to the upper mantle and consequently crossing, and then distributed below the Moho. If so, this confirms the previous suggestion of Tatsumi et al. (2008) that the sub-arc Moho is not a rigid material boundary between crust and mantle materials but is chemically transparent and permeable to crustal components.

It may be reasonable to assume a basaltic composition for the initial arc crust, as magmas produced in the mantle wedge are basaltic in composition (e.g., Tatsumi and Eggins, 1995), except in some unusual tectonic settings where andesitic primary magmas are generated (Tatsumi, 2006) such as during boninite activity at the very initial stage of IBM evolution (e.g., Meijer, 1980). Therefore, in order to create a continental crust with intermediate compositions in arc settings, the extraction of mafic components from the basaltic arc crust is necessary. A transfer of the restitic crustal component, which has a more mafic composition than the original basaltic composition of the arc crust, to the mantle may provide such a mechanism (see Table 1), causing the arc crust to evolve and possibly form continental crust. Kodaira et al. (2007b) calculated the average seismic velocity excluding the restite layer and suggested that the velocity was close to that of a typical continental crust. The complementary restite may thus be referred to as "anti-continent" (Tatsumi et al., 2008, 2014).

One characteristic seismic feature of the IBM system is the distribution of a seismic reflector at depths of 20–40 km in the upper mantle, which defines the base of the low-V uppermost mantle layer (Sato et al., 2009), the layer inferred to be the restite or anti-continent. It may thus be interesting to compare the thickness of this low-V layer with that predicted by the modeling for the anti-continent.

As a simple model, we assumed that a 3:2 mixing of a rhyolite and a basalt melt containing 75 and 50 wt% SiO_2, respectively, forms a magma with 60 wt% SiO_2, consistent with the lithology of the IBM middle crust. Such a rhyolitic magma can be produced by ~10 % partial melting of the basaltic crust (Tatsumi and Suzuki, 2009). As a result, the thickness of the restite (T_R) was expressed as a function of the thickness of the middle crust (T_M): $T_R = 2/5 \times T_M \times 9$. Applying this simple equation to the along-arc seismic structure (Kodaira et al., 2007a; Sato et al., 2009), the distribution of the restite or anti-continent layer was obtained (Fig. 6). The amount of anti-continent required to generate the IBM middle crust resulted in a much thicker layer than that observed for the low-V uppermost mantle layer.

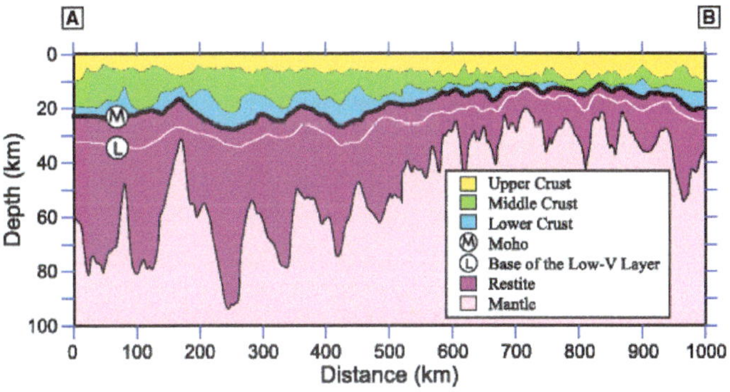

Figure 6: Depths to the base of the upper, middle, and lower crusts and the inferred restite layer along an arc-parallel section of the IBM (shown in Fig. 2) based on the seismic data of Kodaira et al. (2007a) and the model presented here. They require a restite layer that is much thicker than the low-V uppermost mantle layer, suggesting that most of the restite has delaminated.

One possible reason for this inconsistency could be the density contrast between the anti-continent and the normal upper mantle (harzburgite). The inferred density of the uppermost mantle, composed of a restite, anti-continent layer, was higher than that for the underlying normal mantle (Fig. 3) due to garnet stabilization in the restite layer (Fig. 5 and Table 2). This would have caused gravitational instability and may have resulted in the "delamination" of the overlying dense anti-continent. This process has been repeatedly emphasized as a possible physical consequence of crustal evolution (England & Houseman, 1989; Turcotte, 1989; Kay & Kay, 1993; Jull & Kelemen,2001; Tatsumi et al., 2008). Interestingly, high-pressure experiments and geochemical modeling have demonstrated that a sinking anti-continent is, in contrast to a subducting oceanic crust, always denser than the surrounding mantle, suggesting that it can penetrate through the upper-lower mantle boundary without stagnation and accumulate at the base of the mantle where it could be recycled in mantle plumes (Tatsumi et al., 2014).

While the IBM arc has been creating a buoyant continental crust, the anti-continent has been simultaneously delaminating from the arc crust. Although the process of the delamination of the anti-continent is unclear, extensive magma generation in the mantle wedge may reduce the viscosity of the upper mantle, triggering the delamination of the dense anti-continent from the arc crust.

CONCLUSIONS

In our solar system, the continents, their highland topography, and the andesitic crust that forms them are unique to the Earth. On other terrestrial planets, the crust is entirely basaltic. Andesitic magmatism and continent crust-forming processes are caused by the subduction of oceanic plates, i.e., the operation of plate tectonics. Plate tectonics was initiated by the foundering of a surface plate along a large crack or fault within a plate, which could only develop in the presence of liquid water at the surface. Water also plays a key role in the generation of andesitic magma, lowering the solidus temperature of the basaltic crust, causing partial melting of the basaltic crust, and producing differentiated, i.e., rhyolitic to andesitic, magmas.

The basaltic crust has evolved into an andesitic continental crust via the delamination of the SiO_2-deficient and dense anti-continent that was created complementary to the continental crust from the initial basaltic crust. The Earth has been simultaneously creating continents at the top of the mantle and the anti-continent at the base of the mantle, and consequently has evolved into a unique planet.

DECLARATIONS

Acknowledgements

We would like to thank Toshihiro Suzuki and Koji Shukuno for productive discussions and Alex Nicholls, editor Toshi Yamazaki, and two anonymous reviewers for their constructive comments on the manuscript.

Competing interests

The authors declare that they have no competing interests.

Authors' contributions

All authors participated in the design of the study and data analysis. All authors read and approved the final manuscript.

REFERENCES

1. Aitta A (2012) Venus' internal structure, temperature and core composition. Icarus 218:967–974
2. Albarède F (1998) The growth of continental crust. Tectonophys 296:1–14
3. Behn MD, Kelemen PB (2006) Stability of arc lower crust: insights from the Talkeetna arc section, south central Alaska, and the seismic structure of modern arcs. J Geophys Res 111: doi:10.1029/2006JB004327
4. Christensen NI, Mooney WD (1995) Seismic velocity structure and composition of the continental crust: a global view. J Geophys Res 100:9761–9788
5. Clift P, Vannucchi P (2004) Controls on tectonic accretion versus erosion in subduction zones: implications for the origin and recycling of the continental crust, Rev Geophys 42: doi: 10.1029/2003RG000127
6. Coleman RG (1977) Ophiolites. Springer Verlag, New York, p 220
7. Connolly JAD (1990) Multivariable phase-diagrams—an algorithm based on generalized thermodynamics. Am

J Sci 290:666–718

8. Connolly JAD (2005) Computation of phase equilibria by linear programming: a tool for geodynamic modeling and its application to subduction zone decarbonation. Earth Planet Sci Lett 236:524–541
9. Crawford WC, Hildebrand JA, Dorman LM, Webb SC, Wiens DA (2003) Tonga Ridge and Lau Basin crustal structure from seismic refraction data. J Geophys Res 108:2195. doi:10.1029/2001JB001435
10. Dewey JF, Bird JM (1971) Origin and emplacement of the ophiolite suite: Appalachian ophiolites in Newfoundland. J Geophys Res 76:3179–3206
11. Dilek Y, Furnes H (2011) Ophiolite genesis and global tectonics: geochemical and tectonic fingerprinting of ancient oceanic lithosphere. Geol Soc Am Bull 123:387–411
12. England P, Houseman G (1989) Extension during continental convergence, with application to the Tibetan plateau. J Geophys Res 94:17561–17579
13. Ewart A (1982) The mineralogy and petrology of Tertiary-Recent orogenic volcanic rocks: with special reference to the andesitic-basaltic compositional range. In: Thorp RS (ed) Andesites: orogenic andesites and related rocks. John Wiley and Sons, New York, pp 25–95
14. Gill JB (1981) Orogenic andesites and plate tectonics. Springer, Berlin
15. Hacker BR, Abers GA, Peacock SM (2003) Subduction factory; 1, theoretical mineralogy, densities, seismic wave speeds, and H2O contents. J Geophys Res 108: doi:10.1029/2001JB001127
16. Hopkins MD, Harrison TM, Manning CE (2010) Constrains on Hadean geodynamics from mineral inclusions in >4 Ga zircons. Earth Planet Sci Lett 298:367–376
17. Ishii T, Sato H, Haraguchi S, Fryer P, Fujioka K, Bloomer SH, Yokose H (2000) Petrological characteristics of peridotites from serpentinite seamounts in the Izu-Ogasawara-Mariana Forearc. J Geogra 109:517–530
18. Ishizuka O, Uto K, Yuasa M, Hochstaedter AG (2002) Volcanism in the earliest stage of back-arc rifting in the Izu-Bonin Arc revealed by laser-heating $^{40}Ar/^{39}Ar$ dating. J Volcanol Geotherm Res 120:71–85
19. Ishizuka O, Tani K, Reagan MK, Kanayama K, Umino S, Harigane Y, Sakamoto I, Miyajima Y, Yuasa M, Dunkley DJ (2011) The timescales of subduction initiation and subsequent evolution of an oceanic island arc. Earth Planet Sci Lett 306:229–240
20. Jull M, Kelemen PB (2001) On the conditions for lower crustal convective instability. J Geophys Res 106:6423–6446
21. Karato S, Wu P (1993) Rheology of the upper mantle: a synthesis. Science 260:771–778
22. Kawate S, Arima M (1998) Petrogenesis of the Tanzawa plutonic complex, central Japan: exposed felsic middle crust of the Izu-Bonin-Mariana arc. Island Arc 7:342–358
23. Kay RW, Kay SM (1993) Delamination and delamination magmatism. Tectonophys 219:177–189
24. Kelemen PB (1995) Genesis of high Mg-andesites and the continental crust. Contrib Mineral Petrol 120:1–19
25. Kodaira S, Sato T, Takahashi N, Ito A, Tamura Y, Tatsumi Y, Kaneda Y (2007a) Seismological evidence for variable growth of crust along the Izu intra-oceanic arc. J Geophys Res 112:B05104. doi:10.1029/2006JB004593
26. Kodaira S, Sato T, Takahashi N, Miura S, Tamura Y, Tatsumi Y, Kaneda Y (2007b) New seismological constraints on growth of continental crust in the Izu-Bonin intra-oceanic arc. Geology 35:1031–1034
27. Komiya T, Maruyama S, Masuda T, Nohda S, Hayashi M, Okamoto K (1999) Plate tectonics at 3.8–3.7 Ga: field evidence from the Isua accretionary complex, southern West Greenland. J Geol 107:515–554
28. Korenaga J (2009) Scaling of stagnant-lid convection with Arrhenius rheology and the effects of mantle melting. Geophys J Int 179:154–170
29. Meijer A (1980) Primitive arc volcanism and a boninite series: Examples from western Pacific island arcs. In: Hayes DE (ed) The tectonic evolution of Southeast Asian seas and islands, vol 23, American Geophysical Union Monograph. AGU, Washington, pp 269–282
30. Moresi L, Solomatov V (1998) Mantle convection with a brittle lithosphere: thoughts on the global tectonic styles of the Earth and Venus. Geophys J Int 133:669–682
31. Nakanishi A, Kurashimo E, Tatsumi Y, Yamaguchi H, Miura S, Kodaira S, Obana K, Takahashi N, Tsuru T, Kaneda Y, Kawasaki T, Hirata N (2009) Crustal evolution of the southwestern Kuril arc, Hokkaido, Japan, deduced from seismic velocity and geochemical structure. Tectonophys 472:105–123
32. Nutman AP, Friend CRL, Bennett VC (2002) Evidence for 3650–3600 Ma assembly of the northern end of the Itsaq Gneiss Complex, Greenland: implication for early Archean tectonics. Tectonics 21:1–28
33. Rudnick RL (1995) Making continental crust. Nature 378:571–578
34. Sakamoto I, Hirata D, Fujioka K (1999) Description of basement rocks from the Izu-Bonin arc. Res Rep Kanagawa Pref Mu 9:21–39
35. Sato T, Kodaira S, Takahashi N, Tatsumi Y, Kaneda Y (2009) Amplitude modeling of the seismic reflectors in the crust-mantle transition layer beneath the volcanic front along the northern Izu-Bonin island arc. Geochem Geophys Geosys 10: doi:10.1029/2008GC001990
36. Schubert G, Turcotte DL, Olson P (2001) Mantle convection in the earth and planets. Cambridge Univ Press, Cambridge
37. Shirey SB, Kamber BS, Whitehouse MJ, Mueller PA, Basu AR (2008) A review of the isotopic and trace element evidence for mantle and crustal processes in the Hadean and Archean: implications for the onset of plate tectonic subduction. Geol Soc Am Special Paper 440:1–29
38. Smith DE, Zuber MT, Phillips RJ, Solomon SC, Hauck SA II, Lemoine FG, Mazarico E, Neumann GA, Peale SJ, Margot JL, Johnson CL, Torrence MH, Perry ME, Rowlands DD, Geossens S, Head JW, Taylor AH (2012)

Gravity field and internal structure of Mercury from NESSENGER. Science 336:214–217

39. Solomatov VS (1995) Scaling of temperature- and stress-dependent viscosity convection. Phys Fluids 7:266–274
40. Stern RJ (2005) Evidence from ophiolites, blueschists, and ultrahigh-pressure metamorphic terranes that the modern episode of subduction tectonics began in Neoproterozoic time. Geology 33:557–560
41. Stern RJ, Tamura Y, Ishizuka O, Shukano H, Bloomer SH, Embley RW, Leybourne M, Kawabata H, Nunokawa A, Nichols AL, Kohut E, Pujana I (2013) Volcanoes of the Diamante Cross-Chain: evidence for a mid-crustal felsic magma body beneath the southern Izu-Bonin-Mariana arc. In: Tuena A et al. (eds) Orogenic andesites and crustal growth, Geol Soc London Spec Pub 385:235-255
42. Stevenson DJ (2001) Mars' core and magnetism. Nature 412:214–219
43. Suyehiro K, Takahashi N, Ariie Y, Yokoi Y, Hino R, Shinohara M, Kanazawa T, Hirata N, Tokuyama H, Taira A (1996) Continental crust, crustal underplating, and low-Q upper mantle beneath an oceanic island arc. Science 272:390–392
44. Takahashi N, Kodaira S, Klemperer LS, Tatsumi Y, Kaneda Y, Suyehiro K (2007) Crustal structure and evolution of the Mariana intra-oceanic island arc. Geology 35:203–206
45. Takahashi N, Kodaira S, Tatsumi Y, Kaneda Y, Suyehiro K (2008) Structure and growth of the Izu-Bonin-Mariana arc crust: I. Seismic constraint on crust and mantle structure of the Mariana arc–back-arc system. J Geophys Res 113:B01104. doi:10.1029/2007JB005120
46. Tamura Y, Gill JB, Tollstrup D, Hawabata H, Shukuno H, Chang Q, Miyazaki T, Takahashi T, Hirahara Y, Hodaira S, Ishizuka O, Suzuki T, Kido Y, Fiske RS, Tatsumi Y (2009) Silicic magmas in the Izu-Bonin oceanic arc and implications for crustal evolution. J Petrol 50:685–723
47. Tani K, Dunkley DJ, Kimura J-I, Wysoczanski RJ, Yamada K, Tatsumi Y (2010) Syncollisional rapid granitic magma formation in an arc-arc collision zone: evidence from the Tanzawa plutonic complex, Japan. Geology 38:215–218
48. Tatsumi Y (2000) Continental crust formation by crustal delamination in subduction zones and complementary accumulation of the enriched mantle I component in the mantle. Geochem Geophys Geosys 1, doi : 10.1029/2000GC000094
49. Tatsumi Y (2006) High-Mg andesites in the Setouchi volcanic belt, southwestern Japan: analogy to Archean magmatism and continental crust formation? Ann Rev Earth Planet Sci 34:467–499
50. Tatsumi Y, Eggins S (1995) Subduction zone magmatism. Wiley, Boston
51. Tatsumi Y, Suzuki T (2009) Tholeiitic vs. calc-alkalic differentiation and evolution of arc crust: constraints from melting experiments on a basalt from the Izu-Bonin-Mariana arc. J Petrol 50:1575–1603
52. Tatsumi Y, Sakuyama M, Fukuyama H, Kushiro I (1983) Generation of arc basalt magmas and thermal structure of the mantle wedge in subduction zones. J Geophys Res B88:5815–5825
53. Tatsumi Y, Shukuno H, Tani K, Takahashi N, Kodaira S, Kogiso T (2008) Structure and growth of the Izu-Bonin-Mariana arc crust: 2. Role of crust-mantle transformation and the transparent Moho in arc crust evolution. J Geophys Res: 113, doi:10.1029/2007JB005121
54. Tatsumi Y, Suzuki T, Ozawa H, Hirose K, Hanyu T, Ohishi Y (2014) Accumulation of 'anti-continent' at the base of the mantle and its recycling in mantle plumes. Geochim Cosmochim Acta 143:23–33
55. Taylor B (1992) Rifting and the volcanic-tectonic evolution of the Izu-Bonin-Mariana arc system. In: Taylor B. et al. (eds) Proceedings of ocean drilling program, scientific results, Texas A&M Univ., College Station 126: 627–650
56. Taylor SR (1967) The origin and growth of continents. Tectonophys 4:17–34
57. Taylor SR, McLennan SM (1995) The geochemical evolution of the continental crust. Rev Geophys 33:241–265
58. Taylor SR, McLennan SM (2009) Planetary crusts: their composition, origin and evolution. Cambridge Univ Press, Cambridge
59. Turcotte DL (1989) Geophysical processes influencing the lower continental crust. In: Mereu RF, Mueller S, Fountain DM (eds) Properties and processes of Earth's lower crust. Am Geophys Union, Washington, pp 321–329
60. Valencia D, Sasselov DD, O'Connell RJ (2007) Radius and structure models of the first super-Earth planet. Astrophys J 656:545–551

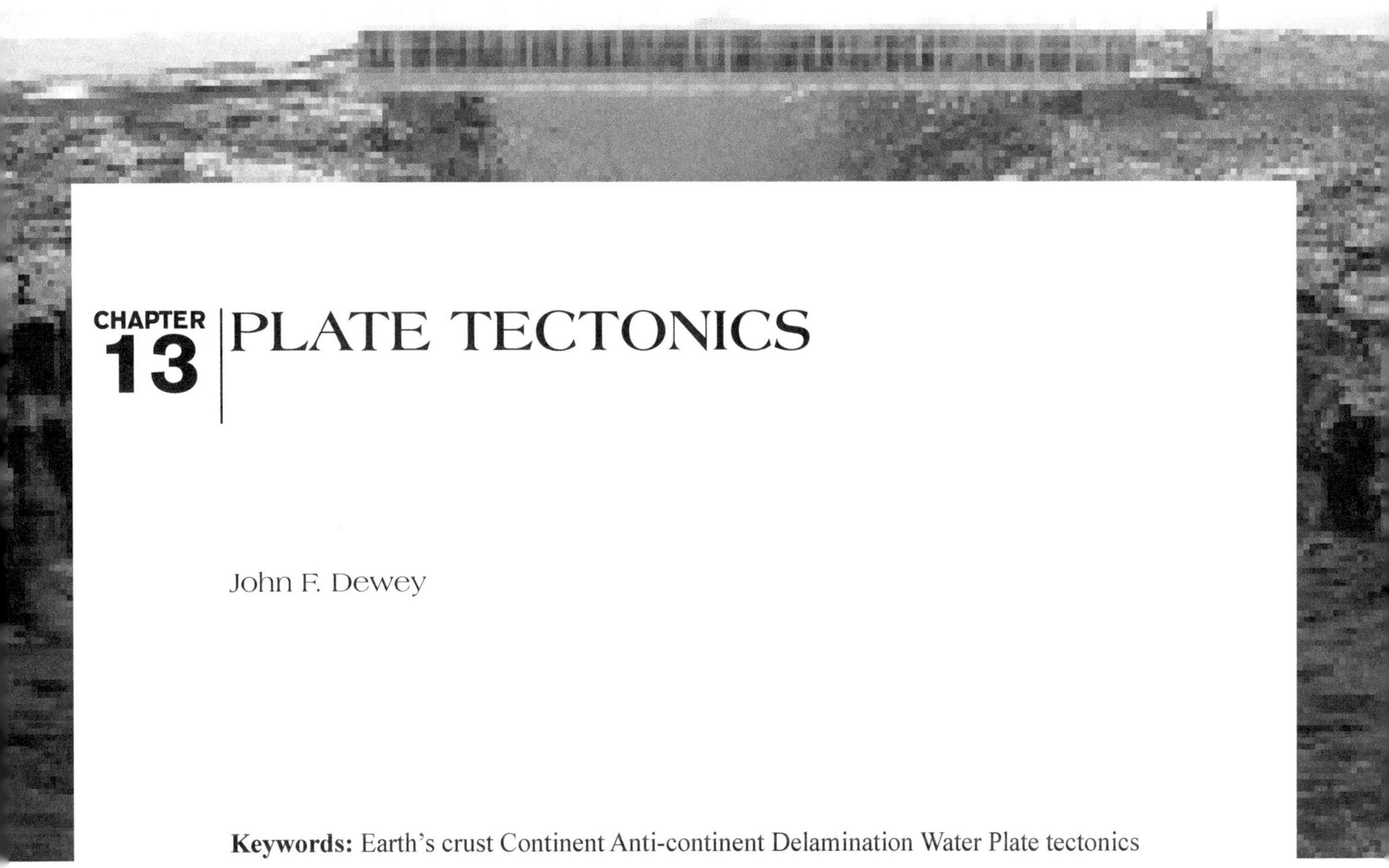

CHAPTER 13 PLATE TECTONICS

John F. Dewey

Keywords: Earth's crust Continent Anti-continent Delamination Water Plate tectonics

THE PRE-1965 HARBINGERS

The general similarity of the shapes of the eastern and western shorelines of South America and Africa, respectively, was noticed and recorded by Francis Bacon at a time when the detailed geography of the shorelines was ill-determined and by Antonin Snider [**1**] when the existence of the continental shelf and its margin were unknown. Bacon did not claim or even mention that this implied that Africa and South America once fitted together. Alfred Wegener [**2**] and Alexander Du Toit [**3**] proposed the idea of continental drift and the opening of the South Atlantic from the excellent geometric fit of the South Atlantic continental margins (the edge of the continental shelf not the shorelines), the close similarity of South American Upper Palaeozoic faunas and floras, and the continuity of geological zones and structures when the two continents were restored to contiguity. Implicit but not explicit was that the two continental margins and, by inference, the continents must have been torsionally rigid to maintain the fitting edge. In retrospect, this seems to be compelling evidence for continental drift yet it was a subject for lively, often vituperative, debate until the 1960s. There was substantial division between geologists working in the Southern Hemisphere, where the evidence for continental drift is so clear, and the Northern Hemisphere where the evidence is less clear. Lester King (e.g. [**4**]), of the University of Cape Town, travelled the globe after World War II proselytizing for continental drift. As an undergraduate in London in 1956, I heard him talk in Imperial College, clearly and convincingly, when I became convinced that continental drift must be a reality. Stanley Hollingworth and Stanley Westoll, professors of geology at University College London and Newcastle, respectively, were firm adherents of drift in the 1950s. Westoll enjoyed taking a sheet of newsprint, tearing it along a jagged edge and then defying onlookers to claim that the fit of the two margins was a coincidence, a simple but salutary lesson to the anti-drifters. The conflict was exacerbated by Harold Jeffreys [**5,6**], who argued that continental drift is impossible from the theoretical standpoint that there was no known mechanism to explain it. This was a classic case not only of theory outrunning observation but theory constructed upon the incorrect notion that weak continents, dominated by the strength of quartz, ploughed through a strong oceanic basement

Citation: John F. Dewey, Plate Tectonics, .http://dx.doi.org/10.1098/rsta.2014.0227.

dominated by the strength of olivine. Jeffreys' assumptions were incorrect but his influence was profound until the mid-1960s, especially in the USA. It is said that, if one espoused the ideas of continental drift in the USA before the 1960s, an academic career in geology was doomed. From 1970, the reverse has been true. Major figures of great influence such as Walter Bucher, Charles Drake and Maurice Ewing strongly opposed continental drift, the last until his death in 1974. Meyerhoff [7] waged warfare against plate tectonics to his death in 1994. As a young lecturer in Cambridge in the late 1960s, Jeffreys and R. A. Lyttleton, an expanding Earth enthusiast, suggested that I abandon the mobilist 'nonsense' that was gaining ground and credence because it would ruin my career. Opposition to mobilism could be construed as strange in that Keith Runcorn [**8**] and Ted Irving [9] had already demonstrated, with great precision and clarity, the relative motion of continents from palaeomagnetic studies.

All this, until the 1960s, was concerned with continental edge fitting because the geology of the ocean floor was poorly known, although Arthur Holmes [**10**] and David Griggs [**11**] had proposed a convective model involving ocean growth and destruction. By the mid-1960s, the geology of the ocean floor was becoming better known and the concept of sea-floor spreading along the axes of mid-ocean ridges and subduction in oceanic trenches had been proposed by Bob Dietz [**12**] and Harry Hess [**13**]. This, with the explanation of Vine & Matthews [**14**] of the zebra-like stripes of positive and negative magnetic anomalies as a 'sea-floor spreading magnetic tape recorder', laid much of the basis for plate tectonics. 'Sam' Warren Carey [**15**,**16**] developed several important contributions to 'mobilism' including the fast rotation of micro-continents and the opening of oceanic sphenochasms such as Iberia and the Bay of Biscay, and the growth of nemataths (hotspot tracks). Carey, like Bruce Heezen [**17**], failed to embrace and contribute to the plate tectonics revolution by hoisting his flag upon the mast of an expanding Earth, in spite of the clear evidence, from palaeomagnetic triangulation, of a constant size Earth at least during the Phanerozoic. By the mid-1960s, the contrast between old silicic/intermediate continents and a young mafic/ultramafic ocean floor was becoming clear.

A little known but profoundly prescient and original harbinger of the concept of a rotation pole that describes the relative motion of continental masses was Quennell [**18**]. Quennell recognized that the general trend of the Dead Sea fault system forms part of a small circle, normal to which great circles intersect at a pole near Gibraltar. From this, Quennell argued that the relative motion of Arabia with respect to Africa can be described as an anticlockwise rotation whose rate is given by the slip-rate (about 20 mm yr^{-1}) on the sinistral Dead Sea Fault, now termed a transform fault, which also accounts for the opening of the Red Sea. This was a remarkable discovery of the use of a rotation pole to describe the relative motion of torsionally rigid continents. Another related idea before its time was Wellman's [**19**] recognition of the role of the Alpine fault in New Zealand as a dextral, crustal scale strike-slip fault that transferred motion between two zones of convergence, now termed a trench–trench transform.

Tuzo Wilson entered, enthusiastically, the mobilism saga in the early 1960s, before which he had been a fixist in the North American tradition. His first key paper [**20**] compared the Cabot Fault of the Maritime Provinces of Canada with the Great Glen Fault of Scotland supporting the opening of the North Atlantic. Similarly, Dewey & Kay [**21**] showed, in a reconstruction of the North Atlantic, the clear continuity of the Appalachian and Caledonian Belts. Wilson [**22**] used Carey's nemataths as hotspot track lines to demonstrate the movement of the ocean floor across plumes in the mantle and the relative motion of the ocean floor with respect to oceanic ridges. Especially persuasive is the V-shaped relationship of the Rio Grande Rise and Walvis Ridge to the Tristan da Cunha Hotspot. Wilson [**23**] not only explained the transform offsets of ridge axes and subduction zones, but also laid out a comprehensive global model of sea-floor spreading, major strike-slip and subduction along a network of connected earthquake boundaries, showing how transforms that connect ridges and subduction zones and subduction zones with different polarities cause the earthquake-bounded areas (plates) to change size and shape. Tuzo Wilson was, tectonically, almost there. In the early 1970s, he confided to me and others that, had he known of Euler's theorem, he would have nailed the theory of plate tectonics. Like Wegener, Du Toit and Carey, the torsional rigidity of 'plates' was implicit in his 1963 and 1965 papers, but he missed the essential point of the relative motion of plates described by rotations around axes passing through the centre of Earth intersecting the surface of Earth at poles of rotation. The scene was now set for Bullard *et al.* [**24**], which has become the central and most durable paper from a 1964 Royal Society Discussion Meeting on continental drift organized by Blackett, Bullard and Runcorn. The volume, published in 1965 [**25**], provides a fascinating snapshot of the point at which majority opinion was poised to swing from 'fixist' to 'mobilist' interpretations of oceans and continents, prominently driven by the rapidly increasing quality of palaeomagnetic data.

BULLARD, EVERETT AND SMITH, 1965

Bullard *et al.* [**24**] was the key paper in the paradigm change that led from continental drift to plate tectonics. The essential, critical and original features of the paper are as follows:

- The complete separation of the kinematic certainty of continental drift from concerns about the mechanism by which it was achieved.
- Restoration of the continents around the Atlantic was achieved by minimizing the misfit between the true continental margins (edge of the continental shelf at the 50 fathom line, about 900 m) not the shorelines. Snug and convincing fits were achieved.
- Torsional rigidity of the continents was assumed because distortion of continents and their margins would prohibit fitting. The superb fit demonstrates torsional rigidity.
- Fitting of continental margins was done by using finite difference angular rotations around axes passing through the centre of Earth intersecting Earth's surface at poles of rotation. Euler's theorem was used to make the finite difference rotations.
- Rifted continental margins may be very sharp with a rapid transition from continent to ocean, the ideal margin that fits cleanly. Overlaps were explained mainly as the growth of deltas since continental separation and voluminous basaltic lava fields.

- Underlaps are mainly the result of continental stretching rather than a sharp break.
- The recognition that smaller complicated areas like the Caribbean and Gulf of Mexico need special geological analyses to unravel the behaviour of small blocks.
- When the continents around the Atlantic are restored to a 'Pangea configuration' (all continents fitted together), there appears the huge, westward-narrowing and terminating gap of the Tethyan ocean complex, which closed as a result of the Central and North Atlantic opening at different times around different poles of rotation.
- Whereas most Atlantic margins were the site of extensional rifting, several, such as the equatorial Atlantic and the boundary between Greenland and Svaalbard, enjoyed hundreds of kilometres of strike-slip relative motion along giant intracontinental transform faults.
- Some small continental blocks, such as Iberia, have rotated by up to 40° during periods as short as 10 Myr around nearby rotation poles to open small, wedge-shaped oceans (Carey›s sphenochasms) such as the Bay of Biscay.

These were profoundly important reconstructions, results and conclusions that led directly to the formulation of the theory of plate tectonics shortly afterwards, and then to its geological implications and corollaries. Reception of the paper was enthusiastic and enhanced by the very high quality of the drafting of key figures in well-chosen geographical projections with publication in colour; overlap and gaps were shown, respectively, in red and blue. The fitting was achieved using a least-squares misfit between digitized continental margins on a sphere, which avoided the problems of earlier attempts that used moulded caps on a spherical last or simply ignored the issues of sphericity. Early fit diagrams were semi-schematic, whereas those of Bullard *et al.* [**24**] were based upon a simple and clear mathematical construction and a fitting algorithm on Earth's spherical surface based upon explicit rotations. The Earth science community saw the paper as a major breakthrough in tectonics; it was used widely in structure and tectonics courses, including mine, as an advance in our quantitative understanding of tectonics. It led to similar studies using the same techniques (e.g. [**26–28**]).

PLATE TECTONICS

Plate tectonics was formulated by McKenzie & Parker [**29**]; they showed how the tectonics of the North Pacific can be described, very simply and elegantly, by the relative motion of a small number of torsionally rigid lithospheric plates (Pacific, Juan de Fuca, North America around poles of rotation). Three basic types of plate boundary were recognized: ridges where new oceanic lithosphere is created; trenches where lithosphere is subducted; and transform faults where the relative motion is accommodated along lithosphere-penetrating transcurrent faults. This paper transformed tectonics; within a few years, dozens of papers elaborated many aspects of plate tectonics. McKenzie [**30**] speculated on the consequences and causes of relative plate motion and McKenzie & Morgan [**31**] wrote the definitive paper on plate triple junctions. Sykes [**32**] supported Wilson's hypothesis of ridge to ridge transforms using earthquake first motions. Comprehensive global plate tectonic models were developed by Morgan [**33**], Isacks *et al.* [**34**] and Le Pichon [**35**]. Error-minimizing global vector nests for instantaneous relative plate rotations were developed by Chase [**36**], Minster *et al.* [**37**] and De Mets [**38**]; these provided, for the first time, a comprehensive quantitative solution for the relative motion of the global plate system.

Plate tectonics had an immediate and massive effect upon the geological world in a way that continental drift never had. Plate tectonics provided the basis for understanding most of tectonic and historical geology (e.g. [**39–42**]) such as rift valleys, mountain belts, major strike-slip faults, deep ocean trenches, volcanic arcs and the evolution of the continental crust and large regions of Earth. Essentially, this was the qualitative geological 'clothing' for plate tectonics. The term 'Wilson cycle' was adopted for the opening and closing of oceans from continental splitting through arc–continent collision, to continent–continent collision in honour of his paper that first explained the Appalachian–Caledonian Orogen in terms of ocean opening and closing [**43**]. It was soon realized, from earthquake distribution, that plate boundaries in oceans are generally very sharp and well defined in contrast to the more complicated and diffuse plate boundary zones of continents. Soon after the oceanic lithosphere is generated by sea-floor spreading, it acquires great torsional rigidity.

Of especial importance was the development of the quantitative relationship between plate tectonics and geology. The first paper of this genre was Atwater [**44**] in which the basic tectonic evolution of the US Cordillera was related to the relative motion of the Cordilleran margin to the plates of the eastern Pacific. Smith [**45**] and Dewey *et al.* [**28**] related the opening of the Central and North Atlantic around different poles of rotation to the closing of the Tethyan oceanic 'gap' and the development of plate boundaries and tectonics in the Alpine orogenic system. Sengor [**46**] introduced the concepts of Palaeo-Tethys, the larger ocean(s) that formed the gap on the Bullard *et al.* reconstruction, and Neo-Tethys oceans that were formed by the detachment of continental strips during the closure of Palaeo-Tethys. Pindell & Dewey [**47**] described the tectonic evolution of the Gulf of Mexico and Caribbean region by analysing the history of the relative motion of North and South America. Dewey [**48**] provided a basic framework for the complex changing evolution of plate boundary zone rock systems from the inherent complexities of plate boundary evolution. He showed that poles of rotation are of three types: finite difference (Euler poles), finite motion and instantaneous. In a three-plate system, only two poles can be finite motion; the third must be only instantaneous with respect to which its plate boundary must move. In other words, a three-plate vector triangle cannot close except instantaneously at a point. It is likely that most poles are only instantaneous.

FURTHER IMPLICATIONS

The quantification of continental drift [**24**] leading to plate tectonics [**29**], in the context of the 1960s overall, was the greatest paradigm change in twentieth century geology. There are few areas of geology that it has not touched from tectonics to palaeontology; for the last 47 years, it has formed the basis for most of our understanding of Earth's evolution. There are, however, a substantial numbers of corollaries of plate tectonics of which we have less than a full picture and understanding. There is still disagreement about the rheology of the lithosphere, where its strength lies, how its strength varies, and the cause of its torsional rigidity [**49**,**50**]. Perhaps the biggest challenge that tectonic and structural geology faces is the detailed relationship between relative plate motion and rock structures and strain in continental plate boundary zones, and as a result of evolving plate boundaries across which the slip vector changes, and triple junctions that change their position and type (**figure 1**). Analyses have been attempted, fairly successfully, in transtensional zones between blocks with plate boundary zones such as the Eastern California Shear Zone, where Dewey & Taylor [**51**,**52**] used GPS, the inversion of large earthquake arrays, palaeomagnetism, volcanic systems and the detailed kinematics of fault systems to deduce the evolution of bulk strain facies patterns in area and depth in the Coso region. This approach can be conducted only in superbly exposed regions and involves a huge amount of field mapping and structural analysis. Also, as yet, it has been done only in a few regions at less than whole plate boundary zone scale. GPS has been a powerful tool in measuring instantaneous relative plate motion and relative block motions within continental plate boundary zones. Analysis of earthquakes alone gives a useful but incomplete picture because a substantial component of bulk displacement and strain occurs aseismically or accompanied by very small earthquakes. A further problem is why continents rift along older orogens, for example, the Greenland/Baltica split along the Caledonides and the Central Atlantic along the Appalachians/Mauritanides [**53**].

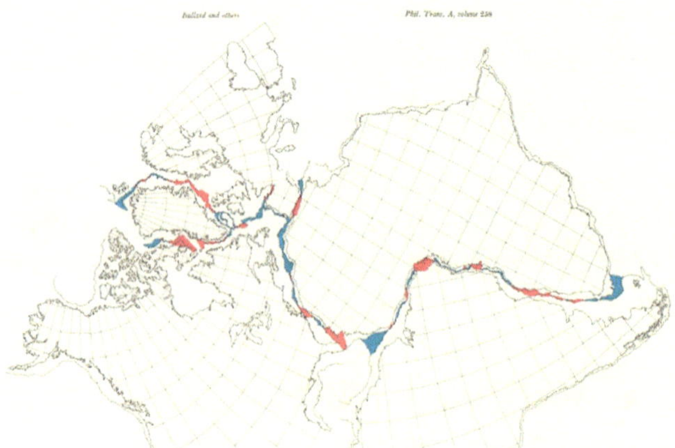

Figure 1. The fit of the continents around the Atlantic [**24**]. Copyright The Royal Society.

A final word about Teddy Bullard the man. He was a larger-than-life, wonderfully exuberant and enthusiastic person of great charm and friendliness. He was extraordinarily curious about everything and infected all those around him with a desire to find things out. As a young lecturer in Cambridge during the 1960s, he both enthused me and gave me profound advice, especially not to get stuck in a small niche of geology permanently but to expand the scale and range of one's research, to follow one's own interests not necessarily the flavour of the month and certainly not the proactive demands of senior figures, indeed any others, especially committees, who claim, falsely, to know what is important, and to always stay with science, rejecting arid administrative paths. His influence on the scientific world was profound. All who knew him will remember him with great affection for his wisdom, cleverness, originality, advice, guidance, friendship and the sheer fun of being with him.

ACKNOWLEDGEMENTS

Throughout my career, I have been encouraged and helped by numerous people, especially John Hayward, Jack Kirkaldy, John Ramsay, Janet Watson, Teddy Bullard, Walter Pitman, Celal Sengor and Paul Ryan.

Footnotes

- One contribution of 17 to a theme issue 'Celebrating 350 years of Philosophical Transactions: physical sciences papers'.
- The featured article can be viewed at http://dx.doi.org/10.1098/rsta.1965.0020.

REFERENCES

1. Snider A. 1858 La Creation et ses mysteres. Paris, France: Frank et Dentu.
2. Wegener A. 1924 The origin of continents and oceans. London, UK: Methuen.
3. Du Toit AL. 1937 Our wandering continents. Edinburgh, UK: Oliver and Boyd.
4. King LC. 1951 South African scenery. Edinburgh, UK: Oliver and Boyd.
5. Jeffreys H. 1924 The Earth: its origin, history, and physical constitution, 1st edn. Cambridge, UK: Cambridge University Press.
6. Jeffreys H. 1976 The Earth: its origin, history, and physical constitution. Cambridge, UK: Cambridge University Press.
7. Meyerhof AA. 1970 Continental drift, II: high-latitude evaporite deposits and geologic history of Arctic and North Atlantic oceans. J. Geol. 78, 406–444. (doi:10.1086/627538)
8. Runcorn SK. 1962 Palaeomagnetic evidence for continental drift and its geophysical cause. Cont. Drift Int. Series 3, 1–40. (doi:10.1016/B978-1-4832-2982-9.50006-1)
9. Irving E. 1964 Palaeomagnetism and its application to geological and geophysical problems. New York, NY: John Wiley.
10. Holmes A. 1944 Principles of physical geology. Edinburgh, UK: Nelson.
11. Griggs D. 1939 A theory of mountain building. Am. J. Sci. 237, 611–650. (doi:10.2475/ajs.237.9.611)
12. Dietz RS. 1961 Continent and ocean basin evolution by spreading of the ocean floor. Nature 190, 854–857. (doi:10.1038/190854a0)
13. Hess HH. 1955 Serpentine, orogeny and epeirogeny. Geol. Soc. Am. Spec. Paper 62, 391–407. (doi:10.1130/SPE62-p391)
14. Vine FJ, Matthews DH. 1963 Magnetic anomalies over oceanic ridges. Nature 199, 947–949.

(doi:10.1038/199947a0)

15. Carey SW. 1954 The rheid concept in geotectonics. J. Geol. Soc. Aust. 1, 67–117. (doi:10.1080/14400955308527848)
16. Carey SW. 1958 The tectonic approach to continental drift; continental drift; a symposium, pp. 177–358. Hobart, Australia: University of Tasmania, Geology Department.
17. Heezen BC. 1960 The rift in the ocean floor. Sci. Am. 205, 99–111.
18. Quennell AM. 1958 The structural and geomorphic evolution of the Dead Sea Rift. Geol. Soc. Lond. Q. J. 114, 1–24. (doi:10.1144/gsjgs.114.1.0001)
19. Wellman HW. 1955 New Zealand quaternary tectonics. Geol. Rundsch. 43, 248–257. (doi:10.1007/BF01764108)
20. Wilson JT. 1962 Cabot Fault, an Appalachian equivalent of the San Andreas and Great Glen Faults and some implications for continental displacement. Nature 195, 135–138. (doi:10.1038/195135a0)
21. Dewey JF, Kay G. 1968 Appalachian and Caledonian evidence for drift in the north Atlantic. In The history of the Earth's crust (ed. RA Phinney), pp. 161–167. Princeton, NJ: Princeton University Press.
22. Wilson JT. 1963 Evidence from islands on the spreading of ocean floors. Nature 197, 536–538. (doi:10.1038/197536a0)
23. Wilson JT. 1965 A new class of faults and their bearing on continental drift. Nature 207, 343–347. (doi:10.1038/207343a0)
24. Bullard E, Everett JE, Smith AG. 1965 The fit of the continents around the Atlantic. Phil. Trans. R. Soc. Lond. A 258, 41–51. (doi:10.1098/rsta.1965.0020)
25. Blackett PMS, Bullard EC, Runcorn SK. 1965 Continental drift. Phil. Trans. R. Soc. Lond. A 258, 1–333. (doi:10.1098/rsta.1965.0016)
26. Smith AG. 1971 Alphine deformation and the oceanic areas of the Tethys, Mediterranean, and Atlantic. Geol. Soc. Am. Bull. 82, 2039–2070. (doi:10.1130/0016-7606(1971)82[2039:ADATOA] 2.0.CO;2)
27. Pitman WC, Talwani M. 1972 Sea-floor spreading in the North Atlantic. Geol. Soc. Am. Bull. 83, 619–646. (doi:10.1130/0016-7606(1972)83[619:SSITNA]2.0.CO;2)
28. Dewey JF, Pitman WC, Ryan WBF, Bonnin J. 1973 Plate tectonics and the evolution of the Alpine system. Geol. Soc. Am. Bull. 84, 3137–3180. (doi:10.1130/0016-7606(1973) 84<3137:PTATEO>2.0.CO;2)
29. McKenzie DP, Parker RL. 1967 The North Pacific, an example of tectonics on a sphere. Nature 216, 1276–1280. (doi:10.1038/2161276a0)
30. McKenzie DP. 1969 Speculations on the consequences and causes of plate tectonics. Geophys. J. Int. 18, 1–32. (doi:10.1111/j.1365-246X.1969.tb00259.x)
31. McKenzie DP, Morgan WJ. 1969 Evolution of triple junctions. Nature 224, 125–133. (doi:10.1038/224125a0)
32. Sykes LR. 1967 Mechanism of earthquakes and nature of faulting on the mid-ocean ridges. J. Geophys. Res. 72, 2131–2153. (doi:10.1029/JZ072i008p02131)
33. Morgan WJ. 1968 Rises, trenches, great faults, and crustal blocks. J. Geophys. Res. 73, 1959–1982. (doi:10.1029/JB073i006p01959)
34. Isacks B, Oliver J, Sykes LR. 1968 Seismology and the new global tectonics. J. Geophys. Res. 73, 5855–5899. (doi:10.1029/JB073i018p05855)
35. Le Pichon X. 1968 Sea floor spreading and continental drift. J. Geophys. Res. 73, 3661–3697. (doi:10.1029/JB073i012p03661)
36. Chase CG. 1972 The N-plate problem of plate tectonics. Geophys. J. Int. 29, 117–122. (doi:10.1111/j.1365-246X.1972.tb02202.x)
37. Minster JB, Jordan TH, Molnar P, Haines E. 1974 Numerical modeling of instantaneous plate tectonics. Geophys. J. Int. 36, 541–576. (doi:10.1111/j.1365-246X.1974. tb00613.x)
38. DeMets C. 1993 Earthquake slip vectors and estimates of present-day plate motions. J. Geophys. Res. 98, 6703–6714. (doi:10.1029/92JB02868)
39. Dewey JF. 1969 The evolution of the Appalachian/Caledonian orogen. Nature 222, 124–129. (doi:10.1038/222124a0)
40. Dewey JF, Bird JM. 1970 Mountain belts and the new global tectonics. J. Geophys. Res. 75, 2625–2647. (doi:10.1029/JB075i014p02625)
41. Dickinson WR. 1970 Relation of andesites, granites, and derivative sandstones to arc-trench tectonics. Rev. Geophys. 8, 813–860. (doi:10.1029/RG008i004p00813)
42. Hamilton W. 1969 Mesozoic California and the underflow of Pacific mantle. Geol. Soc. Am. Bull. 80, 2409–2430. (doi:10.1130/0016-7606(1969)80[2409:MCATUO]2.0.CO;2)
43. Wilson JT. 1966 Did the Atlantic close and then reopen? Nature 211, 676–681. (doi:10.1038/211676a0)
44. Atwater T. 1970 Implications of plate tectonics for the Cenozoic tectonics of western North America. Geol. Soc. Am. Bull. 81, 3513–3536. (doi:10.1130/0016-7606(1970)81 [3513:IOPTFT]2.0.CO;2)
45. Smith AG. 1971 Alpine deformation and the oceanic areas of the Tethys, Mediterranean and Atlantic. Geol. Soc. Am. Bull. 82, 2039–2070. (doi:10.1130/0016-7606(1971)82 [2039:ADATOA]2.0.CO;2)
46. Sengor AMC. 1979 Mid-Mesozoic closure of Permo-Triassic Tethys and its implications. Nature 279, 590–593. (doi:10.1038/279590a0)
47. Pindell J, Dewey JF. 1982 Permo-Triassic reconstruction of Western Pangea and the evolution of the Gulf of Mexico/Caribbean region. Tectonics 1, 179–211. (doi:10.1029/ TC001i002p00179)
48. Dewey JF. 1975 Finite plate evolution: some implications for the evolution of rock masses at plate margins. Am. J. Sci. 275-A, 260–284.
49. Jackson J. 2002 Strength of the continental lithosphere: time to abandon the jelly sandwich. GSA Today 12, 4–10. (doi:10.1130/1052-5173(2002)012<0004:SOTCLT>2.0.CO;2)
50. Watts AB, Burov EB. 2003 Lithospheric strength

and its relationship to the elastic and seismogenic layer thickness. Earth Planet. Sci. Lett. 213, 113–131. (doi:10.1016/S0012- 821X(03)00289-9)

51. Dewey JF, Taylor T, Monastero FC. 2008 Transtension in the brittle field: the Coso region, southern California. Int. Geol. Rev. 50, 193–217. (doi:10.2747/0020-6814.50.3.193)

52. Taylor T, Dewey JF, Monastero FC. 2008 Transtensional deformation of the brittle crust: field observations and theoretical applications in the Coso-China Lake region, eastern margin of the Sierra Nevada Microplate, southeastern California. Int. Geol. Rev. 50, 218–244. (doi:10.2747/0020-6814.50.3.218)

53. Ryan PD, Dewey JF. 1997 Continental eclogites and the Wilson cycle. Geol. Soc. Lond. J. 154, 437–442. (doi:10.1144/gsjgs.154.3.0437)

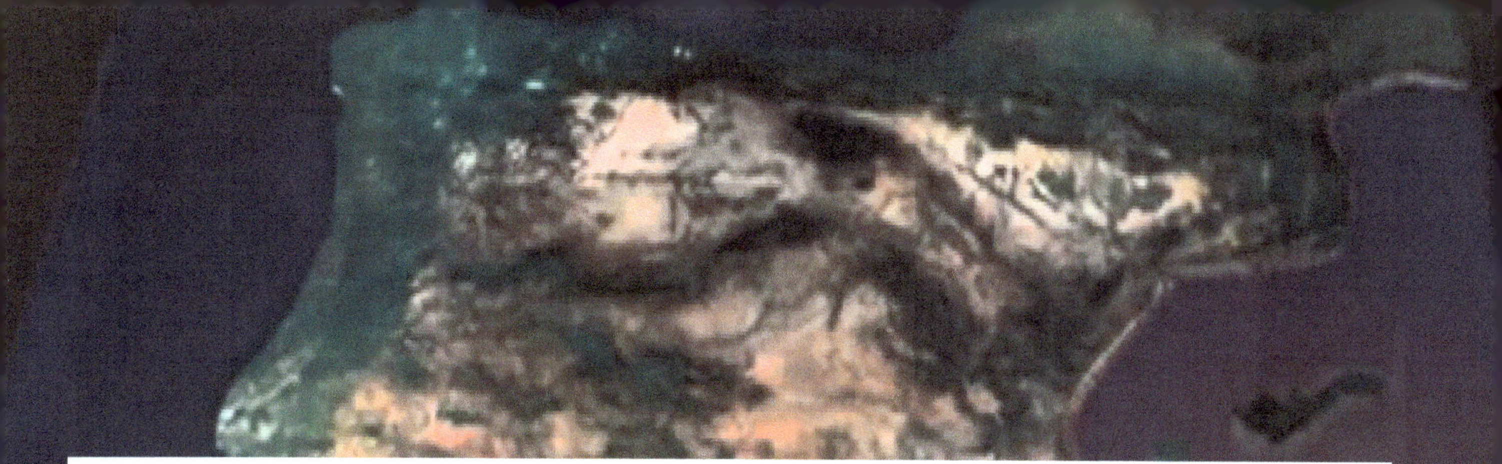

CHAPTER 14 PLATE TECTONIC EVOLUTION

Jan Golonka and Aleksandra Gawęda

INTRODUCTION

The role of an active margin of Eurasia during Mesozoic and Cenozoic times was well defined (Golonka, 2004). The trench-pulling effect of the north dipping subduction, which developed along the new continental margin caused rifting, creating the back-arc basin as well as transfer of plates from Gondwana to Laurasia. The present authors applied this model to the southern margin of Laurussia during Paleozoic times. The preliminary results of their work were presented during the Central European Tectonic Group (CETEG) in 2011 in Czech Republic. The supercontinent of Laurussia, defined by Ziegler (1989), included large parts of Europe and North America. The southern margin of this supercontinent stretched out between Mexico and the Caspian Sea area. The present authors attempted to characterize the entire margin, paying the special attention to Central and Eastern Europe.

Citation: Jan Golonka and Aleksandra Gawęda (2012). Plate Tectonic Evolution InTech, DOI: 10.5772/50009.

METHODS

The present authors were using a plate tectonic model, which describes the relative motions plates and terranes during Paleozoic times. This model is based on PLATES, GPLATES and PALEOMAP software (see Golonka *et al.* 1994, 2003, 2006a,b, Golonka 2000, 2002, 2007a,b,c, 2009a,b). The plate tectonic reconstruction programs generated palaeocontinental base maps. It takes tectonic features in the form of digitised data files, assembles those features in accordance with user specified rotation criteria (Golonka *et al.* 2006a).

The rigid, outer part of the earth divided into many pieces known as lithospheric plates, comprising both the continental landmasses and oceanic basins These plates are in motion relative to each other and to the earth itself. Assuming the earth is a sphere, the motion of a plate across the earth's surface can be described as motion about the axis of a pole of rotation that goes through the centre of the earth. The intersection of the pole's axis with the earth's surface is referred to by its latitude/longitude coordinates. The distance the plate travels about the pole is an angular distance and is recorded in degrees. A stage pole of rotation describes the distance a plate moved from one time to the next time (i.e. from 20 Ma to 10 Ma). A finite pole of rotation describes the total distance a plate moved from some time in the past to the present day (i.e. from 20 Ma to 0 Ma). A rotation file contains a list of poles of rotation for various plates.

The rotation files used by applied software contain finite poles of rotation. Thus for each plate, there are several finite poles of rotation for different times in the past. Plate models that use rotation files that describe the motion of plates relative to other plates are called relative framework models. Among, the data that show the relative motions between plates are fracture zones. Fracture zones are essentially flowlines between plates. For example, in the South Atlantic, the fracture zones show the motion of South America relative to (or away from) Africa. The finite pole of rotation describing this motion is a relative pole. South America is referred to as the moving plate and Africa as the fixed plate (Golonka *et al.* 2006a). The rotation file contains a list of finite rotations between pairs of tectonic elements, at different episodes of time, with brief bibliographic notes or general comments for each individual rotation.

PALEOZOIC MAJOR CONTINENTAL PLATES AND OCEAN RELATED TO THE LAURUSSIA SUPERCONTINENT

The break-of the supercontinent Pannotia (Dalziel at al., 1994) during the latest Precambrian times (Golonka, 2000, 2002) lead to the formation of the new continents. Fig 1. depicts the position of these continents at the beginning of the Paleozoic.

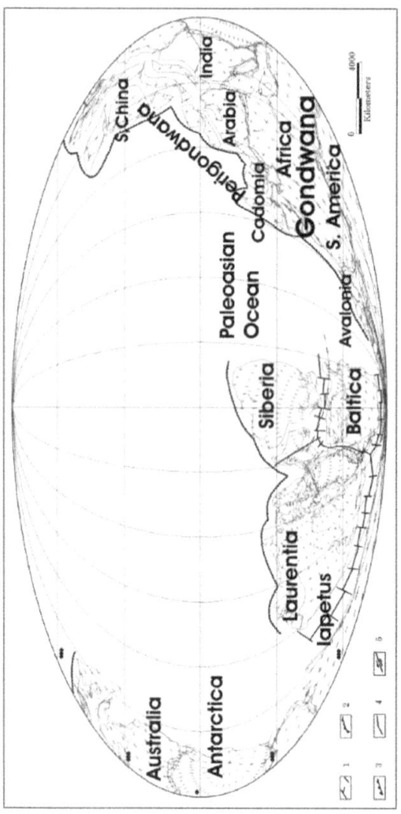

Figure 1. Plate tectonic global map of Early Cambrian (plates position as of 544 Ma). Mollweide projection. Modified from Golonka (2012). 1 - oceanic spreading center and transform faults, 2 - subduction zone, 3 - thrust fault, 4 - normal fault, 5 - transform fault.

Baltica

The continent Baltica was named after the Baltic Sea. It consisted of a major parts of northern and eastern Europe. It was bounded by the north by the border of the shelf of Norway, southern Barents Sea and Novaya Zemlya, on the east by the Ural suture, and on the southwest by a suture located close, but not quite along the Teisseyre-Tornquist line (Scotese & McKerrow, 1990; Golonka *et al.*, 1994). The southern boundary is more controversial. The Ukrainian shield certainly belonged to Baltica (Zonenshain et al, 1990). Perhaps also fragments of the North European platform like the Malopolska block, Bruno-Vistulicum, Moesia and other small blocks located now around the Baltic Sea belonged to Baltica (Kalvoda & Bábek 2010, Żelaźniewicz et al., 2009, Besutiu, 2001). In Central and southwestern Europe the possible boundary of Baltica is marked by the extent of North European plate below the Carpathian and Balkan nappes (Figs. 1-3)

Gondwana

The supercontinent Gondwana, also known as Gondwanaland (Vevers, 2004) was named after the ancient Indian tribes Gonds; in Sanskrit Gondwana means "the forest of Gonds". The continents forming the core of Gondwana include South America, Africa, Madagascar, India, Antarctica and Australia. The location of numerous smaller continental blocks that bordered Gondwana is less certain. The following were adjacent to Gondwana at some time during the Paleozoic: Yucatan, Florida, Avalonia, central European (Cadomian) terranes between the Armorica and Bohemian Massif, Moesia, Iberia, Apulia and the smaller, southern European terranes, central Asian terranes (Karakum and others), China (several separate blocks), and the Cimmerian terranes of Turkey, Iran, Afghanistan, Tibet and Southeast Asia (Figs, 1, 2).

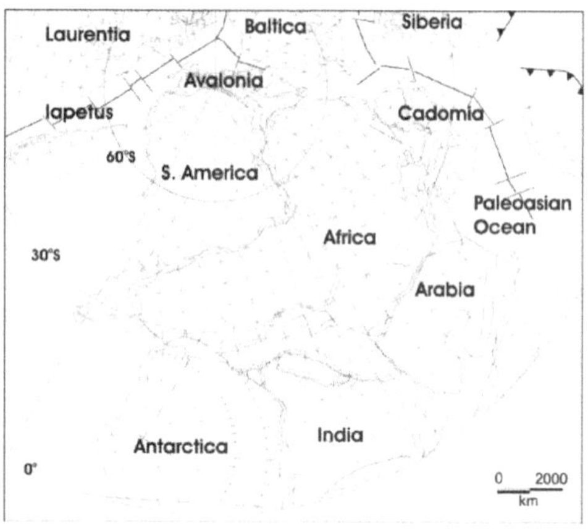

Figure 2. Plate tectonic map of Gondwana. Late Vendian - Early Cambrian (plates position as of 544 Ma). Stereographic polar projection. Modified from Golonka (2012). Legend as in Fig. 1

Laurentia

The continent Laurentia was named after the Laurentia Shield, in turn after St. Lawrence (Laurentius in Latin) River. North America was a major component of Laurentian plate. This plate also included Greenland, Chukotka peninsula, Svalbard and large part of the Barents Sea (Barentsia) fragment of Alaska (North Slope), northwest Ireland, and Scotland, (Golonka, 2000, 2002, Ford & Golonka, 2003,Golonka *et al.*, 2003). (Figs 1-3). Its southern (present day eastern boundary is located within Appalachians, its northern (present day western) boundary is located within Rocky Mountains. The relationship between Laurentia and Siberia remains speculative (Figs 1-5).

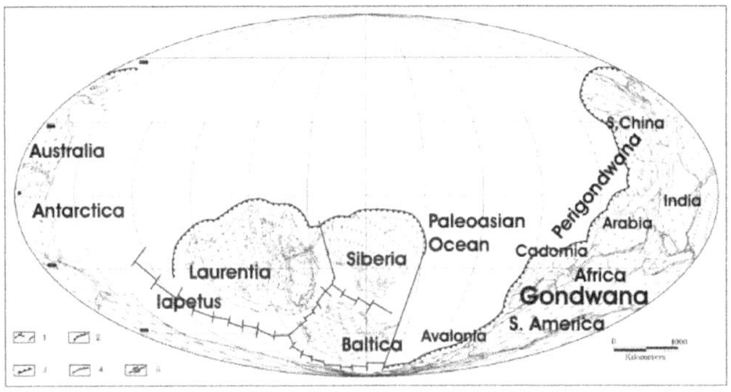

Figure 3. Plate tectonic map of Middle Cambrian (plates position as of 510 Ma). Mollweide projection. Modified from Golonka (2012). 1 - oceanic spreading center and transform faults, 2 - subduction zone, 3 - thrust fault, 4 - normal fault, 5 - transform fault.

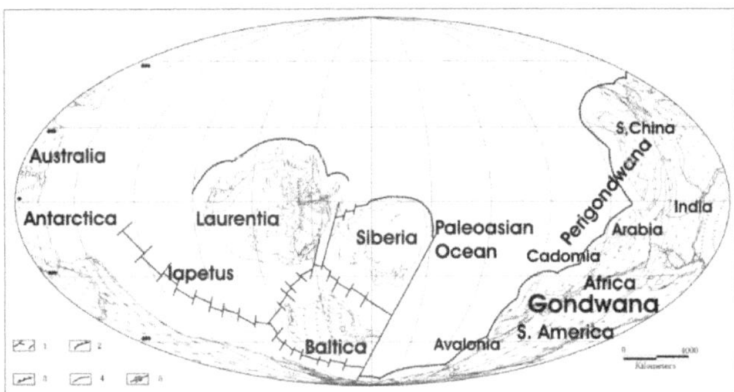

Figure 4. Plate tectonic map of Late Cambrian (plates position as of 498 Ma). Mollweide projection. Modified from Golonka (2012). 1 - oceanic spreading center and transform faults, 2 - subduction zone, 3 - thrust fault, 4 - normal fault, 5 - transform fault.

Avalonia

The name Avalonia is derived from the Avalon peninsula, Newfoundland, eastern Canada. The names 'Avalon Composite Terrane' (Keppie, 1985), or Superterrane were also used. Western Avalonia included terranes in northern Germany, the Ardennes in Belgium and northern France, England, Wales, southeastern Ireland, eastern Newfoundland, much of Nova Scotia, southern New Brunswick and some coastal parts of New England (Golonka 2009, McKerrow et al. 1991). The inclusion of terranes in the eastern part of Avalonia is more speculative (Fig. 5).

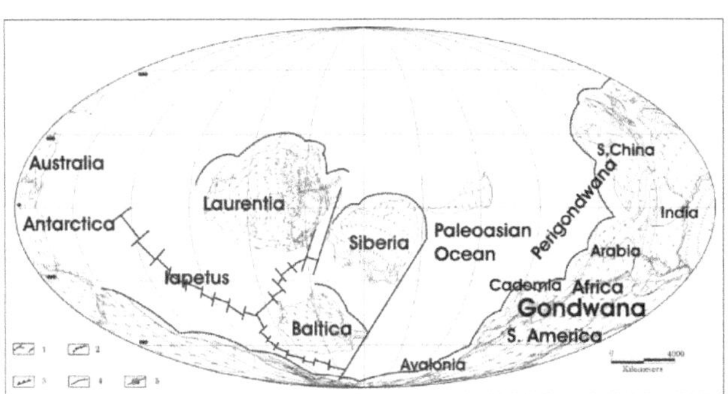

Figure 5. Plate tectonic map of Early Ordovician (plates position as of 485 Ma). Mollweide projection. Modified from Golonka (2012). 1 - oceanic spreading center and transform faults, 2 - subduction zone, 3 - thrust fault, 4 - normal fault, 5 - transform fault.

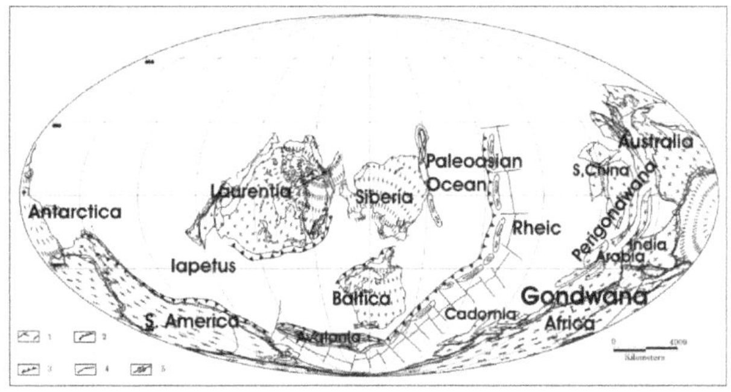

Figure 6. Plate tectonic map of Middle Ordovician (plates position as of 472 Ma). Mollweide projection. Modified from Golonka (2012). 1 - oceanic spreading center and transform faults, 2 - subduction zone, 3 - thrust fault, 4 - normal fault, 5 - transform fault.

Perhaps eastern Avalonia consisted of northwestern Poland fragments now included into Sudety Mountains and Bohemian Massif, terranes accreted in Carpathian-Balkan and Pannonian regions, containing fragments consolidated during Cadomian orogeny, and also Serbo-Macedonian massif, Rhodopes and Pontides in eastern Europe and adjacent part of Asia (Golonka, 2012). Avalonia originated after Gondwana break-up during Early-Middle Ordovician times (Fig.6).

Cadomia

Cadomia is named after city of Caen, Roman/Latin Cadomia in northern France also known as Armorica or Armorica Group Terrane (e.g. Lewandowski 2003 and references therein) or Gothic terranes (Stampfli, 2001). It consisted of the fragments of western and Central Europe between northwestern France (Brittany) and CzechRepublic consolidated during Cadomian orogeny (Figs.1-6). It includes Saxoturingian zone/terrane in Germany and northwestern CzechRepublic. Cadomia belonged to Perigondwana during Early Paleozoic times. Perhaps it was detached from Gondwana during Silurian times (Golonka et al. 2006a, Lewandowski 2003), the existence and nature of this detachment as well as extension of Cadomia eastward and westward remain quite speculative.

Iapetus Ocean

The IapetusOcean is named after Titan, son of Uranus, the sky, and Gaia, the Earth, father of Atlas. It preceded Atlantic off the American coast, therefore Atlantic was derived from Atlas and proto-Atlantic was named after Atlas' father. The IapetusOcean was located between Gondwana and Laurentia (Figs 1-6), later also between Avalonia and Laurentia (Fig. 6). It started to open as a rift between Laurentia and Rodinia in late Neoproterozoic, about 760-700 Ma, and the maximum obtained at about 600-520 Ma (Kamo et al. 1989; Cawood et al. 2001). Its closure is connected with the rotation and collision of Baltica, Laurentia and the fragment of Gondwana – Avalonia (Hartz & Torsvik 2002). The Iapetus suture was formed Caledonian Orogeny was formed in the time interval 480-440 Ma in the west while the eastern part (in Europe) closed at 440-420 Ma. The eastern extension of Iapetus is not so certain (Golonka, 2002, 2006a, b). Perhaps it was located between Baltica and Gondwana. Part of the IapetusOcean is known as TornquistSea, the oceanic basin located between the southwest and southern margin of Baltica and Gondwana, and later also Avalonia plate. The name is derived from the Tornquist zone in central Europe, already mentioned border of Baltica. The position of TornquistSea is speculative, because Baltica rotated during Early Paleozoic times.

Rheic Ocean

The RheicOcean was named after Rhea, Titaness-Goddess, daughter of Uranos, the sky, and Gaia, the Earth, wife of Cronos and mother of Zeus.RheicOcean originated between Gondwana and Avalonia (Fig. 6), later between Gondwana and Laurussia, during Early Paleozoic times (Nance et al. 2010 and references therein). It was opened during Late Cambrian – Early Ordovician times, around 500 Ma. System of back-arc basins developed along the northern branch of RheicOcean after formation of Laurussia during Devonian times (von Raumer & Stampfli, 2008). These basins are considered either as a part of the Rheic Ocean (e.g. Nance et al. 2010, Golonka, 2007b, McKerrow et al., 1991) or as a separate entity known as Rheno-Hercynian or Moldanubian basin in Central Europe (e.g. Golonka et al. 2006b, Golonka 2002, Schulmann et al., 2009). RheicOcean narrowed during Devonian and was closed during Carboniferous times as a result of collision of Laurussia with Gondwana.

GLOBAL EARLY PALEOZOIC PLATE TECTONIC EVOLUTION LEADING TO THE ASSEMBLY OF LAURUSSIA

Gondwana supercontinent was located around the South Pole at the beginning of Paleozoic (Figs 1-2). Baltica and Laurentia were located at the high latitude in the southern hemisphere, their southern margins close to the South Pole. They drifted apart from Gondwana during the Late Vendian times (Torsvik et al., 1996, Golonka et al, 2002, 2009b, 2012). Their breakup led to the formation of new oceans, including Iapetus (Figs. 1-2). Continued seafloor spreading occurred in this ocean during Cambrian times (Figs. 2-4). The fragmentation of northern margin of Gondwana was also marked by magmatic activity in the 550-500 Ma (Dörr et al., 1998, Turniak et al., 2000, Tichomirowa, 2002, Burda and Klötzli, 2011). Laurentia drifted rapidly northward and rotated counter-clockwise, reaching low latitudes (Golonka, 2000, 2002). Seafloor spreading also occurred within the PleionicOcean between East Siberia and Baltica. The relationship between Laurentia and Siberia remains quite speculative. Latest Cambrian – earliest Ordovician was the time of maximum dispersion of continents during the Paleozoic. Baltica, Laurentia and Siberia drifted further northward (Fig. 5). The subduction along the central margin of Gondwana caused the onset of rifting of the Avalonian terranes (Golonka et al., 1994, Golonka, 2000, 2002). The subduction along the northern margin of Baltica was perhaps related to the Ordovician rotation of this plate (see Golonka et al., 1994, 2006b, Torsvik et al., 1996,McKerrow et al., 1991, Golonka, 2000, 2002, cocks & Torsvik, 2011). The distance between Gondwana and Laurentia, which was situated on equator reached5000 km (Golonka, 2002, Golonka et al. 2006).

Early-Middle Ordovician were the times of a major plate reorganization (Golonka, 2000, 2002, 2009b,2012, Golonka et al. 2006b). Avalonia probably started to drift from Gondwana and move northward toward Baltica and Laurentia (Golonka, 2000, 2002, 2009b, 2012). This movement was related to the origin of RheicOcean. (Fig. 6).The IapetusOcean had begun to narrow.

During Late Ordovician times (Fig. 7 the Rheic Ocean between Gondwana and Avalonia widened significantly (Golonka, 2000, 2002, 2009b, 2012). The position of Cadomia remains uncertain. On the presented reconstruction, the Cadomian blocks are positioned relatively close to Gondwana. This is mainly based on the paleobiogeographical data (Scotese and McKerrow, 1990, Robardet et al., 1993,Golonka, 2002, 2009b, 2012). The alternative reconstructions assumed that these terranes were rifted away and formed separate Cadomia plate floating within the RheicOcean (Lewandowski, 1993, see also Golonka et al., 2006a). Latest Ordovician- Early Silurian were the times of collision between Avalonia and Baltica (Fig. 8).

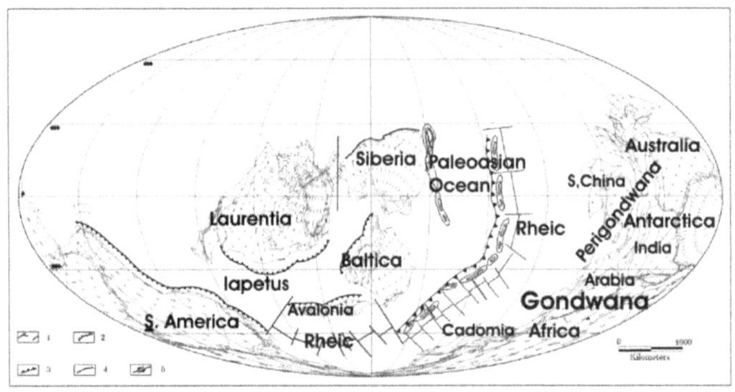

Figure 7. Plate tectonic map of Late Ordovician (plates position as of 452 Ma). Mollweide projection. Modified from Golonka (2012). 1 - oceanic spreading center and transform faults, 2 - subduction zone, 3 - thrust fault, 4 - normal fault, 5 - transform fault.

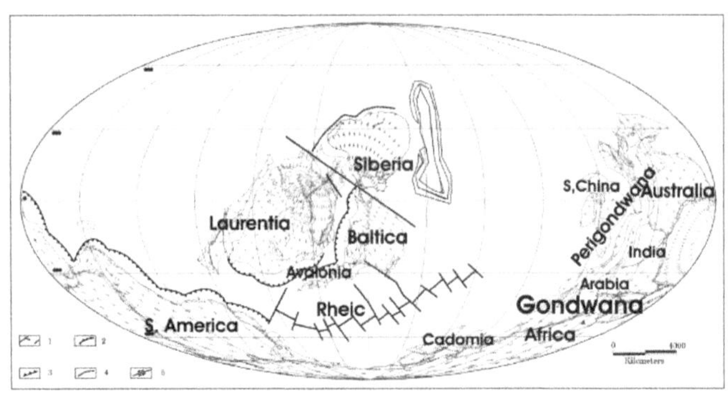

Figure 8. Plate tectonic map of Early Silurian (plates position as of 435 Ma). Mollweide projection. Modified from Golonka (2012). 1 - oceanic spreading center and transform faults, 2 - subduction zone, 3 - thrust fault, 4 - normal fault, 5 - transform fault.

This convergence was dominated by a strike-slip suturing of the two continents, rather than by full-scale continent-continent collision(Golonka, 2000, 2002). Northwestern Poland and adjacent part of Germany was joined with Baltica, along a strike-slip fault zone known as the Tornquist-Teisseyre line forming new continent - Balonia. Perhaps the Brunovistulicum and Malopolska terranes of southern Poland also belonged to Avalonia and joined Baltica (Moczydłowska, 1997, Bełka et al., 2000,Golonka, 2002, 2009b). In the Central Western Carpathians Late Ordovician – Early Silurian tonalitic gneisses of calc-alkaline character, associated with meta-gabbros, revealed the presence of magmatic episodes at 470-435 Ma (Kohut et al. 2008, Janák et al. 2002, Gaab et al. 2003, Gawęda

& Golonka, 2011). These rocks and granites present results of docking of Avalonia to Baltica. Also in the East Carpathians in Romania the intrusion of 459-470 Ma granitoids (Munteanu & Tatu, 2003, Pana et al., 2002, Ballintoni et al., 2010) document the collision-related tectono-magmatic effects of docking of eastern prolongation of Avalonia to Baltica. The Scandian Orogeny was the result of the collision between Balonia and Laurentia. the onset of the orogeny occurred during the Early Silurian times and by late Silurian the orogeny was concluded (Golonka et al., 1994, Golonka, 2000, 2002, 2009b). The main phase of the Scandian orogeny is marked by nappes in Norway and Greenland as well as large crustal thickening (Dewey & Burke 1973, Torsvik et al., 1996, Golonka, 2000, 2002, 2009b). During the Mid-Silurian Avalonia collided with Laurentia (Fig. 9). This collision is known as the Caledonian orogeny, named after Roman name of Scotland – Caledonia. This name is also extending into other related orogenic events in Scanidinavia, Greenland and Ventral Europe. After the complete closure of the IapetusOcean, the continents of Baltic, Avalonia, and Laurentia formed the continent of Laurussia (Ziegler, 1989).

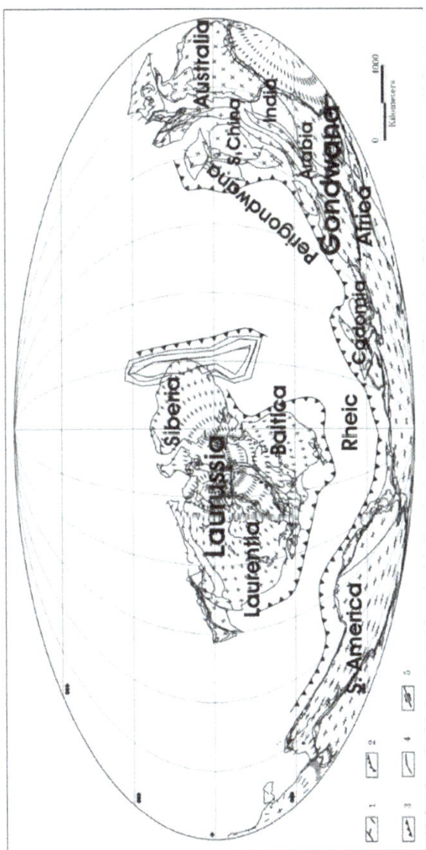

Figure 9. Plate tectonic map of Late Silurian (plates position as of 425 Ma). Mollweide projection. Modified from Golonka (2012). 1 - oceanic spreading center and transform faults, 2 - subduction zone, 3 - thrust fault, 4 - normal fault, 5 - transform fault.

LAURUSSIA DURING DEVONIAN TIMES

Supercontinent Laurussia existed during Late Silurian and Devonian times. Late Silurian Global paleogeography depicted on Fig. 9 id showing separation of Gondwana and Laurussia as well as subduction along the southern margin of Laurussia. Fig 10 depicts the global paleogeography during Early Devonian times. It is showing a possibility of Early Devonian collision between South and North America according to Golonka (2002, 2007b, 2012 see also Keppie 1989, McKerrow *et al.* 1991,Dalziel *et al.* 1994, Keppie *et al.* 1996). This collision was marked by orogenic events in Venezuela, Columbia, Peru, and northern Argentina (Gallagher & Tauvers 1992, Williams 1995). Paleomagnetic data (Kent & Van der Voo 1990, Van der Voo 1993, Lewandowski 1998, 2003) and paleobiogeography (Young 1990) also support the hypothesis about the proximity of South and North America (Golonka, 2007, 20012 b). This proximity is also shown by recent maps of Cocks & Torsvik (2010), however without the collision. The Carolina region (Rast & Skehan 1993) also contains an element of collision and transpression between South and North America, acting as an indenter, with a dextral strike-slip component (Golonka, 2012).

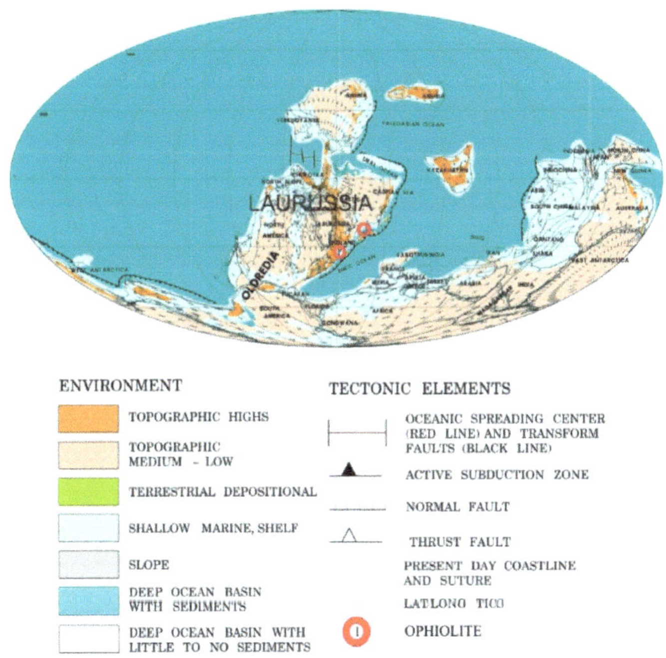

Figure 10. Paleogeographic c map of Early Devonian (plates position as of 401 Ma). Mollweide projection. Modified from Golonka 2007b, 2012) Ophiolites – 1 Lizard ophiolite, 2 – central and eastern European ophiolites (depicted in details on Fig. 11).

The accretion of Avalonia was followed by the development of north dipping subduction, along the new continental margin of Laurussia. That subduction caused rifting and creating the new back-arc basin. That basin was first recognized in central Europe as Rheno-Hercynian zone (Ziegler, 1989,Franke, 1992, Franke et al., 1995, Golonka, 2007b and references therein). Its paleogeography resembled present-day marginal Seas of East Asia. Fragment of Rheno-Hercynian zone, with oceanic crust, was also named Lizard-GiessenOcean (Zeh & Gerdes, 2010). The name Lizard is derived from the well known Lizard Ophiolitic Complex in Cornwall, United Kingdom (Bromley, 1975, 1976,Kirby, 1976), marked as number 1 on Fig.10. This Devonian complex contain peridotites, serpentinites dolerite dikes as well as amphibolites (Bromley, 1975, 1976, Kirby, 1976, Cook et al., 2002, Clark et al., 2003). Probably El Castillo volcanic rocks from Spain are related to the Lizard ophiolites (Gutiérez-Alonso et al., 2008). They yield Devonian, 394,7±1,4 Ma age of magmatism.

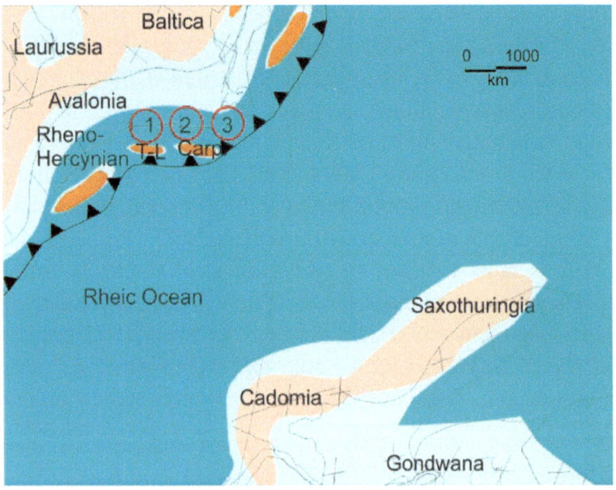

Figure 11. Paleogeography of Laurussia margin in central-eastern Europe during Early Devonian). Mollweide projection. Legend as on Fig. 10. Abbreviations: T – L - Tepla Barrandien – Lugia terranes, Carp. – Central Carpathian and Balkan terranes. Ophiolites: 1- Central-Sudetic ophiolite, 2 - Western Carpathian ophiolite, 3 - Balkan-South Carpathian ophiolite.

The eastern extension of the Rheno-HercynianBasin is marked by ophiolites in Sudety area in Poland and in the Carpathian-Balkan area (Figs. 10, 11). During Devonian times the Central-Sudetic ophiolites were located between Avalonian terrenes, sutured to Laurussia, and several microplates/terranes now included in the mosaic structure of Bohemian Massif and SudetyMountains. These terranes are marked on the Fig. 11 as T – L – Tepla-Barrandien – Lugia. They include several small blocks like Tepla-Barrandien, Góry Sowie and others, with uncertain, speculative position between Cadomia-Saxoturungia and Laurussia (see. e.g. Aleksandrowski et al., 2000, Franke & Żelaźniewicz, 2000, Winchester et al., 2002, Mazur, et al., 2006, Schulmann et al., 2009, Kryza & Pin, 2010, Nanece et al., 2010). According to Kryza & Pin (2000) the SHRIMP zircon data supplied evidence for Devonian age of the Central-Sudetic ophiolites, around 400 Ma.

The age of the ophiolitic remnants within the Carpathians from Tatric and Gemeric units Unit yield various ages: 371 Ma - 385-383 Ma (Putiš et al. 2009), 391 Ma (Gawęda, 2008), and 394 Ma (Kohut et al., 2006). The recorded events are related to the development of a back-arc basin with ocean crust as a result of the slab roll-back and derivation of ribbon-like Proto-Carpathian Terrane from Laurussia (see Gawęda & Golonka, 2011*)*. The associated granitoid magmatism post-dated that event (387 Ma; Burda & Klötzli 2011, Putiš et al., 2008). The Balkan-South Carpathian ophiolite belt yield the isotopic age of 406-399 Ma (Zakariadze et al, 2007), comparable with the Central-Sudetic ophiolites age being around 400 Ma (Kryza, Pin, 2010).

Gondwana drifted northward and rotated clockwise during Devonian times (Scotese and McKerrow, 1990, Scotese and Barret, 1990, Golonka, 2000, 2002, 2007b). At the same time, Laurussia was rotating clockwise (Torsvik et al., 1996) at a somewhat faster rate. Figure 12 depict global paleogeography during the Late Devonian times.

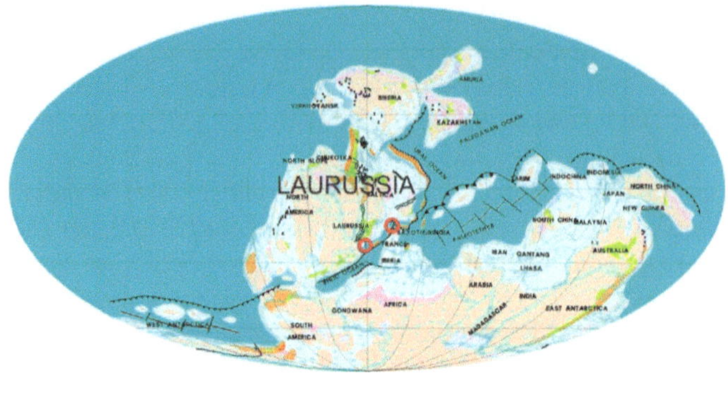

Figure 12. Paleogeographic c map of Late Devonian (plates position as of 370 Ma). Mollweide projection. Legend as on Fig. 10. Modified from Golonka 2007b, 2012).Ophiolites – 1 Lizard ophiolite, 2 – central and eastern European ophiolites (depicted in details on Fig.13).

The first contact between Laurussia and the Cadomian promontory of Gondwana occurred in Central Europe This contact marks the onset of Hercynian orogeny. The Saxoturingian part of the Cadomian plate collided with small terranes in Germany, CzechRepublic and Poland (Fig. 13). Collisional events were marked also in Carpathians and Balkans. The Rheno-HercynianBasin changed its character from extensional into compressional one. By latest Devonian – Early Carboniferous it displayed synorogenic features, filling with turbiditic flysch and culm facies (Golonka, 2007b).

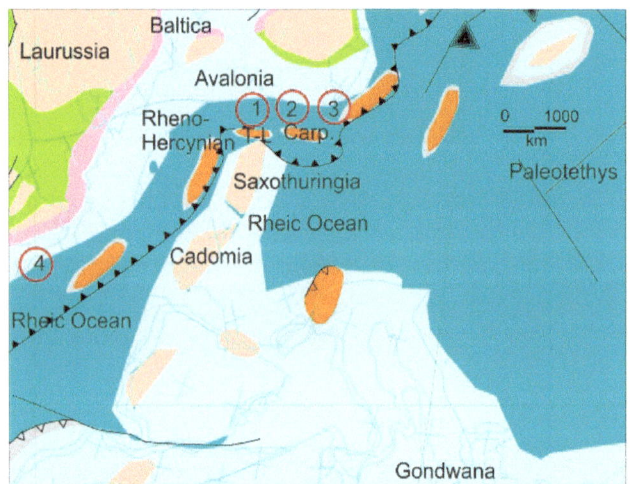

Figure 13. Paleogeography of Laurussia margin in central-eastern Europe during Late Devonian). Mollweide projection. Legend as on Fig. 10. Abbreviations: T – L - Tepla Barrandien – Lugia terranes, Carp. – Central Carpathian and Balkan terranes. Ophiolites: 1- Central-Sudetic ophiolite, 2 - Western Carpathian ophiolite, 3 - Balkan-South Carpathian ophiolite, 4 – Lizard ophiolite.

The new subduction zone on the southern margin Laurussia and the formation of a back-arc resulted in the intensive partial melting and magmatic activity from 370 to 340 Ma (Burda & Gawęda 2009,Schulmann et al. 2009). The continuous subduction along both old and new subduction zones resulted in docking of the Cadomia- Saxoturingia Terrane to Laurussia and significant consumption of the RheicOcean (Fig. 10). Finally the subduction caused formation of the Variscan Orogenic Suture, extending from Turkey to Mexico. That event resulted in voluminous granitoid magmatism and amphibolite-granulite facies metamorphism – the most prominent features of all crystalline cores, present in all the Variscan complexes (Stipska et al., 1998; Schulmann et al. 2009). These Variscan granitoid magmas were formed and intruded in the interval of 370-340 Ma, contemporaneously with Variscan nappes formation (Dallmayer et al., 1996; Burda et al., 2011) and synchronous with uplift of the continental blocks during prolonged collision (Janak et al., 1999, Gawęda et al., 2000). The resulting collision was possibly associated with basaltic underplating or slab break-off (Broska & Uher, 2001; Finger et al., 2009). Most of the granitoid bodies show hybrid character, with both mantle and crustal components involved (e.g. Słaby & Martin, 2005, Burda et al. 2011). Their VAG and CAG affinities suggest that melted metasediments, representing crustal component of the magma, were originally deposited both in the volcanic arc and intracontinental basins during subduction at the active, Andean-type continental margin (Schulmann et al., 2009). The melted material represented mainly recycled Proterozoic components, with addition of Paleozoic ones and influence of DM (depleted mantle) component as a source of heat (Poller et al., 2000). The subsequent uplift caused the exhumation of eclogitic remnants (Janak et al., 1996) and caused their retrograde metamorphism in granulite – amphibolite facies regime.

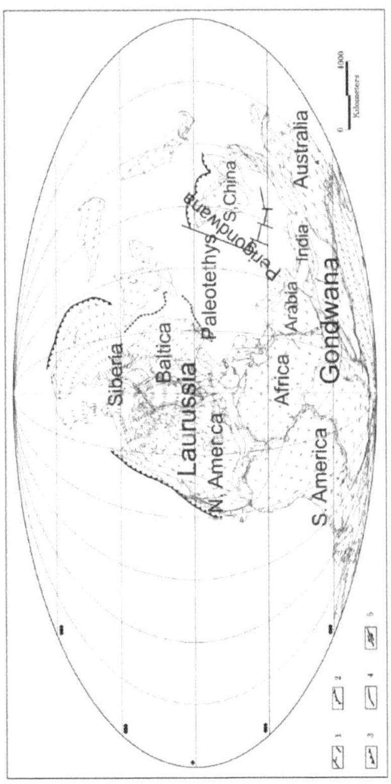

Figure 14. Plate tectonic map of Early Carboniferous (late Visean – Serpukhovian), plates position as of 328 Ma). Mollweide projection. Modified from Golonka (2012). 1 – oceanic spreading center and transform faults, 2 – subduction zone, 3 – thrust fault, 4 – normal fault, 5 – transform fault.

The ongoing Hercynian convergence in Europe led to large scale dextral shortening, overthrusting and emplacement of parts of the accretionary complexes (Edel & Weber, 1995). The amount of convergence was modified by large, dextral and sinistral transfer faults. The thrusting took place in the Tatra Mts. in the Carpathians (Gawęda et al., 2000, Golonka, 2000, 2002, 2007b). The collision between Gondwana and Laurussia continued to develop during Carboniferous times (Figs. 14, 15). The intercontinental collision began to affect the northwestern part of Africa, developing Mauretinides, Bassarides, Rokelides orogens is. The Alleghenian orogeny in North America continued (Hatcher et al., 1989, Rast & Skehan, 1993), prograding westwards to the Ouachita fold belt in Arkansas, Oklahoma, Texas and adjacent part of Mexico (Arbenz, 1990, Golonka, 2000, 2002). The clockwise rotation of Gondwana resulted in the involvement of the deformation. This Gondwanian influence resulted in the convoluted shape of the Hercynian orogen, strike-slip zones (Franke et al., 1995) and Hercynian deformation at the eastern end in Poland. The European foreland basin was elevated or changed its sedimentation regime from flysch to molasse. The central Pangean mountain range was formed, which extended from Mexico to Poland (Golonka, 2000, 2002, 2007b). The Laurussia continent ceased its independent existence becoming a part of the supercontinent Pangea.

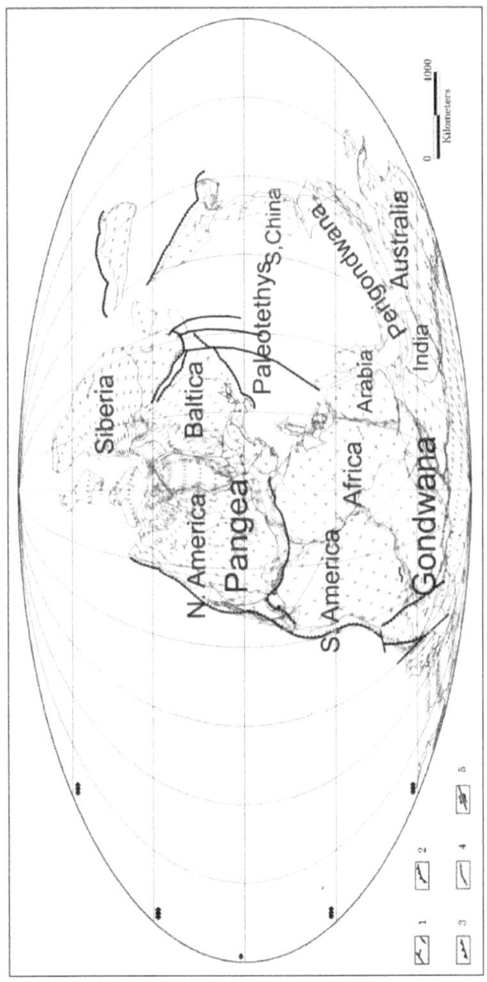

Figure 15. Plate tectonic map of Late Carboniferous (plates position as of 302 Ma). Mollweide projection. Modified from Golonka (2012). 1 – oceanic spreading center and transform faults, 2 – subduction zone, 3 – thrust fault, 4 – normal fault, 5 – transform fault.

CONCLUSIONS

- The Late Precambrian (Vendian) to present plate tectonic processes contributed to the complex structure of Western and Central Europe.
- The supercontinent of Laurussia, originated as a result of a closure of Iapetus Ocean and collision of Baltica, Avalonia and Laurentia.
- Laurussia originated during Late Silurian times
- The accretion of Avalonia was followed by the

development of north dipping subduction, forming the new Rheno-Hercynian Basin back-arc basin during Devonian times

- The oceanic crust of the Rheno-Hercynian Basin is recognised by ophiolites in Western, Central and Eastern Europe. The Lizard ophiolite in U.K. and Central-Sudetic ophiolite in Poland represent the best developed Devonian oceanic complexes.
- The new geological research (including dating) in the Carpathian-Balkan area supports the idea about the prolongation of the Rheic Suture to the east.
- The new geological research (including dating) in the Carpathian-Balkan area supports the idea about the prolongation of the Rheic Suture to the east.
- The tectonic events and associated magmatism point out the development of the back-arc at the southern margin of Laurussia in the time interval 406-371 Ma.

Laurussia cease to exist during Hercynian orogeny in Carboniferous times and was included into supercontinent Pangea.

ACKNOWLEDGEMENT

This research has been partially financially supported by Ministry of Science and Higher Education grants (Statutory Activities) through the AGH University of Science and Technology in Krakow grants no. 11.11.140.447 (J. Golonka) and Ministry of Education and Science grant No N307 027837 (A. Gawęda).

REFERENCES

1. Aleksandrowski, P., Kryza, R., Mazu,r S., Pin, C. & Zalasiewicz, J.A,. (2000). The Polish Sudetes: Caledonian orVariscan? Transactions of the Royal Society, Edinburgh, Earth Sciences (1999), Vol.90, 127-146.
2. J. K. Arbenz, 1990The Ouachita system, The Geology of North America, Bally, A. W. & Palmer, A. R., (Eds.,) Vol. A., 371396Boulder, Geological Society of America,
3. I. Ballintoni, C. Balica, A. Seghedi, M. N. Ducea, 2010Avalonian and Cadomian terranes in North Dobrogea, Romania. Precambrian Research, 1823
4. Z. Bełka, H. Ahrendt, W. Franke, K. Wemmer, 2000The Baltica-Gondwana suture in central Europe: Evidence from K-Ar ages of detrital muscovites and biogeographical data, Geological Society, London, Special Publications, 179
5. L. Besutiu, (2001, Baltica. Moesia-a, terrane?. derived, 20. Pancardi, I. I. , Abstracts, CP29Ádám, A., Szarka, L., Szendröi, J. (Eds.) Hungarian Academy of Sciences, Sopron, Humgary.
6. A. V. Bromley, 1975Is the Lizard Complex, South Cornwall, a fragment of Hercynian oceanic crust?. The Lizard: A Magazine of Field Studies, Vol., 33211
7. A. V. Bromley, 1976A new interpretation of the Lizard Complex, S. Cornwall, in the light of the ocean crust model. Proceedings. Geological. Society London, 132
8. I. Broska, P. Uher, 2001Whole-rock chemistry and genetic typology of the West-Carpathian Variscan granites. Geologica Carpathica, 522
9. J. Burda, A. Gawęda, 2009Shear-influenced partial melting in the Western Tatra metamorphic complex: geochemistry and geochronology. Lithos, 110
10. A. H. Clark, H. A. I. Sandeman, A. P. Nutman, D. H. Green, A. C. Cook, 2003Discussion on SHRIMP U-Pb zircon dating of the exhumation of the Lizard Peridotite and its emplacement over crustal rocks: constraints for tectonic models. Journal of the Geological Society, London, 160
11. L. R. M. Cocks, T. H. Torsvik, 2011The Palaeozoic geography of Laurentia and western Laurussia: a stable craton with mobile margins. Earth Science Reviews, 106
12. C. A. Cook, R. E. Holdsworth, M. T. Styles, 2002The emplacement of peridotites and associated oceanic rocks from the Lizard Complex, southwest England, Geological Magazine, 139
13. D. Dallmeyer, F. Neubauer, R. Handler, H. Fritz, W. Muller, D. Pana, M. Putis, 1996Tectonothermal evolution of the internal Alps and Carpathians: evidence from 40Ar-39Ar mineral and whole-rock data. In:Schmid S.M., Frey M., Froitzheim N., Heilbronner R., Stuenitz, H. (eds.): Alpine geology, proceedings of the second workshop: 2nd workshop on Alpine geology, Eclogae Geologica Helvetica, 89203227
14. I. W. D. Dalziel, Salda. L. H. Dalla, L. M. Gahagan, 1994Paleozoic Laurentia-Gondwana interaction and the origin of the Appalachian-Andean mountain system, Geological Society of America Bulletin, 106
15. J. F. Dewey, V. B. S. Burke, 1973Tibetan, Variscan, and Precambrian basement reactivation, products of continental collision, Journal of Geology, 81
16. W. Dörr, J. Fiala, Z. Vejnar, D. Zulauf, 1998U-Pb zircon ages and structural development of metagranitoids of the Tepla crystalline complex: evidence for pervasive Cambrian plutonism within the Bohemian Massif (Czech Republiv). International Journal of Earth Sciences, 87
17. J. B. Edel, K. Weber, 1995Cadomian terranes, wrench faulting and thrusting in the central Europe Variscides: geophysical and geological evidence, Geologische Rundschau, 84
18. F. Finger, A. Gerdes, M. Rene, G. Riegler, 2009The Saxo-Danubian Granite Belt: magmatic response topost-collisional delamination of mantle lithosphere below the south-west sector of the Bohemian Massif (Variscan Orogen). Geologica Carpathica, 603
19. D. Ford, J. Golonka, 2003Phanerozoic paleogeography, paleoenvironment and lithofacies maps of the circum-Atlantic margins, Thematic set on paleogeographic reconstruction and hydrocarbon basins: Atlantic, Caribbean, South America, Middle East, Russian Far East, Arctic. Golonka J. (ed.), Marine and Petroleum Geology, 20
20. W. Franke, 1992Phanerozoic structures and events in central Europe, The European Geotraverse, A continent revealed, Blundell, D., Freeman, R. & Mueller, S. (Eds.), 164180University of Cambridge, Cambridge.

21. W. Franke, A.. Żelaźniewicz, (2000, 2000The eastern termination of the Variscides: terrane correlation and kinematic evolution, Orogenic processes:quantification and modelling in the Variscan belt, Geological Society Special. Publication, 179

22. W. Franke, R. D. Dallmeyer, K. Weber, 1995Geodynamic Evolution, Pre-Permian geology of Central and Eastern Europe, IGCP 233 international conference: Gottingen, Federal Republic of Germany: Berlin, Dallmeyer, R. D., Franke, W. & Weber, K., (Eds.), 579593Springer-Verlag.

23. A. Gaab, U. Poller, W. Todt, M. Janák, 2003Geochemical and isotopic characteristics of the Murán Gneiss Complex, Veporic Unit (Slovakia), Journal of the Czech Geological Society, 481-252

24. J. J. Gallagher, P. R. Tauver,s, 1992Tectonic evolution of northwestern South America, Basement tectonics. Mason, R. (Ed), 123137Kluver Academic Publishers.

25. A. Gawęda, 2008An apatite-rich enclave in the High Tatra granite (Western Carpathians): petrological and geochronological study. Geologica Carpathica, 594

26. A. Gawęda, J. Golonka, (2011, Variscan plate dynamics in the Circum-Carpathian area. Travaux Géophysiques XL (2011), Abstracts of the 9th Central European Tectonic Groups meeting, Hotel Skalský Dvůr, Czech Republic, 1317April 2011, 19. Prague. Institute of Geophysics Academy of Sciences of the Czech Republic,.

27. A. Gawęda, K. Kozłowski, K. Piotrowska, 2000Early-Variscan collision and generation of leucogranite melts in the Western Tatra Mountains (S-Poland, W-Carpathians). Journal of the Czech Geological Society, 45

28. J. Golonka, 2000Cambrian-Neogene Plate Tectonic Maps. 1125Wydawnictwa Uniwersytetu Jagiellońskiego, Kraków.

29. J. Golonka, 2002Plate-tectonic maps of the Phanerozoic. Phanerozoic reef pattern, Kiessling W., Flügel E. & Golonka J. (eds.), SEPM (Society for Sedimentary Geology) Special Publication, 72

30. J. Golonka, 2004Plate tectonic evolution of the southern margin of Eurasia in the Mesozoic and Cenozoic, Tectonophysics, 381

31. J. Golonka, 2007aLate Triassic and Early Jurassic paleogeography of the world. Palaeogeography, Palaeoclimatology, Palaeoecology, 244

32. J. Golonka, 2007bPhanerozoic Paleoenvironment and Paleolithofacies Maps. Late Paleozoic. Mapy paleośrodowiska i paleolitofacje fanerozoiku. Późny paleozoik. Kwartalnik AGH. Geologia, 332

33. J. Golonka, 2007cPhanerozoic Paleoenvironment and Paleolithofacies Maps. Mesozoic. Mapy paleośrodowiska i paleolitofacje fanerozoiku. Późny mezozoik. Kwartalnik AGH. Geologia, 332

34. J. Golonka, 2009aPhanerozoic Paleoenvironment and Paleolithofacies Maps.Cenozoic. Mapy paleośrodowiska i paleolitofacje fanerozoiku. Kenozoik. Kwartalnik AGH. Geologia,, 354

35. J. Golonka, 2009bPhanerozoic Paleoenvironment and Paleolithofacies Maps.Early Paleozoic. Mapy paleośrodowiska i paleolitofacje fanerozoiku. Wczesny paleozoik. Kwartalnik AGH. Geologia,, 354

36. J. Golonka, 2012Paleozoic paleoenvironment and paleolithofacies maps of Gondwana, WydawnictwaAGH Publishing House. Kraków. In press

37. J. Golonka, M. I. Ross, C. R. Scotese, 1994Phanerozoic paleogeographic and paleoclimatic modeling maps. Pangea: Global environment and resources, Embry A. F., Beauchamp, B. & Glass, D. J., (Eds.),. Canadian Society of Petroleum Geologists Memoir, 17147

38. J. Golonka, M. Krobicki, N. Oszczypko, A. Ślączka, T. Słomka, 2003Geodynamic evolution and palaeogeography of the Polish Carpathians and adjacent areas during Neo-Cimmerian and preceding events (latest Triassic- earliest Cretaceous), Tracing tectonic deformation using the sedimentary record, McCan, T. & Saintot, A. (Eds.),Geological Society Special Publications. 208138158

39. J. Golonka, L. Gahagan, M. Krobicki, F. Marko, N. Oszczypko, A. Slaczka, 2006aPlate Tectonic Evolution and Paleogeography of the Circum-Carpathian Region. In: Golonka, J. & Picha, F. (Eds.) The Carpathians and their foreland: Geology and hydrocarbon resources. American Association of Petroleum Geologists Memoir, 841146

40. J. Golonka, M. Krobicki, J. Pająk, Giang. Nguyen Van, W. Zuchiewicz, 2006bGlobal plate tectonics and paleogeography of Southeast Asia. Faculty of Geology, Geophysics and Environmental Protection, AGH University of Science and Technology, Arkadia, Kraków.

41. G. Gutiérrez-Alonso, J. B. Murphy, J. Fernández-Suárez, M. A. Hamilton, 2008Rifting along the northern Gondwana margin and the evolution of the Rheic Ocean: A Devonian age for the El Castillo volcanic rocks (Salamanca, Central Iberian Zone). Tectonophysics, 461

42. E. H. Hartz, T. H. Torsvik, 2002Baltica upside down: a new plate tectonic model for Rodinia and Iapetus Ocean. Geology, 303

43. R. D. Hatcher, W. A. Jr Thomas, P. A. Geiser, A. W. Snoke, S. Mosher, D. V. Wiltschko, 1989Alleghenian orogen, The Appalachian-Ouachita Orogen in the United States, Hatcher R. D., Jr, Thomas, W. A. & Viele, G. W., (Eds.),V. F,. 233318Boulder, Geological Society of America, The Geology of North America,

44. M. Janák, V. Hurai, L. Ludhova, Brien. P. J. O`, E. E. Horn, 1999Dehydration melting and devolatilization during exhumation of high-grade metapelites: the Tatra Mountains, Western Carpathians. Journal of Metamorphic Geology, 17

45. M. Janák, F. Finger, D. Plašienka, I. Petrik, B. Humer, S. Meres, B. Luptak, 2002Variscan high P-T recrystallization of Ordovician granitoids in Veporic Unit (Nizke Tatry Mountains, Western Carpathians): new petrological and geochronological data. Geolines, 14

46. J. Kalvoda, O. Bábek, (2010, 2010The Margins of Laurussia in Central and Southeast Europe and Southwest Asia. Gondwana Research, 17

47. S. L. Kamo, C. F. Gowar, T. E. Krogh, 1989Birthdate for the Iapetus Ocean ? A precise U-Pb zircon and

baddeleyite age for Long Range dikes, southeast Labrador. Geology, 177

48. D. V. Kent, R. Van der Voo, 1990Palaeozoic palaeogeography from palaeomagnetism of the Atlantic-bordering continents, Palaeozoic palaeogeography and biogeography, McKerrow, W. S. & Scotese, C. R., (Eds.), Geological Society Memoir, 12

49. J. D. Keppie, 1985The Appalachian collage, The Caledonide orogen, Scandinavia, and related areas, 12171226Gee, D.G., & Sturt, B. (Eds)., New York, J. Wiley and Sons.

50. J. D. Keppie, 1989Northern Appalachian terranes and their accretionary history, Terranes in the Circum-Atlantic Paleozoic orogens. Dallmeyer, R. D. (Ed), Geological Society of America, Special Paper, 230

51. J. D. Keppie, J. Dostal, J. B. Murphy, R. D. Nance, 1996Terrane transfer between eastern Laurentia and western Gondwana in the Early Paleozoic: Constraints on global reconstructions, Avalonia and Related Peri-Gondwanan Terranes of the Circum-North Atlantic, Nance R. D. & Thompson M. D. (Eds), Geological Society of America Special Paper, 304

52. G. A. Kirby, 1979The Lizard Complex as an ophiolite. Nature, 282

53. M. Kohut, P. Konecny, P. Siman, 2006The first finding of the iron Lahn-Dill mineralization in the Tatric Unit of the Western Carpathians. Mineralogia Polonica- Special Papers, 28

54. M. Kohut, U. Poller, Ch. Gurk, W. Todt, 2008Geochemistry and U-Pb detrital zircon ages of metasedimentary rocks of the Lower Unit, Western Tatra Mountains (Slovakia), Acta Geologica Polonica, 58

55. R. Kryza, C. Pin, 2010The Central Sudetic ophiolites (SW Poland): petrogenetic issues, geochronology and paleotectonic implications. Gondwana Research, 17

56. M. Lewandowski, 1998Assembly of Pangea: Combined Paleomagnetic and Paleoclimatic Approach, Circum-Arctic Palaeozoic Faunas and Facies, Ginter, M. & Wilson, M. H. (Eds), Ichtyolith Issues Special Publication, 4

57. M. Lewandowski, 2003Assembly of Pangea: Combined Paleomagnetic and Paleoclimatic Approach. Advances in Geophysics, 46

58. S. Mazur, P. Aleksandrowski, R. Kryza, T. Oberc-Dziedzic, . , 2006The Variscan Orogen in Poland, Geological Quarterly, 50

59. W. S. Mc Kerrow, J. F. Dewey, C. R. Scotese, 1991The Ordovician and Silurian development of the Iapetus Ocean The Murchison symposium; proceedings of an international conference on the Silurian Systemin Bassett, M. G., Lane, P. D. & Edwards, D. (Eds.), Special Papers Palaeontology, 44

60. M. Moczydłowska, 1997Proterozoic and Cambrian successions in Upper Silesia: an Avalonian terrane in southern Poland, Geological Magazine, 134

61. M. Munteanu, M. Tatu, 2003The East-Carpathian Crystalline-Mesozoic Zone (Romania): Paleozoic amalgamation of Gondwana- and East European Craton-derived terranes. Gondwana Research, 62

62. R. D. Nance, G. Gutiérrez-Alonso, J. D. Keppie, U. Linnemann, J. B. Murphy, C. Quesada, R. A. Strachan, N. H. Woodcock, 2010Evolution of the Rheic Ocean. Gondwana Research, 17

63. D. Pana, I. Ballintoni, L. Heaman, R. Creaser, 2002The U-Pb and Sm-Nd dating of the main lithotectonic assemblages of the East Carpathians, Romania. Geologica Carpathica, 53

64. M. Putiš, P. Ivan, M. Kohút, J. Spišiak, P. Siman, M. Radvanec, P. Uher, S. Sergeev, A. Larionov, Š. Méres, R. Demko, M. Ondrejka, 2009Meta-igneous rocks of the West-Carpathian basement, Slovakia: indicators ofEarly Paleozoic extension and shortening events. Bulletin Societe Géolologique France, 1806

65. N. Rast, J. W. Skehan, 1993Mid-Paleozoic orogenesis in the North Atlantic, the Acadian orogeny. The Acadian Orogeny, recent studies in New England, Maritime Canada, and the autochtohonous foreland., Roy C. & Skehan J. W. (Eds), Geological Society of America Special Paper, 275

66. M. Robardet, J. Blaise, E. Bouyx, R. Gourvennec, H. Lardeux, A. Le Hérissé, J. Le Menn, M. Melou, F. Paris, Y. Plusquellec, J. Poncet, S. Régnault, M. Rioult, M. Weyant, 1993Paléogeographie de l'Europe occidentale de l'Ordovicien au Dévonien; Paleogeography of Western Europe from the Ordovician to the Devonian, Bulletin Societe géologique de France, 164

67. K. Schulmann, J. Konopásek, V. Janoušek, O. Lex, J. Lardeaux, M. , J. Ede,l, B. , Štípská, P. , S. Ulrich, 2009An Andean type Palaeozoic convergence in the Bohemian Massif. Comptes Rendus- Geoscience, 341

68. C. R. Scotese, W. S. Mc Kerrow, 1990Revised world maps and introduction, Palaeozoic palaeogeography and biogeography, McKerrow W. S. & Scotese C. R. (Eds), Geological Society of London Memoir, 12121

69. C. R. Scotese, S. F. Barret, 1990Gondwana's movement over the South Pole during the Paleozoic: evidence from lithologic indicators of climate. In: W.S. McKerrow and C.R. Scotese (Eds.) Paleozoic Paleogeography and Biogeography,, Geological Society of London, Memoir 127585

70. J. W. Sears, 2012Transforming Siberia along the Laurussian margin. Geology, doi:G32952.1

71. E. Słaby, H. Martin, 2005Mafic and felsic magma interaction in granites: the karkonosze Hercynian pluton (Sudetes, Bohemian Massif). Journal of Petrology, 49

72. P. Štipská, K. Schulmann, A. Kroener, 1998From Cambro-Ordovician rifting to Variscan collision at the NE margin of the Bohemian Massif: petrological, geochronological and structural constraints. Paleozoic orogenesis and crustal evolution of the European lithosphere- post-conference excursion, 2431

73. M. Tichomirova, 2002Zircon inheritance in diatexite granodiorites and its consequence on geochronology- a case study in Lusatia and Erzgebirge (Saxo-Thuringia, eastern Germany). Chemical Geology, 191

74. T. H. Torsvik, M. A. Smethurst, J. G. Meert, R. Van der

Voo, W. S. Mc Kerrow, M. D. Brasier, B. A. Sturt, H. J. Walderhaug, 1996Continental break-up and collision in the Neoproterozoic and Palaeozoic; a tale of Baltica and Laurentia, Earth-Science Reviews, 403-4

75. K. Turniak, S. Mazur, R. Wysoczański, 2000SHRIMP zircon geochronology and geochemistry of tyhe Orlica snieznik gneisses (Variscan Belt of Central Europe) and their tectonic implications. Geodinamica Acta, 13

76. J. J. Veevers, 2004Gondwanaland from 650-500 Ma assembly through 320 Ma merger in Pangea to 185-100 Ma breakup: Supercontinental tectonics via stratigraphy and radiometric dating. Earth-Science Reviews, 68

77. Raumer. J. F. von, G. M. Stampfli, 2008The birth of the Rheic Ocean- Early Paleozoic subsidence patterns and subsequent tectonic plate scenario, Tectonophysics, 461

78. K. E. Williams, 1995Tectonic Subsidence Analysis and Paleozoic Paleogeography of Gondwana, Petroleum basins of South America, Tankard, A.J., Suarez, S. & Welsink, H. J. (Eds),American Association of Petroleum Geologists Memoir, 62

79. J. A. Winchester, P. A. Floyd, Q. G. Crowley, M. A. J. Piasecki, M. K. Lee, T. C. Pharaoh, P. Williamson, D. Banka, J. Verniers, J. Samuelsson, U. Bayer, A. Marotta, M. , J. Lamarche, W. Franke, W. Dörr, P. Valverde-Vaquero, U. Giese, M. Vecoli, H. Thybo, M. Laigle, M. Scheck, H. Maluski, D. Marheine, S. R. Noble, R. R. Paarish, J. Evans, H. Timmerman, A. Gerdes, A. Gutcrch, M. Grad, S. Cwojdzinski, Z. Cymerman, W. Kozdroj, R. Kryza, P. Alexandrowsk, S. Mazur, V. Stedrá, J. Kotková, Z. Belka, F. Patoćka, V. Kachlik, 2002Palaeozoic amalgamation of Central Europe: New results from recent geological and geophysical investigations. Tectonophysics, 360

80. G. C. Young, 1990Devonian vertebrate distribution patterns and cladistic analysis of paleogeographic hypothesis, Palaeozoic palaeogeography and biogeography. McKerrow W. S. & Scotese C. R. (Eds), Geological Society of London Memoir, 12

81. G. S. Zakariadze, S. O. Karamata, Y. Dilek, (2007, of. E. Significance, Paleozoic Paleo-Tethyan Ophiolites in the Balkan Terrane and the Greater Caucasus for the Cadomian-Hercynian Continental Growth of Southern Europe. American Geophysical Union, Fall Meeting 2007, Abstract #43BAmerican Geophysical Union, Washigton D.C.

82. A. Zeh, A. Gerdes, 2010Baltica- and Gondwana-derived sediments in the Mid-German Crystalline Rise (Central Europe): Implications for the closure of the Rheic ocean, Gondwana Research, 17

83. P. A. Ziegler, 198, 1989Evolution of Laurussia. Kluwer Academic Publishers, Dordrecht.

84. L. P. Zonenshain, M. L. Kuzmin, L. N. Natapov, 1990Geology of the USSR: A Plate-Tectonic Synthesis. Page, B. M. (Ed), Geodynamics Series, American Geophysical Union, 21

85. A. Żelaźniewicz, A. Seghedi, J. Żaba, M. . Fanning, Z. Buła, 2009More evidence on Neoproterozoic terranes in Southern Poland and southeastern Romania, Geological Quarterly, 51

CHAPTER 15 TECTONIC GEOMORPHOLOGY AND VOLCANO-TECTONIC INTERACTION

Engielle M. R. Paguican and Marcus I. Bursik

Keywords: Volcano-tectonic, Tectonic-geomorphology, Hat Creek Graben region, California, Southern Cascades, Morphometry, Spatial distribution, Volcano alignments

The eastern boundary of the Southern Cascades (Hat Creek Graben region), California, USA, is an extensively faulted volcanic corridor between the Cascade Range and Modoc Plateau. The morphology of the region is a result of plate motions associated with different tectonic provinces, faulting, and recurring volcanic activity, making it an ideal place to study the interrelationship between tectonics, volcanoes, and geomorphology. We use the morphometry and spatial distribution of volcanoes and their interaction with regional structures to understand how long term regional deformation can affect volcano evolution. A database of volcanic centers and structures was created from interpretations of digital elevation models. Volcanic centers were classified by morphological type into cones, sub-cones, shields and massifs. A second classification by height separated the larger and smaller edifices, and revealed an evolutionary trend. Poisson Nearest Neighbor analysis showed that bigger volcanoes are spatially dispersed while smaller ones are clustered. Using volcano centroid locations, about 90 lineaments consisting of at least three centers within 6 km of one another were found, revealing that preferential north-northwest directed pathways control the transport of magma from the source to the surface, consistent with the strikes of the major fault systems. Most of the volcano crater and collapse scar openings are perpendicular to the north northwest-directed maximum horizontal stress, expected for extensional environments with dominant normal faulting. Early in the history of a volcano or volcano cluster, melt propagates to the surface using the easiest and most efficient pathway, mostly controlled by the pre-existing normal faults and near-surface stress fields, as indicated by the pervasive vent alignments. Volcano growth continues to be dependent on the regional structures as indicated by the opening directions, suggesting structural control on the growth of the volcanic edifices. The results present a particularly well-defined case in which extension of a volcanic region is accommodated mostly by faulting, and only partly by intrusion to form volcanoes. This is attributed to a low magma supply rate.

Citation: Engielle M. R. Paguican and Marcus I. Bursik, Tectonic Geomorphology and Volcano-Tectonic Interaction http://dx.doi.org/10.3389/feart.2016.00076.

INTRODUCTION

Volcanoes occur in a wide variety of shapes and sizes as a result of the interaction of constructive and destructive geological and environmental processes. The quantitative description of volcano morphology, called volcano morphometry, is a systematic way to characterize the morphology of volcano edifices, allowing for the extraction of information on the processes that interact and factors that control growth history and evolution (Grosse et al., 2012, 2014). At the same time, describing a volcanic field in which volcanoes grow is essential for characterizing key controls on volcano growth in space, which reflects the pathway used by magma to reach the surface, and the existence of any dominantly controlling factor that may have created preferential pathways (Le Corvec et al., 2013). At the surface and through time, the growth or destruction of a volcano is influenced by the volcano-tectonic setting of the region. Studies looking at the influence of regional stress on the direction of sector-collapse and debris-avalanche amphitheater opening have been done for Southeast Asian volcanoes in tectonic settings dominated by strike-slip faults, particularly those in the Philippines, by Lagmay and Valdivia (2006), Japan by Ui et al. (1986), and Indonesia by Bahar and Girod (1983), which showed that the opening direction of such features is at an acute angle relative to the regional maximum horizontal stress. Their results are in disagreement with earlier models by Moriya (1980), Siebert (1984), and Nakamura (1977), which show volcano openings perpendicular to the maximum horizontal stress being more prevalent in extensional regions (Tibaldi, 1995), where tabular conduits and dikes that propagate along normal faults promote collapses perpendicular to the regional maximum horizontal stress. This difference in results may be due to differences in the underlying fault kinematics, the resulting geometry of large-scale sliding and debris avalanching, original crater opening direction due to temporal and topographic stress-field control within a volcanic body (Ui et al., 1986), a dipping substrate, or a spreading basement.

The eastern boundary of the Southern Cascades (herein referred to as the Hat Creek Graben region) in California is an extensively faulted volcanic corridor with spectacular, high, steep scarps in a bedrock of late Tertiary and Quaternary volcanic and sedimentary deposits (Wills, 1991; Muffler et al., 1994; Blakely et al., 1997). It is at the boundary between two distinct geologic and geomorphic provinces (Jenkins, 1938), the Cascade Range on the west and the Modoc Plateau on the east, between Mt. Shasta and Lassen Peak on the north and south, respectively. It is in the transition zone between several significant volcanic and tectonic provinces, including the east-west convergence and subduction of the Gorda Plate underneath the North American Plate, the current north-south shortening within the Klamath Mountain Region on the northwest, and the region of east-west extensional faulting that produced the dominant north- to northwest- striking normal faults, forming the horst and graben topography and voluminous basaltic volcanism (White and Crider, 2006). Volcano morphology is a result of recurring volcanic activity within this diverse tectonic setting, from more than 500 vents over the past 7 Ma (Muffler et al., 1994), making it an important place to study the long-term interrelationship between tectonics, geomorphology, and volcanic activity. Many studies have recently explored the importance of magma in shaping divergent plate boundaries (Wright et al., 2006; Ebinger et al., 2010; Sigmundsson et al., 2015;Acocella and Trippanera, 2016), and here we seek to discern whether some of these ideas also apply to intraplate magmatic provinces experiencing extension, such as the Hat Creek Graben region. This study aims to provide a tectonic-geomorphic interpretation of the eastern boundary of the southern Cascades—the Hat Creek Graben region. It is hoped that the results will aid in producing a more refined understanding of regional volcano-tectonic interaction and history, which may help guide future hazard assessment. We explore the overall geomorphology and activity of individual volcanoes and the potential effects of regional stress regimes. We characterize local structures within each volcano, and compare the trend of volcano erosional and collapse scars with features in the basement faults, and regional maximum horizontal stress. We furthermore attempt to refine our understanding of the probable kinematics of underlying faults that may cause volcano instability.

REGIONAL TECTONIC SETTING AND FAULT SYSTEMS

The tectonics of the Hat Creek Graben region (Figure 1A) is dominated by the convergence of the Pacific, Juan de Fuca, Gorda and North American Plates; the rigid block of the Sierran Microplate; and the dominantly extensional western Basin and Range (Hammond and Thatcher, 2005). Subduction of the Juan de Fuca and Gorda Plates underneath the North American Plate forms the Cascadia Subduction Zone and the associated Cascade Arc volcanoes. To the south, the Sierran Microplate behaves as a rigid block that, due to the faster rate of right-lateral shear on its western side, accommodated by the right-lateral San Andreas Transform Fault and associated structures (Dixon et al., 2000; Faulds et al., 2005), is translated northward. This causes compression (north-south shortening) in the Klamath Mountains region on the northwest (McCaffrey, 2005). Furthermore, the motion of the Sierran Microplate contributes to the clockwise rotation of the Oregon forearc, where the Sierran Microplate and Cascadia forearc blocks meet (McCaffrey, 2005). On the eastern boundary of the Sierran Microplate, the Walker Lane accommodates some 15–25% of the total relative plate motion (Faulds et al., 2005), with GPS measured strain rates of about 11 mm yr^{-1} (Dokka and Travis, 1990; Argus and Gordon, 1991; Sauber et al., 1994; Bennett et al., 1999; Dixon et al., 2000; Hammond and Thatcher, 2007). Some of the right-lateral shear related to relative Pacific and North American plate motion is accommodated within the Hat Creek Graben region, and connects with poorly defined zones of deformation between the Gorda, Pacific, and North American plates near the Klamath Mountains (Miller et al., 2001). Crustal stresses related to the complex block motions and volcanic processes drive regional seismicity, and the observed patterns of faulting of the Hat Creek Graben region.

Fault scarps are prominent in the Hat Creek Graben region, especially in places where they displace resistant Pliocene to early Pleistocene basalt, with up to millions of years of accumulated offset (Blakeslee and Kattenhorn, 2013). Major fault systems (Figure 1B) identified and described by previous workers include the Hat Creek, Rocky Ledge, McArthur, Soldier Creek, and Pittville Fault Systems.

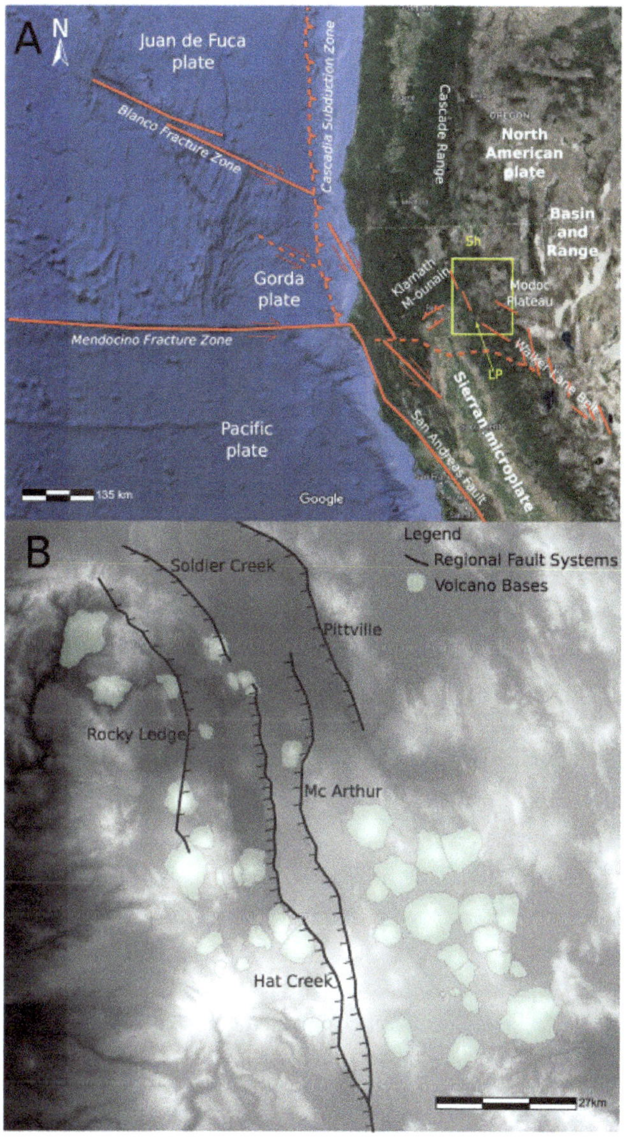

Figure 1. (A) Regional tectonic setting of the eastern boundary of the Southern Cascades (Hat Creek Graben region), California, is dominated by the convergence of the Pacific, Juan de Fuca, Gorda, and North American Plates resulting in the formation of the Cascadia Subduction Zone and its associated Cascade Arc Volcano with Lassen Peak (LP) as its southernmost volcano; the rigid block of the Sierran Microplate that acts like a piston compressing the Klamath Mountain on the northern boundary and creating the Ink Creek Fold Belt southwest of Lassen Peak; the Oregon forearc where the Sierran Microplate and Cascadia forearc blocks meet; the dominantly extensional western Basin and Range Region and Walker Lane; and the Hat Creek Graben region (yellow box) that includes the area south of Mt. Shasta (Sh), Pit River Region, Mushroom Rock Anticline, Cinder Butte, Butte Creek, Bald Mountain, Big Lake, Brushy Butte, and Fall River Valley. Modified from Faulds et al. (2005) and Austin (2013). **(B)** General distribution of the regional fault systems and volcanoes.

The Hat Creek Fault System is a late Pleistocene to Holocene (Parrish, 2006; U.S. Geological Survey and California Geological Survey, 2006; Muffler et al., 2010), 47-km long, up to 490-m high, west-facing escarpment along the east side of Hat Creek Valley (Blakeslee and Kattenhorn, 2013). It has a slip rate of about 1-3.6 mm yr^{-1}, and is capable of producing an earthquake of moment magnitude about 6.7, with a recurrence interval of 667 ± 167 yr (Wills, 1991; Blakeslee and Kattenhorn, 2013). The Hat Creek Fault has a component of right-lateral offset along with normal offset (Wills, 1991), with segments connected by relay ramps and monoclines (Muffler et al., 1994). Due to sporadic volcanic activity, surface rupture along the Hat Creek fault has progressively migrated westward since the Late Pleistocene, with older scarps successively abandoned (Blakeslee and Kattenhorn, 2013). North of the Pit River, slip of the Hat Creek Fault probably transfers to the Soldier Creek Fault, which displaces a 1.2 Ma basalt by more than 260 m, giving a long-term vertical separation rate of 0.22 mm yr^{-1}. On the western edge of Hat Creek valley is the 9 km long, north-trending and east-facing Rocky Ledge Fault System. It has a component of right-lateral motion on a dominantly normal structure. The strike-slip component is consistent with the northwest propagation of the Walker Lane Belt, with suggested Holocene activity (Wills, 1991; Austin, 2013). The Rocky Ledge Fault System has an average slip rate of 0.4 ±0.2 mm yr^{-1}, an order magnitude less than the late Quaternary slip rate on the Hat Creek Fault (Sawyer and Ramelli, 2012). East of and parallel to the Hat Creek fault zone, there are normal faults with a right-lateral component, as well as the west-facing scarps of the McArthur Fault Zone. A steep scarp in alluvium with very weak soil development, a large closed depression that flanks a west-facing scarp in alluvium south of Big Lake, and tonal lineaments in plowed fields indicate Holocene offset along this fault (Wills, 1991). The more northwesterly Pittville Fault extends for about 48 km to the southeast, partly controlling the course of the Pit River and offsetting Brushy Butte, the volcanic center north of Big Lake (Peterson and Martin, 1980). Recent movements are indicated North of the Pit River Valley, where the fault forms a 1–3 m scarp in Late Pleistocene basalt, with open fissures up to 2 m deep on the upthrown side.

METHODOLOGY

Shuttle Radar Topography Mission (SRTM), higher-resolution, local LiDAR, and 10-m spacing National Elevation Data (NED) (Gesch et al., 2002) were used in the present study. NED data, covering the area of the Southern Cascades from 40.22 to 41.33 N and 122.08 to 120.22 W in decimal degrees, were acquired from the USGS National Map website (http://ned.usgs.gov), and were the primary dataset used for mapping and interpreting the overall tectonics, geomorphology and volcano distribution. In addition, Landsat satellite images, which have been used to study remote active volcanoes and to map basaltic volcanic fields (Francis and Wells, 1988), were downloaded. The elevation data were loaded in ArcGIS® software and saved as an ENVI compatible file for processing and topographic modeling.

Topographic Modeling, and Structural and Geomorphological Interpretation

Topographic modeling techniques were applied to the DEM using RSI ENVI® software generated surface parameters, including shaded relief, slope, and aspect maps. Landsat 7 Thematic Mapper Plus (ETM+) and panchromatic band 8 images were processed to combine bands and obtain maximum contrast to check surface features and deposits observed in the DEM. Generated images were compiled using ArcGIS® software.

Geomorphic features, including volcanic, tectonic, and volcano-tectonic features such as volcano edifices, cone and lava fields, and faults, as well as features resulting from gravitational instability and erosion were visually interpreted and described from the images. Interpretations were checked in the field. Structural analysis of the elevation dataset was based on the interpretation of rectilinear features or lineaments of regional scale. Lineaments consist of sharp tonal differences and alignments of geomorphological features like cones, streams and rivers, ridges and crests. Within tectonically active areas such as the Hat Creek Graben region, lineaments usually coincide with the fracture network and the major fracture patterns. Volcanoes are identified from descriptions in previous literature and visually interpreted by their positive relief with a circular or elongated shape, and may have a small crater near the summit. Volcano bases were manually delineated, by following the break in slope around the edifice base. Thus, only the objectively definable, visible edifice was traced and far-reaching fall and flow products were not considered in the delineation. Main lineament directions and volcano base elongations were then defined by the azimuth frequency distribution plotted on rose diagrams. Terrain modeling, geomorphological interpretation, and geomorphometry were then combined to interpret the relationships between the regional stress, basement faults, and volcanoes.

Volcano Morphometry

Individual volcano base and crater shapefiles were extracted as regions of interest from the DEM and processed using MORVOLC to generate a quantitative characterization of each individual edifice, and to develop a systematic way to compare the edifices. MORVOLC (Grosse et al., 2009) is an IDL-based code that generates a database of quantitative descriptions of the size (height, width, and volume), planview shape (ellipticity and irregularity), and profile shape (height-width ratio, summit width to basal width ratio, and slope) of stratovolcanoes and cones. Morphometric analysis can provide clues to the processes that interact during the growth history, development and evolution of each volcano, and can be used to draw an evolutionary trend for volcanoes in a region. MORVOLC calculates a 3D basal surface from the edifice outline using a least-squares criterion, which was used to estimate the volume and height parameters. Elevation contour lines were then generated from the DEM, and a summit region was defined at the elevation where the edifice starts flattening out.

Volcano Spatial Distribution and Alignments

MORVOLC generated the geographic coordinates of volcano summits that we used as the volcano centroids. These point data sets were used as input information for running the Geological Image Analysis Software (GIAS, www.geoanalysis.org) code in Matlab developed by Beggan and Hamilton (2010). GIAS evaluates the amount of randomness or any patterns of a spatial distribution of points by comparing the mean distance between all the nearest neighbors in the population with those defined by an idealized statistical model (Clark and Evans, 1954), Poisson Nearest Neighbor Analysis is preferred, in the present case. The nearest neighbor distances are determined both for the observed and expected distribution and then their mean distances are calculated to obtain the population-dependent R_0 and R_e (Baloga et al., 2007), respectively. Our observed population consists of the number of volcanoes per subgroup based on their morphometric class. The sample area is a convex hull generated by connecting the outermost points of the population (Hamilton et al., 2010; Le Corvec et al., 2013). To overcome the sample-dependent bias and properly assess the suitability of the PNN analysis, the R and c values were plotted with a confidence interval of 2σ. GIAS outputs a plot of the volcanic centers and the convex hull of the volcanic area, the R and $c$statistical values, and the skewness vs. kurtosis.

Alignment patterns found by PNN provided important constrains for understanding controls on regional volcanic evolution. Using the area of the convex hull from PNN analysis computed by GIAS, and the observed volcano population based on morphological characterization by MORVOLC, we calculated the normalized distance between the volcanoes used to create alignments, based on the density of the volcanoes. This takes into account the problem of volcanic fields covering a large area even though the mean distance between volcanic centers is generally small, due to magma propagating through new fractures or re-activated pre-existing structures (Valentine and Hirano, 2010). Due to differences in densities and mean distances of volcanic centers for each volcanic field within the Hat Creek region, we applied the relationship between the maximum length for the generation of alignments and the density of volcanic centers originally determined by Le Corvec et al. (2013) for monogenetic volcanic fields. This relationship was used to calculate the maximum length to generate lineaments in the Hat Creek volcanic region. A Three-point Alignments Matlab code used for volcanic alignment analysis (Le Corvec et al., 2013) was used to identify groups of points in a population of volcanic centers that form straight lines (Wadge and Cross, 1988; Connor et al., 1992; Baloga et al., 2007; Bleacher et al., 2009; Le Corvec et al., 2013), and to generate a list of alignment azimuths. Directions of alignments are presented in a rose diagram to find out if there is a a strong controlling factor that results in preferred alignment during volcano growth.

Volcano Morphology and Regional Stress Relationship

We compared crater morphology and regional stress relationships only for bigger volcanoes, to draw from a homogeneous population. Amphitheaters, breaches, scars or craters were our primary indicator for describing a part of the volcano's history that could be related to regional stress (Lagmay and Valdivia, 2006). We identified volcanoes that show amphitheaters, volcanic scars and breaches. A scar was used to describe craters deformed by major collapse events, including debris avalanches and large landslides. A crater breach is a scar-like deformation brought about by erosion such as by glaciers (cirques) or by meteoric and hydrological events forming gullies and gully heads.

To determine regional stress, we obtained the orientation of the nearest regional maximum horizontal stress node to each bigger volcano from the World Stress Map, a standard global compilation of contemporary tectonic stress (Heidbach et al., 2008). Stress orientation in the map is from various types of stress indicators, such as earthquake focal mechanisms; well bore breakout and drilling induced features; *in-situ* stress measurement such as overcoring, hydraulic fracturing, and borehole slotter; and young geologic data from fault slip analysis and volcanic vent alignments (Heidbach et al., 2008). We then measured the direction at which scars and breaches open using the angle measurement tool in GNU Image Manipulation Program®. The absolute difference between these two angles is the angle difference.

RESULTS

Topographic Modeling, and Structural and Geomorphological Interpretation

Elevation Modeling

The elevation range in the Hat Creek region is from 150 to 3200 m with a mean elevation of 1400 ± 400 m. To simplify initial exploration of terrain development, elevation density slicing was applied to the DEM, generalizing the topography. This process enhanced the regional detail of the DEM. Based on the original and density sliced DEM (Figures 2A,B), there are two major elevation profiles: the low elevation areas in cool colors and high areas in warm colors. The low elevation areas are subdivided into four areas based on the continuity of the low elevations compared to adjacent, higher areas: A, B1, B2, and C. The low areas decrease in mean elevation from east to west. The lowest mean elevation (A), darkest blue in Figure 2A and magenta in Figure 2B, are the Sacramento Valley and Sacramento River and its tributaries from the Klamath Mountains in the north to the western side of Lassen Volcano National Park in the south. This area is separated from the lower area of Lake Britton (B1) on the east by Chalk and Hatchet Mountains, which are both part of the Klamath Mountains. The Pit River, which is the northern main tributary of the Sacramento River, carves the divide between these two mountains. In lightest blue and highest in mean elevation among the three low areas is an area that includes lakes in the southern part of Brushy Butte lava field, and McArthur and Fall River Mills (B2). It is separated from the lower western area (B1), of Lake Britton and Hat Creek Valley, by the Hogback ridge and Saddle and Haney Mountains. The green, flat area consists of Lookout and Bieber areas and Ash Creek State Wildlife Area (C), possibly the easternmost tributary of the Sacramento River.

The high elevation areas are subdivided based on the texture and intensity of warm colors in the DEM and the geographic location (D1, D2, E1, E2). The warm, geographically continuous colors in the southern part of the study area are divided into an area of smooth contours on the western side that coincides with the Cascade Range (D1) and the relatively irregular-edged Modoc Plateau (D2) on the east and northeast sides. Two other high elevation areas are geographically separated and thus are identified as separate profiles: the Klamath Mountains (E1) and Taylor, Jimmerson and Widow Mountains (E2). The elevation density map (Figure 2B) highlights the edges of high elevation areas (in green: D1, D2, E1, E2), and the break in slope that marks the base of volcano edifices (in red: insets F, G, H, I, J).

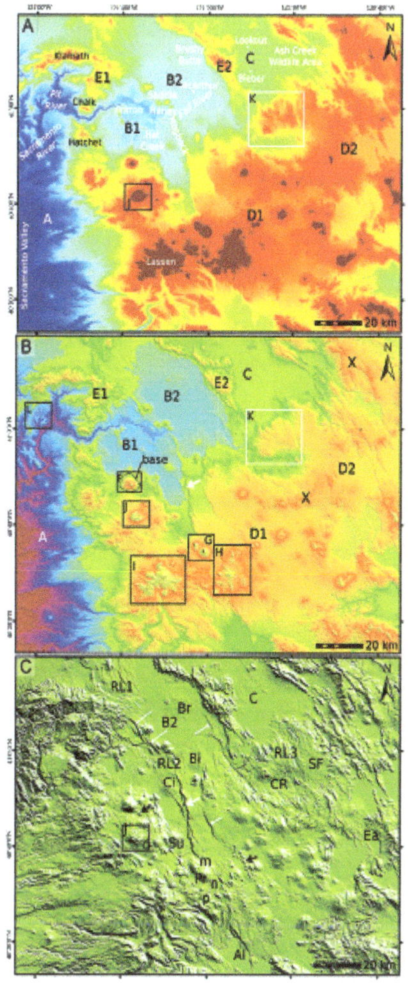

Figure 2. Density-sliced DEM and Shaded Relief Maps of the Hat Creek Graben and its vicinity (Gesch et al., 2002). (A) Highlights the different elevation regions high in warm colors and low in cold colors and **(B)** green as the boundary between the two main regions and red as the break in slope that marks the bases of volcanoes. Based on mean elevation and continuity of elevation vs. adjacent areas, the low elevation areas are subdivided into A: Sacramento Valley and Sacramento River including its main tributary on the north, the Pit River; B1: Hat Creek Valley and Lake Britton, B2: McArthur and Fall River, and C: Lookout, Bieber and Ash Creek Wildlife Area. High elevation areas are subdivided based on texture and intensity of warm colors and geographic location. D1: Cascade Range; D2: Modoc Plateau; E1: Klamath Mountains; and E2: Taylor, Jimmerson and Widow Mountains. Red linear areas marked with x and K (inset) are ridges, features in insets F, G, and J denote stratovolcanoes, and insets H and I denote complex volcano edifices. Inset L denotes areas of river erosion. **(C)** Shaded relief image with water bodies: lakes Almanor (Al) and Eagle (EA); reservoirs: Silva Flat (SF) and Catfish (CR); flood plains: B2 and C; volcanoes: Sugarloaf Mountain (Su), Crater Peak (inset J), Prospect Peak (Pr); lava fields: Cinder Butte (Ci), Big Cave (Bi), Brushy Butte (Br), m and n; lava flows (o and p); cones (black arrows); faults (white arrows); and possible lineaments as localized faults (RL1, RL2, RL3). Arrows are drawn perpendicular to the general strike of the lineaments.

Features observed and mapped include the morphology of the edifices. Mountain ridges that form narrow and linear features (x) are contrasted against volcano edifices, observed as being more equant in planview (insets F, G, J). Stratovolcano bases are identified by the outermost, circular or equant, boundaries (insets F, G:1-2, J). The non-circular and very irregular edges of the base also identify the more complex volcanoes (such as those in insets H and I). "Piedmont" volcanoes (such as in inset G) are formed when the base of one volcano (1) is truncated by or buried under an adjacent edifice (2). Big volcanoes or broad areas of higher elevation relative to adjacent areas (insets H, I) are comprised of concentrations of (sometimes linear) cones (inset H), or are heavily eroded stratovolcanoes (inset I). Smaller stratovolcanoes (insets F, G, J) can also show irregularities of their bases, which can be evidence for collapse, sliding or unequal deposition and erosion. In inset J is an area that has experienced a deep-seated erosional or mass-wasting process.

Other high elevation areas (insets K, L) with differences in the regularity and smoothness of the edges (in green) can indicate areas that may be experiencing varying degrees of erosion. Inset K, showing a volcano base with a regular planview edge as indicated by smoother color contour, may indicate a relatively lesser amount and lower rate of erosion, while the volcano in inset L may be experiencing a higher and more stable rate of erosion by a river tributary.

Shaded Relief

The shaded relief map (Figure 2C) shows a 3D depiction of the earth's surface. The flat and very smooth surfaces represent: the water bodies such as lakes Almanor (Al) and Eagle (Ea); reservoirs such as Silva Flat (SF) and Catfish (CR); and flood plains such as B2 and C. Rough and sloping areas represent either mountains or mountain ridges if they are elongated, and volcanoes if they are more circular in planview.

Volcanoes having rough surface textures but very few gullies or other linear erosional features, such as Sugarloaf Mountain (Su), have relatively fresh lava (lava is not easily erodible), which may translate into a young relative age of the volcano. Those with rough surface texture or lower slopes, such as Cinder Butte (Ci), Big Cave (Bi), and Brushy Butte (Br), or those being non-circular in planview, such as those in"m" and "n," are identified as lava fields, as long as there is a recognizable source. Lava flows ("o" and "p") are also recognizable as having levees on the narrower sides, and subsidiary flows on the edges of the main flow. Cones (black arrows) are easily recognizable as small mounds with relatively smooth and steep slopes. Relative ages of mountains are marked by the existence and density of gullies.

Other surface feature easily recognized on a shaded relief image are faults and linear features. Linear features that show continuity, and coincide with regional fault systems described in the literature, as well as those that show obvious vertical displacement of other features (white arrows), are identified as faults. Identification of linear arrays of cones (black arrows) and faults (white arrows) is a starting point in recognizing more subtle structures. Repeating lineaments (RL1, RL2, RL3) in the same area, and with the same strike, are also likely faults. In recognizing faults, the most obvious ones may not have the biggest offset, but may have the best preserved scarp, which is due to their age, offset, and erosion and deposition rates.

The hillshade image enhances some of what can be picked out in color mapping (Figure 2C). The addition of a 3D depiction behind the DEM enhances the surface features observed, while the simplified and generalized version of the shaded relief image confirms the existence of both the surface features and structures that may have been otherwise over-interpreted. Although (Figure2C) is just an enhancement of Figure 2A, one of its advantages is that shadows generated by the particular sun angle considered in the generation of the shaded relief image are not emphasized. Shadow emphasis might be helpful for delineating faults (thick white arrow), but this is not always the case for delineating volcano bases like that of Crater Peak (inset J), as the surfaces are of much the same texture, generating the same types of shadows, in all directions.

Slope Modeling

Topographic modeling was also used to generate a slope map (Figures 3A,B). The dataset for the study area ranges from 0° to 81.5° with a mean slope of 11°. Steep slopes are emphasized in Figure3A. This color mapping of the slope map is useful for confirming the delineations of faults.

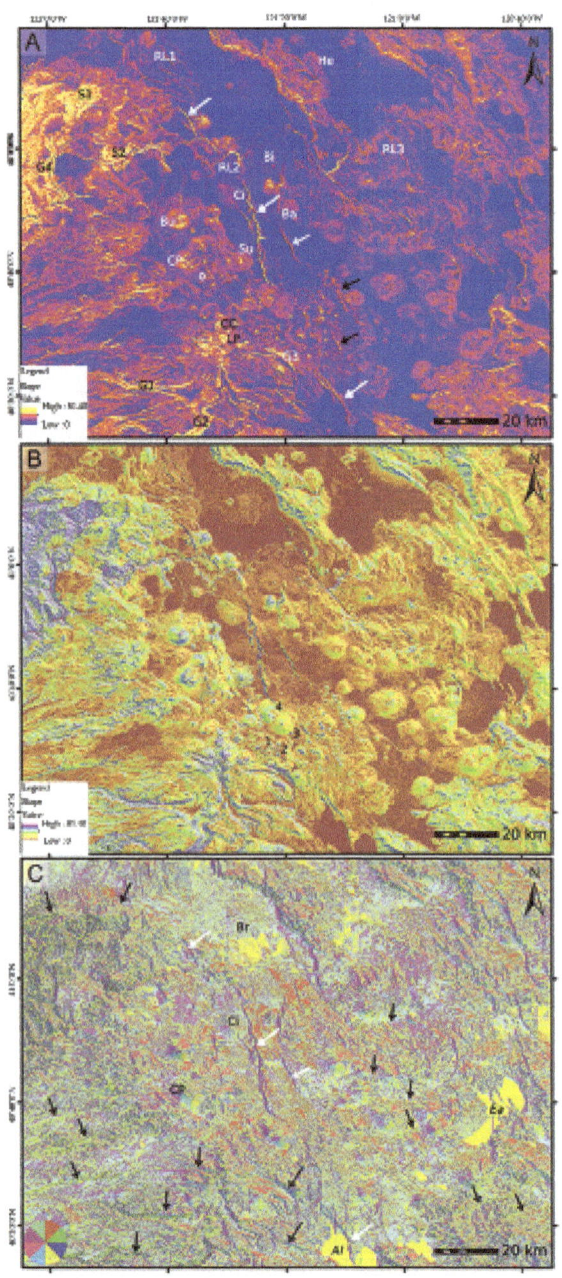

Figure 3. Slope and aspect maps (Gesch et al., 2002). (A) Density sliced slope map where steep areas including volcano and cone summits and fault scarps are yellow, edifices are pink, and relatively flat surfaces and water bodies are blue. The boundaries between the blue

and pink areas are the break in slopes that mark a volcano base. Arrows are drawn perpendicular to the general strike of the lineaments. Fault groups based on orientation and strike, degree of curvature and sinuosity are RL1, RL2, RL3. Hat Creek Fault and its projected continuation is shown by thick white arrow. McArthur fault is indicated by a thin white arrow. Steep gorges are marked G1, G2, G3, and G4. Steeply sloping, wide areas are S1 and S2. Stratovolcanoes with known and observed deformed edifices are Crater Peak (CP), Chaos Crags (CC) and Lassen Peak (LP). Other edifices are Sugarloaf Peak (Su), Cinder Butte (Ci), Big Cave (Bi), Bald Mountain (Ba), and Burney Mountain (Bu) and lava fields (o) and Hermit Butte (He). **(B)** Emphasizes lava fields and their source areas labeled 1–11. **(C)**Aspect map of the Hat Creek Graben and adjacent areas. Polygon colors and azimuths: red (0–45°), green (45–90°), blue (90–135°), yellow (135–180°), cyan (180–225°), magenta (225–270°), maroon (270–315°), and sea green (315–360°). Water bodies: lakes Almanor (Al) and Eagle (Ea) are in yellow. White and black arrows point at west- and south- facing fault controlled geomorphic features and linear discontinuities.

Faults are identified as relatively straight lineaments, often members of a set with consistent strike. Faults in the study area have a considerable amount of vertical movement, as measured by the extent of the area colored yellow. The faults were classified according to the degree of curvature. Faults in RL1, RL2, RL3, for example, are grouped together. Faults that belong to the same set generally have the same orientation, maximum along-strike throw as indicated by maximum cross-strike width, planview curvature, and thus approximately the same age, as should be the case.

The main Hat Creek fault may well have a southern continuation (thick white arrow), which occurs along the northeastern bank of the northwestern lobe of Almanor Lake (Al). To the east of the Hat Creek Fault is the trace of the McArthur fault (thin white arrow), which may continue to the south or southeast as evidenced by the linear cones directly north of AL.

Steep gorges, G1, G2, G3, and G4, are identifiable; they often indicate in this region that the stream segment has been uplifted. Steeply sloping, wide areas, such as S1 and S2 indicate landsliding or a high rate of erosion. A closer look at the steep summits of some volcanoes and domes, such as Chaos Crags (CC) and Crater Peak (CP), shows that these are relatively smooth in texture, compared to stratovolcanoes that have been deformed by violent activity, such as Lassen Peak.

Cones (black arrows) are defined by yellow circular regions that indicate steepness from the base to the summit, with a lower-slope angle, cool colored (blue) region in the middle. Because slope maps emphasize areas of steepness, they are good for delineating lava flow extent, since the edges of lava flows tend to be steep, irregular and narrow relative to adjacent areas. Because slope steepness is a factor in landslide susceptibility mapping, steeper regions on volcanic edifices are at increased risk for landslides and collapses.

Slope maps also indicate relative ages of volcanic edifices. If we compare Burney Mountain, Bald Mountain (Ba) and Hermit Butte (He), for example, Burney Mountain seems to have been covered by younger surface deposits and less affected by erosion than the other two volcanoes.

Another variation in slope map presentation (Figure 3B) emphasizes lava fields (labeled 1–11) and their source areas. Areas covered by lava flows, such as those near Sugarloaf Peak (Su), Cinder Butte (Ci), Big Cave (Bi), the area north of Burney Mountain (Bu) and the area southeast of Crater Peak (o), have rough texture in the slope map, but a mean slope near zero.

Aspect Modeling

Slope aspect mapping is classified as a continuous terrain segmentation scheme (Miliaresis and Kokkas, 2004). In the current approach, aspect was computed for every DEM grid point (Figure3C). Then the aspect was standardized to eight major geographic directions: N, NE, E, SE, S, SW, W, and NW. Linearity in colored polygon in the figure is another type of evidence for faults (black arrows). Faults with dip-slip component of different orientation can be distinguished with this technique, e.g., those with the western block moving down relative to the eastern block (white arrows). Areas of deposition of rough materials, such as emplaced lava flows or avalanche and slide toes tend to have highly variable, specular color patterns, meaning that the area has slopes facing in many directions on a relatively fine spatial scale, such as at Cinder Butte (Ci), Brushy Butte (Br), and Crater Peak (CP).

Structural and Lineament Mapping

Regional fault systems (Figure 4A) were drawn and identified in consultation with previous work as: the Hat Creek, Rocky Ledge, McArthur, Soldier Creek, and Pittville Fault Systems. These fault systems are predominantly normal, with right-lateral components and evidence of activity up to Holocene time (Wills, 1991; Muffler et al., 1994, 2010; Parrish, 2006; U.S. Geological Survey and California Geological Survey, 2006; Sawyer and Ramelli, 2012; Blakeslee and Kattenhorn, 2013). We traced the possible extent of these structures, as well as other shorter, more curvilinear, and relatively minor, faults such as those that affect Burney Mountain (Bu), Crater Peak (CP), West Prospect Peak (WP), Prospect Peak (PP), Bogard Butte (Bo), Whaleback (Wh), and Fox Mountain (Fo) (Figures 4B–F). Some of the smaller faults, with strike directions that coincide with faults identified by U.S. Geological Survey and California Geological Survey (2006) to be <130 ka, <1600 ka and >1600 ka, are assumed to be of the same age as the larger faults, and are grouped with them. The smaller faults so mapped are possibly parts of the Walker Spring fault system, and the unnamed faults in the Susanville lake and Big Valley areas, while Other Primary Regional 1 faults are unnamed. Some of our other Primary Regional 2 traces also belong to the <1600 ka faults in the USGS Fault Map.

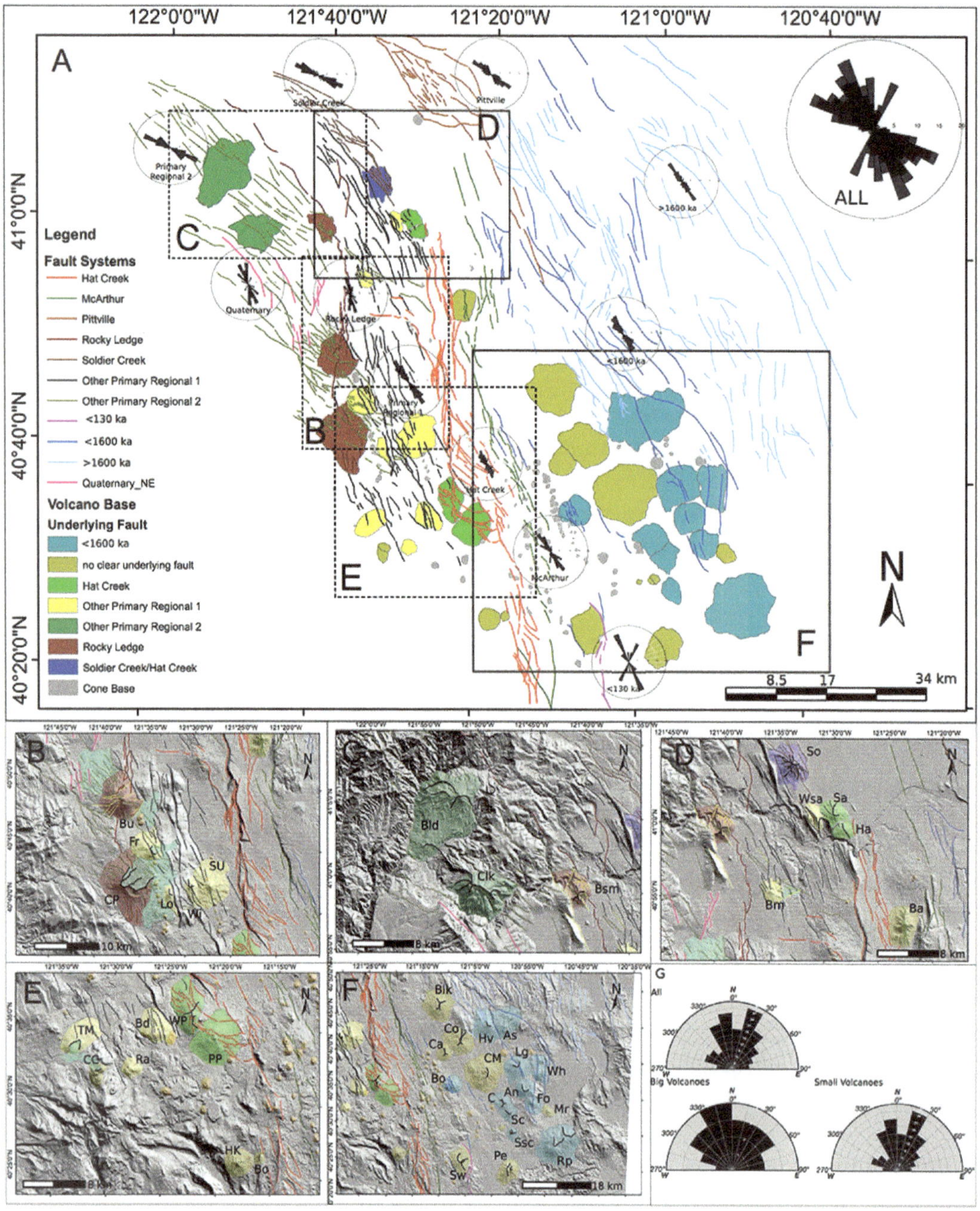

Figure 4. Volcano-tectonic relationships in Hat Creek Graben region. **(A)** Base map showing fault systems, volcano bases color coded on fault system that mostly affects edifice, and water bodies; rose diagrams of fault strikes by fault system; and locations (squares) of hill-shaded maps **(B–F)** showing fault traces and effects on edifice: edifice instability and collapse, edifice disintegration, and splitting. **(G)** Rose diagrams of volcano base elongations for both big (cones, sub-cones, shields, and massifs) and small (small cones, small sub-cones, and small shields) volcano sub-groups, and all volcanoes.

Most of the faults trend northwest (Figure 4A). However, in more detail, the fault strike evolves from north-northwest to more nearly west-northwest, from east to west near the northern edge, while it goes from west-northwest toward north in the central to southern part of the study area. The small faults are defined as short lineaments on volcanoes without clear continuity into nearby areas. Such lineaments include the trace just to the northeast of Crater Peak (CP) (Figure 4B), which continues northwest up to the northwest side of Burney Mountain (Bu). These traces can contain multiple, short lineaments, but are constrained areally. Another area of localized faulting is the corridor of northwest trending short faults that affect the northern sectors of Prospect (PP) and West Prospect Peaks (WP) (Figure 4E). On some volcanic edifices, such as Bogard Butte (Bo), Whaleback (Wh), and Fox Mountain (Fo) (Figure 4F), small faults seem to split the edifices. In the case of Crater Peak, the small fault seems to have at least partly resulted from a major collapse. In these cases, the lineaments developed in response to a gravity-driven event that affected the volcano. However, in the case of Bogard Butte, Whaleback and Fox, it appears that the faults have influenced the shape and development of the volcano, suggesting that they are basement features that have propagated into the edifice.

Volcano Morphometry

Based on the MORVOLC output, we classified the volcanoes of the Hat Creek Graben region using primarily height to basal width ratio (Figure 5A). Those with a height to basal width ratio greater than 0.15 were classified as *cones*; those with intermediate values between 0.1 and 0.15 as *sub-cones*, and those with values less than 0.1 as *shields* or *massifs*. These last two were classified further based on height. A second classification based on height was done to differentiate small and large *cones, sub-cones* and *shields*. This classification was used to develop an evolutionary trend for volcanoes in the Hat Creek Graben region (Figure 5B).

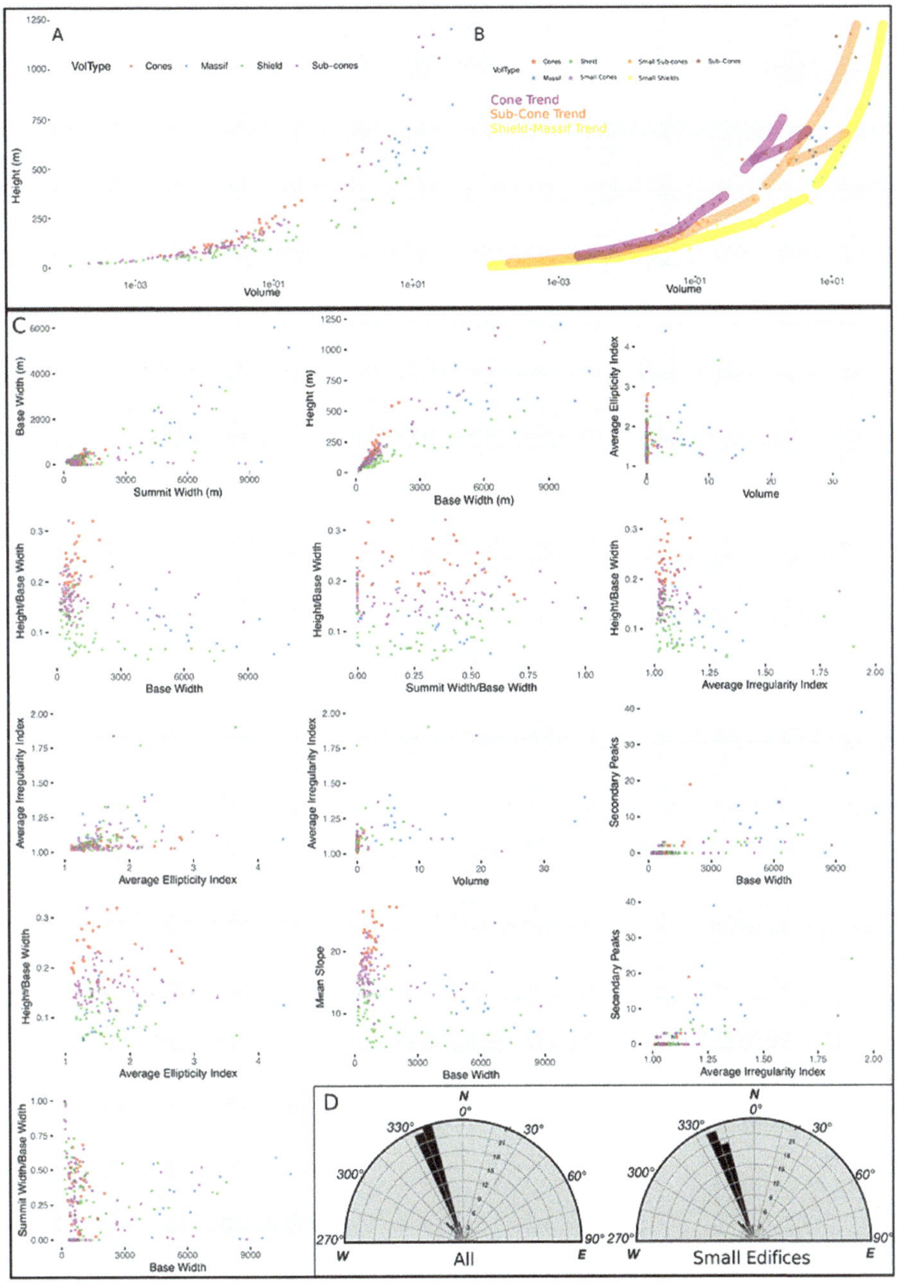

Figure 5. Volcano morphometry. **(A)** Volume vs. height plots for all volcanoes grouped by morphological classification: cones, sub-cones, shields, and massifs based on their height and basal width ratio; **(B)** A second morphological classification based on height separating bigger and smaller edifices on a volume vs. height plot to draw a volcano evolutionary trend;**(C)** X-Y plots of selected morphometric parameters grouped by morphological type (cones, sub-cones, shields and massifs); and **(D)** A rose diagram showing the preferred direction of alignments that can be formed by at least 3 volcano centroids within 6 km.

The evolutionary trend begins with cones. Cones can either maintain their height to basal width ratio to increase in volume into larger cones, or enlarge in basal width to become sub-cones. Sub-cones can continuously increase in volume to become a bigger sub-cone, or evolve in basal width into a large but low volcano, a shield, and ultimately into a massif. Lastly, small volcanoes with height and basal width ratios of a shield can grow into massifs. All *morphologically big* volcanoes in the region are those with identified names, as discussed in the geomorphological classification of features in the region (Part 1 of this work). However, some of these previously named, *big*volcanoes needed to be reclassified as small cones, small sub-cones, or small shields, because their heights were less than 350 m (Figure 5B). Of the total of 185 drawn volcanic centers in the Hat Creek Graben region, 33 were classified as bigger stratovolcanoes (massifs), and 152 are morphologically smaller, mostly monogenetic cones (Tables 1, 2).

Table 1. Morphometry and crater morphology of the Hat Creek Graben region stratovolcanoes.

Name	Type	HeightMax	VolMax	WidthBase	WidthSummit	σH_{max}	Crater direction
0-crater	C	84.61	0.05	1142.83	0.00		
1-crater	A	204.39	0.08	1041.97	374.30		
2-crater	B	183.54	0.09	1135.70	528.73		
3-crater	A	306.50	0.13	1055.52	449.16		
4-crater	C	139.37	0.06	1013.24	26.91		
5-crater	A	126.70	0.02	714.21	0.00		
6-crater	A	315.51	0.19	1308.55	500.73		
7-crater	B	242.07	0.13	1137.36	0.00		
8	A	117.98	0.01	507.83	206.89		
9-crater	A	224.08	0.07	877.83	464.97		
10-crater	B	229.48	0.15	1447.89	509.98		
11-crater	B	219.39	0.11	1060.86	692.81		
12-crater	B	214.63	0.12	1291.19	0.00		
13-crater	A	132.87	0.02	661.04	323.29		
14-crater	B	149.46	0.03	818.74	0.00		
15-crater	A	175.64	0.04	723.62	419.76		
16	A	114.79	0.01	447.11	51.27		
17-crater	B	101.96	0.01	457.38	256.21		
18-crater	A	261.68	0.10	939.22	410.03		
19-crater	B	102.62	0.02	623.90	267.48		
20-crater	A	171.83	0.06	994.90	682.40		
21-crater	B	177.72	0.07	1026.93	574.44		
22	B	52.45	0.00	290.10	83.82		
23-crater	A	154.56	0.04	733.61	378.59		
24-crater	B	76.64	0.01	576.39	79.77		
25-crater	C	47.43	0.00	487.56	0.00		
26-crater	C	55.34	0.00	474.03	227.70		
27-crater	A	135.25	0.02	625.51	0.00		
28	B	187.48	0.03	584.53	226.79		
29	B	108.22	0.02	643.34	47.04		
30-crater	C	139.33	0.07	1227.03	588.47		
31-crater	B	111.39	0.03	765.83	0.00		
32-crater	B	102.78	0.02	673.14	370.83		
33-crater	B	102.82	0.02	674.84	0.00		
34-crater	C	151.75	0.04	783.18	0.00		
35	C	88.47	0.01	629.07	111.51		
36	A	147.91	0.03	775.96	207.36		
37-crater	C	134.52	0.05	1069.24	484.42		
38-crater	B	112.97	0.03	863.77	383.09		
39-crater	B	165.27	0.06	1170.85	484.79		
40-crater	B	143.32	0.06	1125.81	487.76		
41-crater	A	105.69	0.01	528.68	352.49		
42	B	114.39	0.01	534.37	131.56		
43-crater	B	130.47	0.03	799.15	392.52		
44-crater	C	64.24	0.00	477.99	0.00		
45-crater	C	93.98	0.03	786.42	424.89		
46	B	126.67	0.02	591.44	175.94		
47-crater	B	93.58	0.01	508.92	0.00		
48	B	80.06	0.01	450.60	132.66		
49-crater	B	109.34	0.02	628.01	261.94		

50-crater	C	120.30	0.02	636.28	0.00
51-crater	B	182.88	0.07	936.56	360.90
52-crater	B	167.79	0.05	822.82	0.00
53-crater	B	108.30	0.02	675.58	0.00
54-crater	B	224.42	0.11	1104.87	0.00
55-crater	C	187.71	0.05	958.14	0.00
56	B	158.98	0.03	664.97	97.53
57	B	104.07	0.01	563.44	215.91
58-crater	A	187.47	0.06	864.39	486.56
59	C	51.58	0.01	852.92	322.74
60	C	94.23	0.05	1310.23	215.51
61	C	40.92	0.01	743.28	68.89
62	C	37.97	0.00	430.59	221.30
63	C	40.09	0.01	583.25	426.13
64	B	77.89	0.01	682.35	86.88
65	C	65.40	0.01	825.39	215.03
66	C	56.35	0.01	885.72	343.24
67	C	139.26	0.18	2107.28	401.35
68	C	138.39	0.14	1721.25	251.66
69	C	91.09	0.02	933.47	247.91
70	C	113.38	0.04	1179.01	334.75
71-crater	C	179.44	0.15	1823.35	96.44
72	B	108.63	0.02	945.42	125.35
73-crater	C	154.42	0.20	1766.18	0.00
74	C	74.18	0.03	1040.01	395.14
75	A	64.44	0.00	362.00	165.14
76-crater	A	101.59	0.01	434.40	226.27
77-crater	A	117.40	0.01	477.54	0.00
78	B	102.70	0.02	605.24	356.17
70-crater	B	127.52	0.02	564.57	0.00
80	B	77.40	0.01	524.39	145.68
81-crater	C	211.84	0.21	1425.68	0.00
82-crater	A	157.53	0.04	818.09	52.04
83-crater	B	114.51	0.02	650.72	357.50
84	B	106.04	0.02	619.50	131.41
85	C	57.50	0.00	565.37	239.89
86	A	103.03	0.01	528.98	143.81
87	B	49.49	0.00	263.73	164.77
88	B	36.95	0.00	254.93	150.73
89	A	77.47	0.00	390.64	68.46
90	B	59.02	0.00	342.86	127.66
91	B	65.75	0.01	455.40	169.45
92	B	63.53	0.00	424.76	28.38
93	B	54.93	0.00	340.69	160.95
94	B	123.61	0.02	695.64	288.28
95	B	89.32	0.01	387.70	170.94
96	B	64.96	0.00	397.58	101.11
97	B	60.23	0.00	455.70	126.14
98-crater	B	79.23	0.01	475.31	0.00
99	B	77.13	0.00	349.00	191.65

100	B	150.58	0.05	887.18	39.84		
101	B	63.28	0.00	393.48	77.79		
102	B	92.74	0.01	599.06	70.93		
103	C	87.09	0.01	507.15	300.71		
104	B	112.22	0.01	634.24	113.11		
105	B	99.97	0.02	705.61	193.97		
106	C	74.63	0.01	600.43	303.98		
107	C	25.31	0.00	204.68	201.25		
108	B	26.47	0.00	182.46	182.47		
109	B	57.98	0.00	382.88	184.81		
111	C	26.39	0.00	165.84	84.97		
112	C	25.69	0.00	179.67	115.05		
113	C	57.08	0.01	486.19	324.70		
114	C	13.17	0.00	136.52	58.86		
115	C	28.59	0.00	257.31	85.42		
116	C	25.56	0.00	178.49	65.47		
117	B	75.55	0.00	412.84	70.80		
118	C	79.40	0.01	606.86	212.14		
119	B	78.02	0.01	609.21	192.07		
120	A	248.03	0.09	1001.44	154.58		
121	A	245.45	0.06	776.61	38.08		
122	A	227.57	0.06	828.54	29.72		
123	A	200.54	0.06	919.19	287.44		
124	A	63.50	0.00	296.52	82.74		
125	B	23.34	0.00	160.64	160.65		
126	B	70.80	0.00	363.09	174.56		
127	B	35.82	0.00	215.30	208.05		
128	C	34.88	0.00	245.70	145.70		
129-crater	C	56.29	0.00	410.13	0.00		
130	B	40.44	0.00	185.63	148.57		
131-crater	B	106.93	0.01	506.88	0.00		
132	C	54.39	0.01	618.51	309.63		
133-crater	C	39.82	0.00	251.05	176.48		
134-crater	B	81.29	0.01	430.47	0.00		
135-crater	C	46.17	0.00	341.62	0.00		
136	B	27.79	0.00	141.67	122.89		
137-crater	B	56.99	0.00	331.60	0.00		
138	B	32.73	0.00	208.66	178.63		
139	B	57.79	0.00	352.41	258.45		
140-crater	B	76.94	0.01	547.62	4.20		
Antelope Mountain	Massif	557.75	5.55	5378.46	1851.37	SS = 4°	27
Ashurst Mountain	Shield	499.62	11.53	7851.94	3228.31		
Badger Mountain	Massif	650.39	5.33	4386.47	2371.60		
Bald Mountain	Sub-cones	626.85	3.42	4730.09	2298.75		
Bald Mountain	Massif	1202.07	35.02	9565.24	110.92	NF = 153°	259, 238, 207, 113
Blacks Mountain	Massif	605.71	15.40	8488.95	51.96		
Bogard Buttes	Massif	587.63	4.94	5158.42	1837.69		
BontePeak	A	338.02	0.23	1610.41	38.37		
Brush Mountain	C	236.37	0.63	2887.93	1578.00		
Burney Mountain-crater	Sub-cones	1110.31	13.67	6552.93	2566.33	NF = 156°	307, 111

Burney Spring Mountain	Massif	536.63	3.10	4297.78	73.10	NF = 156°	293, 111
Cal Mountain	C	295.35	1.38	4423.38	963.10	NF = 156°	247, 66
Campbell Mountain	C	342.72	2.48	5110.78	1168.56	SS = 4°	10, 240
Chalk Mountain	Sub-cones	1174.27	19.33	6685.52	3465.11	NF = 153°	195, 193, 182, 230
Chaos Crags north	Cones	543.46	0.48	1698.41	262.63	NF = 156°	300
Chaos Crags south	Cones	569.94	1.07	2015.65	669.66		
Cone Mountain	C	324.32	4.99	7417.35	2446.71	SS = 4°	278
Crater Mountain-crater	Massif	578.82	14.22	10240.00	6012.92	SS = 4°	68,97
CraterPeak-crater	Sub-cones	1054.51	23.26	8809.16	21.25	NF = 156°	66
Fox Mountain	Massif	632.63	6.11	4850.22	1100.21		
FreanerPeak	Massif	869.30	6.99	4686.70	864.89	NF = 156°	76
Haney Mountain	B	337.23	0.71	2624.56	344.25	NF = 156°	44
Harkness Mountain	Sub-cones	521.40	2.39	3760.78	959.82		
Harvey Mountain	Massif	592.13	10.12	7563.22	3408.96	SS = 4°	25
Logan Mountain	Shield	450.59	7.50	7194.79	2435.96	SS = 4°	321
Logan Mountain	Sub-cones	600.12	1.81	2612.66	291.51	NF = 156°	65
Merrill Mountain	C	215.82	0.79	3194.44	751.84		
Mudleft	C	103.07	0.04	938.09	66.45		
Mudright	C	92.44	0.08	1501.66	410.10		
Pegleg Mountain	Massif	577.98	6.00	6148.08	1209.84	SS = 4°	89
ProspectPeak-crater	Massif	699.11	8.53	6352.69	2226.61		98
RakerPeak	Sub-cones	395.72	0.88	2637.22	746.77		
Roop Mountain	Massif	820.53	36.66	10924.98	5125.81		61
Saddle Mountain	Sub-cones	511.10	1.61	3332.40	492.58		14
SCampbell Mountain	C	203.86	1.36	4282.21	1491.79		345
SLongLake	C	67.64	0.02	912.56	255.52		221
SofSCampbell Mountain	C	202.66	0.75	3691.73	551.48		54
Soldier Mountain	Sub-cones	662.69	4.38	5045.36	84.61		40
SugarloafPeak	Sub-cones	847.47	8.96	6274.42	170.30		
Swain Mountain	Shield	429.74	5.62	6355.22	557.52	SS = 4°	1,290
Table Mountain	Shield	487.35	3.92	4590.87	2526.60	NF = 156°	7, 232, 195, 240, 306
WestProspectPeak	Sub-cones	1163.52	11.90	5306.43	46.43		
Whaleback Mountain	Massif	518.45	6.12	6219.10	2998.09	SS = 4°	62, 99
WilcoxPeak	Sub-cones	583.33	1.76	2985.07	205.76	NF = 156°	27, 108
WSaddle Mountain	Sub-cones	414.88	1.08	2674.36	199.76		

*Height and width are in m; Volume is in m^3; σH_{max} is the maximum horizontal stress orientation (azimuth, in degrees) and tectonic regime (SS = strike-slip or NF = normal faulting) nearest to each volcano edifice with crater opening or collapse scar (stress data from Heidbach et al., 2008); crater directions are in degrees; volcano types A, B, and C refer to small cones, small sub-cones, and small shields that fall within the lower purple, orange, and yellow lines in **Figure 5B**. Dataset available in Supplementary Materials.*

Table 2. Properties and results of the statistical analyses for each morphological subgroup (big and small), and for the all the volcanoes in the Hat Creek Graben.

		Smaller	Bigger	ALL
Measured nearest neighbor properties	Number of volcanic centers (N)	152	33	185
	Area Convex Hull (m^2)	3.26E+09	3.54E+09	4.28E+09
	Density (nb volcanic centers/m^2)	4.66E−10	9.33E−11	4.32E−10
	Min distance NN (m)	4.92E+02	4.29E+03	4.92E+02
	Max distance NN (m)	1.83E+05	1.65E+05	1.23E+05
	Mean distance NN (m)	1.60E+04	7.01E+04	2.01E+04
	Standard deviation (1σ)	2.17E+04	4.60E+04	2.36E+04
	Standard deviation (2σ)	4.33E+04	9.19E+04	4.72E+04
	Skewness	4.28	0.62	2.43
	Kurtosis	25.75	−0.80	6.40
Nearest neighbor results relative to the Poisson model	Poisson Re (m)	2.32E+04	5.18E+04	2.40E+04
	R	0.69	1.35	0.83
	Ideal R given N_i	1.04	1.08	1.03
	R negative thresholds at 1σ	1.08	1.19	1.07
	R positive thresholds at 1σ	1.13	1.29	1.12
	R negative thresholds at 2σ	0.99	0.98	0.99
	R positive thresholds at 2σ	0.94	0.87	0.95
	c	−7.25	3.89	−4.33
	Ideal c given N_i	0.84	0.90	0.84
	c negative thresholds at 1σ	1.93	2.07	1.92
	c positive thresholds at 1σ	3.02	3.24	3.00
	c negative thresholds at 2σ	−0.25	−0.27	−0.24
	c positive thresholds at 2σ	−1.34	−1.44	−1.33
Alignment analysis	Distance maximal for the generation of lineaments (m)	5800	9900	6000
	Number of lineaments (m)	85	0	93

NN, nearest neighbor.

We present x-y scatter plots of volcanoes in the region based on morphological classification parameters (Figure 5C). Cones have narrow summit widths, while shields and massifs tend to have broader summits, wider bases and more peaks, and are sometimes more irregular in outline. Basically, shields and massifs are high volcanoes with wide bases. Sub-cones, shields and massifs are voluminous volcanoes compared to cones. Cones are mostly identifiable by their height and high mean slope angle, in which sub-cones follow them closely. Sub-cones, massifs and some shields show the most irregular or elliptical planview shapes (Figure 6A).

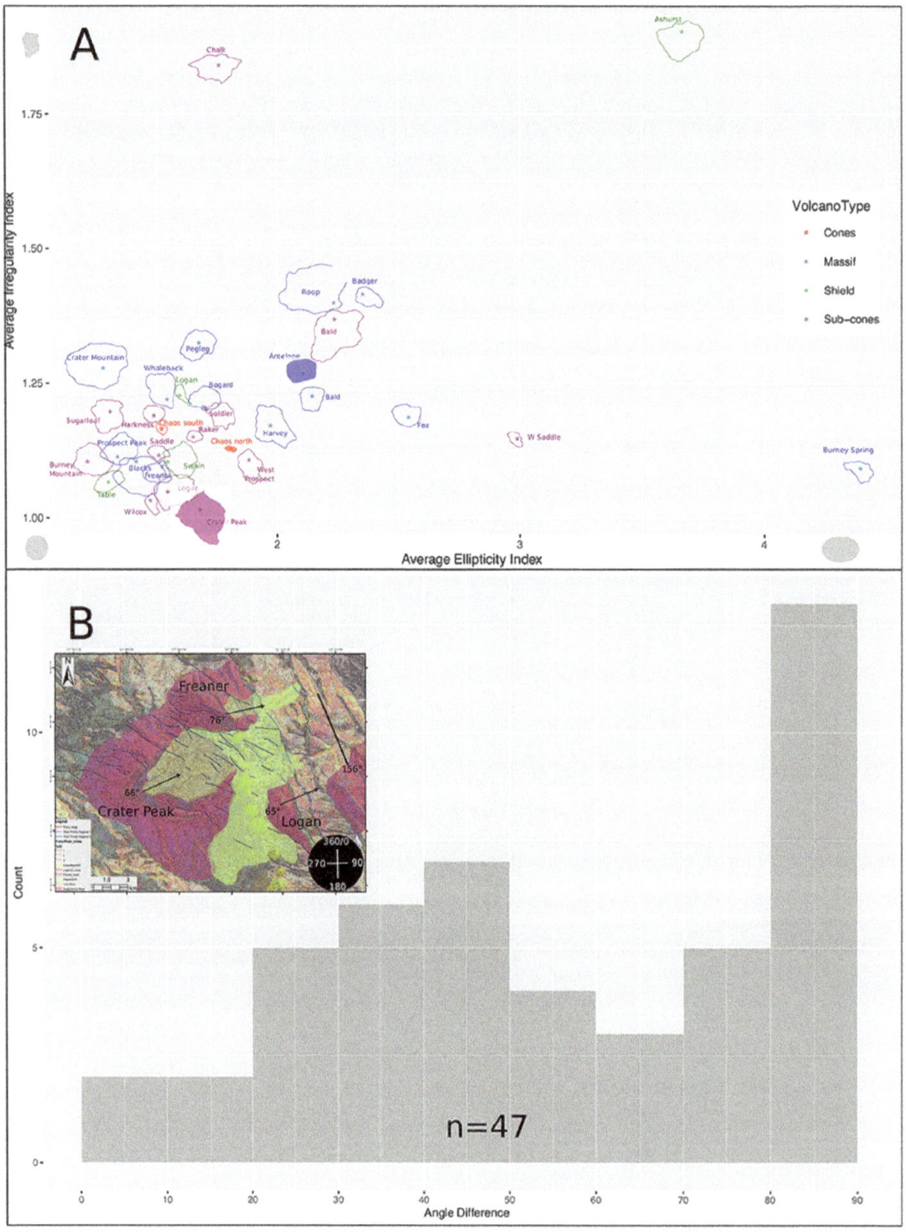

Figure 6. Volcano morphology and breaches. (A) Ellipticity and Irregularity Indices (not to scale) of bigger volcanoes in the Hat Creek Graben region grouped by morphological type as cones, sub-cones, shields and massifs. Filled polygons are those with evidence of collapse scars. **(B)** Frequency diagram showing the relationship between the opening direction of craters and the maximum horizontal stress. Inset is an example of how angle difference is calculated.

Volcano bases are elongated (Figure 4G), with two modes, north-northwest and northeast; most of this scatter is influenced by the trend of the large population of smaller volcanoes. Big volcanoes also generally trend north-northwest, but this trend is less clear. Volcano bases seem to be elongated consistent with the trend of the directly underlying local structures.

Volcano Spatial Distribution and Alignments

The Hat Creek Graben region appears to be a volcanic field with distinctive subgroups of volcanoes, having different spatial distributions. Poisson Nearest Neighbor Analysis was applied to the morphologically smaller and bigger volcanoes as subgroups, and separately, to all volcanoes (Table2). The smaller volcano subgroup and the group of all volcanoes have c values outside the $\pm 2\sigma$ significance level but R less than -2σ, which means they are spatially clustered. The bigger volcano subgroup, however, has c values outside the $\pm 2\sigma$ significance level and R greater than $+2\sigma$, which means that volcanoes in this group are dispersed relative to the Poisson distribution. Based on the density of volcanoes per subgroup, and using the relationship established using the global database of monogenetic volcanic fields (Le Corvec et al., 2013), we used 5800, 9900, and 6000 m as the maximal distance needed for the recognition of lineaments for smaller and bigger subgroups, and the group of all volcanoes, respectively.

The three-point MATLAB script was used to automatically extract the different alignments of the volcanoes, and plot for each the direction of the alignment on a rose diagram (Figure 5D). Bigger volcanoes generated no preferred alignment, while the smaller volcano subgroup and the group of all volcanoes have a north-northwest preferred direction of alignments. Again, this quantitatively determined strike is consistent with the strike of local, tectonic faults.

Crater Morphology and Stress Field

Only the subgroup of bigger volcanoes was considered for investigating crater opening, either a volcano collapse scar or an erosional breach, because of the prevalence of these features on the bigger volcanoes. The regional maximum horizontal stress and crater opening directions, both in azimuth, were noted for each bigger volcano with crater openings in Table 1. For example, as in Figure 6B (inset), the maximum horizontal regional stress has an azimuth of 156° near Freaner, Logan and Crater Peak volcanoes, which have summit openings (erosional breaches for Freaner and Logan, and collapse scar for Crater Peak) of 76°, 65°, and 66°, respectively. The angle difference for each of these three volcanoes is the absolute value of the difference between 156°and the summit opening direction, which is 80°, 89°, and 90°, respectively (Table 1). A frequency diagram (Figure 6B) of the angular difference between the crater opening direction for the bigger volcanoes and the maximum horizontal stress has a mode at about 90°, although crater openings at an acute angle are also found, albeit at lower frequencies. Most of the crater openings are thus perpendicular to the maximum horizontal stress, which is consistent with the observation ofTibaldi (1995) and models by Moriya (1980) and Siebert (1984).

DISCUSSION

Several regional fault systems are oriented northwest-southeast; these are comprised of normal faults with right-lateral components. Individual faults within these systems pass through or underneath most of the volcanoes. The stratovolcanoes contain clear traces of the regional faults, or manifestations of collapse or erosional features on their edifices that can be related to the faults. Cones are more prevalent in the southeastern part of the study area.

Magma Propagation

The Hat Creek Graben region consists of volcano subgroups that have different spatial distributions; smaller volcanoes tend to cluster, while bigger volcanoes are dispersed. The density-dependent, preferred alignments reveal a single north-northwest direction for the small volcano subgroup, and all volcanoes considered together. These observations are consistent with the control of melt production at depth and magma propagation by pre-existing regional structures, and with the control of magma propagation in the near surface by the local stress regime. Melt production is related to the oblique subduction of the Gorda Plate underneath the Northern American Plate, and is manifested in the volcanism at the surface. Regional and near-surface stresses are influenced by Walker Lane and Cascades tectonism. Extension is accommodated by the pervasive faulting and magmatism. As the area extends, it produces weak regions that provide energy-efficient paths along which magma propagates to the surface.

Normal faults, dominant in the area, influence dike orientation and the generation of alignments at the surface, because dikes are more likely to intercept and follow dipping faults than vertical fractures (i.e., Le Corvec et al., 2013). Near-surface stress also affects magma propagation (i.e.,Caputo, 1995) it is possible that high differential stress did not allow stress swap, therefore the intrusions followed the orientation of the earlier dikes. The generation of multiple orientations of volcanic alignments is perhaps attributable to the interaction of near-surface stress field influences and pre-existing fractures.

Magma pressure, which is also a factor in dike propagation (i.e., Jolly and Sanderson, 1997), may be low, so that magma only follows pre-existing fractures that are favorable for reactivation in the prevailing stress field. Although pathways may vary slightly according to stress field at time of propagation and eventual emplacement (i.e., Valentine and Krogh, 2006), in general, they have been consistent through the lifetime of Hat Creek region volcanism.

The propagation of magma that formed the bigger volcanoes is also structurally controlled, because these volcanoes stand atop major regional structures, and the faults affect the stability of the volcanic edifices. The random, dispersed spatial distribution of the bigger volcanoes may be attributed to their long-lived plumbing systems, which are capable of capturing or re-focusing nearby dikes as they propagate to the surface, most of the time erupting from a single vent. Changes in the local stress field caused by the growing edifice may also have enhanced dike capture.

Edifice Evolution

Most volcanoes in the study area are small, with maximum heights about 250–300 m, and volumes generally less than 1 km^3. Based on the height and basal width ratio, some volcanoes can be classified as cones, which are relatively regular and more nearly circular in planview, have relatively smaller summit areas and narrower basal widths compared to sub-cones and to the bigger and more voluminous shields and massif volcanoes. Quantitative morphology furthermore suggests that volcanoes evolve along several paths: small cones continue to grow as cones, or they grow into sub-cones; small sub-cones continue to grow into bigger sub-cones, or they develop into shields and massifs; and shields can grow into massifs. Underlying major faults systems have greatly influenced the growth, partial destruction, and general evolution of volcano morphology.

Quantitative evaluation furthermore revealed volcanic basal elongations in two main directions, north-northwest and northeast. The existence and magnitude of the northeast mode in the elongations is consistent with spreading parallel to displacements on regional normal faults with right-lateral components, either by preferred feeder dike orientation or by deformation.

Deformed and Collapsing Volcanoes

Several types of volcanic edifice deformation are observed in the Hat Creek Graben region. Some volcanoes show collapse features, others erosional breaches, and some are being split by faulting.

Chaos Crags is a stratovolcano with a known collapse scar (Figure 4E). The collapse of the unstable, very steep, soft-cored, and unconsolidated plug dome of Chaos North resulted in a debris avalanche, called the Chaos Jumbles. The jumbles were emplaced as a high yield strength material capable of deforming and shearing in three separate rockfalls about 300 years ago Eppler et al. (1987), damming Manzanita Creek and forming Manzanita Lake. A projection of a northwest striking fault, here part of a Primary Regional 1 group of faults (Figures 4A,E), which splits Table Mountain to the northwest, may be masked underneath the Crags and Jumbles. Post-eruptive mass wasting and sedimentation on the slopes of Chaos Crags have masked any evidence for the mechanism of generation of the debris avalanche (Perez, 1998). Activity on the fault could have triggered instability and collapse.

Crater Peak is another stratovolcano that shows morphological evidence of collapse (Figure 4B). Crater Peak has a large and deep-seated scar, and a possible runout deposit elongated toward the northeast marked by a rough surface texture that would be expected in a debris avalanche deposit. The scar was glacially eroded after the collapse, as shown by the smoother, uppermost parts of the amphitheater, which thus resembles, or is in part, a cirque. Comparing the structures and scar elongation direction of Crater Peak to Figure 9E of Wooller et al. (2009), we speculate that the collapse could have been caused by a basement normal fault, with a right-lateral component, possibly the Rocky Ledge Fault (red line in Figures 4A,B, thicker black line Figure 7A), which passes through the center of the edifice. In this scenario, the Rocky Ledge Fault destabilized and eventually caused a collapse approximately normal to its strike direction.

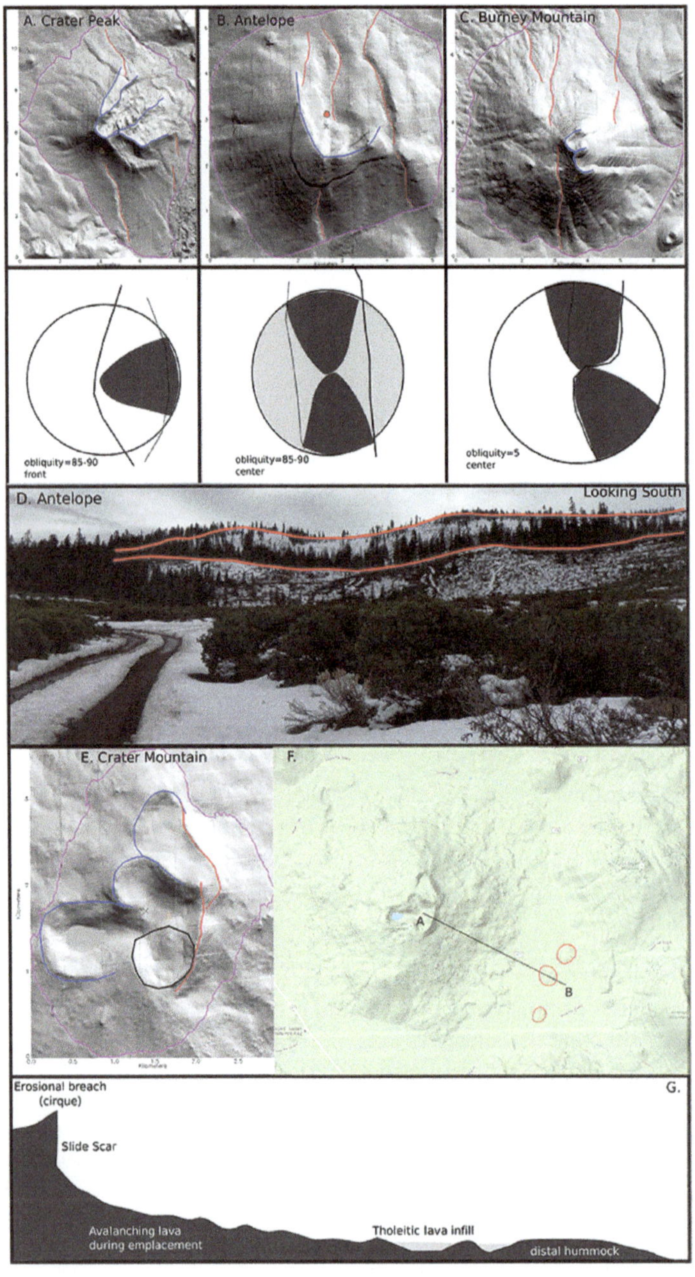

Figure 7. Deformed volcanoes in the Hat Creek Graben region. (A–C) Shaded relief delineated with volcano base (purple), collapse scars or erosional breaches (blue) and structural lineaments (red) for Crater Peak **(B)**, and Antelope **(C)**, and Burney **(D)**Mountains compared with their

predictive model cones by Wooller et al. (2009)underneath, where the gray circles represent the edifice; solid lines are main normal faults; dotted polylines are antithetic and minor faults; and gray polygons are unstable sectors; **(D)**View looking at Antelope Mountain Slide scar from the northern sector of the volcano marked as red dot in **(B)**. The red lines are the slide scar (top), and terrace (bottom) that mark the top of the major listric faults that accommodate the sliding of the whole area. **(E)**Shaded relief and **(F)** google earth images, and **(G)** cross section of Crater Mountain Slide. Location of cross section A-B is drawn over google earth image **(F)**. Volcano base (purple); cirques (blue); scar and hummocks (red); possible crater trace (black) are marked.

Relations, confirmed in the field, showed that Antelope Mountain has also collapsed (Figures 4F,7B,D), with deposits emplaced toward the northeast. From remote sensing, lineaments on the eastern base of the volcano can be traced (Figures 4A,F), of the same set that split Whaleback volcano to the northeast. Another set that perhaps influenced the stability of Antelope Mountain more strikes northwest, splits Logan Mountain to the north, and affects the western flank of Fox Mountain. The structures affecting Antelope Mountain are grouped with the <1600 ka faults, are possibly part of the Walker Spring fault system. These may be related to the faults in the Susanville and Big Lake Valley areas, which may have been active during the Holocene.

Based on remote sensing analysis, four more volcanoes notably show evidence of edifice collapse: Crater, Burney and Table Mountains; and Raker Peak. Burney and Table Mountains are deformed, and show morphological traces of edifice collapse structures (Figures 4B,E). The structures at Burney Mountain appear to be manifestations of ongoing deformation of the edifice by the Rocky Ledge Fault, the underlying basement fault (Figure 4B), which may ultimately result in sector collapse. In the field, Burney Mountain has two small craters that both open to the east. These are likely to have evolved into cirques by small, summit glaciers flowing to the east, leaving behind till, possibly in the form of a moraine, below the northern cirque (Figure 7C). The location of lineaments on the northwest and southeast sectors of Burney Mountain is consistent with underlying normal faults with little strike-slip component (Figure 7C). The lineaments (in yellow) perpendicular to the Rocky Ledge Fault trace may not be tectonic in origin, but rather lava flow ridges (Figure 4B).

Crater Mountain (Figures 4E–G) is a relatively "strong" and stable volcano, built mostly of lava flows, with no apparent hydrothermal alteration at the surface, nor any indication that it is built on deformable and weak substrata. At the summit, three craters have been traced, all elongated to the east, which may be modified by glaciers. There may be another, fourth, crater as well. The head-scarp morphology in the summit area, where the north-south scar is drawn (Figures 7E–G), suggests some type of gravity failure on the east flank. The scarp could have been made by multiple slides, affecting slightly different sectors. It is possible that a series of slides were emplaced without a thin, weak or wet sliding base, which resulted in slumping or rotation of relatively coherent blocks, forming a toe now exposed as three low mounds or distal hummocks (Paguican et al., 2014). Spaces between hummocks have been infilled by tholeiite lavas.

Raker Peak (Figure 4E) is another edifice with an apparent northeastward opening. This sector deformed relative to the others. Although there is no basement fault near the edifice, a river flows across the eastern piedmont, which could have cut the base of the edifice, causing its eventual collapse. It is also possible that a segment of a Primary Regional 2 fault is buried by the volcano.

Last, the southwestern sector of Table Mountain (Figure 4E) has a crater-like structure that opens toward the south. The difference in slope steepness between the upper and lower portions of the edifice within the scar is evident.

Volcanoes with Erosional Breaches and their Underlying Faults

Volcanoes with an erosional breach or particular sectors affected by relatively more gullies include Pegleg, Swain, Roop Mountain, Logan, Campbell, South of Campbell, South of South Campbell, and Harvey Mountain (Figure 4F); Chalk and Bald Mountains in the north (Figure 4C); Raker Peak (Figure 4E), Cone and Cal Mountains (Figure 4F), Soldier, Haney and, Saddle Mountains (Figure4D), and Burney Spring Mountain (Figure 4C) in the east.

Harvey Mountain is bounded to the east by Ashurst volcano. These two stratovolcanoes are separated by a segment of a <1600 ka regional fault that also affects Logan, Antelope and Fox Mountains. On the northeast is a possible erosional feature most likely to develop under the influence of an active normal fault with a strike-slip component. Logan, Campbell and South of Campbell all have erosional features opening toward the northeast or northwest. Of these, only Logan Mountain is clearly traversed by a <1600 ka fault, the same fault that splits Fox Mountain and is related to the slide on Antelope Mountain. The volcano to the south of South Campbell has no apparent nearby fault to which we can relate the north and northwest scars. However, a river flows near the western base, which might have laterally undermined the volcano base, causing instability on the northwestern flank. The western flank of Roop Mountain has gullies. The northeastern sector is the most incised by deep gullies, with major gully heads. Based on its relatively bigger size and volume compared to other volcanoes, and the density and size of gullies, Roop Mountain is probably an old volcanic edifice.

Swain Mountain has traces of two fault on the southwestern and northeastern piedmonts. Two erosional scars are shaped like cirques and have gullies within their walls. Breaks in slope that create benches or terraces within the scar may indicate that this is the most deformed part of the edifice, causing it to be more easily affected by erosional processes. The locations and opening directions of the scars are consistent with deformed sectors of volcanoes affected by normal faults with a strike-slip component occurring at the base and facing toward the edifice (Wooller et al., 2009). Pegleg Mountain also has two scar traces but unlike Swain Mountain, there is no nearby fault that can be explicitly connected to it. Thus, all sectors of the volcano are affected by gullies.

In the northern portion of the Hat Creek Graben region, where both the Rocky Ledge and Hat Creek faults terminate, and more dense Primary Regional 1 and 2 faults, and other Quaternary faults, are delineated, Chalk, Bald, Soldier, Haney, Saddle and Burney Spring Mountains are observed to be heavily traversed by the regional structures. In addition, the volcanoes occur along a major river (the Pit River), so they have many more erosional features and gullies in almost all sectors. They are therefore more affected by both frequent sliding and gully formation than other volcanoes.

On the eastern side of the study area, two volcanic edifices, Cone and Cal Mountains, each, have two erosional scars opening in opposite directions. There is no clear trace of a regional fault or any manifestation of movement of possible faults except for these scars.

Splitting Volcanoes, and the Faults Splitting Them

A few volcanoes are either very young, composed of strong and consolidated materials, or constructed on relatively stable basements. Even though they stand atop regional faults, may remain intact, except for splitting, in which some parts of the edifice move vertically relative to adjacent areas. Splitting volcanoes are affected by through-going faults, deforming yet not causing major instability or collapse. Whaleback, Fox, and Bogard (Figure 4F) Mountains are examples of this. Freaner and Wilcox (Figure 4B) are splitting elliptical volcanoes, as shown by vertical differences between adjacent parts of the edifices. Some volcanoes, such as West Prospect Peak, Prospect Peak (Figure 4E), and Sugarloaf (Figure 4B), have retained their ellipticity and regularity, even though faults split them.

Fault vs. Dike Dominated Regional Extension

In magmatic slow and ultraslow spreading plate boundaries and rift zones, extension is often accommodated by magmatic activity, or diking with associated faulting (Acocella and Trippanera, 2016, and references therein). Normal faults that propagate from the surface downwards and create graben-like structures on the surface are recurrent structural features. Symmetric, minor grabens are observed in Lakagigar and Bardarbunga in Iceland, and Dallol in Afar, induced by single diking episodes. Asymmetric, larger grabens can be seen in Krafla, Iceland, and Fantale, Ethiopia, formed by repeated or distributed diking. At these spreading plate boundaries, extension at depth seems to be accommodated mostly by diking, with normal faulting playing a negligible role. In shallower layers, normal faults and extensional tectonic structures accommodate the deformation, alongside a few feeder dikes (Acocella and Trippanera, 2016, and references therein).

While we cannot rule out the idea that dikes play a major role in extension at depth in the Hat Creek Graben region, and that some of the shorter faults might have been generated by diking episodes, it seems that at shallower levels, regional normal faults accommodate most of the extension. The extent of the Hat Creek Graben region, its regional fault systems with total displacements as high as a few hundred meters, recurrence intervals of a few 100 years, and the influence of the regional fault systems on the development of the volcanoes furthermore suggest that it is unlikely that most accommodation is by diking episodes, even deeper in the crust. The Hat Creek Graben region and the East African Rift are spreading at similar rates of 2.9 ± 0.8 mm yr^{-1}(Zeng and Shen, 2014) and 3–4 mm yr^{-1} (Stamps et al., 2008), respectively. However, it appears that the Hat Creek Graben region has a lower magma supply rate because it has smaller volcanoes than does the East African Rift (Grosse et al., 2014). The Hat Creek Graben region presents as an example of a different mechanism of extension in a small rift system associated with a transform plate boundary.

CONCLUSIONS

Topographic mapping and modeling, based mostly on NED data, were used to describe the geomorphology and structures of the Hat Creek Graben region. In addition to the obvious, known major geomorphological features, such as Lassen Peak, Pit River and Hat Creek valleys, smaller-scale geomorphological features, including mountain ridges, volcanic edifices, and lava fields, were characterized in detail using different topographic modeling techniques. Initial assessment of the shape and regularity of the volcano bases gave an overview of the processes that have affected or are currently affecting the volcanoes. Erosional features and the density of structures delineated on volcanoes were also noted. The study can serve as a standard for other volcanic areas in the world with strong tectonic interaction.

Using the height and basal width ratios from the morphometric analysis allowed us to classify the individual volcanic edifices of the Hat Creek Graben region as cones, sub-cones, shields and massifs. Separating those less than 350 m high as smaller cones, sub-cones and shields revealed an evolutionary trend of volcanic forms in the region: smaller cones can grow into bigger cones if they increase their height in proportion to their basal width, or evolve into sub-cones if their basal width increases faster than their height. Smaller sub-cones can grow into bigger sub-cones, or grow into the wide variety of shields or massifs; and smaller shields can grow into larger shields or massifs. As volcanoes develop and grow on top of major regional faults, their activity and growth pattern are influenced by regional extension accommodated on north-northwest striking faults that are mostly normal with strike-slip component. The stress regime and fault orientation can cause edifice deformation by collapse, erosion, or splitting. The majority of the deformational features, often manifested as collapse scars or erosional breaches, have opening directions perpendicular to the maximum horizontal stress, similar to those in other extensional settings (Tibaldi, 1995) and may, therefore, be related to the normal component of movement on the underlying faults. A considerable number, however, open at an acute angle and therefore are affected by the right-lateral component of movement on the underlying faults.

Spatial distribution varies for the volcano subgroups in the Hat Creek Graben region. The smaller volcanoes subgroup, and the combined smaller and bigger volcanoes, are clustered with northwest-southeast alignments formed by at least three volcanoes within 6 km of one another, while the bigger volcanoes subgroup is random and dispersed. This implies that the initial and clustered growth of volcanoes in the Hat Creek Graben region is controlled by the pre-existing major structures. Thus, the regional structures and volcanoes within the Hat Creek Graben region are interacting. The spatial dispersion of the bigger volcanoes can be attributed to volcanogenic, near-surface pressure changes—as volcanoes grow and long-lived plumbing systems develop—capable of refocusing local dikes. Thus, in the case of the Hat Creek Graben region, the regional, pre-existing major faults within the crust greatly influenced the initial alignment and distribution of volcanoes, and their continued growth is closely related to local as well as regional strain. Faults (by slip) and volcanoes (by increasing total crustal volume through dilation during intrusion and propagation of magma to the surface) both accommodate regional extension in the Hat Creek Graben region. However, extension is accommodated more by faulting than by dikes, which along with the small size of the stratovolcanoes, suggests that the magma supply rate is low, with low-pressure, as magma propagates to the surface using the most efficient, easiest, pre-existing fractures. Thus, the propagation of magma to the surface through a complex plumbing system has been insufficient to accommodate much of the extension, which is therefore dominated by normal and other extensional, purely tectonic structures. This is manifested by the dominantly fault-related topography of the Hat Creek Graben region, and the overwhelming control of the tectonic structures on the growth, development, and evolution of volcanoes. The tectonic structures should continue to have a great influence on the stability and deformation of the volcanoes into the future.

We recognize that a more comprehensive factoring in of temporal data, such as description of erosion rates, and dating of collapse events relative to the age of fault activity, will be important for a complete characterization of the morphological and morphometric evolution of the Hat Creek Graben region.

Author Contributions

EP and MB both contributed to the conception and design of the work, acquired, analyzed, and interpreted the data for the work. EP executed the detailed analyses, and drafted the manuscript. Revisions were made by both EP and MB. We both approved the version of the manuscript we submitted and we will both be accountable for the accuracy and integrity of this work.

Conflict of Interest Statement

The authors declare that the research was conducted in the absence of any commercial or financial relationships that could be construed as a potential conflict of interest.

ACKNOWLEDGMENTS

We thank Laura Pioli and Gianluca Groppelli, and the Associate Editor, Dr. Roberto Sulpizio for their comprehensive and constructive reviews. We also thank the Editor in Chief, Dr. Valerio Acocella for his helpful comments. Dr. G. N. Fabbro provided essential help in formatting technicalities and improving the drafts. We are grateful for field assistance by Mr. G. Bursik and Ms. Leandra Marshall. This research was funded in part by NASA grant NNX12AQ10G, and by the State University of New York.

REFERENCES

1. Acocella, V., and Trippanera, D. (2016). How diking affects the tectonomagmatic evolution of slow spreading plate boundaries: overview and model. *Geosphere* 12:GES01271.1. doi: 10.1130/GES01271.1
2. Argus, D. F., and Gordon, R. G. (1991). Current Sierra Nevada-North America motion from very long baseline interferometry: implications for the kinematics of the western United States. *Geology* 19, 1085–1088.
3. Austin, L. J. (2013). "Fault kinematics from LiDAR analysis of the Rocky Ledge Fault, Shasta County, California," in *Geological Society of America Abstracts with Programs* (Denver, CO: Geological Society of America), 546.
4. Bahar, I., and Girod, M. (1983). Controle structural du volcanisme indonesien (Sumatra, Java-Bali); application et critique de la methode de Nakamura. *Bull. Soc. Geol. Fr.* XXV, 609–614. doi: 10.2113/gssgfbull.S7-XXV.4.609
5. Baloga, S. M., Glaze, L. S., and Bruno, B. C. (2007). Nearest-neighbor analysis of small features on Mars: applications to tumuli and rootless cones. *J. Geophys. Res. E Planets* 112, 1–17. doi: 10.1029/2005JE002652
6. Beggan, C., and Hamilton, C. W. (2010). New image processing software for analyzing object size-frequency distributions, geometry, orientation, and spatial distribution. *Comput. Geosci.* 36, 539–549. doi: 10.1016/j.cageo.2009.09.003
7. Bennett, R. A., Davis, J. L., and Wernicke, B. P. (1999). Present-day pattern of Cordilleran deformation in the western United States. *Geology* 27, 371–374.
8. Blakely, R. J., Christiansen, R. L., Guffanti, M., Wells, R. E., Donnelly-Nolan, J. M., Muffler, L., et al. (1997). Gravity anomalies, Quaternary vents, and Quaternary faults in the southern Cascade Range, Oregon and California: implications for arc and backarc evolution. *J. Geophys. Res.* 102:22513. doi: 10.1029/97JB01516
9. Blakeslee, M. W., and Kattenhorn, S. A. (2013). Revised earthquake hazard of the Hat Creek fault, northern California: a case example of a normal fault dissecting variable-age basaltic lavas. *Geosphere* 9, 1397–1409. doi: 10.1130/GES00910.1
10. Bleacher, J. E., Glaze, L. S., Greeley, R., Hauber, E., Baloga, S. M., Sakimoto, S. E. H., et al. (2009). Spatial and alignment analyses for a field of small volcanic vents south of Pavonis Mons and implications for the Tharsis province, Mars. *J. Volcanol. Geotherm. Res.* 185, 96–102. doi: 10.1016/j.jvolgeores.2009.04.008
11. Caputo, R. (1995). Evolution of orthogonal sets of coeval extension joints. *Terra Nov.* 7, 479–490. doi: 10.1111/j.1365-3121.1995.tb00549.x
12. Clark, P. J., and Evans, F. C. (1954). Distance to nearest neighbor as a measure of spatial relationships in populations. *Ecol. Soc. Am.* 35, 445–453. doi: 10.2307/1931034
13. Connor, C. B., Condit, C. D., Crumpler, L. S., and Aubele, J. C. (1992). Evidence of regional structural controls on vent distribution: springerville volcanic field, Arizona. *J. Geophys. Res.* 97, 12349–12359. doi: 10.1029/92JB00929
14. Dixon, T. H., Miller, M., Farina, F., Wang, H., and Johnson, D. (2000). Present-day motion of the Sierra Nevada block and some tectonic implications for the Basin and Range province, North American Cordillera. *Tectonics* 19:1. doi: 10.1029/1998TC001088
15. Dokka, R., and Travis, J. (1990). Role of the eastern california shear zone in integrated shear along the to historic measurements mm yr present rate mm yr 4

determined. *Geophys. Res. Lett.* 17, 1323–1326. doi: 10.1029/GL017i009p01323

16. Ebinger, C., Ayele, A., Keir, D., Rowland, J., Yirgu, G., Wright, T., et al. (2010). Length and timescales of rift faulting and magma intrusion: the afar rifting cycle from 2005 to present. *Annu. Rev. Earth Planet. Sci.* 38, 439–466. doi: 10.1146/annurev-earth-040809-152333
17. Eppler, D. B., Fink, J., and Fletcher, R. (1987). Rheologic properties and kinematics of emplacement of the chaos jumbles rockfall avalanche, Lassen Volcanic National Park, California. *J. Geophys. Res.* 92:3623. doi: 10.1029/JB092iB05p03623
18. Faulds, J. E., Henry, C. D., and Hinz, N. H. (2005). Kinematics of the northern Walker Lane: An incipient transform fault along the Pacific-North American plate boundary. *Geology* 33, 505–508. doi: 10.1130/G21274.1
19. Francis, P., and Wells, A. (1988). LANDSAT Thematic Mapper observa- tions of debris avalanche deposits in the Central Andes.*Bull. Volcanol.* 50, 258–278. doi: 10.1007/BF01047488
20. Gesch, D., Oimoen, M., Greenlee, S., Nelson, C., Steuck, M., and Tyler, D. (2002). The national elevation dataset. *Photogramm. Eng. Remote Sens.* 68, 5–32.
21. Grosse, P., Euillades, P. A., Euillades, L. D., and van Wyk de Vries, B. (2014). A global database of composite volcano morphometry. *Bull. Volcanol.* 76:784. doi: 10.1007/s00445-013-0784-4
22. Grosse, P., van Wyk de Vries, B., Euillades, P. A., Kervyn, M., and Petrinovic, I. A. (2012). Systematic morphometric characterization of volcanic edifices using digital elevation models. *Geomorphology* 136, 114–131. doi: 10.1016/j.geomorph.2011.06.001
23. Grosse, P., van Wyk de Vries, B., Petrinovic, I. A., Euillades, P. A., and Alvarado, G. E. (2009). Morphometry and evolution of arc volcanoes. *Geology* 37, 651–654. doi: 10.1130/G25734A.1
24. Hamilton, C. W., Fagents, S. A., and Thordarson, T. (2010). Explosive lavawater interactions II: self-organization processes among volcanic rootless eruption sites in the 17831784 Laki lava flow, Iceland. *Bull. Volcanol.* 72, 469–485. doi: 10.1007/s00445-009-0331-5
25. Hammond, W. C., and Thatcher, W. (2005). Northwest Basin and Range tectonic deformation observed with the Global Positioning System, 1999-2003. *J. Geophys. Res. Solid Earth* 110, 1–12. doi: 10.1029/2005JB003678
26. Hammond, W. C., and Thatcher, W. (2007). Crustal deformation across the Sierra Nevada, northern Walker Lane, Basin and Range transition, western United States measured with GPS, 2000-2004. *J. Geophys. Res. Solid Earth* 112, 1–26. doi: 10.1029/2006JB004625
27. Heidbach, O., Tingay, M., Barth, A., Reinecker, J., Kurfeß, D., and Müller, B. (2008). *The World Stress Map Database Release 2008*. doi: 10.1594/GFZ.WSM.Rel2008
28. Jenkins, O. (1938). *Geomorphic Map of California: California Division of Mines and Geology Map Sheet, Scale Approximately 1:2,000,000.*
29. Jolly, R., and Sanderson, D. (1997). A Mohr circle construction for the opening of a pre-existing fracture. *J. Struct. Geol.* 19, 887–892. doi: 10.1016/S0191-8141(97)00014-X
30. Lagmay, A. M. F., and Valdivia, W. (2006). Regional stress influence on the opening direction of crater amphitheaters in Southeast Asian volcanoes. *J. Volcanol. Geotherm. Res.* 158, 139–150. doi: 10.1016/j.jvolgeores.2006.04.020
31. Le Corvec, N., Spörli, K. B., Rowland, J., and Lindsay, J. (2013). Spatial distribution and alignments of volcanic centers: clues to the formation of monogenetic volcanic fields. *Earth-Sci. Rev.* 124, 96–114. doi: 10.1016/j.earscirev.2013.05.005
32. McCaffrey, R. (2005). Block kinematics of the Pacific-North America plate boundary in the southwestern United States from inversion of GPS, seismological, and geologic data. *J. Geophys. Res. Solid Earth* 110, 1–27. doi: 10.1029/2004JB003307
33. Miliaresis, G., and Kokkas, N. (2004). Segmentation and terrain modeling of extraterrestrial chasmata. *J. Spat. Sci.* 49, 89–99. doi: 10.1080/14498596.2004.9635024
34. Miller, M., Johnson, D. J., Rubin, C. M., Dragert, H., Wang, K., Qamar, A., et al. (2001). GPS-determination of along-strike variation in Cascadia margin kinematics: implications for relative plate motion, subduction zone coupling, and permanent deformation. *Tectonics* 20, 161–176. doi: 10.1029/2000TC001224
35. Moriya, I. (1980). *Bandaian Eruption and Landforms Associated with It: Collection of Articles in Memory of Retirement of Prof. K. Hishimura*. Technical Report, Tohoku University.
36. Muffler, L., Blakely, R. J., and Clynne, M. (1994). Late Quaternary normal faulting of the Hat Creek Basalt, Northern California.*Geol. Soc. Am. Bull.* 106, 195–200.
37. Muffler, L., Robinson, J., Felgar, T., Dutton, D., and Clynne, M. (2010). *Database for the Geologic Map of Lassen Volcanic National Park and Vicinity, California.* Data to accompany Scientific Investigations Map 2899. Available online at:http://pubs.usgs.gov/sim/2899/database.html
38. Nakamura, K. (1977). Volcanoes as possible indicators of tectonic stress orientation-principle and proposal. *J. Volcanol. Geotherm. Res.* 2, 1–16. doi: 10.1016/0377-0273(77)90012-9
39. Paguican, E. M. R., van Wyk de Vries, B., and Lagmay, A. M. F. (2014). Hummocks: how they form and how they evolve in rockslide-debris avalanches. *Landslides* 11, 67–80. doi: 10.1007/s10346-012-0368-y
40. Parrish, J. G. (2006). *Simplified Geologic Map of California.* California Geological Survey, Map Sheet 57. Available online at:http://www.conservation.ca.gov/cgs/information/publications/ms/documents/ms057.pdf
41. Perez, F. L. (1998). Talus fabric, clast morphology, and botanical indicators of slope processes on the chaos crags (California cascades), U.S.A. *Geogr. Phys. Quat.* 52, 1–22. doi: 10.7202/004861ar
42. Peterson, J., and Martin, L. (1980). *Geologic Map of the Baker-Cypress BLM Roadless Area and Timbered Crater*

Rare II Areas, Modoc, Shasta, and Siskiyou Counties, Northeastern California. U.S. Geological Survey, Miscellaneous Field Studies Map 1214-A. Available online at: https://pubs.er.usgs.gov/publication/mf1214A

43. Sauber, J., Thatcher, W., Solomon, S., and Lisowski, M. (1994). Geodetic slip rate for the Eastern California shear zone and the recurrence time of Mojave Desert earthquakes. *Nature* 367, 264–266. doi: 10.1038/367264a0
44. Sawyer, T., and Ramelli, A. (2012). *Structural Analysis, Evaluation, and Quaternary Activity of the Hat Creek and Rocky Ledge Faults, Shasta County, Northeastern California.* Technical Report, Piedmont Geosciences Report, October 2012.
45. Siebert, L. (1984). Large volcanic debris avalanches: characteristics of source areas, deposits, and associated eruptions. *J. Volcanol. Geotherm. Res.* 22, 163–197. doi: 10.1016/0377-0273(84)90002-7
46. Sigmundsson, F., Hooper, A., Hreinsdóttir, S., Vogfjörd, K. S., Ófeigsson, B. G., Heimisson, E. R., et al. (2015). Segmented lateral dyke growth in a rifting event at BárÃřarbunga volcanic system, Iceland. *Nature* 517, 191–195. doi: 10.1038/nature14111
47. Stamps, D. S., Calais, E., Saria, E., Hartnady, C., Nocquet, J. M., Ebinger, C. J., et al. (2008). A kinematic model for the East African Rift. *Geophys. Res. Lett.* 35, 1–6. doi: 10.1029/2007GL032781
48. Tibaldi, A. (1995). Morphology of pyroclastic cones and tectonics. *J. Geophys. Res.* 100, 24521–24535. doi: 10.1029/95JB02250
49. Ui, T., Yamamoto, H., and Suzuki-Kamata, K. (1986). Characterization of debris avalanche deposits in Japan. *J. Volcanol. Geotherm. Res.* 29, 231–243. doi: 10.1016/0377-0273(86)90046-6
50. U.S. Geological Survey California Geological Survey (2006). *Quaternary Fault and Fold Database for the United States.* Available online at: http://earthquake.usgs.gov/hazards/qfaults/
51. Valentine, G., and Krogh, K. (2006). Emplacement of shallow dikes and sills beneath a small basaltic volcanic center The role of pre-existing structure (Paiute Ridge, southern Nevada, USA). *Earth Planet. Sci. Lett.* 246, 217–230. doi: 10.1016/j.epsl.2006.04.031
52. Valentine, G. A., and Hirano, N. (2010). Mechanisms of low-flux intraplate volcanic fields - Basin and Range (North America) and northwest Pacific Ocean. *Geology* 38, 55–58. doi: 10.1130/G30427.1
53. Wadge, G., and Cross, A. (1988). Quantitative methods for detecting aligned points: an application to the volcanic vents of the Michoacan-Guanajuato volcanic field, Mexico. *Geology* 16, 815–818.
54. White, I., and Crider, J. (2006). Extensional fault-propagation folds: mechanical models and observations from the Modoc Plateau, Northeastern California. *J. Struct. Geol.* 28, 1352–1370. doi: 10.1016/j.jsg.2006.03.028
55. Wills, C. J. (1991). Active faults north of Lassen Volcanic National Park, Northern California. *Calif. Geol.* 44, 49–72.
56. Wooller, L., van Wyk de Vries, B., Cecchi, E., and Rymer, H. (2009). Analogue models of the effect of long-term basement fault movement on volcanic edifices. *Bull. Volcanol.* 71, 1111–1131. doi: 10.1007/s00445-009-0289-3
57. Wright, T. J., Ebinger, C., Biggs, J., Ayele, A., Yirgu, G., Keir, D., et al. (2006). Magma-maintained rift segmentation at continental rupture in the 2005 Afar dyking episode. *Nature* 442, 291–294. doi: 10.1038/nature04978
58. Zeng, Y., and Shen, Z. K. (2014). Fault network modeling of crustal deformation in California constrained using GPS and geologic observations. *Tectonophysics* 612–613, 1–17. doi: 10.1016/j.tecto.2013.11.030

CHAPTER 16 IMPACT OF 2004 TSUNAMI IN THE ISLANDS OF INDIAN OCEAN: LESSONS LEARNED

Georges Ramalanjaona

INTRODUCTION

Tsunami is a series of ocean waves typically caused by large undersea earthquakes or volcano eruptions at tectonic plate boundaries. These surges of water may reach 100 feet and cause widespread destruction when they crash ashore. They race across the sea at a speed up to 500 miles per hour and cross the entire Pacific Ocean in less than one day. Their long wavelength means that they lose very little energy along the way.

Tsunami of December 2004, caused by a 9.0 magnitude earthquake, is the most infamous tsunami of modern times with disastrous consequences in many areas [1](i)humanitarian toll: it affected more than 18 countries from Southeast Asia to Southern Africa, killing more than 250,000 people in a single day and leaving more than one million homeless,(ii)economic toll: it left several million of dollars of economic loss affecting fishing and tourist industries,(iii)environmental and medical threats including water pollution and flooding and endemic diseases.

The rationale for writing this paper is to report the tsunami events in the eleven nations bordering the Indian Ocean, as they received less publicity than their Southeast Asian countries counterpart although the 2004 tsunami had real humanitarian, economic, and environmental impact in these regions more than 1,000 miles away from the epicenter [2].

Furthermore, these regions are at risk from the devastating effects of future tsunami due to the presence of a tectonic interactive plate [3], absence of a tsunami warning system in the Indian Ocean, and lack of established communication network providing timely information to that region.

METHODOLOGY

This paper is a review of documents collected by WHO and other organizations/authors involved in disaster management during the 2004 tsunami.

Citation: Georges Ramalanjaona, "Impact of 2004 Tsunami in the Islands of Indian Ocean: Lessons Learned," Emergency Medicine International, vol. 2011, Article ID 920813, 3 pages, 2011. doi:10.1155/2011/920813.

RESULTS AND DISCUSSION

Impact of Tsunami in the Islands of the Indian Ocean

These eleven countries bordering the Indian Ocean are Mauritius, Madagascar, Reunion Island, and Seychelles, Comoros islands and by geographical extension include countries in southern borders of Africa such as Somalia, Tanzania, Mozambique, and South Africa.

These individual countries suffered humanitarian loss with more than 3,000 people killed and left more than 10,000 homeless about 1,000 miles away from epicenter. In terms of economic toll, several million dollars were reported accompanied by environmental threat due to flooding.

Specifically included is a country by country report [4] with other south-Asian countries.

(i) Mauritius

Large waves completely submerged one village in north of the island. Although there was no death published, a significant economic loss in millions of dollars was reported.

(ii) Madagascar

Waves up to 10 meters were seen in southeastern region of the island. There was one death and more than 1,000 people homeless. Furthermore, there were considerable economic damages inflicted in touristic and fishing industries and infrastructure disruptions due to flooding and beach erosion (http://savannah.gatech.edu/cee/groups/tsunami/madagascar.html).

(iii) Reunion Island

It suffered mostly economic damages over one million dollars involving fishing industries with more than 200 boats sunk. No deaths were reported.

(iv) Seychelles

Ten people were reported killed, and flooding destroyed a major bridge between the capital Port Victoria and main airport. Also, the island reported devastating economic loss in millions of dollars due to hotels, housing, public utilities, and fishing damages.

(v) Somalia

More than 300 deaths were reported and 5,000 displaced.

(vi) Tanzania

Tsunami killed ten people with unknown number missing along with significant economic damages.

(vii) Kenya

Two deaths and two injured people were reported.

(viii) South Africa

8 people were killed about 8,000 km away from the epicenter.

(ix) Indonesia

122,232 deaths and 113,937 missing.

(x) Shri Lanka

30,974 killed and 4,698 missing.

(xi) Thailand

5,395 killed and 2,993 missing.

(xii) Maldives

82 deaths and 26 missing.

(xiii) Malaysia

68 deaths reported.

(xiv) Myanmar

59 killed.

(xv) Bangladesh

2 killed.

(xvi) Burma

90 killed.

Lessons Learned from 2004 Tsunami

To prevent the devastating effects of future tsunami, these islands of Indian Ocean have set their priorities in achieving 3 goals [5]:(i)development of disaster tsunami program which include implementation of tsunami program at national level, regional, and international levels and coordination of all these programs,(ii)development of an Indian Ocean early warning system,(iii)development of tsunami research program.

National Level

The most studied plans are the Madagascar plan, the tsunami early warning and response system in Mauritius, and the creation of the Department of Risk and Disaster Management in Seychelles.

(i) Madagascar Plan

It was developed in 2006 and is the most exhaustive of all the other national plans and should serve as a model for other islands. It includes 5 objectives:(a)development of national evacuation plan on tsunami,(b)establishment of early warning system in conjunction with regional system,(c)increase public and community awareness through publication and training of media and local authorities,(d)conduct mock exercises on tsunami,(e)strengthen the operational capacity of national meteorological service to include national warning system.

Regional Level

Disaster management is a regional priority in the Indian Ocean due to permanent threat of cyclones, floods, and tsunamis. The stated two goals set by a series of regional meetings in 2005 and 2006 are [6] the following:(i)implementation of Indian Ocean tsunami warning and mitigation system (IOTWS), which focused on defining disaster management and reduction (prevention, mitigation, response and relief) of disaster by all the participating countries,(ii)development of integrated regional information network (IRIN) with the goals of creation of an early warning system for the islands in the Indian Ocean and ensuring adequate equipment to manage natural disasters including tsunamis.

The important issues are the cost of establishing such warning system in the Indian Ocean, the transfer of existing technology versus improving, old one, global warming and extreme weather events in that region.

International Level

A series of international meetings have been convened to discuss the role international organizations [7]:(i) international coordination meeting sponsored by UNESCO intergovernmental oceanographic commission in Mauritius in 2005 with a dual goal:(a)development of tsunami warning and mitigation system,(b)coordination of national tsunami warning center with regional centers,(ii)international strategy for disaster reduction (UN/ISDR) attended by representatives from the Indian Ocean countries and international experts on early warning system in 2006 with two objectives:(a) funding of projects and rehabilitation of roads and bridges(b) increase public awareness and training of key staff in tsunami preparedness and warning at all levels.

Future and Challenges

The main challenge for all the islands of the Indian Ocean is to coordinate all the national efforts with existing regional and international endeavors to meet their stated priorities before the next tsunami events.

The role of one special group of physicians should be mentioned at all these levels.

Emergency physicians are knowledgeable on the risks of tsunami and are trained in the field of disaster management, thus they are true expert. They should get involve as leaders in local, national, and international organizations as resources in disaster management as well as humanitarian institutions such as Red Cross.

CONCLUSION

This paper is the first peer-reviewed paper on the impact of the 2004 tsunami on the islands bordering the Indian Ocean and the lessons learned from this event from national, regional, and international organizations to prevent such events from occurring again in the future.

Tsunami is an ever-present and real threat for the these islands of the Indian Ocean due to the presence of a tectonic interactive plate.

Their disaster management priority is the development of an early tsunami warning system in order to effectively and timely communicate with all the people in that region.

Disaster management should involve national, regional, and international organizations at all levels in order to develop tsunami program, fund tsunami projects, and continue research program.

REFERENCES

1. "WHO- Situation report on tsunami," http://www.who.int/hac/crises/international/asia_tsunami/en/.
2. "The world wide web virtual library-2004 tsunami disaster –www.vl," scholarly and factual analyses, March 2005.
3. D. E. James, "Imaging crust and upper mantle beneath southern Africa: the southern Africa broadband seismic experiment," Leading Edge, vol. 22, no. 3, pp. 238–249, 2003. View at Google Scholar · View at Scopus
4. "Indian Ocean tsunami disaster December 26,2004 and reconstruction-Coordinates," Online Journal of Map and Geographic Round Table, American Library Association February 2011, http://www.stonybrook.edu/libmap/coordinates.htm.
5. E. A. Okal, et al., "Field survey of the 2004 Indonesian Tsunami in Madagascar," Earthquake Spectra, vol. 22, no. S3, pp. S263–S283, 2006. View at Google Scholar
6. M. P. Escaleras and C. A. Register, "Mitigating natural disasters through collective action: the effectiveness of Tsunami early warnings," Southern Economic Journal, vol. 74, no. 4, pp. 1017–1034, 2008.View at Google Scholar · View at Scopus
7. K. Suter, "After the tsunami: the politics of international relief," Contemporary Review, vol. 286, March 2005. View at Google Scholar

CHAPTER 17

PLATE BOUNDARY DEFORMATION FROM A 3-D VELOCITY FIELD INVERSION OF INSAR AND GPS

Carolina Pagli, Hua Wang, Tim J. Wright, Eric Calais, and Elias Lewi

Keywords: plate spreading;magma processes;postrifting

INTRODUCTION

Established plate spreading is cyclical; the rates and spatial distribution of strain vary greatly during each phase of the cycle, as different tectonic and magmatic mechanisms control the deformation [*Sigmundsson*, **2006**; *Wright et al.*, **2012**]. The cyclicity of plate spreading has been appreciated following diking events at a few places: in Krafla (Iceland, 1975–1984), in Asal–Ghoubbet (Djibouti, 1978), and Dabbahu (Afar, 2005–2010) [*Sigmundsson*, **2006**; *Vigny et al.*, **2007**; *Wright et al.*, **2012**]. Three phases have been identified during a full plate spreading cycle: interdiking, codiking, and postdiking. During the interdiking phase, extension across the plate margin is steady, gradually building up tensile stresses in the elastic layer. These stresses are then released through repeated dike injections and eruptions (codiking phase) [e.g., *Wright et al.*, **2006**]. A transient postdiking period follows, with extension across the plate margin occurring at a rate several times higher than the long-term spreading rate [e.g., *Foulger et al.*, **1992**].

The transient postdiking deformation has been explained assuming two different mechanisms. In Iceland, it has been largely explained by the slow relaxation of stresses induced in the ductile part of the crust and mantle by earlier dike intrusions in the elastic layer [*Foulger et al.*, **1992**;*Heki et al.*, **1993**; *Hofton and Foulger*, **1996**]. Conversely, both viscoelastic stress relaxation [*Vigny et al.*, **2007**] and magma flow into the deep crust [*Cattin et al.*, **2005**] have been invoked as the cause of the accelerated deformation following the 1978 Asal–Ghoubbet episode. In Afar, the onset of the rifting episode in 2005 was marked by a large magma intrusion (2–2.5 km^3) such that has caused the initiation of a postdiking phase. At the same time, repeated intrusions have also continued occurring during the 2005–2010 time period, further contributing to the postdiking deformation. The postdiking deformation has been modeled assuming either viscoelastic stress relaxation or continued dike injection at depth [*Grandin et al.*, **2010**; *Nooner et al.*, **2009**] as well as a combination of stress relaxation and inflation/deflation of magma chambers [*Hamling et al.*, **2014**].

Afar has been created by the separation of the Nubian, Arabian, and Somalian Plates during the past 30 Ma (Figure 1a) [*Barberi and Varet*, **1977**; *Courtillot*, **1980**; *Ebinger et*

Citation: Pagli, C., H. Wang, T. J. Wright, E. Calais, and E. Lewi (2014), plate boundary deformation from a 3-D velocity field inversion of InSAR and GPS doi:10.1002/ 2014JB011391.

al., **2010**]. Extension occurs along three rift branches, the Red Sea, the Gulf of Aden, and the Main Ethiopian Rift. Current full spreading velocities are 18 mm/yr between Nubia and Arabia, 16 mm/yr for the Somalian–Arabian Plates, and 6 mm/yr for Nubia–Somalia [*McClusky et al.*, **2010**;*Saria et al.*, **2014**; *Vigny et al.*, **2006**]. Since ~3 Ma, the central axis of the plate boundary in Afar is marked by a series of rift segments, some of which have been volcanically active (red outlines in Figure 1a), while others are purely tectonic structures (black lines in Figure 1a). South of 15°N, the onland continuation of the Red Sea Rift in Afar is marked by a lineament of active volcanoes [*Field et al.*, **2012**; *Nobile et al.*, **2012**; *Pagli et al.*, **2012**; *Wright et al.*, **2012**] that can be traced to ~12°N. However, the mode by which a connection between the Red Sea and the Gulf of Aden rifts is realized in southern Afar remains unclear. A possible model is one of rift propagation together with block rotations [*Kidane et al.*, **2003**; *Manighetti et al.*, **1998**;*Tapponnier et al.*, **1990**].

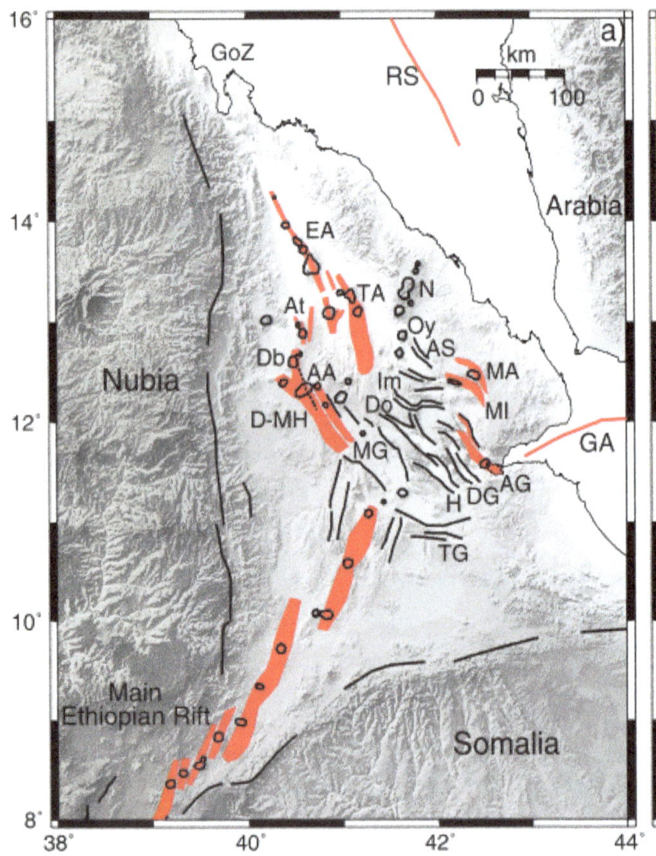

Figure 1. Afar region. (a) The red fillings are the magmatic rift segments: EA (Erta Ale), TA (Tat Ale), At (Alayta), D-MH (Dabbahu–Manda Hararo), AG (Asal–Ghoubbet), MI (Manda-Inakir), and MA (Moussa-Alli). The black lines are amagmatic rift segments: AS (Akulle–Sarodda), Im (Immino), Do (Dobi), MG (Manda Gargori), H (Hanle), and DG (Derela–Gaggade). The black outlines mark the volcanoes: N (Nabro), Db (Dabbahu), AA (Ado Ale), and Oy (Oyma). The dashed line along D-MH marks the area intruded by the 2005–2010 dikes. TG is the Tendaho-Gobha fault, and the Afar region is bounded by border faults (black lines). RS is the Red Sea ridge and GA is the Gulf of Aden ridge. GoZ indicates the gulf of Zula. (b) Data set and mesh used in the 3-D velocity field inversion. The black rectangles mark the InSAR tracks; the black triangles are the GPS sites operated by NSF and the Afar Rift Consortium [*McClusky et al.*, **2010**; *Kogan et al.*, **2012**; ***Vigny et al.*, 2006, 2007**; *Saria et al.*, **2013**].

Interferometric synthetic aperture radar (InSAR) has the ability to measure displacements at high accuracy (subcentimeter) and over large areas (~100 km), but measurements are one-dimensional, and extracting the three-dimensional displacement field remains challenging. Previous attempts to obtain the across-rift horizontal and the vertical components in Dabbahu have involved combining InSAR images from different orbits [*Hamling et al.*, **2014**]. The limitations of this method are that two components of the displacement field can be separated only in the overlapping area of the two sets of images, losing the InSAR spatial coverage, and that assumptions have to be made about the orientation of deformation, for example that there is no north-south (or rift parallel) motion [*Wright et al.*, **2004**]. This assumption is not ideal for concentric deformation at volcanoes where the north-south component of the displacement will be mapped in apparent vertical motion.

In this paper we combine all available GPS data with InSAR from eight tracks in a common reference frame, using the method of *Wang and Wright* [**2012**] to derive the first high-resolution continuous 3-D velocity field of the Afar rift (Ethiopia). We are able to capture the kinematic complexity of the plate boundary deformation as well as resolve the current strain distribution at regional scale. In particular, we can capture the full spatial extent of the transient postdiking deformation that follows the onset of the Afar diking event in 2005 in Dabbahu. Furthermore, we identify broad deformations in the rift segments of central Afar and a focused extension in southern Afar, at Asal–Ghoubbet, where the segment is experiencing an interdiking phase.

InSAR AND GPS DATA ANALYSIS

We analyzed GPS and InSAR data from 2005 to 2010 to reconstruct the plate boundary kinematics of Afar (Figure 1b). Following a two-step approach, we combined InSAR data from multiple tracks together with all available GPS measurements to obtain the three-dimensional velocity field for Afar. First, we obtain the line of sight (LOS) surface velocities for each InSAR track using a multi-interferogram method, π-rate software [*Wang et al.*, **2012**]. Second, we combine these with the GPS observations, using the velocity-field method developed by *Wang and Wright* [**2012**].

Data

We created about 350 interferograms using images from eight Envisat tracks (006, 028, 049, 071, 235, 278, 300, and 321) in both descending and ascending orbits, in image and wide-swath modes, spanning the period between 2005 and 2010 and covering the whole Afar region (Figure 1b and Figure S1 in the supporting information). The data cover ~350,000 km^2, which is the largest area ever imaged continuously with InSAR.

We used the Jet Propulsion Laboratory (JPL)/California Institute of Technology ROI_PAC software [*Rosen et al.*, **2004**] and a 3 arc sec Shuttle Radar Topography Mission digital elevation model (SRTM DEM) to create all the interferograms. We used a standard 20 azimuth and 4 range looks and filtered interferograms using a power spectrum filter before phase unwrapping with the branch-cut method. Phase unwrapping errors were initially identified in each interferogram using the phase closure technique on minimum spanning trees [*Kruskal*, **1956**]. They were then manually corrected when the absolute sum of the closure was equal to or larger than 2π. The data were geocoded to the 3 arc sec SRTM DEM (~90 m resolution).

The GPS data set analyzed here includes a total of 112 sites. The GPS sites near Dabbahu–Manda Hararo and in central Afar were measured between 2007 and 2010 as a part of the Afar Rift Consortium and National Science Foundation (NSF) projects [*Nooner et al.*, **2009**], while some GPS velocities from the Red Sea coast, the Gulf of Aden, and the Main Ethiopian Rift are from other sources [*Kogan et al.*, **2012**; *McClusky et al.*, **2010**; *Saria et al.*, **2013**; *Vigny et al.*, **2006**, **2007**]. We combined all GPS velocities in the common International Terrestrial Reference Frame 2008 and then converted the velocities with respect to Nubia using common sites between the different solutions and the Nubia frame as defined by *Saria et al.* [**2013**]. Horizontal velocities at sites located on stable Nubia fit a single-plate model with a weighted root-mean-square residual of 0.6 mm/yr. Offset corrections were estimated and subtracted from the GPS time series in order to remove the contributions from dike intrusions at Dabbahu so that GPS velocities in this area are representative of postdiking displacements. The examination of GPS position time series corrected for dike intrusions show that for the time interval considered here, displacements are linear with time and can be described by a constant velocity at each site.

INSAR TIME SERIES ANALYSIS

We obtained time series of cumulative displacements, average velocities, and their associated uncertainties for each InSAR track using the π-rate software. The π-rate software uses a network-based approach on the geocoded interferograms to mitigate common sources of errors in the interferograms, such as orbital and atmospheric delay errors [*Biggs et al.*, **2007**;*Elliott et al.*, **2008**]. In all our time series analyses, we removed contributions from sudden events (see description in the following) such as those caused by dike intrusions and eruptions.

In this study, the focus is on the regional-scale deformation pattern in Afar. Therefore, we averaged 10 geocoded neighboring pixels to ~900 m resolution in all interferograms, which reduced both the noise and the number of data points. Any unwrapping errors that may have not been manually corrected, as described in section 3.1, were again identified, using the phase closure technique on minimum spanning trees, as implemented in the π-rate software [*Wang et al.*, **2012**], and these pixels were automatically excluded from the analysis. Quadratic orbital corrections were applied to all the InSAR tracks, as these have lengths of several hundred kilometers. Linear functions were also tested but provided a worse fit to the data, as have been found elsewhere in large regions [*Wang and Wright*, **2012**]. Topographically correlated atmospheric delay errors were estimated and removed using a linear function to the height [*Elliott et al.*, **2008**; *Wang et al.*, **2012**].

We removed sudden displacement steps in the time series using a novel cross-correlation method in the time series analysis that we developed for this study. The method has three steps. (i) We invert the interferograms for the raw time series, without smoothing, and then form at least two independent interferograms spanning a given sudden event. (ii) In order to identify the area of the interferogram affected by a sudden deformation, we calculate the cross correlation between the independent interferograms. Areas affected by a sudden deformation are highly correlated, while other areas have low correlation. (iii) For the final step, our aim is to subtract from those interferograms spanning the time of a sudden event, the deformation corresponding to the highly correlated areas but leaving all other areas unchanged (the correlation ranges from 0 to 1 for poorly to highly correlated areas). For this purpose, we used an arctangent function,

$$f = 0.5 + \frac{1}{\pi} \cdot \tan^{-1}[D \cdot (C - C_T)] \qquad (1)$$

where C is the correlation value as calculated in step (ii), CT is a correlation threshold, and $D$is the slope of the function. We then obtained a residual interferogram (without sudden deformation) according to $\Delta\phi = \phi - f \cdot \phi$, where ϕ is the incremental shortest-interval interferogram spanning the event. For values of correlation greater than the threshold, f will approach 1 (for high values of D). Conversely, for low values of correlation, f will be close to 0. Therefore, in the residual interferogram, $\Delta\phi$, most of the correlated deformation will be subtracted, while poorly correlated area will be practically unchanged. Unlike a step, an arctangent function can avoid unrealistic sharp boundaries in the residual interferogram, where C is close to CT. We found that a CT between 0.6 and 0.7 was able to reliably identify the areas of coherent sudden deformation present in all interferograms spanning the event. The sharpness of the boundary is controlled by the slope *D,* with high values removing the correlated signal more strongly than lower values. Different values of correlation coefficient threshold and slope were tested, and the residual interferograms were inspected to ensure that sudden displacements had been removed; we found that *D* in the range of 20 to 25 gave good results. Moreover, the value of *f* will be somewhat smaller than 1 even for the strongest correlated signal, as we avoid using too large values of *D*, which would leave sharp boundaries where *C* is close to *CT*. This implies that after the correction, some sudden deformation is left in the residual interferogram. Therefore, the correlation procedure is applied twice in order to fully remove sudden deformations.

We show in Figure 2 a detailed example of the cross-correlation method, including the individual steps that lead to the removal of the sudden deformation caused by the November 2007 dike intrusion in Dabbahu, using two independent interferograms (Figures 2a and 2b). In Figures 2c and 2d, we show the correlation, *C*, and the arctangent function, *f*, respectively, and in Figures 2e and 2f, the result estimated sudden deformation and residual. Since the correlation is applied twice, in Figures 2g–2i, we also show the parameters of the second correlation procedure. Finally, in Figures 2j–2l, we show the final estimated sudden deformation and the residuals for the two independent interferograms. Furthermore, in Figure S2 in the supporting information, we show the results of the cross-correlation method using three independent interferograms covering the March 2008 Dabbahu intrusion. These examples demonstrate the efficiency of the cross-correlation method in removing sudden dike deformations while leaving other signals, such as time-progressive deformations near Hararo, unchanged. In both examples, most of the deformation caused by the intrusions is removed, and we observe no significant improvement in the sudden deformation removal when correlating three rather than two independent interferograms. Although, correlating an increasingly large number of independent interferograms should provide a better identification of the sudden deformation, it might be that the difference between correlating two or three independent interferograms is not significant. Dike intrusions in Dabbahu have occurred at closely spaced time interval such that only in a few cases, it is possible to correlate up to three independent interferograms. More dense satellite acquisitions and creation of larger number of independent interferograms will be possible in the future to further test the selection and number of independent interferograms to be used in the correlation.

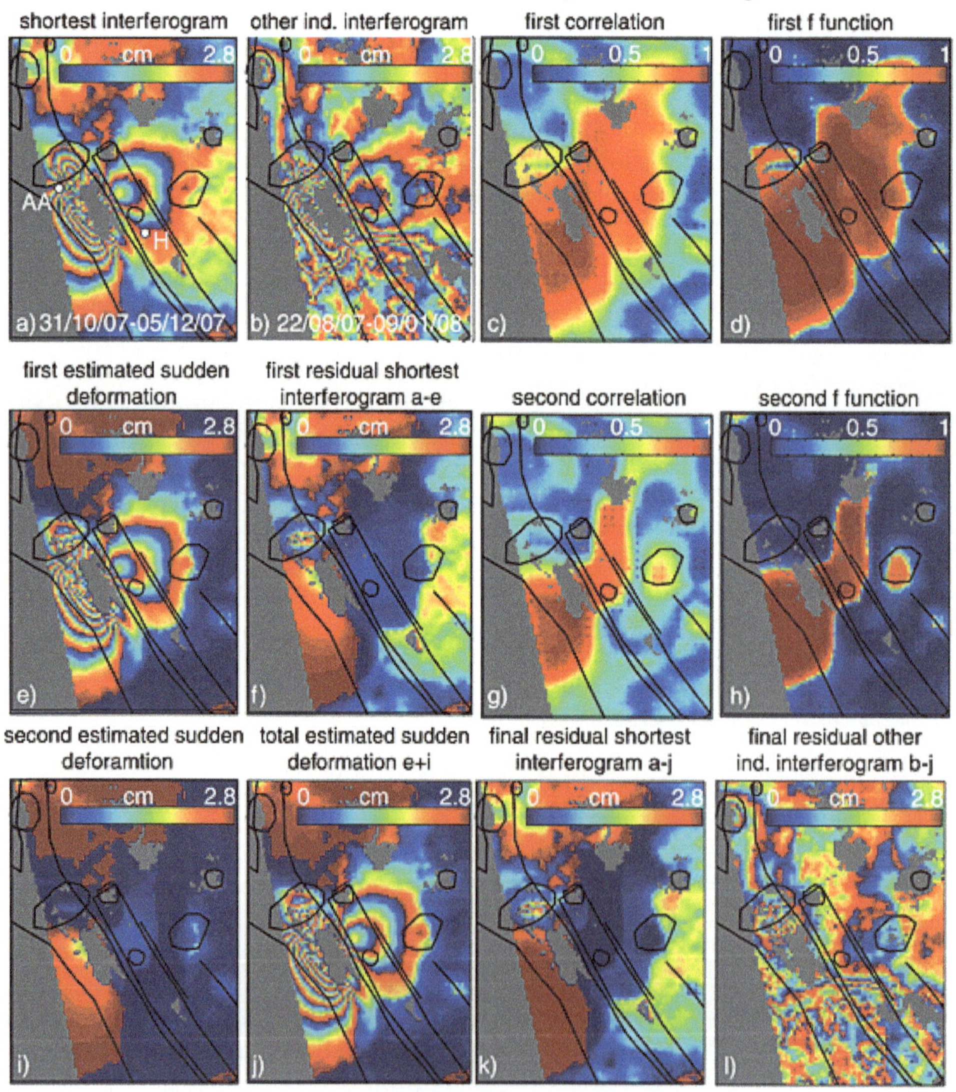

Figure 2. Removal of the displacement caused by the November 2007 intrusion in Dabbahu using the cross-correlation method and two independent interferograms of track 028. (a and b) The two shortest independent interferograms. (c) The correlation, (d) the *f* function, (e) the estimated sudden deformation, and (f) the residual, respectively, after the first correlation are shown. (g–i) The parameters of the second cor-

relation and (j) the total estimated sudden deformation. (k–l) The final residuals for the two shortest independent interferograms. Acquisition dates of the interferograms are marked in Figures 2a and 2b. The letters AA and H mark the positions of Ado Ale and Hararo.

A set of observed model and residual interferograms spanning the times of all dike intrusions in Dabbhau is shown in Figure S3 in the supporting information. It must be noticed that this method will remove only episodic displacements when at least one short time spanning interferogram across a given episode is available. If only long-interval interferograms are used, then both sudden displacements and any other signal exceeding the noise level are likely to correlate and hence be removed. Our time series include interferograms spanning the shortest time period, in most cases about a month, across an episodic event. Furthermore, atmospheric noise can be significant in Afar, but the cross-correlation correction is applied before any atmospheric cleaning such that only strong deformation signals that exceed the noise level will give a high correlation and be subtracted. Therefore, our average velocity maps can preserve the stable deformation rates, while deformations due to dike intrusions and eruptions are removed. The main advantages of separating steady and sudden deformations with the cross-correlation method are that a priori models of such deformations are not needed nor assumptions about the trend of the steady deformation. Future SAR missions will provide scientists with dense satellite acquisitions making cross-correlation method a potentially efficient way to separate steady and sudden deformation.

Furthermore, we applied atmospheric phase screen (APS) filtering to mitigate atmospheric disturbances. The APS consists of two steps: a temporal high-pass filter followed by a spatial low-pass filter similar to *Ferretti et al.* [**2001**]. Here we used a Gaussian temporal filter with 1σ of 0.5 years to minimize temporal variations shorter than a year as we focus on long-term, steady velocities. We used a Butterworth spatial filter and estimate the variance and covariance using a best fit 1-D covariance function [*Hanssen*, **2001**].

Finally, we estimated a least square best fit velocity and its RMS misfit to the interferograms on a pixel-by-pixel basis, using a variance-covariance matrix to account for the spatial and temporal correlation between interferograms [*Biggs et al.*, **2007**; *Wang et al.*, **2009**]. We exclude unreliable pixels from the velocity maps by removing pixels that are not coherent for at least a set number of independent epochs (between 15 and 20). We also removed pixels with RMS misfits larger than an a priori value, here set to 3–4 mm/yr.

InSAR TIME SERIES RESULTS

Initially, we obtained time series of incremental displacement together with average LOS velocities and the associated uncertainties for each InSAR track spanning the longest time period possible (Figure S1 in the supporting information). We show the time-evolving cumulative displacements for points close to the Ado Ale volcano and the Hararo subsidence zone for tracks 028 and 300 (Figure 3 and Figure S4 in the supporting information), which offer long and frequent acquisition histories (Figure S1 in the supporting information). The patterns of cumulative displacements in Figures 3a and 3b show that Ado Ale experienced continuous uplift while Hararo subsided. Our time series show a transient deformation with fastest displacement rates (Figures 3c and 3d) both at Ado Ale and Hararo in early 2006, followed by a rapid decay in late 2006 and approximately linear trends in 2007–2010. More importantly, our velocity time series show the first compelling evidence that the velocity changes at Ado Ale and Hararo were coeval, supporting a possible connection between the two magma systems [*Grandin et al.*, **2010**]. Higher melt flow from Hararo to Ado Ale in 2006 may have occurred in response to pressure changes caused by the large initial intrusion (2–2.5 km^3) in October 2005. The cumulative volume of the 14 subsequent intrusions (1 km^3) is less than half of the initial dike, requiring less magma flowing into the system in the later part of the diking event. How the connection between Hararo and Ado Ale is accomplished remains unclear. However, the subsidence at Hararo does not appear to be caused by magma moving from the Hararo system to feed the Dabbahu dike intrusions directly, as our time histories indicate that no significant change in subsidence rate occurred during the repeated intrusions (Figure 3b and Figure S4b in the supporting information).

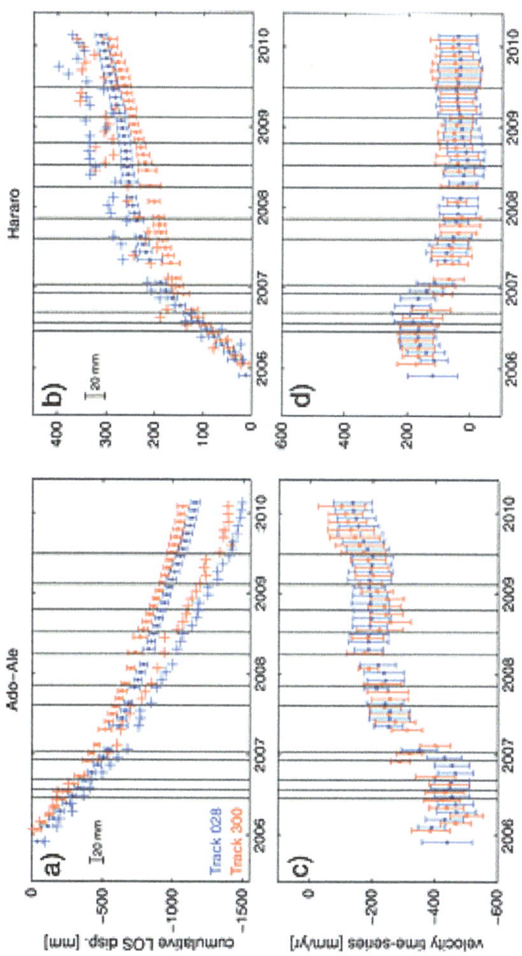

Figure 3. Time series of (a and b) cumulative displacements and (c and d) incremental velocities in the satellite line of sight (LOS) at Ado Ale and Hararo from tracks 028 (blue) and 300 (red). The plus signs show the raw time series (including sudden deformations), and the dots with 2 sigma error bars are the final smoothed time series after the removal of offsets and APS. In Figures 3a and 3b, the error bars are scaled, and the scale is given. The black vertical lines give the time of dike intrusions and eruptions. The positive values indicate the range increase.

The only significant offset displacement is identified at Hararo during the November 2007 intrusion (plus signs in

Figure S24b in the supporting information), but this is the signal from the dike opening that intruded in the southern part of the rift, close to Hararo (Figure 2). Likewise, uplift at Ado Ale follows a smooth trend without any fast replenishment following the intrusions (Figure 3a and Figure S4a in the supporting information). We conclude that velocity changes at Ado Ale and Hararo magmatic systems are coeval and that an inverse correlation exists in the long-term (approximately years) velocity trends at the two sites, with uplift at Ado Ale coupled with subsidence at Hararo. These signals are decoupled from the dike intrusion processes occurring at shallow crustal depths and are likely caused by magma flow at deep crustal levels.

In this paper we aim at combining average InSAR velocity maps with GPS to obtain a 3-D velocity field. Regular GPS measurements near Dabbahu started only in 2007, and InSAR velocities have been approximately linear since that time (Figure 3). Therefore, we further calculated maps of average LOS InSAR velocities and associated RMS misfits for the 2007–2010 time period (Figure 4 and Figures S5–S9 in the supporting information) and use these to combine with the GPS. The InSAR velocities (Figure 4) show a consistent pattern with the largest deformation concentrated at the Dabbahu segment. There is a range decrease signal along the segment in all the tracks in spite of the different viewing geometries, indicating that most of this deformation is consistent with uplift. Subsidence at Hararo (range increase) is also clearly seen in different velocity maps. Another deformation signal consistent with subsidence is observed at the southern tip of the Dabbahu–Manda Hararo rift and is likely related to the exploitation of an aquifer [*Tessema*, **2013**]. Although some obvious vertical motions can be identified in the InSAR velocity maps, horizontal motions remain challenging to extract using InSAR only. For this purpose, we combined our InSAR 2007–2010 velocity maps with the GPS to extract the three-dimensional velocity field.

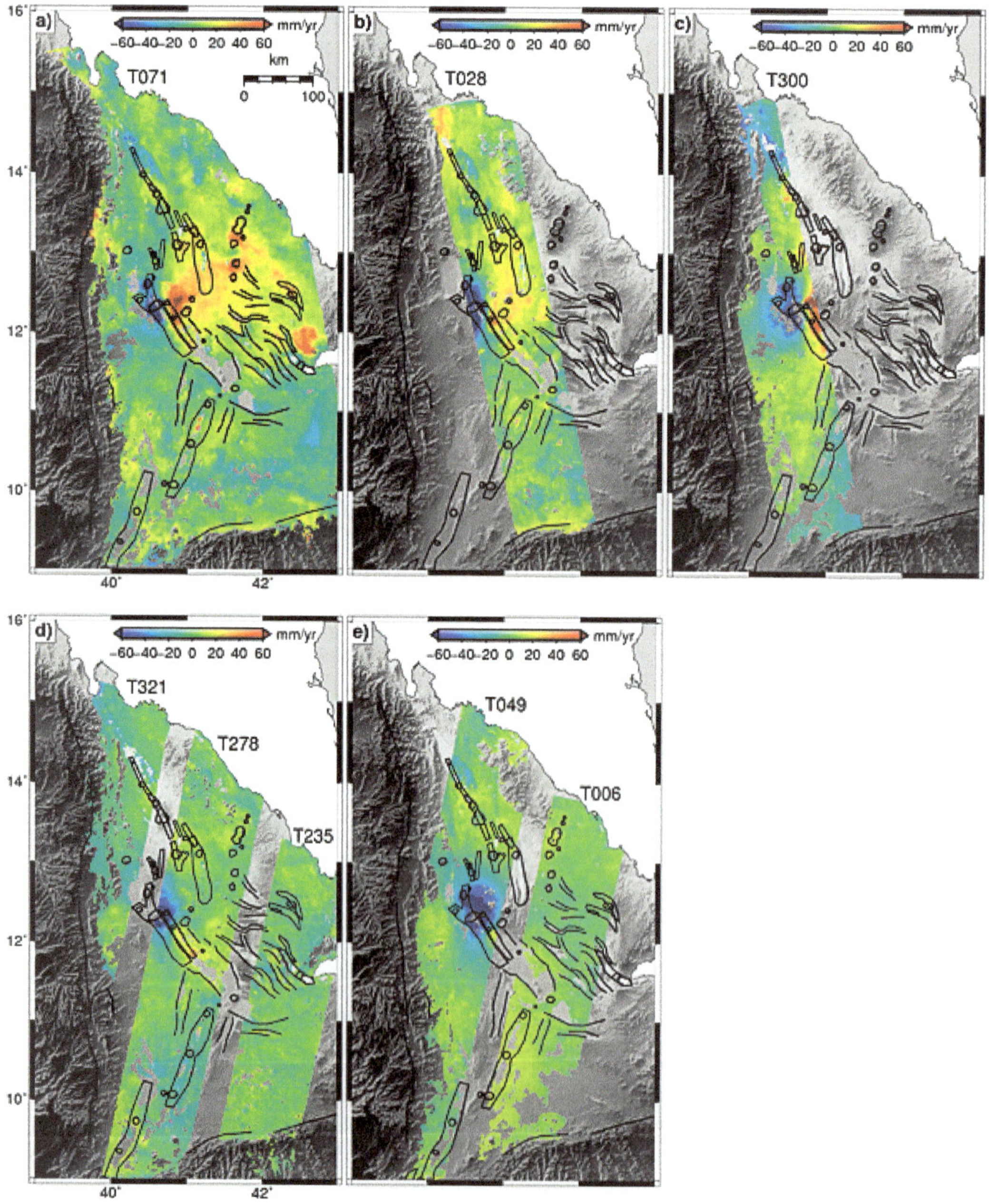

Figure 4. Average 2007–2010 LOS surface velocities for each InSAR track. The positive values indicate the range increase.

4 3-D CONTINUOUS VELOCITY FIELD

We combined the eight InSAR velocity maps (Figure 4) with the available GPS data to invert for the three-dimensional velocity field, using a method originally applied to GPS [*England and Molnar*, **2005**] and adapted to GPS and InSAR by *Wang and Wright* [**2012**]. We divided the Afar region into a mesh of triangular elements with varying size and assumed that the velocity varies linearly with latitude and longitude within each triangle. Therefore, the geodetic observations (InSAR and GPS) within each triangle are related to the velocities of their vertices by an interpolation function. A denser mesh with nodes at ~10 km spacing is used at the rift axis and over volcanic segments, while progressively larger elements with node spacing of ~20 km are used in the far field. Prior to the inversion, we applied further multilooking to the line of sight velocity data sets, giving them a final pixel size of ~4.5 km. We used GPS velocities with respect to Nubia and assumed that velocities at a few points in the stable part of Nubia are equal to zero (black dots in Figure 5a). Since only a few GPS points are available on the Somalian Plate, the motion of this plate was poorly defined. We fixed the velocities at two additional points on the Somalian Plate (white inverted triangles in Figure 5a). At these points, we assumed velocities equal to the GPS points on the Somalian Plateau. As our fixed points in Somalia are far from active rift segments and the Nubia–Somalia extension is less than 10 mm/yr, this does not impact on the velocity field obtained within the Afar region. We inverted for the velocities of the triangular vertices using the system of equation described in *Wang and Wright* [**2012**] and including a quadratic model to fit orbital and long-wavelength atmospheric delay errors. The system of equations was solved using a least square method that included full variance-covariance matrices as well as smoothing with a Laplacian operator. We selected a smoothing factor that minimizes the trade-off between the solution roughness and the weighted RMS misfit of the models (Figure S10 in the supporting information). The RMS residual between the InSAR rate maps and the predictions from the velocity field is 7 mm/yr, while the GPS has a total residual of 9 mm/yr, with 4 mm/yr of residual in the east and north components and 15 mm/yr in the vertical. The RMS residuals are comparable to the uncertainty of the input data; therefore, the solution from our 3-D inversion finds a relatively good fit to the data. InSAR LOS observed, fitted, and residual velocities are shown in Figures S11–S15 in the supporting information. Some residual deformation mainly near Ado Ale and Dabbahu volcanoes is attributed to mesh size that is likely too coarse to accurately reproduce focused signals at volcanoes. Also, using a single smoothing factor for the whole region does not allow the inversion to fit every spatial wavelength of the deformation field, in particular at volcanoes, causing some residuals.

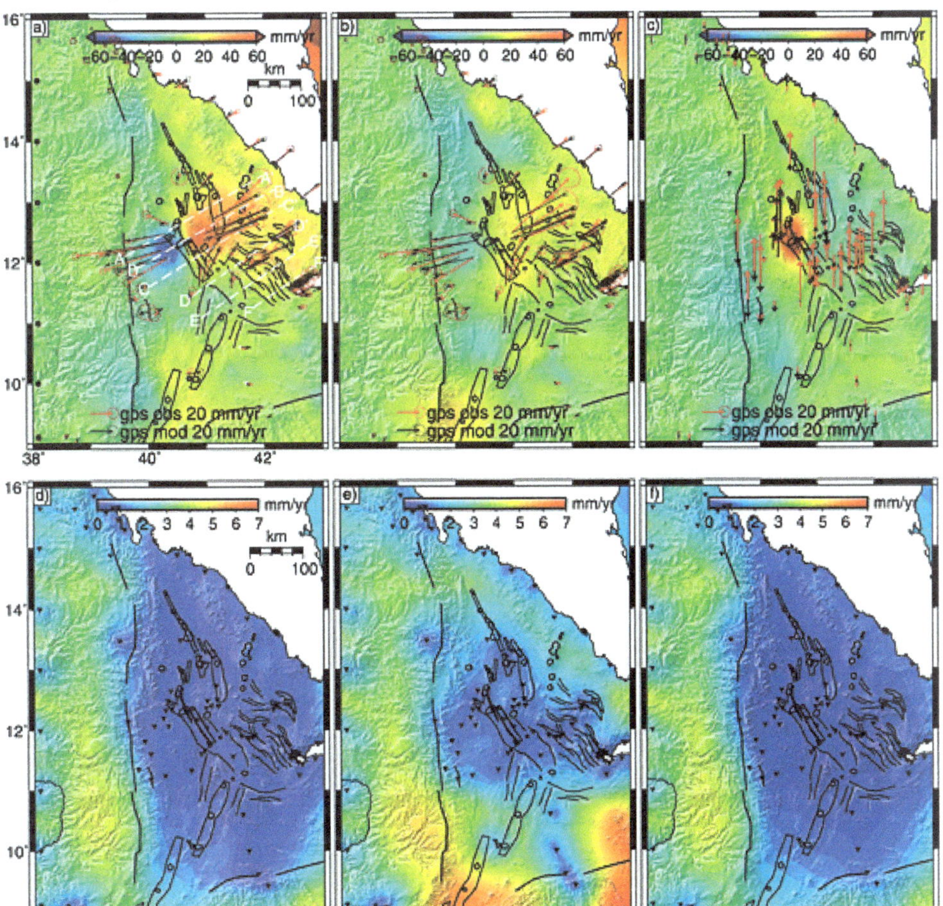

Figure 5. Continuous 3-D velocity field of Afar 2007–2010 and uncertainties. (a) Map of east-west surface velocity and overlapped GPS horizontal velocities relative to stable Nubia. The positive values indicate the movement toward east. (b) Map of north-south surface velocity and overlapped GPS horizontal velocities relative to stable Nubia. The positive values indicate the movement toward north. (c) Map of vertical surface velocity and overlapped GPS vertical velocities. The positive values indicate the uplift. (d) Map of the uncertainty on the east-west surface velocity. (e) Map of the uncertainty on the north-south surface velocity. (f) Map of the uncertainty on vertical surface velocity. The black inverted triangles mark the GPS stations.

After finding the velocities at the vertices of the triangles, we obtained a continuous velocity field by interpolation (Figures 5a–5c). Uncertainties on each component of the velocity field have also been calculated and are shown in Figures 5d–5f. For the velocity field inversion, we used GPS velocities with respect to Nubia; therefore, our final velocities are also in this reference system. Finally, we calculate the horizontal strain rates, coherent rotations, and their uncertainties at each vertex using spherical approximation equations (A6 in *Savage et al.* [**2001**]). Uncertainties are small compared to the displacement rates (Figures 5d–5f). The uncertainties on the east-west and vertical velocities are small over the entire Afar depression and do not exceed 4 mm/yr, with the largest values observed in the southern and western plateaus, where we lack InSAR observations due to loss of coherence. Larger uncertainties, up to 7 mm/yr, are observed in the north-south velocity. This is expected as the north-south motion is the least-well resolved by InSAR. The pattern of north-south uncertainties is alsolargely dependent on the GPS station distribution, with smaller uncertainties in central and southern Afar, where most GPS stations are located but larger uncertainties where GPS stations are sparse, such as in the Danakil Alps, the Somalian Plateau, the Main Ethiopian Rift, and the western plateau. Another reason for the generally smaller uncertainties in central and southern Afar is that most independent ascending and descending InSAR acquisitions are available in these areas.

The most prominent signals are the exceptionally fast extension across Dabbahu (Figure 5a) and the uplift centered at Ado Ale (Figure 5c). The east-west plate motions are also well resolved in southern and central Afar, allowing us to clearly define the location of the axis of the plate boundary, the width of the plate boundary deformation zone, as well as the rift segmentation. The highest extension rates and strain of Afar are ~110 mm/yr at Dabbahu and 20×10^{-7} strain/yr (Figure 5a, Figures 6a–6c, Figure 7a, and Figure 8a), as a result of postdiking processes. Extension is higher in the central part of the segment, at Ado Ale, and decays symmetrically at the segment tips (Figures 6a–6c). An analogous pattern is observed in the vertical velocities with values as high as 80 mm/yr in the center of the segment, at Ado Ale (Figure 5c and Figures 6a–6c). Since the horizontal velocities are not uniform along Dabbahu but decay toward the segment tips, this causes some shear (Figures 7b and 8b) as well as coherent rotations in the horizontal plane (Figure 9a). Coherent rotations about the y and $x$axes are also observed due to the vertical uplift in the center of the segment (Figures 9b and9c). The strain and rotation rates (Figures 7-9) are significantly larger than the estimated uncertainties (Figures S16–S18) over the entire Afar depression and follow a pattern as also observed for the velocity field inversion (Figures 5c and 5d), with higher uncertainties in the southern and western plateaus.

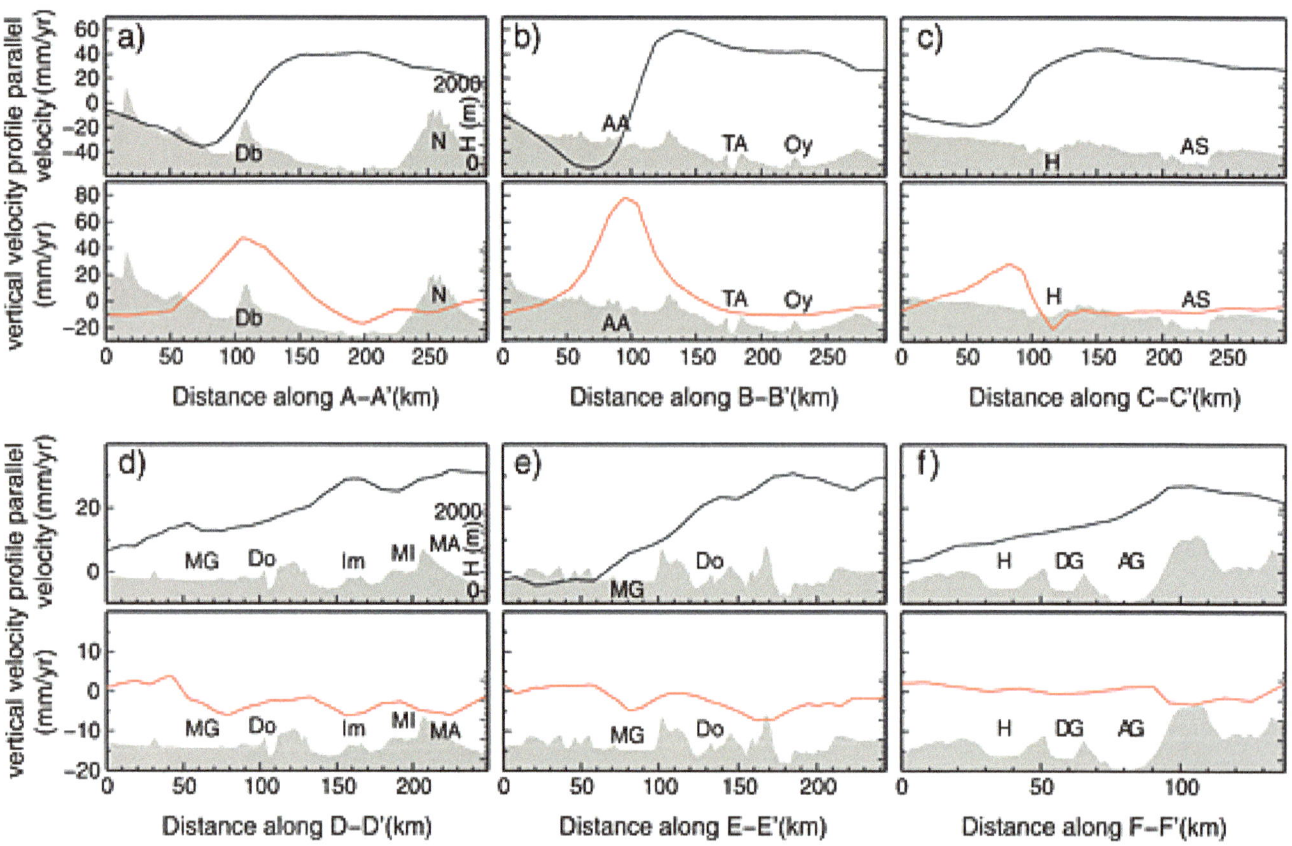

Figure 6. (a–f) Cross sections of surface velocities. Profile-parallel horizontal velocities (black lines) and vertical velocities (red lines). Cross section locations are shown in Figure 5a. The grey shading shows the topography, and the scale is read on the right-hand vertical axis of the figure, as in Figure 6a. The letters mark the rift segments and volcanoes as in Figure 1a.

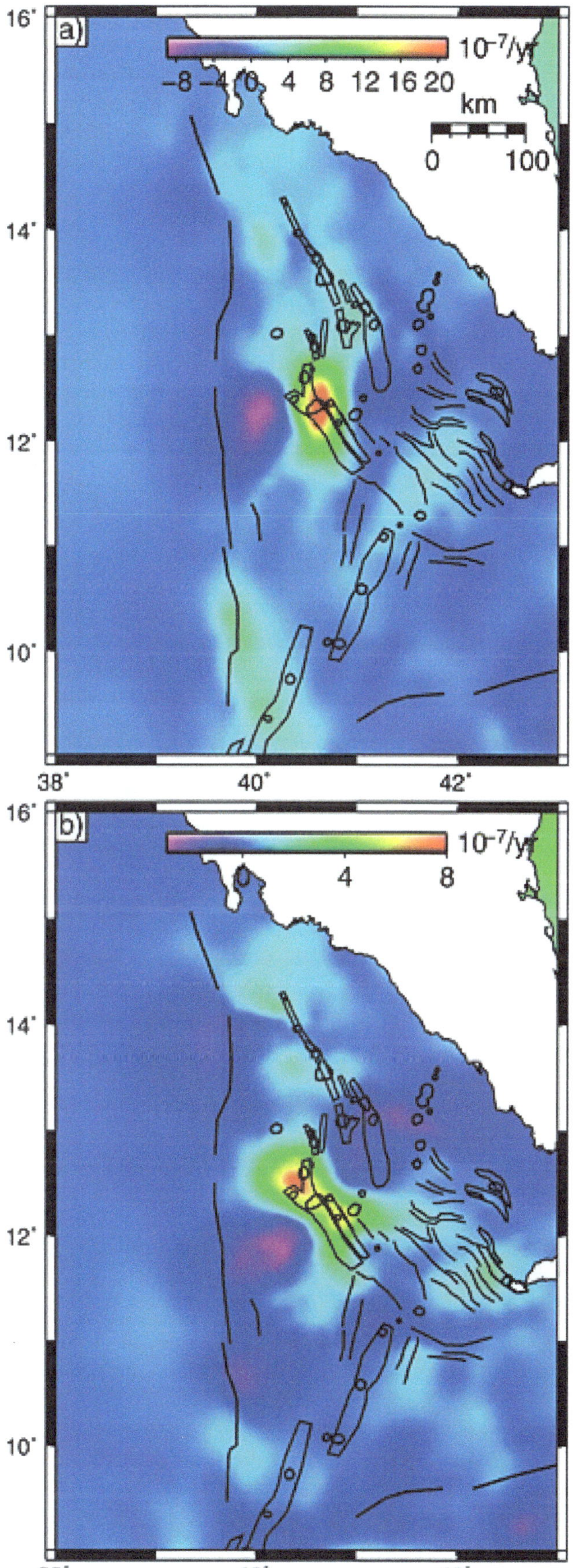

Figure 7. (a) Normal strain rate, $\dot{\varepsilon}_{xx}$. (b) Shear strain rate, $\dot{\varepsilon}_{xy}$.

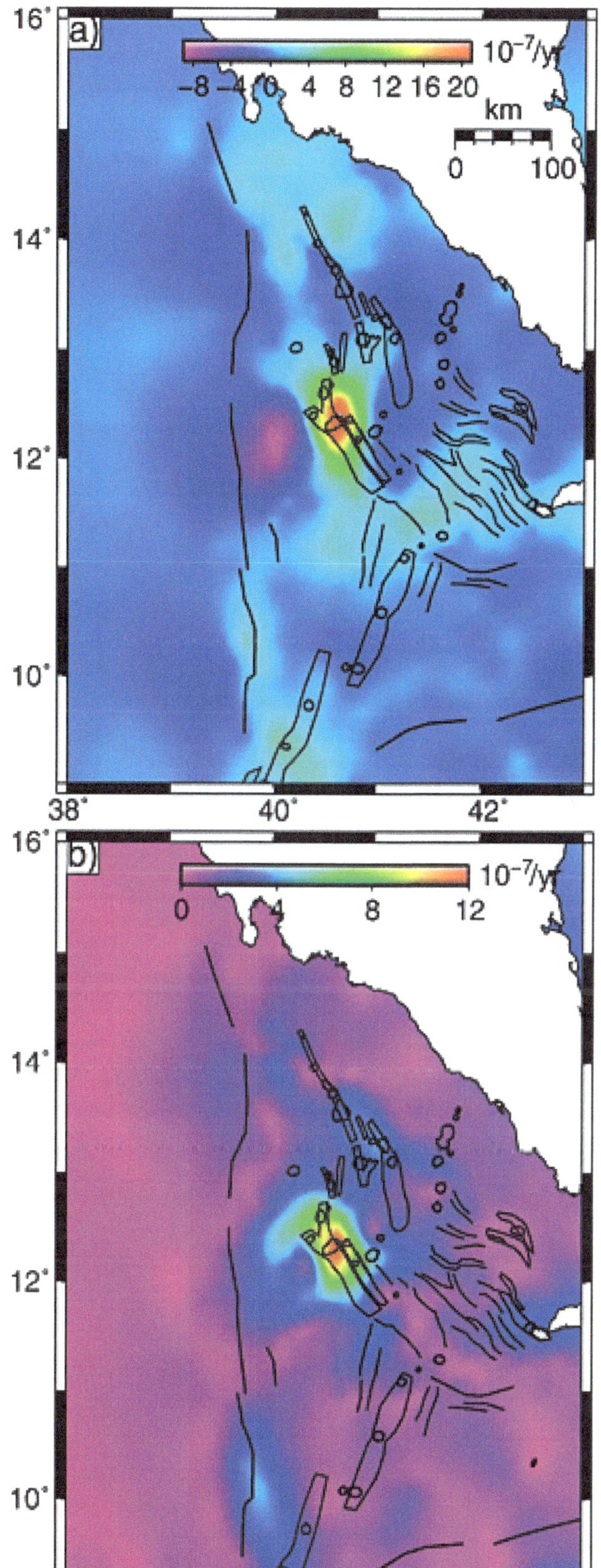

Figure 8. (a) First invariant of the horizontal strain rate tensor, also called dilatation or areal strain, positive values are extension. (b) Maximum horizontal shear strain rate.

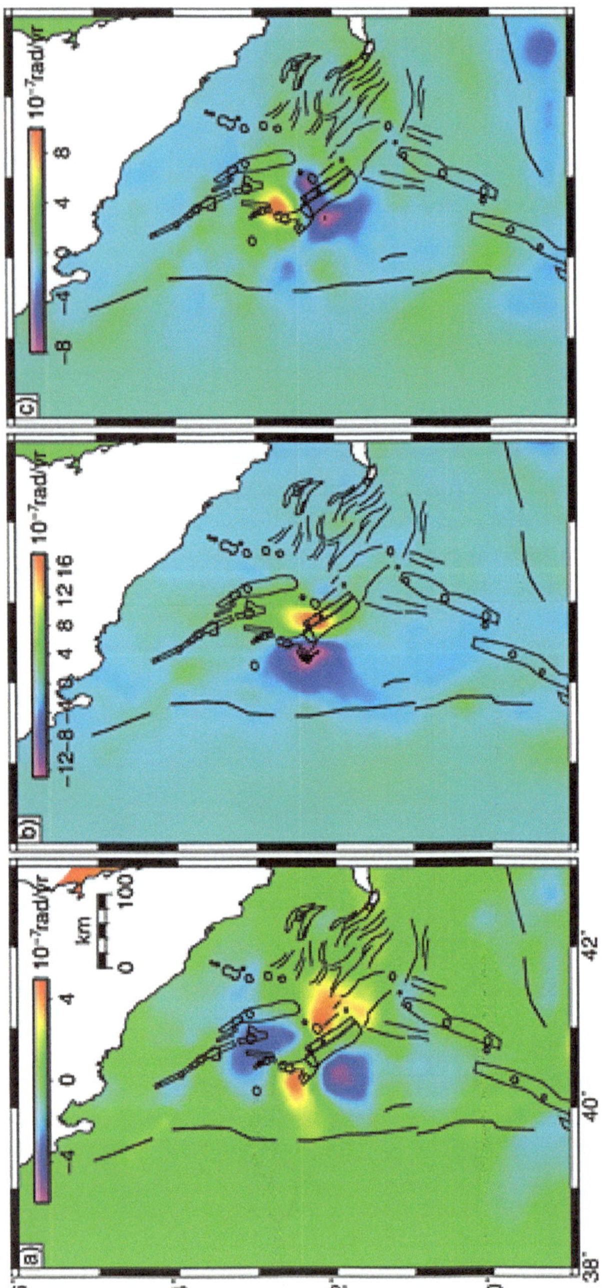

Figure 9. Coherent rotations: (a) Rotations about vertical axis, positive clockwise. (b) Rotations about *y* axis, positive toward east. (c) Rotations about *x* axis, positive toward north.

To the north of Dabbahu, the rift axis steps over eastward across two zones of extension, the Erta Ale Ridge and Tat Ale (Figure 7a). Extension is observed across the Alayta segment as well (Figure 7a), but strains here are dominated by the nearby postdiking deformation at Dabbahu. Farther north along the Erta Ale Ridge (Figure 7a), the localization of the rift axis is less clear as extension rates decay due to the partitioning of the spreading between the onland segment and the offshore Red Sea Rift. Furthermore, the paucity of GPS sites close to the rift axis does not allow us to fully describe the distribution of tectonic motions.

South of Dabbahu–Manda Hararo, extension is accommodated across a 150–200 km wide plate boundary deformation zone (PBZ) encompassing the Manda Gargori, Dobi, and Immino rifts (Figures 6d and 8a). A similarly broad accommodation zone is observed farther south with spreading distributed across Manda Gargori and Dobi in a PBZ about 150 km wide (Figures 6e and 8a). Strain is distributed over a much narrower area in the southeastern part of the plate boundary, where extension is mainly localized in Asal–Ghoubbet (Figures 6f and 8a) with only minor extension in Hanle and Derele–Gaggade and a PBZ of about 50 km width. Furthermore, no significant strain is observed at the Tendaho fault (Figures 7-9), suggesting that this structure is not currently active. However, motions of only a few mm/yr would not be resolved by our strain maps as the InSAR data spans a limited time period, 2007–2010. The row of volcanoes between Nabro and Oyma caldera also lacks significant strain (Figures 7 and 8), suggesting that these volcanoes do not currently mark an active portion of the plate boundary. Central volcanic activity, like the 2010 Nabro eruption [*Hamlyn et al.*, **2014**], where we do not observe any precursory deformation, is likely associated with local magmatic processes similar to off-rift volcanoes.

DISCUSSION AND CONCLUSIONS

The novel cross-correlation method described in this work allows us to estimate and remove sudden deformation from intrusions and eruptions without using a priori source models or assuming a linear deformation rate. Another advantage of this method is that the residual signal can then be effectively cleaned from atmospheric noise during time series analysis, using both APS and smoothing. The minimization of atmospheric noise in the time series allows us to identify a sharp decrease in uplift rate in 2006 at Ado Ale coinciding to a subsidence decrease at Hararo. This observation has important implications for the mechanism of rifting as it implies that a direct connection between Ado Ale and Hararo likely exists. We suggest that a continuous magma flow from a distant part of the segment toward the Ado Ale chamber in the rift center occurs, contributing to the observed uplift. The inflation/deflation at Ado Ale and Hararo and the superimposed accelerated spreading have been modeled as viscoelastic stress relaxation and individual magma chambers [*Hamling et al.*, **2014**]. However, a physical connection between the two magma sources has not been proposed yet. We observe a segment-centered uplift of up to 80 mm/yr, indicating that magma flow is still a primary mechanism of deformation during the postdiking phase. On the other hand, no vertical displacements are currently observed in central and southern Afar, indicating lack of significant magmatic activity at shallow levels in these regions at present time.

Our three-dimensional velocity field provides for the first time a picture of the current plate boundary deformation in Afar, illustrating contrasted patterns and processes at work. The dominant signal is the transient postdiking deformation around Dabbahu, where melt migration between connected magma chambers and likely viscous relaxation occur. There, spreading velocities are about 6 times larger than the long-term averages of 18 mm/yr, and postdiking deformation affects a region that is about 200 km wide. In central Afar, extension is distributed over a broad (150–200 km) plate boundary zone including several overlapping rift segments. Interestingly, this region is devoid of active magmatic segments (Figure 1) and has the thickest crust in Afar, ~26 km [*Hammond et al.*, **2011**; *Makris and Ginzburg*, **1987**]. In contrast, we find that spreading is accommodated across a very narrow (<50 km) zone in southern Afar (Asal–Ghoubbet), where previous diking events occurred

[*Cattin et al.*, **2005**;*Vigny et al.*, **2007**]. The width over which the plate boundary deformation is accommodated, as well as whether magmatic or tectonic processes are dominant, appears to vary along strike in Afar, probably reflecting the amount of crustal thinning and subcrustal melt available.

This differs from the more mature Iceland spreading center, where a PBZ ranging between 50 and 100 km is observed across all the spreading rift segments [*Árnadóttir et al.*, **2009**;*LaFemina et al.*, **2005**]. In Afar, less strain focus occurs at rift segments that are not known to have experienced diking events, like the Manda Gargori, Dobi, and Immino rifts. This is likely the result of the less mature stage of rifting of Afar, where continental rupture has not completely evolved to oceanic spreading. Plate spreading in Afar is achieved not only through episodic intrusions (i.e., diking events in Dabbahu and Asal–Ghoubbet) but also through diffuse crustal extension and normal faulting in areas of thicker and colder crust that have not experienced diking events, like the Manda Gargori, Dobi, and Immino rifts. This is also consistent with the seismicity of the area showing normal faulting mechanisms [*Keir et al.*, **2013**; *Sigmundsson*, **1992**] as well as thick crust and lower V_p/V_s ratio, suggesting lower degrees of melt than in other Afar rift segments [*Hammond et al.*, **2011**]. These results should be an impetus for developing advanced deformation models explaining these coexisting behaviors over distinct rift segments, possibly representative of different evolutionary phases of the transition from continental rupture to oceanic spreading. Figure 10 shows a schematic view of the different modes of strain accommodation in the Afar rift segments. Figure 10a depicts a typical rift segment that has evolved through repeated rifting events and magma intrusions, such as Dabbahu but also Asal–Ghoubbet, while Figure 10b shows two overlapping rift segments with a widespread strain accommodation area, where deformation is dominated by faulting, and it is lacking a system of intrusions as created during rifting events. This mode of deformation is observed at the Manda Gargori, Dobi, and Immino rifts.

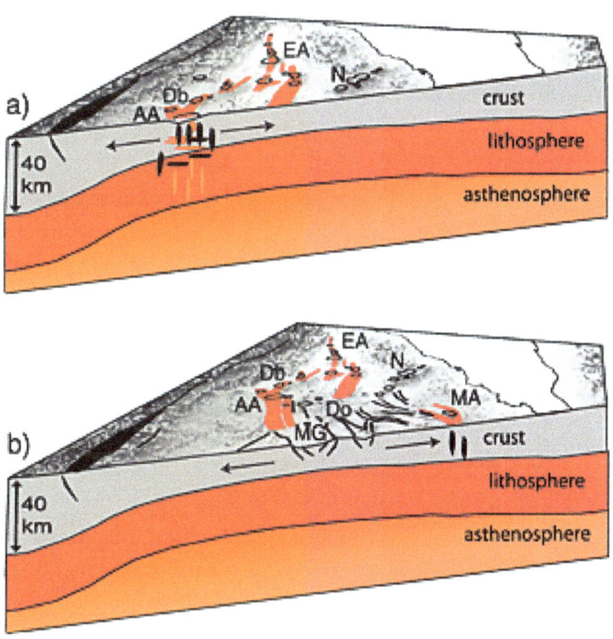

Figure 10. Cartoon of the different modes of strain accommodation in the Afar rift segments. (a) Rift segment that has evolved through rifting events and repeated magma intrusions, such as Dabbahu. (b) Extension accommodated by faulting, lacking a system of intrusions as created during rifting events, but causing a widespread strain accommodation area as in the Manda–Gargori and Dobi grabens.

Our strain maps also show coherent horizontal rotations about a vertical axis in the area around Dabbahu due to the segment-centered accelerated extension (Figure 9a). Although we cannot reconstruct the long-term rotation pattern in Afar as we lack detailed knowledge about past diking events, our results show that the pattern of postdiking deformation can induce coherent rotations, providing another possible way of explaining the paleomagnetic block rotations observed in Afar [*Kidane et al.*, **2003**].

Using our three-dimensional velocity field and strain maps, we identified the main areas of plate boundary deformation as well as the rift segmentation in Afar. As the Gulf of Aden rift becomes onland in Djibouti, extensions are confined to the Asal–Ghoubbet rift. The plate boundary then steps over westward across the Manda Gargori, Dobi, and Immino grabens, and then reconnects to the Dabbahu–Manda Hararo rift. The absence of significant strain along the Nabro and Oyma row of calderas but also at Moussa–Alli and Manda–Inakir Inakir suggests that these structures are less active than previously thought and that extension and magmatic activity have shifted toward the western rift segments. This is also in agreement with seismic tomography results showing areas of low velocities interpreted as zones of passive mantle upwelling underneath the western rifts, while no significant anomalies are observed in the eastern rifts [*Hammond et al.*, **2013**]. A similar scenario has also occurred in Iceland where rifting jumped eastward from the Snaefellsnes, western and eastern rifting zones [*Sigmundsson*, **2006**].

ACKNOWLEDGMENTS

C.P. thanks Peng Li and Zhenhong Li for their support with the wide swath processing software. C.P. gratefully acknowledges the support she receives through her Rita Levi Montalcini fellowship (Nota MIUR Montalcini 26259_21/12/2013). This work was supported by the UK National Environment Research Council NERC NE/D008611/1, NE/D01039X/1, and NE/E007414/1 grants and the Centre for the Observation and Modelling of Earthquakes, Volcanoes, and Tectonics. H.W. was supported by the National Natural Science Foundation of China grants 41104016 and 41372221. E.C. acknowledges support from NSF grants EAR-0635789 and EAR-0613651 and from JPL subcontract 148113. The SAR data used to produce all interferograms are copyright to ESA from CAT 1 3435. All processed interferograms are archived at the University of Leeds (contact T.J.Wright@leeds.ac.uk), and all the GPS data and station velocities are archived at Ecole Normale Superieure (contact eric.calais@ens.fr). The π-rate software developed by Hua Wang and used for the time series analysis is freely available online (http://homepages.see.leeds.ac.uk/~earhw/software/pi-rate/index.html).

REFERENCES

1. Árnadóttir, T., B. Lund, W. Jiang, H. Geirsson, H. Björnsson, P. Einarsson, and T. Sigurdsson (2009), Glacial rebound and plate spreading: Results from the first countrywide GPS observations in Iceland, *Geophys. J. Int.*,**177**(2), 691–716.

2. Barberi, F., and J. Varet (1977), Volcanism of Afar - Small-Scale Plate Tectonics Implications, *Geol. Soc. Am. Bull.*,**88**(9), 1251–1266.
3. Biggs, J., T. Wright, Z. Lu, and B. Parsons (2007), Multi-interferogram method for measuring interseismic deformation: Denali fault, Alaska, *Geophys. J. Int.*, **170**(3), 1165–1179.
4. Cattin, R., C. Doubre, J. B. de Chabalier, G. King, C. Vigny, J. P. Avouac, and J. C. Ruegg (2005), Numerical modelling of quaternary deformation and post-rifting displacement in the Asal-Ghoubbet rift (Djibouti, Africa),*Earth Planet. Sci. Lett.*, **239**(3–4), 352–367.
5. Courtillot, V. E. (1980), Opening of the Gulf of Aden and Afar by progressive tearing, *Phys. Earth Planet. Inter.*,**21**(4), 343–350.
6. Ebinger, C., A. Ayele, D. Keir, J. Rowland, G. Yirgu, T. Wright, M. Belachew, and I. Hamling (2010), Length and Timescales of Rift Faulting and Magma Intrusion: The Afar Rifting Cycle from 2005 to Present, *Annu. Rev. Earth Planet. Sci.*, **38**(1), 439–466.
7. Elliott, J. R., J. Biggs, B. Parsons, and T. J. Wright (2008), InSAR slip rate determination on the Altyn Tagh Fault, northern Tibet, in the presence of topographically correlated atmospheric delays, *Geophys. Res. Lett.*, **35**, L12309, doi:10.1029/2008GL033659.
8. England, P., and P. Molnar (2005), Late Quaternary to decadal velocity fields in Asia, *J. Geophys. Res.*, **110**, B12401, doi:10.1029/2004JB0003541.
9. Ferretti, A., C. Prati, and F. Rocca (2001), Permanent scatterers in SAR interferometry, *IEEE Trans. Geosci. Remote*, **39**(1), 8–20.
10. Field, L., T. Barnie, J. Blundy, R. A. Brooker, D. Keir, E. Lewi, and K. Saunders (2012), Integrated field, satellite and petrological observations of the November 2010 eruption of Erta Ale, *Bull. Volcanol.*, **74**(10), 2251–2271.
11. Foulger, G. R., C. H. Jahn, G. Seeber, P. Einarsson, B. R. Julian, and K. Heki (1992), Post-Rifting Stress-Relaxation at the Divergent Plate Boundary in Northeast Iceland, *Nature*, **358**(6386), 488–490.
12. Grandin, R., A. Socquet, M. P. Doin, E. Jacques, J. B. de Chabalier, and G. C. P. King (2010), Transient rift opening in response to multiple dike injections in the Manda Hararo rift (Afar, Ethiopia) imaged by time-dependent elastic inversion of interferometric synthetic aperture radar data, *J. Geophys. Res.*, **115**, B09403, doi:10.1029/2009JB006883.
13. Hamling, I. J., T. J. Wright, E. Calais, E. Lewi, and Y. Fukahata (2014), InSAR observations of post-rifting deformation around the Dabbahu rift segment, Afar, Ethiopia, *Geophys. J. Int.*, **197**(1), 33–49.
14. Hamlyn, J. E., D. Keir, T. J. Wright, J. W. Neuberg, B. Goitom, J. O. S. Hammond, C. Pagli, C. Oppenheimer, and J. M. Kendall (2014), Seismicity and subsidence following the 2011 Nabro eruption, Eritrea: Insights into the plumbing system of an off-rift volcano, *J. Geophys. Res. Solid Earth*, doi:10.1002/2014JB011395.
15. Hammond, J. O. S., J. M. Kendall, G. W. Stuart, D. Keir, C. Ebinger, A. Ayele, and M. Belachew (2011), The nature of the crust beneath the Afar triple junction: Evidence from receiver functions, *Geochem. Geophys. Geosyst.*, **12**, Q12004, doi:10.1029/2011GC003738.
16. Hammond, J. O. S., et al. (2013), Mantle upwelling and initiation of rift segmentation beneath the Afar Depression, *Geology*, **41**(6), 635–638.
17. Hanssen, R. F. (2001), Radar Interferometry: Data Interpretation and Error Analysis, 328 pp., Kluwer Acad. Publ., Dordrecht, Netherlands.
18. Heki, K., G. R. Foulger, B. R. Julian, and C. H. Jahn (1993), Plate Dynamics near Divergent Boundaries - Geophysical Implications of Postrifting Crustal Deformation in Ne Iceland, *J. Geophys. Res.*, **98**(B8),14,279–14,297.
19. Hofton, M. A., and G. R. Foulger (1996), Postrifting anelastic deformation around the spreading plate boundary, north Iceland. 1. Modeling of the 1987–1992 deformation field using a viscoelastic Earth structure, *J. Geophys. Res.*, **101**(B11), 25,403–25,421.
20. Keir, D., I. D. Bastow, C. Pagli, and E. L. Chambers (2013), The development of extension and magmatism in the Red Sea rift of Afar, *Tectonophysics*, **607**, 98–114.
21. Kidane, T., V. Courtillot, I. Manighetti, L. Audin, P. Lahitte, X. Quidelleur, P. Y. Gillot, Y. Gallet, J. Carlut, and T. Haile (2003), New paleomagnetic and geochronologic results from Ethiopian Afar: Block rotations linked to rift overlap and propagation and determination of a similar to 2 Ma reference pole for stable Africa, *J. Geophys. Res.*, **108**(B2), 2102, doi:10.1029/2001JB000645.
22. Kogan, L., S. Fisseha, R. Bendick, R. Reilinger, S. McClusky, R. King, and T. Solomon (2012), Lithospheric strength and strain localization in continental extension from observations of the East African Rift, *J. Geophys. Res.*, **117**, B03402, doi:10.1029/2011JB008516.
23. Kruskal, J. B. (1956), On the shortest spanning subtree of a graph and the traveling salesman problem, *Proc. Am. Math. Soc.*, **7**(1), 48–50.
24. LaFemina, P. C., T. H. Dixon, R. Malservisi, T. Arnadottir, E. Sturkell, F. Sigmundsson, and P. Einarsson (2005),Geodetic GPS measurements in south Iceland: Strain accumulation and partitioning in a propagating ridge system, *J. Geophys. Res.*, **110**, B11405, doi:10.1029/2005JB003675.
25. Makris, J., and A. Ginzburg (1987), The Afar Depression: Transition between continental rifting and sea-floor spreading, *Tectonophysics*, **141**(1–3), 199–214.
26. Manighetti, I., P. Tapponnier, P. Y. Gillot, E. Jacques, V. Courtillot, R. Armijo, J. C. Ruegg, and G. King (1998),Propagation of rifting along the Arabia-Somalia plate boundary: Into Afar, *J. Geophys. Res.*, **103**(B3), 4947–4974.
27. McClusky, S., et al. (2010), Kinematics of the southern Red Sea-Afar Triple Junction and implications for plate dynamics, *Geophys. Res. Lett.*, **37**, L05301, doi:10.1029/2009GL041127.
28. Nobile, A., C. Pagli, D. Keir, T. J. Wright, A. Ayele, J.

Ruch, and V. Acocella (2012), Dike-fault interaction during the 2004 Dallol intrusion at the northern edge of the Erta Ale Ridge (Afar, Ethiopia), *Geophys. Res. Lett.*, **39**, L19305, doi:10.1029/2012GL053152.

29. Nooner, S. L., L. Bennati, E. Calais, W. R. Buck, I. J. Hamling, T. J. Wright, and E. Lewi (2009), Post-rifting relaxation in the Afar region, Ethiopia, *Geophys. Res. Lett.*, **36**, L21308, doi:10.1029/2009GL040502.

30. Pagli, C., T. J. Wright, C. J. Ebinger, S. H. Yun, J. R. Cann, T. Barnie, and A. Ayele (2012), Shallow axial magma chamber at the slow-spreading Erta Ale Ridge, *Nat. Geosci.*, **5**(4), 284–288.

31. Rosen, P. A., S. Hensley, G. Peltzer, and M. Simons (2004), Updated repeat orbit interferometry package released, *Eos Trans. AGU*, **85**(5), 47–47, doi:10.1029/2004EO050004.

32. Saria, E., E. Calais, Z. Altamimi, P. Willis, and H. Farah (2013), A new velocity field for Africa from combined GPS and DORIS space geodetic Solutions: Contribution to the definition of the African reference frame (AFREF), *J. Geophys. Res. Solid Earth*, 1677–1697, doi:10.1002/jgrb.50137.

33. Saria, E., E. Calais, D. S. Stamps, D. Delvaux, and C. J. H. Hartnady (2014), Present-day kinematics of the East African Rift, *J. Geophys. Res. Solid Earth*, 3584–3600, doi:10.1002/2013JB010901.

34. Savage, J. C., W. J. Gan, and J. L. Svarc (2001), Strain accumulation and rotation in the Eastern California Shear Zone, *J. Geophys. Res.*, **106**(B10), 21,995–22,007.

35. Sigmundsson, F. (1992), Tectonic Implications of the 1989 Afar Earthquake Sequence, *Geophys. Res. Lett.*, **19**(9),877–880.

36. Sigmundsson, F. (2006), Iceland Geodynamics, Crustal Deformation and Divergent Plate Tectonics, 209 pp., Praxis-Springer, Chichester, U.K.

37. Tapponnier, P., R. Armijo, I. Manighetti, and V. Courtillot (1990), Bookshelf faulting and horizontal block rotations between overlapping rifts in southern Afar, *Geophys. Res. Lett.*, **17**(1), 1–4.

38. Tessema, T. (2013), Deformation measurnment in Central Afar, Ethiopia uisng InSAR and GPS, Addis Abeba Univ., Addis Abeba, Ethiopia.

39. Vigny, C., P. Huchon, J. C. Ruegg, K. Khanbari, and L. M. Asfaw (2006), Confirmation of Arabia plate slow motion by new GPS data in Yemen, *J. Geophys. Res.*, **111**, B02402, doi:10.1029/2004JB003229.

40. Vigny, C., J. B. de Chabalier, J. C. Ruegg, P. Huchon, K. L. Feigl, R. Cattin, L. Asfaw, and K. Kanbari (2007), Twenty-five years of geodetic measurements along the Tadjoura-Asal rift system, Djibouti, East Africa, *J. Geophys. Res.*,**112**, B06410, doi:10.1029/2004JB003230.

41. Wang, H., and T. J. Wright (2012), Satellite geodetic imaging reveals internal deformation of western Tibet,*Geophys. Res. Lett.*, **39**, L07303, doi:10.1029/2012GL051222.

42. Wang, H., T. J. Wright, and J. Biggs (2009), Interseismic slip rate of the northwestern Xianshuihe fault from InSAR data, *Geophys. Res. Lett.*, **36**, L03302, doi:10.1029/2008GL036560.

43. Wang, H., T. J. Wright, Y. P. Yu, H. Lin, L. L. Jiang, C. H. Li, and G. X. Qiu (2012), InSAR reveals coastal subsidence in the Pearl River Delta, China, *Geophys. J. Int.*, **191**(3), 1119–1128.

44. Wright, T. J., B. E. Parsons, and Z. Lu (2004), Toward mapping surface deformation in three dimensions using InSAR, *Geophys.Res. Lett.*, **31**, L01607, doi:10.1029/2003GL018827.

45. Wright, T. J., C. Ebinger, J. Biggs, A. Ayele, G. Yirgu, D. Keir, and A. Stork (2006), Magma-maintained rift segmentation at continental rupture in the 2005 Afar dyking episode, *Nature*, **442**(7100), 291–294.

46. Wright, T. J., et al. (2012), Geophysical constraints on the dynamics of spreading centres from rifting episodes on land, *Nat. Geosci.*, **5**(4), 242–250.

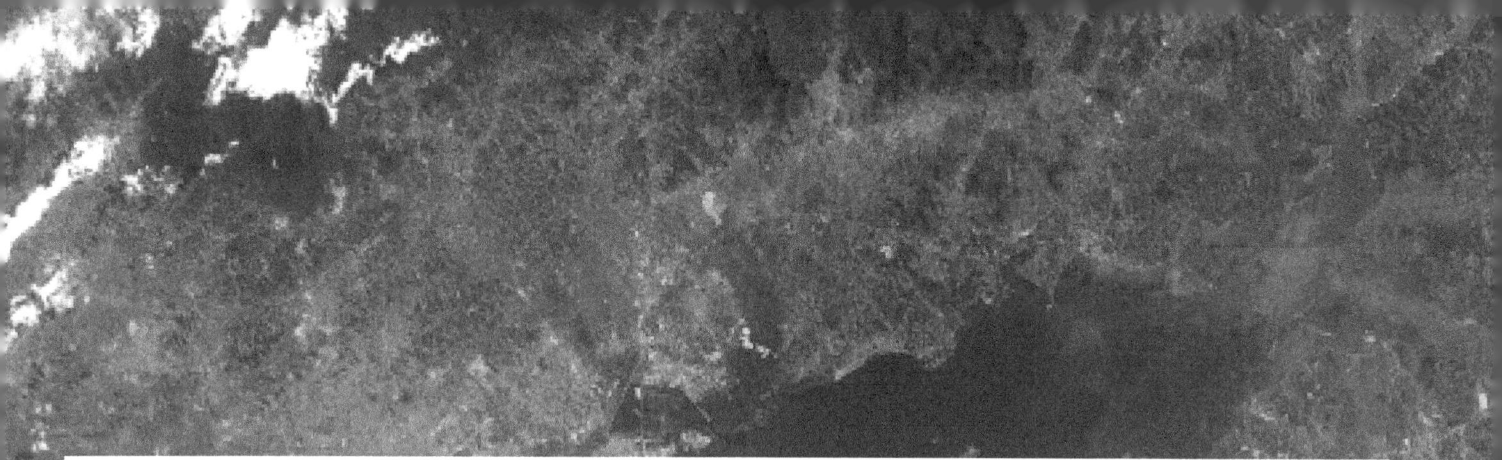

CHAPTER 18

CRUSTAL MOVEMENT AND ITS IMPACT ON CLIMATE CHANGE: A MICRO-SEISMIC STUDY USING GLOBAL POSITIONING SYSTEM (GPS)

Kashyap Mahanta, Jyotirmoy Das Chowdhury, Sangita Baruah, Arun Kumar, Ashok Kumar, Sunil L. Singh, Prakash Barman

Keywords: GPS, Plate Tectonics, Climate Change, Velocity

INTRODUCTION

The whole North-East is geologically an unstable area comprising rocks of Precambrian to Recent alluvium. The affect of thrust and faults in the Himalayan region, the high rising plateau of the Meghalaya, the low lying Brahmaputra Valley at the centre, the mobile subduction zone in the east, the deep basin in Bangladesh bordering Meghalaya, all there had its enervating effect on shaping the seismic character in the region. Its peculiar geologic setting, its plate movements in different directions, the presence of a number of active faults and related pro-active symptoms of surface rejuvenation signify more unstable condition in future. Because of its complex geology and geographical setting, the climate of the North-East India is peculiar. The region falls under high rainfall zone with more than 300 cm of rainfall annually, which makes the region unique. Now a day, some scientists are engaged in correlating rainfall and tectonic movement of plates. They find that the speed of a plate gain movement due to rainfall of the region (Iaffaldano, G.). But, North-East India, which is adjacent to world's highest Himalayan Mountain and encompassed by distinct geologic features, the relationship of rainfall and plate movement is likely to be most prominent in this region. To understand the actual picture, we have installed twelve (12) GPS campaign stations and collected data for four times (2007-2011) in the region. Taking into account the complex geological framework of North-East India, this paper makes attempt to project as to how crustal movement in North-East India is influenced by climate change.

PLATE TECTONICS

The earth is a dynamic plate whose surface is constantly undergoing change. The drifting of continents and sea floor spreading come together as parts of plate tectonics. When the surface of the earth are believed to be composed of a number of thin slabs of plates that moves across the surface. These plates follow a distinct boundary outlined by the location of earthquake epicenters. The plates are formed at the mid-ocean ridges and returned to the earth's interior at convergent plate boundaries such as the volcanic island arcs. Our continents are made of granite, a rock that is less dense than the ocean floor

Citation: Kashyap Mahanta, Jyotirmoy Das Chowdhury, Sangita Baruah, Arun Kumar, Ashok Kumar, Sunil L. Singh, Prakash Barman, Crustal movement and its impact on climate change: a micro-seismic study using Global Positioning System (GPS) ISSN: 2277-1697.

rocks. The continents stand above the ocean floor basalt, and, so, as the plates move they carry the continents with them. In this way the same mechanism that causes sea floor spreading is also the cause of continental drift.

It is known to all that the earth Lithosphere is divided into six large and several small plates which are bounded by three fundamental types of zones that form the plate boundaries (Fig.1). The basic concept of plate tectonic is that Lithosphere is created at spreading axes, slides horizontally part itself along transform faults, and plunges into the mantle at subduction zones. The rate of subduction of Lithosphere must approximately balance the rate of formation of Lithosphere, as there is little evidence to support significant expansion or contraction of the earth during the present episode of sea floor spreading, which extends back at least 180 million years (Burchfiel, et al.).

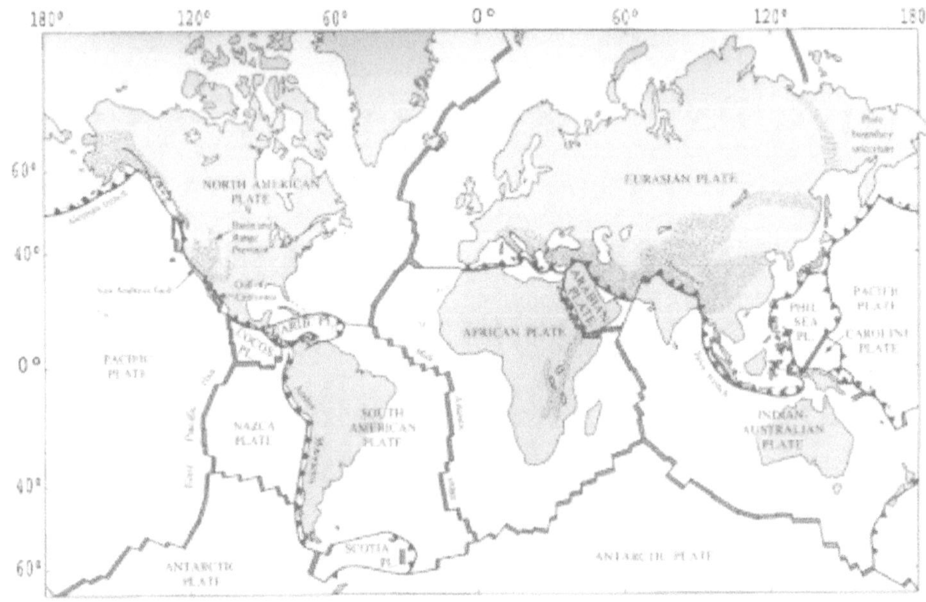

Figure 1: Lithospheric plates of the world.

Human desires to understand the phenomenon of earthquake, the eruption of volcanoes, through various perspectives. Scientists have developed a number of interesting theories on origin of earth and movement of continents. Mention may be made of the theories of Kober, Jeffreys, Hall & Dana and others. But it was Alfred Wegner, who revolutionaries the concept of one ocean-one continent theory and exemplified the reality of continental drift which at present forms the concept of plate tectonics. The matter that plates move relative to each other in convergent or divergent form causing earthquake and sea floor spreading is being viewed in different perspective by a team of scientists from French and Germany. The Australian based study team is of the view that the strengthening Indian monsoon had accelerated movement of the Indian plate over the 10 million years by a factor of about 20 percent (Iaffaldano, G.). For the first time in world's history, scientists have succeeded to bring a linkage between intensifying climate events and tectonic plate movements, which could open up a window for understanding the occurrence of mega thrust earthquake. In the North-East India region, the crustal movement causing enormous strain on the active Kopili Fault region. Added to this, the findings of the Australian scientists that crustal movement is correlated with rainfall implies that this region is actively prone to seismic catastrophic. As such the paper makes a determined attempt to understand the mechanism of fault movement in this tectonically volatile region and also desires to find linkage between the high amount of rainfall in the North-East India (as in Fig. 2) and also the movement of plates.

NORTH-EAST REGION

North-East India with an area of 2,55,089 sq. km. or 7.75% of India's total area encompassing ~3.748% of the country's total population is situated in a corner of the Indian Union with natural frontiers on three sides. The region is located between latitude 29° and 22° North and 89°46′ to 97°45′ East longitude, is literally boxed by China, Myanmar and Bangladesh.

Physiographically, 59.2% of the region is covered by mountains, 28% plain and the rest 12.8% is made up of plateau region. The entire area of Arunachal Pradesh, the biggest of the seven states of the North-East is constituted of a mountainous belt comprising spurs radiating southward from the crest line. Gori Chen (6,339 m or 21,450 ft.) is the highest peak of the territory. South of the Indo-China (Tibet) international boundary in Subansiri district, the altitude of peaks varies between 18,000 ft. to 21,000 ft. (5,487 m to 6,401 m). The Crestline forms India's northern and north-eastern boundary with China (Tibet). But around the north-eastern syntaxial curve, there is a gradual fall in altitude till the crest line, dips as low as 9,000 ft. (2,744 m). Towards the east, along the Indo-Myanmar border, it descends to 15,000 ft. (4,573 m) and even less until it averages about 6,000 ft. (1,828 m) in the northern Patkai Hills. The areas above 6,000 m (19,680 ft) remain snow covered for most part of the year and the presence of a number of waterfalls signifies the youthful topographic nature of the region (Acharyya, 1982). Geologically, North-East India and its adjoining region display tectonically distinct geological domains occurring in intimate spatial association. Rocks representing the entire geological span i.e. from Archaean to the Recent occur in this very small region. Geologically, the Brahmaputra Valley is built by the deposition of more than 1000 m thick alluvium upon sag formed during the period of the rise of the Himalaya. The Brahmaputra Valley is a relatively narrow valley bounded by two mobile young mountain belts, Mishmi block to the northeast and Meghalaya Plateau to the south (Das Chowdhury, 2005).

CLIMATE IN INDIA AND ITS NORTH-EASTERN REGION

According to Koppen, Indian climate is divided into six major climatic subtypes. The world's highest and massive Himalayan mountain system regulates both the South-West and NorthEast monsoon climate of India. India is home to an extraordinary variety of climatic region ranging from tropical in the South to temperate and Alpine in the Himalayan North, the desert in the West to the heaviest rainfall region in the East.

Against the backdrop of the geology and geographical setting of North-East region, the climate also plays an important role. The climate of the North-East region is unique, compared to the rest of the country. In the Northern Arunachal Pradesh the high rising Himalaya determines the pattern of climate there because of velocities in altitude region is very cold whereas in lower altitude and valley areas the climate is highly hot and humid. In Assam, the climate is marked by extreme humidity. From the first week of June, south-west monsoon starts bringing rainfall in Assam. Whereas in Meghalaya, because of its plateau character, the climate is uniquely pleasant all throughout the year. The CherrapunjiMaushingram area records world's heaviest rainfall regions. In Tripura the climate varies from subtropical monsoon climate to temperate condition in the hilly areas with temperature varying from 10°C to 35°C. Nagaland on the other hand experienced heavy monsoon rainfall from June to September.

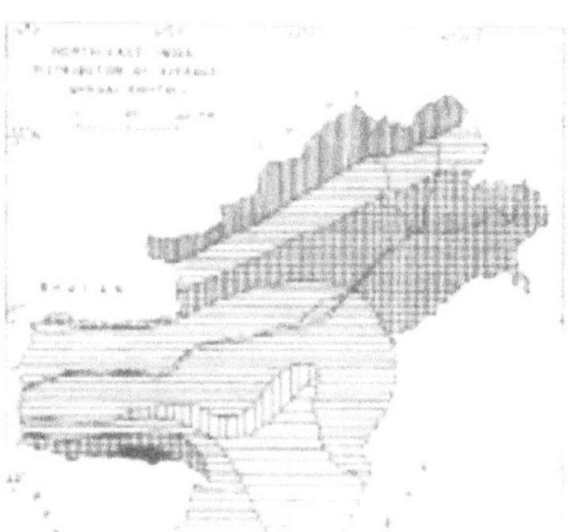

Figure 2: Distribution of average annual rainfall.

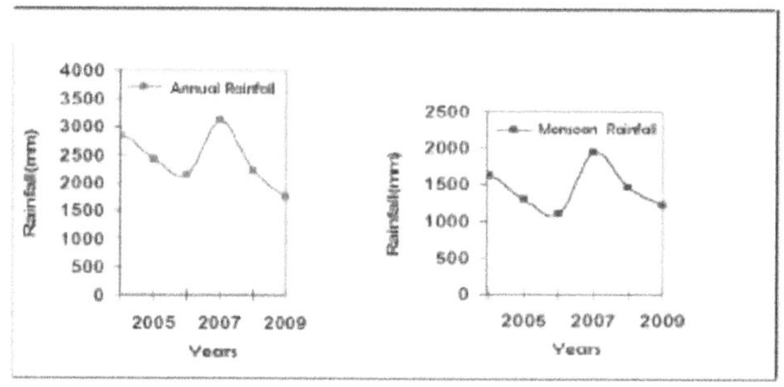

Figure 3: Annual and Monsoon Rainfall Distribution Pattern at Barapani over the last 5 years (2004-2009) (Das et al., 2010).

Monsoon in India gives enormous amount of rainfall from June to September every year. India is also the land of world's heaviest rainfall. Scientists who are recently engaged in correlating rainfall and plate movement on Indian tectonic plate have discover that the speed of the plate gain movement by about 20 percent due to rainfall (Iaffaldano, G.). As the whole North-East India receives more than 300 cm of rainfall annually, it is to be studied whether the plate margin in NorthEast India acquiring faster velocity compared to the whole Indian plate due to heavy rainfall or in vice versa whether rain act as an agent to strengthen the velocity of crustal plate.

METHODOLOGY AND DATA PROCESSING

To understand the crustal deformation in North-East India and its impact on climate change, the work has been carried out on the latest technology using Global Positioning System (GPS). In this study, twelve (12) points in different locations of Kopili Fault region have been selected for campaign study. The first field study was initiated in 2006-07 and then after it was continued for sequent four years up to February, 2011. These GPS campaign stations are studied with Trimble 5700, Leica500 and Leica1200 receivers with choke ring and Zephyr geodetic antennae. The selection of twelve sites was based on granite and feldspar rock formations or on hard stable rocks. All the points were selected in an open area without having any disturbance to the antenna. The machines were installed in all the campaign sites at least for continuous 72 hours, so that we can collect 48 hour data without break. The GPS data so collected has been converted into RINEX observation files and quality check has been conducted using TEQC (Translations, Editing and Quality Checking) software. All these processes of quality check plots were carefully examined and data with high cycle slips, multi-path were removed from the analysis (Sunil and Kumar, 2008). The data collected for the last four epochs (2007-11) processed along with the selected permanent stations like Tezpur (TZPR) and Bomdila (BOMP) and IGS (International GPS Service) stations KUNM, LHAZ, HYDE, IISC, POL2, KIT3. In the processing, we have used GAMIT/GLOBK software developed by Massachusetts Institute of Technology (MIT), USA and a reference frame ITRF 2005 is used for all time series. Ambiguity free and ambiguity fixed solutions were performed with ionosphere free linear combination to account for carrier phase ambiguities and signal delay due to ionosphere (Sunil and Kumar, 2009). The horizontal errors ellipses are calculated by GAMIT/GLOBK software from the uncertainties in the North and East coordinates/velocities using the correlation between the baseline component estimates.

Figure 4: Installation of a GPS machine in Raja Gaon, Nagaon, Assam.

RESULTS

The analysis of data gives a clear picture that the Kopili Fault region is moving in North-NorthEast (NNE) direction traversing the Kopili Fault at an average angle of 84.041°, causing enormous stress and strain at various levels. GPS has revolutionaries the scientific methodology of estimating crustal velocities all over the earth. It is used as scientific and most reliable to determine the crustal movement in the earth's surface, we perceived different in Fig. 4 : Installation of a GPS machine in Raja Gaon, Nagaon, Assam. velocities in different IGS (International GPS Services) stations. The average velocities of KIT3 and POL2 in Eurasian plate is 5.19N (± 0.310mm) and 27.37E (± 0.525mm) which show almost similar velocities. The IGS stations in India IISC and HYDE have the average velocity of 35.06N (± 0.425mm) and 40.54E (± 0.635mm). In NorthEast India the average velocity of the all campaign sites with permanent stations is 27.70N (± 1.091mm) and 40.47E (± 1.092mm) (Table 1 & 2 and Fig. 5).

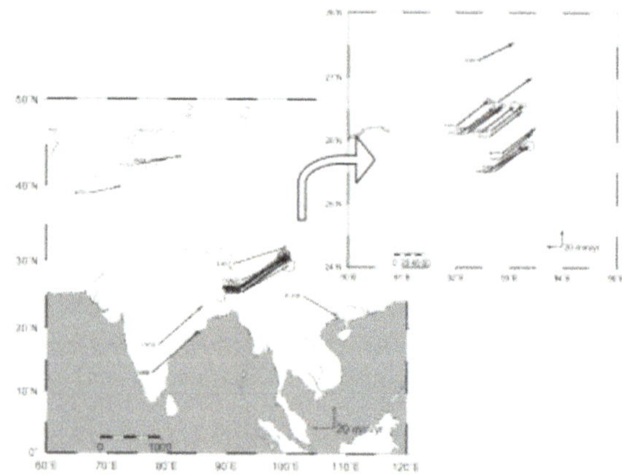

Figure. 5: Velocity plot of various campaign/permanent sites along with IGS stations (Mahanta et al., 2012).

Table - 1: Estimated velocities of campaign and permanent sites of North-East India

Sl. No.	Name of the Stations	Station Code	East (mm/yr)	North (mm/yr)	H Rate (mm/yr)
1	Natun Bazar	NATU	42.42 (±1.06)	29.53 (±1.06)	-0.24
2	Umrangsho	UMRA	38.10 (±.73)	27.51 (±.77)	2.46
3	Sokra Pam	SOKR	42.04 (±2.44)	23.88 (±2.32)	51.16
4	Panimura	PANI	38.39 (±.70)	29.43 (±.75)	2.70
5	Raja Gaon	RAJA	38.93 (±.73)	28.97 (±.79)	-4.24
6	Kheroni	KHER	40.86 (±.67)	28.44 (±.73)	1.85
7	Kumoi	KUMO	41.72 (±.99)	27.21 (±.97)	3.37
8	Jagiroad	JAGI	40.37 (±1.04)	28.73 (±1.03)	2.92
9	Bura Mayong	BURA	38.92 (±1.30)	28.88 (±1.25)	-13.87
10	Amsoi	AMSO	39.73(±2.25)	29.27 (±2.12)	-26.94
11	Kampur	KAMP	40.06(±1.31)	29.74(±1.24)	18.45
12	Khetri	KHET	41.19(±.99)	29.18(±.98)	1.46
13	Tezpur	TZPR	40.77(±.53)	27.31(±.63)	0.58
14	Bomdila	BOMP	43.12(±.55)	19.79(±.64)	4.36

Table 2: Estimated velocities of IGS sites of Indian and Eurasian plates

Sl. No.	Name of the Stations	Station Code	East (mm/yr)	North (mm/yr)	H Rate (mm/yr)
1	Kunming	KUNM	30.71 (±.64)	-17.64 (±.84)	-1.25
2	Lhasa	LHAZ	47.70 (±.50)	14.99 (±.58)	3.36
3	Hyderabad	HYDE	39.82 (±.61)	34.06 (±.42)	-0.97
4	Bangalore	IISC	41.27 (±.66)	36.07 (±.43)	-0.33
5	Bishkek	POL2	26.64 (±.55)	4.64 (±.31)	2.18
6	Kitab	KIT3	28.10 (±.50)	5.75 (±.31)	-0.66

From the above findings we may come to the conclusion that the Eurasian plate is more stable than the Indian plate where as the eastern velocity of the Indian plate and also that of the North-East region is almost same, but Northern velocity of the Indian plate (35.06±0.425mm) is much more higher than the Northern velocity of the NorthEast region (27.70N±1.091mm). Moreover the point that puzzles us is the stability of the Northern Eurasian plate. Because the Indian plate and the North-East region moves 7 times and 5 times more compared to the Eurasian plate respectively.

DISCUSSION AND CONCLUSION

The climate of North-East India will undergo a drastic change in near future, as because of the abnormal plate movement of this region (as mentioned above). Its temperature will rise and also rainfall will increase. According to Indian Network of Climate Change Assessment (INCCA), assessment, the temperature of the region will increase by 1.8°C to 2.1°C during the next two decades. In addition the mean annual rainfall is also likely to increase by 1mm to 6mm per day; however the number of rainy days in a year may decrease from 1-10 days. According to the assessment, the rise of temperature and pattern of changes in rainfall distribution is expected to affect ecology, especially the forests of the NorthEast region.

Moreover, there remains an enormous scope for the scientists to generate a model studying the variations in plate movements in this region and rate of erosion continuing because of prevailing monsoon climate in this region. The enormous monsoonal rain lightening the loads in this region, which affects the plate movement, is also another serious study material for consideration. Urbanisation is gradually rising open cast mining is prevalent in the region, continuous deforestation and construction of roads, railways, airports and more significantly, construction of big dams has its enervating effect on changing the landscape of the region. The report that China is constructing a dam on river Brahmaputra and her likely move to divert water from the river towards interior China would also create devastating ecological effect on this region. The North-East India in the long run would change its physical character. To understand the linkage between impacts of climate change on crustal movement in North-East India, we need to consider the effect of monsoonal rainfall on human changed topography and related crustal movement of the region. As the region demands serious investigation for its complex geodynamic and geophysical characters, we hope to study- the extent up to which climate is responsible for triggering plate movement in North-East India based on our findings.

ACKNOWLEDGEMENT

We acknowledge Ministry of Earth Sciences (MoES) and Department of Science and Technology (DST), Government of India for providing funds vide project no. MoES/P.O.(Seismo)/GPS/48/2005 to Department of Geography, Jagiroad College for carrying out the GPS surveys in the North-East Region.

REFERENCES

1. Acharyya, S. K. 1982 : "Structural framework and tectonics of Eastern Himalaya", Himalayan Geology, Vol. 10, pp. 412-429.
2. Altamimi, Z., Sillard, P. and Boucher, C. 2002, ITRF 2000 : a new release of the International Terrestrial Reference Frame for Earth Sciences application. Journal of Geographical Research Solid Earth 107 (B10), 2214.
3. Burchfiel, B. C., and others : "Physical Geology", Charles Merrill Pub. Co., Columbus, pp. 178-181.
4. Das, A., P. K., Choudhury, B. U., Patel, D. P., Munda, G, C., Ngachan, S. V., Chowdhury, P. 2010 : "Climate Change in North-East India: Recent Facts and events-worry for Agricultureal Management."
5. Das Chowdhury, J. 2005 : Feasibility report on "A delineation of the seismically active Kopili Fault Zone of North-East India", pp. 7-20.
6. DST 1998 : "Report on National Programmed for GPS and Geodetic Studies, National GPS Expert Group." Department of Science and Technology, Govt. of India.
7. Iaffaldano, G. : "Cosmos", Australian National University.
8. Iaffaldano, G. : "Earth and Planetary Science Letters", Australian National University.
9. King, R. W. and Bock, Y. 2002 : Documentation for GAMIT analysis software release 10.0, Massachusetts Institute of Technology (MIT), Cambridge.
10. Mahanta K. and Others 2012 : "Earthquakes and crustal deformation studies of the seismically active Kopili Fault as well as Northeast India : a Scientific field study using GPS", The Clarion, Vol.-1, No. 1, pp. 127-132.
11. Nandy, D. R. 1980 : "Tectonic Pattern in North-Eastern India", Indian J. Earth Sci., pp. 7,103-107.
12. Ravindranath, N. H. and Sukumar, R. 1996 : "Impacts of Climate Change on forest cover in India". Commonwealth Forestry Review 75 (1), pp. 76-78.
13. Sunil, S. L. and Kumar, A. 2008 : "Present day crustal deformation along Indo-Myanmar Arc Region"
14. Sunil, S. L. and Kumar, A. 2009 : "Establishment of a GPS permanent Reference station along IndoMyanmar region at Manipur University, Imphal, Manipur, India".
15. Taher, Ahmed : "Geography of North-East India", pp. 49.

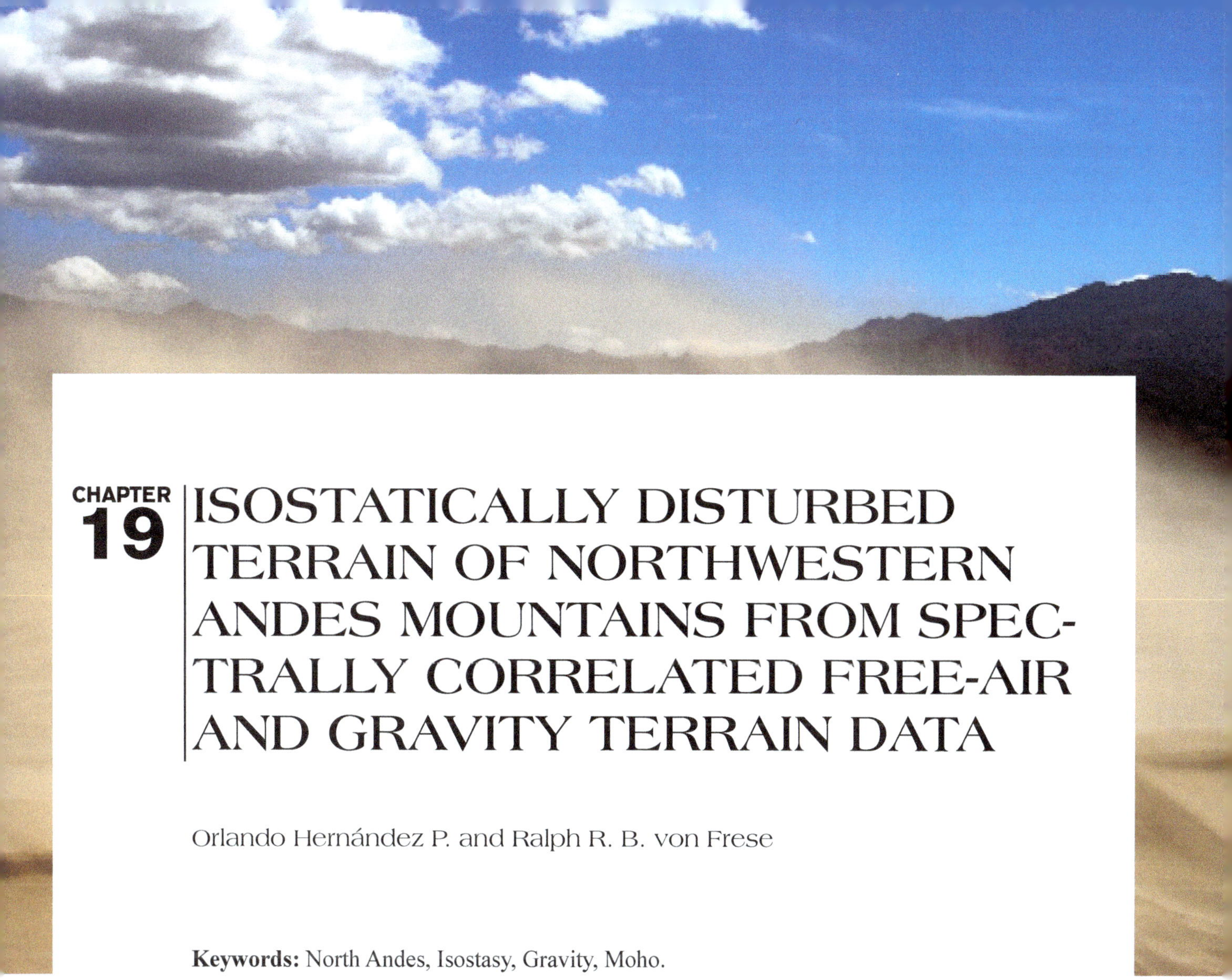

CHAPTER 19

ISOSTATICALLY DISTURBED TERRAIN OF NORTHWESTERN ANDES MOUNTAINS FROM SPECTRALLY CORRELATED FREE-AIR AND GRAVITY TERRAIN DATA

Orlando Hernández P. and Ralph R. B. von Frese

Keywords: North Andes, Isostasy, Gravity, Moho.

INTRODUCTION

The tectonic setting of northwestern South America is poorly understood because of the complex interaction of numerous tectonic plates, and the scarce and incomplete geologic mapping due to the dense vegetation cover and limited accessibility to remote areas. International scientific interest in the northwestern Andes Mountains is intense because it is the key to improve our understanding of the geological evolution of the backbone of America and the Caribbean, and related volcanic and seismic hazards (Bird, 2003; Cediel et al.,2003). The ability of current physical models to predict neotectonic and other intra-plate lithospheric stresses and strain is limited because the structures and geometry of plate boundaries are largely hidden and poorly understood. However, the deployment of geodetic sensors including the GPS-derived ground velocities and high resolution topography have substantially improved the modeling of the lithosphere to analyze surface mass dynamics (Kellogg et al., 1985; Kellogg et al., 1995).

In this paper, spectral correlation theory (e.g., Leftwich et al., 2005; von Frese and Tan, 1999) is applied to terrain and free-air gravity anomalies for new constraints on the boundary zones of the North Andes and Panama Microplates, and the Cocos, Nazca, Caribbean and South American Plates. Normalization and local favorability indexes are implemented to facilitate the visualization and interpretation of gravity anomalies. This methodology has been applied and validated in other crustal studies such as East Asia (Tan and von Frese, 1997), Antarctica (von Frese et al., 1992; 1999), Greenland (Roman, 1999), Ohio (Kim et al., 2000), and Iceland (Leftwich et al., 2005).

Isostatically disturbed terrain produces free-air anomalies that are highly correlated with the gravity effects of the terrain. Spectral correlation of the terrain gravity effects (TGE) and free-air gravity anomalies (FAGA) suggested that the Andes Mountains are isostatically disturbed with an anomalous underlying crust-mantle interface (MOHO). Strong negative terrain-correlated FAGA along western South America and the Greater and Lesser Antilles are consistent with anomalously deepened mantle displaced by subducting oceanic plates.

Citation: Orlando Hernández P. and Ralph R. B. von Frese, Isostatically Disturbed Terrain Of Northwestern Andes Mountains From Spectrally Correlated Free-Air And Gravity Terrain Data, Print version ISSN 1794-6190.

Gravity estimates regionally augment and extend the seismic estimates to highlight crustal deformation from plate collision and subduction in northwestern South America. The gravity anomalies are consistent with a complex evolution for northwestern South America involving the accretion of an oceanic plateau with strong Caribbean affinities, obduction of the oceanic crust, over thrusting and strike-slip faulting. These tectonic characteristics are not readily accounted for by the conventional model of "Andean Type" orogenesis of the Central Andes (Liu et al., 2002).

SPECTRALLY CORRELATED FREEAIR AND TERRAIN GRAVITY

The isostasy of northwestern South America was investigated considering the topography and bathymetry data from National Imagery and Mapping Agency (NIMA) from −8oS to 23.5oN latitude and from −90oW to −58.5oW longitude. Surface and bathymetry elevations from the JGP95E terrain data base (Smith and Sandwell, 1994, 1997) were processed to produce the Digital Elevation Model (DEM) in Figure 1 for the water and rock terrain gravity components at 0.5o nodal spacing. Free-air gravity anomalies (FAGA) were estimated from the EGM96 spherical harmonic Earth Gravity Model to degree and order 360 (Lemoine et al., 1998) at 20 km altitude over the 32o x 32o area at 0.5o nodal spacing in Figure 2. The altitude of 20 km was chosen to help minimizing the effects of local density errors in the terrain gravity modeling (e.g., Leftwich et al., 2005).

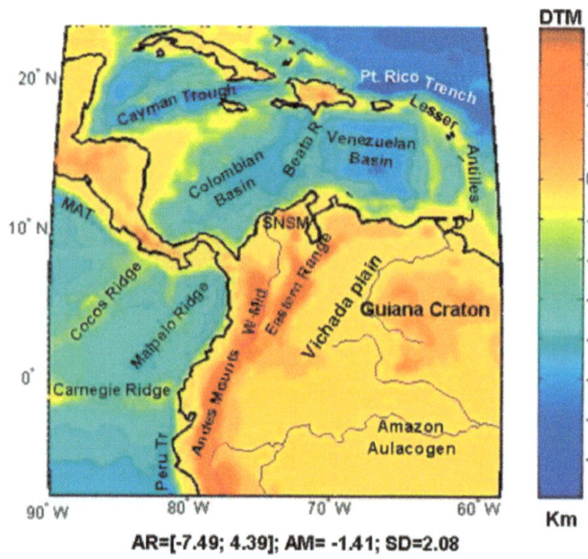

Figure. 1. Topography/bathymetry of northwestern South America with superposed regional tectonic features between −8° S to 23.5° N latitude and from −90° W to −58.5° W longitude. Map annotations include the amplitude range (AR) of (min; max) values, the amplitude mean (AM) and standard deviation (SD). SNSM= Sierra Nevada of Santa Martha, W-Mid = Western-Central ranges. This map was produced using the Albers equalarea conic projection.

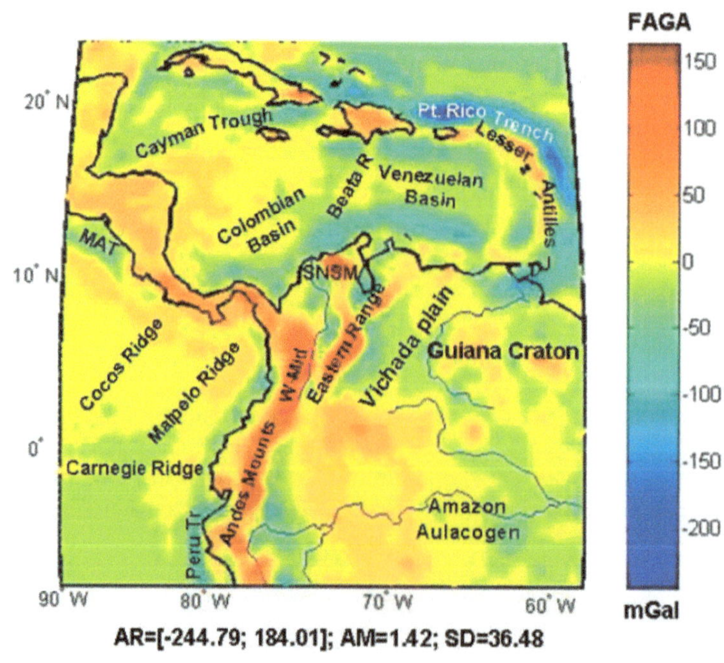

Figure. 2. Free-air gravity anomalies (FAGA) at 20 km elevation for the study region (Lemoine, et al., 1998). Map annotations include the am-

plitude range (AR) of (min; max) values, the amplitude mean (AM) and standard deviation (SD). SNSM= Sierra Nevada of Santa Martha, W-Mid = Western-Central ranges. This map was produced using the Albers equal-area conic projection..

The terrain gravity effects were modeled in spherical coordinates at 20 km altitude by Gauss- Legendre Quadrature integration in Figure 3 (von Frese, 1980). The terrain gravity modeling used densities of 2.8 gm/cm3 for the crust and 1.03 gm/cm3 for oceanic water. Spectral correlation theory was used to analyze the coregistered FAGA and TGE for their anomaly correlations using MatLab (MATHWORKS, 2005). Specifically, the Fourier transforms T and F of TGE and FAGA, respectively, were used to obtain their correlation spectrum (Davis, 1986; von Frese et al., 1997a, Kim et al., 2000) given by:

$$CC(k) = \cos(\Delta\theta k) = Re\left[\frac{F(k)}{T(k)}\frac{|T(k)|}{|Fk|}\right] \quad (1)$$

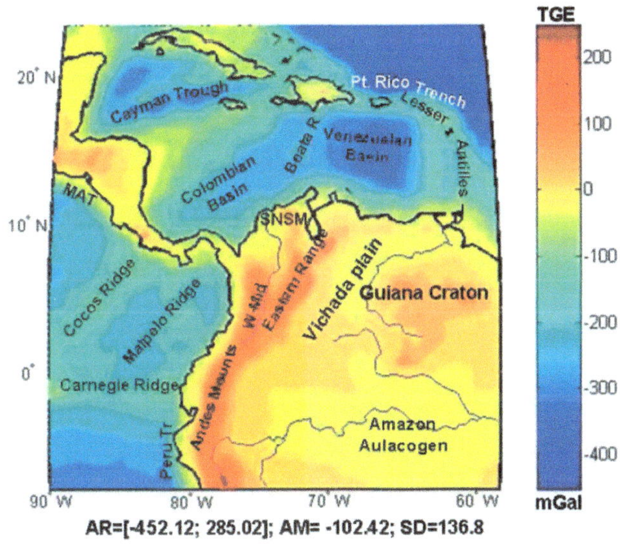

Figure. 3. Terrain Gravity Effects (TGE) at 20 km elevation for the study region by Gauss-Legendre quadrature integration. Map annotations include the amplitude range (AR) of (min; max) values, the amplitude mean (AM) and standard deviation (SD). SNSM= Sierra Nevada of Santa Martha, W-Mid = Western-Central ranges. This map was produced using the Albers equal-area conic projection.

Where CC(k) is the correlation coefficient between the kth wave numbers components F(k) and T(k), and denotes taking the real parts of the wave number components. Usually, CC(k) is evaluated from the cosine of the phase difference (Δθk) between the two k^{th} wave number components.

Using the correlation spectrum between FAGA and TGE, spectral correlation filters were designed to extract terrain-correlated free-air gravity signals. Those wave numbers components showing intermediate to high positive ($CCp(k) \geq 0.3$) and negative ($CCn(k) \leq 0.3$) correlations were identified. The cut off values for the correlation filter were determined to minimize correlative features between the terrain-decorrelated free-air and compensating terrain gravity components. In the other hand, transforming positively and negatively correlated free-air wave number components according to the selected cut off values yielded the terraincorrelated free air gravity anomalies (TCFAGA) in Figure 4. The residual terrain-decorrelated free-air gravity anomalies (TDFAGA) in Figure 5 were calculated by subtracting TCFAGA from FAGA, so that

$$FAGA = TCFAGA + TDFAGA \quad (2)$$

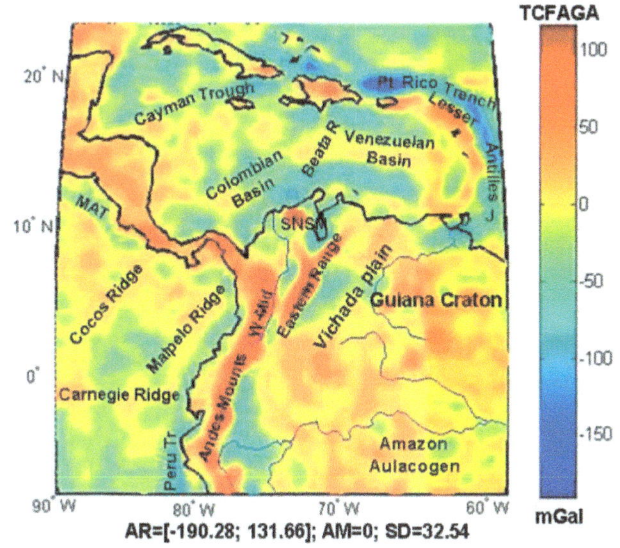

Figure. 4. Terrain-correlated FAGA (TCFAGA) at 20 km elevation for the study area. Map annotations include the amplitude range (AR) of (min; max) values, the amplitude mean (AM) and standard deviation (SD). SNSM = Sierra Nevada of Santa Martha, WMid = Western-Central ranges. This map was produced using the Albers equal-area conic projection.

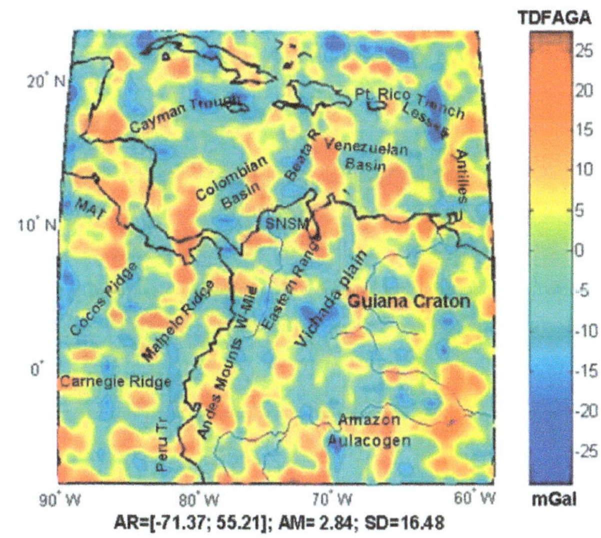

Figure. 5. Terrain-decorrelated FAGA (TDFAGA) at 20 km elevation for the study area. Map annotations include the amplitude range (AR) of (min; max) values, the amplitude mean (AM) and standard deviation (SD). SNSM = Sierra Nevada of Santa Martha, W-Mid = Western-Central ranges. This map was produced using the Albers equal-area conic projection.

TCFAGA are explained by anomalies associated with the topography while TDFAGA include the gravity effects of sources within the crust (e.g., local bodies) and the sub-crust. Similarly, spectral correlation filters were used to extract FAGA-correlated TGE (FCTGE) in Figure 6 and FAGA-decorrelated TGE (FDTGE) in Figure 7. The FCTGE sharpen up the TGE components that may be isostatically disturbed, whereas the FDTGE involves relatively marginal amplitudes that appear to reflect mostly noise in the data and other non-geological effects.

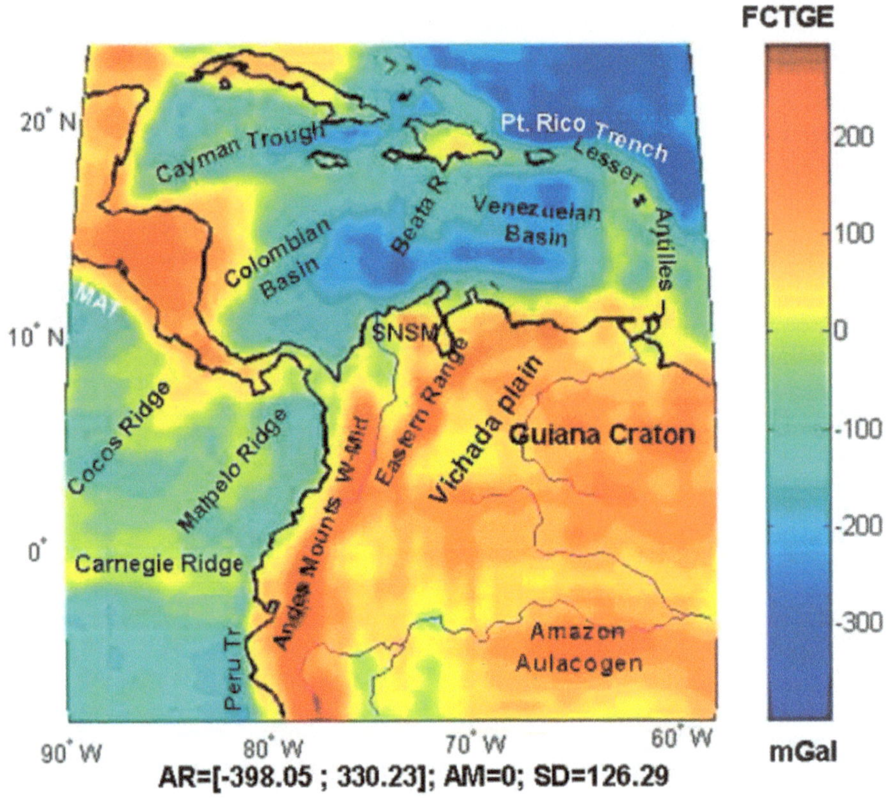

Figure. 6. FAGA-correlated TGE (FCTGE) at 20 km elevation for the study area. Map annotations include the amplitude range (AR) of (min; max) values, the amplitude mean (AM) and standard deviation (SD). SNSM = Sierra Nevada of Santa Martha, W-Mid = Western-Central ranges. This map was produced using the Albers equal-area conic projection.

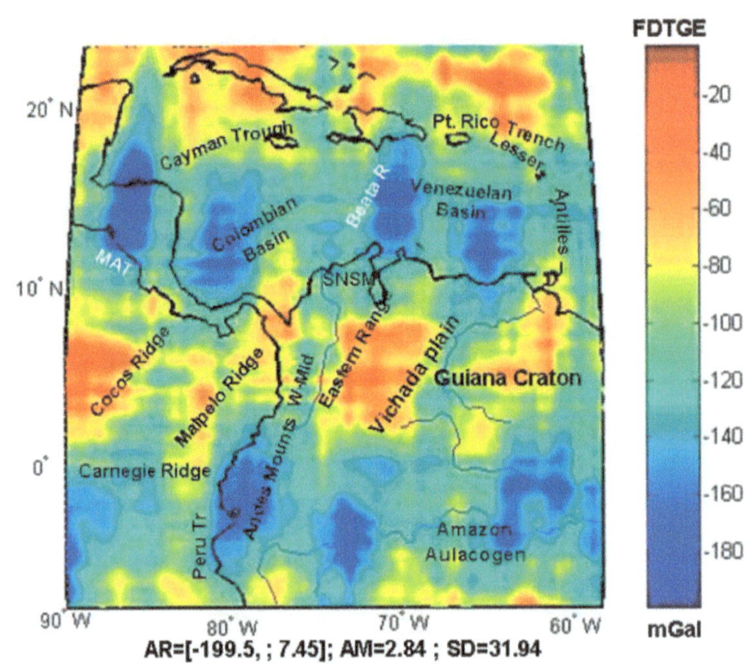

Figure. 7. FAGA-decorrelated TGE (FDTGE) at 20 km elevation for the study area. Map annotations include the amplitude range (AR) of (min; max) values, the amplitude mean (AM) and standard deviation (SD). SNSM = Sierra Nevada of Santa Martha, WMid = Western-Central ranges. This map was produced using the Albers equal-area conic projection.

ANALYSIS OF ANOMALY CORRELATIONS

TCFAGA and FCTGE were normalized to facilitate recognizing the anomaly correlations between them (von Frese et al., 1997). To enhance the visual perception of the anomaly correlations, the following transformation was applied:

$$Z_i(X) = \sigma_Z \left(\frac{x_i - \mu_X}{\sigma_X} \right) + \mu_Z \quad (3)$$

Where µx and x both represent the respective values of the mean and standard deviation of the signal X. The expression in parentheses standardizes the xi coefficients to zero mean and unit standard deviation and is dimensionless. However, the values for mean Z and standard deviation of µZ of the normalized signal Z can be specified by the user to facilitate the visual analysis.

The transformations of TCFAGA and FCTGE used the normalization of Z = 10 to facilitate plotting the two datasets with common plotting parameters (von Frese et al., 1997). Local favorability indices (Merriam and Sneath, 1966) were used to highlight the various anomaly correlations in the normalized TCFAGA and FCTGE. Positively correlated features were mapped out by summed local favorability indexes (SLFI) obtained by (4):

$$SLFI_i = \frac{(z_i(X)) - \mu_Z)}{\sigma_Z} + \frac{(z_i(Y) - \mu_Z)}{\sigma_Z} \quad (4)$$

In equation (4), Z(X) and Z(Y) were both respectively the normalized TCFAGA and FCTGE coefficients. The peak-to-peak correlations between the two data sets were mapped out by SLFIi≥8.7226 in Figure 8, whereas troughto- trough correlations were mapped out by the coefficients satisfying SLFIi≤8.7226 in Figure 9. The SLFI coefficients brought out the positively or directly correlated features, while suppressing the negatively and null correlated features between TCFAGA and FCTGE.

To enhance the perception of inversely correlated features obtained from wave numbers with negative correlation coefficients, the normalized and scaled datasets were subtracted cell by cell for differenced local favorability indexes (DLFI) by (5):

$$DLFI_i = \frac{(z_i(X)) - \mu_Z)}{\sigma_Z} - \frac{(z_i(Y) - \mu_Z)}{\sigma_Z} \quad (5)$$

Positive features in TCFAGA which were correlative with negative FCTGE features (peakto- trough) were mapped out by DLFIi≥4.8904 in Figure 10, whereas negative TCFAGA that were correlative with positive FCTGE were mapped out by DLFIi≤4.8904 in Figure 11. The DLFI coefficients emphasized the inversely correlated features, while suppressing the positively and null correlated features between TCFAGA and FCTGE. Table 1 summarizes the correlations between FAGA and TGE and their correlated filtered components.

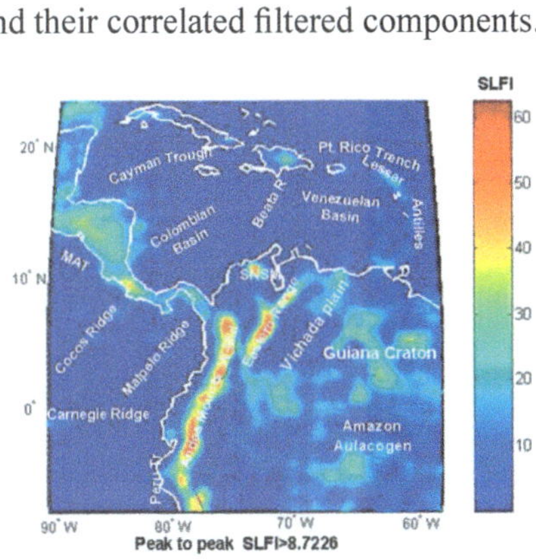

Figure. 8. Summed local favorability indices (SLFI) for TCFAGA and FCTGE at 20 km elevation for the study region showing TCFAGA-peak to FCTGE-peak correlations for SLFI ¸ 8.7226. SNSM= Sierra Nevada of Santa Martha, W-Mid = Western-Central ranges. This map was produced using the Albers equal-area conic projection.

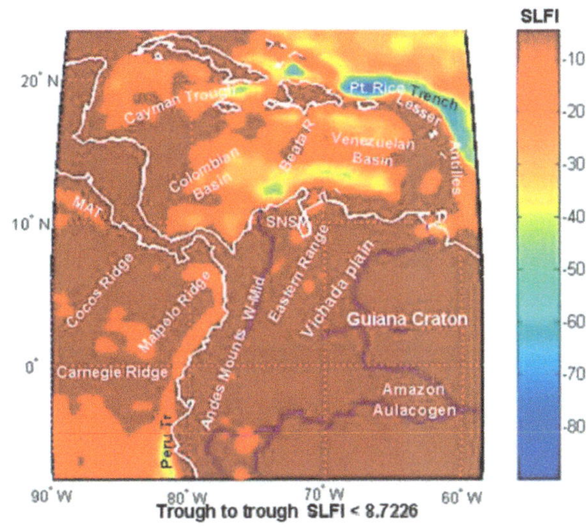

Figure. 9. Summed local favorability indices (SLFI) for TCFAGA and FCTGE at 20 km elevation for the study region showing the TCFAGA-trough to FCTGE-trough correlations for SLFI · −8.7226. SNSM= Sierra Nevada of Santa Martha, W-Mid = Western-Central ranges. This map was produced using the Albers equal-area conic projection

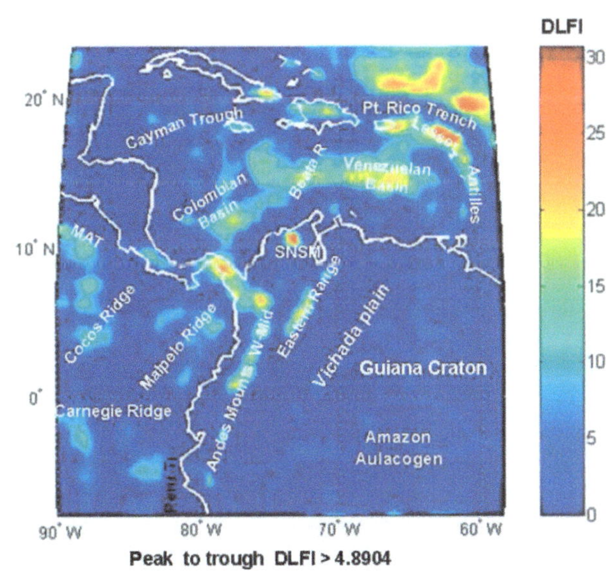

Figure. 10. Differenced local favorability indices (DLFI) between TCFAGA and FCTGE at 20 km elevation for the study region showing the TCFAGA-peak to FCTGE-trough associations for DLFI> 4.8904. SNSM= Sierra Nevada of Santa Martha, W-Mid =Western-Central ranges. This map was produced using the Albers equal-area conic projection.

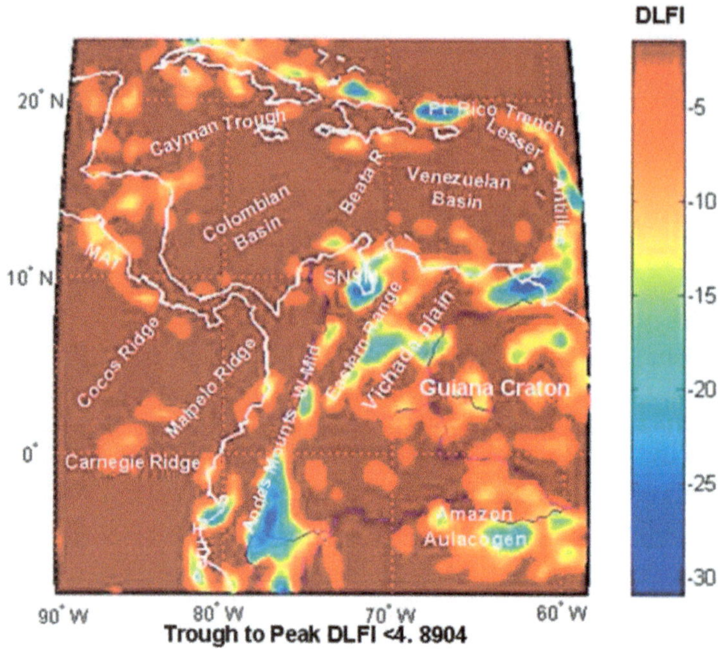

Figure. 11. Differenced local favorability indices (DLFI) for TCFAGA and FCTGE at 20 km elevation for the study region showing the TCFAGA-trough to FCTGE-peak associations for DLFI<- 4.8904. SNSM= Sierra Nevada of Santa Martha, W-Mid = WesternCentral ranges. This map was produced using the Albers equal-area conic projection.

Table. 1. Correlation coefficients (CC) between the gravity anomalies at 20 km altitude for the area from −8oS to 23.5oN latitude and from −90oW to −58.5oW longitude in northwestern South America.

CC	TGE	FCTGE	FDTGE	FAGA	TCFAGA	TDFAGA
TGE	1	0,97	0,24	0,52	0,59	0
FCTGE	0,97	1	0	0,56	0,6	0,05
FDTGE	0,24	0	1	-0,09	0	-0,02
FAGA	0,52	0,56	-0,09	1	0,89	0,45
TCFAGA	0,59	0,6	0	0,89	1	0
TDFAGA	0	0,05	-0,02	0,45	0	1

The increase in the CC between TCFAGA and TGE (CC=0.59) relative to the CC between the raw FAGA and TGE (CC=0.52) facilitates the isostatic analysis of the tectonic features of the study area. However, the increase in the CC between TCFAGA and FCTGE (CC=0.6) with respect to the CC between TCFAGA and TGE (cc=0.59) is negligible and the physical meaning of FCTGE and FDTGE is obscured. The CC between TDFAGA and FAGA (CC= 0.45) is significant in showing the strong influence of the data-grid in the anomaly decomposition of FAGA in TCFAGA and TDFAGA. Therefore, further analyses of TDFAGA are required.

RESULTS

Isostatically disturbed terrains have gravity effects that directly correlate with FAGA. The complementary TDFAGA reflect the effects of lateral density variations in the crust, mantle and core, as well as errors in the data and data processing. In the subsections below, the gravity effects of the DTM and spectral correlations with free-air gravity anomalies are interpreted for constraints on the tectonic features of the study region.

Digital Terrain Model (DTM)

The DTM in Figure 1 presents major morphological and tectonic features that produce strong terrain gravity effects in the gravity anomalies. The topography data are increasingly being integrated into the techniques to study plate boundary zones (Stein & Freymueller, 2002). One approach is to use digital terrain models (DTM) showing the forms of mountain ranges, oceanic trenches, volcanic ridges and island arcs, and oceanic basins to constrain models of the processes that produced them. The DTM of northwestern South America suggests a complex tectonic setting with mountain building and uplift as a consequence of the plate convergences that have created the North Andes Mountains, the Sierra Nevada of Santa Marta and the Cordillera Central Mountains. Flat low lands are associated with the stable cratonic terrains of the Guiana Shield. Deep Pacific Ocean trenches along the Peru - Ecuador - Colombia coastline and the Middle America Trench are associated with subduction zones. The Puerto Rico Trench in the Caribbean is associated with the Caribbean – North American subduction zone. The submarine Carnegie, Cocos and Malpelo volcanic ridges are associated with the Galapagos hot spot. The volcanic arc islands of the Lesser and Greater Antilles reflect intensive geodynamics in the region, including high seismicity and volcanic activity.

FREE-AIR GRAVITY ANOMALIES (FAGA)

The free air gravity anomalies in Figure 2 include the superposed gravity effects of the terrain and subsurface gravity variations. The FAGA amplitudes are smaller than the TGE amplitudes reflecting partial isostatic compensation of the terrain. However, the mean FAGA of 17 mGals indicates that the mean terrain is isostatically disturbed. The largest negative anomalies are over the subduction zones, whereas the strongest positive FAGA are over the Andes Mountains and Sierra Nevada of Santa Marta. Moderate positive and negative FAGA are distributed along the oceanic volcanic ridges, Guiana Craton and oceanic plates. The positive correlation between TGE and FAGA (CC = 0.52) reflects the predominant condition of isostatically disturbed terrain in the tectonically complex study region.

Terrain Gravity Effects

The terrain gravity effects (TGE) in Figure 3 produced by crust - air (2.8 gm/cm3) and crust - water (1.82 gm/cm3) interfaces are the highest density contrasts that can be defined by the lithosphere. Therefore, the minimum and maximum TGE values of -452.12 mGals and 285.02 mGals, respectively, are also the extreme values that FAGA could take in a case of 0% isostatic compensation. Negative TGE (TGEmean = −51.21 mGals) are stronger over the Caribbean Sea than the Pacific Ocean. Negative TGE overlie the Puerto Rico Trench, Cayman Trough, and Colombian and Venezuelan Basins. The Beata Ridge, Lesser and Greater Antilles have relatively positive TGE. In the Pacific, negative TGE overlie the Panama Basin and the Peru Trench, and positive TGE overlie the Carnegie, Cocos and Malpelo Ridges. At the continent, positive TGE reflect the presence of the mountainous Central America, Andes Mountains, Sierra Nevada of Santa Martha and the Guiana Craton. Relatively negative anomalies are associated with the Vichada Plain and Amazon River Aulacogens.

Terrain-Correlated Free-Air Gravity Anomalies

TCFAGA in Figure 4 have a more complex anomaly pattern than TGE inferring a wide range of isostatic disturbances of the terrain. TCFAGA varies from -190.28 to 131.66 mGals, with a mean value of zero. Thus, the TCFAGA reflect both negative and positive disturbances of continental and oceanic features. Volcanic activity and mountain building are associated with positive TCFAGA, whereas negative TCFAGA characterize trenches and troughs. Arcuate positive and negative TCFAGA can be observed along the Greater and Lesser Antilles, along the Caribbean - North American plate boundary. Therefore, the subduction of the North American plate under the Caribbean Plate is well mapped from the TCFAGA. Negative TCFAGA beneath the trenches are due to the negative density contrast (- 0.37 gm/cm3) of the subducting oceanic slab into the mantle. Positive TCFAGA over the recently formed volcanic islands are explained by oceanic crustal thickening which is not fully compensated. The TCFAGA are smaller than TGE, suggesting partial isostatic compensation from the thickening of the oceanic crust into the mantle. Negative TCFAGA are located along the Colombian - Venezuelan Basins and the Caribbean − North Andes subduction zone (Bird, 2003, CASA, 1998). The low TCFAGA amplitudes suggest that the Caribbean - North Andes subduction zone is incipient with respect to the Caribbean - North American or Nazca - North Andes subduction zones. In Central America, the positive TCFAGA from Panama to Costa Rica (60-80 mGals) and from Nicaragua to Honduras (20-40 mGals) are less prominent than the TCFAGA in the Andes (80-100 mGals) suggesting that Central America may be relatively more isostatically compensated than the Andes Mountains. The converging plate boundary between the Panama and North Andes Micro-plates proposed by Kellogg (1985) is not supported by the TCFAGA which shows no distinct anomaly between them.

At the Pacific Ocean, negative TCFAGA clearly define the Middle America Trench. Relatively positive TCFAGA are associated with the flat slab subduction of the Cocos ridge under the Panama − Costa Rica Micro-plate. Negative TCFAGA along the western margin of South America vary from -10 to -60 mGals, being comparatively smaller in the Colombian Trench. These results reflect the lower angle of subduction of the Nazca plate in contrast to the steeper angles of subduction and negative amplitudes at the Chilean - Peruvian subduction zone. Relative TCFAGA highs are associated with the flat slab subduction of the Carnegie Ridge under the North Andes Micro-plate.

At the continent, positive TCFAGA at the Andes Mountains (90 to 130 mGals) show that these mountains are isostatically disturbed. A remarkable isolated positive TCFAGA anomaly is located in northern Colombia, in the Sierra Nevada of Santa Marta (130 mGals). Here, the highest elevations of northwestern South America have also the maximum TCFAGA, showing that these mountains, reaching elevations of 5800 m, are even less compensated than the Andes Mountains suggesting that the Sierra Nevada of Santa Marta has been intensively deformed and uplifted after the Andean orogenesis.

It is not clear how these mountains were formed and why they occupy their present position. These mountains are perhaps allochthonous tectonic blocks that were moved from the east along a left lateral fault (Dewey and Pindell, 1985). The TCFAGA also support this explanation, which implies that lateral stress caused crustal shortening and uplift of the Sierra Nevada of Santa Marta without significant isostatic compensation.

TCFAGA at the Guiana Craton are close to zero suggesting nearly complete isostatic compensation. Local TCFAGA highs and lows may reflect the occurrence of lateral mass variations related to intrusive bodies without topographic expression (TGE = 0). Here, the TCFAGA that are greater than the TGE give an example of overcompensated terrain. The negative TCFAGA along the Eastern Andes Mountains are associated with continent - continent convergence between the North Andes Microplate and the South American Plate with regional thrust faults dipping to the west and intensive compressional and transpressional deformation (Cediel et al., 2003). Therefore, the negative TCFAGA along the eastern cordilleran area appear to reflect the thickening of the continental crust with thick sedimentary sequences that are known to contain oil and gas deposits (Stein & Sella, 2002).

Terrain-Decorrelated Freeair Gravity Anomalies

The TDFAGA in Figure 5 are very poorly correlated with TGE (CC = −0.0575) and mostly show gridded patterns due likely to observational and data processing noise. Thus, we do not consider the geological implications further for the TDFAGA.

TCFAGA AND FCTGE ANOMALY CORRELATIONS

Tectonic analysis of enhanced correlations between the TCFAGA and FCTGE are visualized better using summed local favorability indices with (SFLI $\geq$ 8.7226) in Figure 8. High amplitude peak-to-peak zones along the Andes Mountains overlay mapped mineralized batholithic intrusions (INGEOMINAS, 1987). According to Kerr et al. (1997), subduction related batholiths and extrusive rocks found around the margin of the Caribbean - Colombian cretaceous igneous provinces are of two distinct events: one suite represents the pre-plateau, collision-related volcanism, whereas the other suite, which is slightly younger than the plateau, may be associated with subduction. Therefore, peakto- peak zones along the western Cordillera may be associated with subduction, whereas peakto peak zones along the eastern Cordillera are associated with the pre-plateau collision suite of igneous rocks.

Intermediate amplitude peak-to-peak correlations at the Guiana Craton also indicate isolated topographic highs on relatively denser intrusive rocks of the cratonic basement. These inferred intrusions may be the source rocks of known gold and diamond placer deposits of the Guainia and Orinoco rivers (Romero et al., 1996).

The SLFI ≤ -8.7226 in Figure 9 shows troughto- trough correlations between TCFAGA and FCTGE, respectively. They clearly mark the Puerto Rico, Colombian and Peruvian Trenches, while the Middle American Trench and the Pacific Colombian Trenches are relatively poorly defined. The oceanic trenches are the most isostatically disturbed features in the area and also the most seismically active features. Of special interest are the intermediate troughto- trough correlations of TCFAGA and FCTGE in the Colombian and Venezuela Basins. These areas are undergoing intensive exploration for oil and gas and have productive fields associated with delta deposits of the Magdalena River, the major supplier of sediments to the Caribbean Sea. Therefore, the SLFI anomaly correlations may mark strong prospective areas for locating sedimentary basins in the Caribbean plate.

The differenced local favorability indices with DLFI $\geq$ 4.8904 show TCFAGA-peak to FCTGE-trough correlations in Figure 3.10. They mark topographic depressions that are overcompensated by local denser bodies along the Lesser Antilles, Colombian basin, western and central ranges, Sierra Nevada of Santa Marta and at the North American Plate. The TCFAGAtrough to FCTGE-peak correlations with DLFI $\leq$ 4.8904 in Figure 3.11, on the other hand, show the distribution of low density zones along the eastern Andean Mountains, the upper Amazon River and lower Orinoco river. The negative DLFI zones mark mapped sedimentary basins including the upper, middle and lower Magdalena valley (UMV, MMV and LMV, respectively), and the Orito, Llanos, Maracaibo, Guajira and Orinoco basins with known productive oil and gas fields. Thus, the negative DLFI zone outlined at the upper Amazon river aulacogen may also infer a prospective sedimentary basin for oil and gas exploration.

CONCLUSIONS

The central parts of the Nazca, Cocos and Caribbean Plates are mostly in isostatic equilibrium with TCFAGA close to zero. This condition changes at the plate boundary zones and volcanic ridges. Positive TCFAGA are associated with the Malpelo, Cocos, Carnegie and Beata Ridges. TCFAGA may also be associated with dynamic topography by upwelling mantle. Thickening the oceanic crust by crustal roots into the mantle partially compensates for the TGE. In the case of the Malpelo Ridge, this volcanic range is partially compensated with a crustal root, in agreement with the seamount model proposed by Watts (2001) for isostatic compensation of seamounts and oceanic islands. Depth to the MOHO estimated from gravity data is supported by seismic data (Trummer, 2002). The Andes Mountains are isostatically disturbed. The Airy-type isostatic model is consistent with the thickening of the crust at the Andes Mountains, but lateral density variations between the oceanic and continental crusts must be involved to model TCFAGA. The Pratt-type isostatic model is not appropriate to model TCFAGA of the Andean Mountains because it can not account for the thick continental roots shown in the seismic data (Trummer, 2002).

Locally positive TCFAGA anomalies in the Guiana Craton are associated with relatively denser intrusive bodies affecting the host cratonic rocks. These intrusive bodies do not have a significant surface expression and may be related to mass variations in the crust and upper mantle.

Compressional tectonics has played an important role in the mountain building of the North Andes Micro-plate as a consequence of the relative movements among Nazca, South America and Caribbean Plates and Panama and North Andes Micro-plates. The shortening associated with plate motion, based on GPS data only accounts for less than 50% of the mountain building (Lamb, 2004). Magmatic under-plating and crustal doubling due to over-thrusting may also contribute to the formation of the Andes Mountains and the Sierra Nevada of Santa Marta, as inferred by the positive TCFAGA.

Considering that isostatic equilibrium is the ideal state that any uncompensated mass tends to reach through time, implies that any uncompensated mass of the North Andes Mountains is under pressure to adjust vertically and horizontally to reach equilibrium. Therefore, a direct correlation is expected between disturbed and seismically active terrains, whereas minimal seismic activity is expected for isostatically compensated terrains (Artyushkov, 1973).

SLFI and DLFI enhance TCFAGA and FCTGE anomaly correlation, providing new criteria to locate areas for prospecting mineral and fossil fuel deposits.

The results presented here must be used with care. It has been assumed a simplification of the gravity effects at 20 km altitude, where a compensating mass at 35 km depth has only about 40% of the amplitude of the gravity signal of the surface mass that it compensates; therefore, positive and negative FAGA do not mean, on their own, a lack of compensation. For example, a maximum FAGA of about 184 mGal is greater than the roughly 110 mGal expected for fully compensated terrain, the difference reflects the true lack of compensation.

ACKNOWLEDGMENTS

We thank Mohammad Asgharzadeh and Laramie Potts, from the Geological Sciences Department of the Ohio State University, for their assistance in the inverse gravity modeling. Elements of this research were supported from FULBRIGHTLASPAU, Universidad Nacional de Colombia, COLFUTURO and Sociedad Minera Kedahda S.A.

REFERENCES

1. Artyushkov, E. V. (1973). Stresses in the lithosphere caused by crustal thickness inhomogeneities. Journal of Geophysical Research. 78. 7675-7708.
2. Bird, P. (2003). An Updated digital model of plate boundaries, Geochemistry, Geophysics Geosystems G3. 4 (3).1-52.
3. Cediel, F., Shaw, R. P., and Caceres, C. (2003). Tectonic assembly of the Northern Andes Block. The Circum-Gulf of Mexico and Caribbean Hydrocarbon habitats, basin formation and plate tectonics. AAPG Memoir. 79. 815-848.
4. Davis, J. C. (1986). Statistics and data analysis in geology: John Wiley & Sons Inc. New York, 646 pp.
5. Dewey, J. F. and Pindell, J. L. (1985). Neogene block tectonics of eastern Turkey and Northern South America: Continental applications of the finite difference method. Tectonics. 4. 71-83.
6. von Frese, R. R. B. and Tan, L. (1999). Antarctic crustal modeling from spectral correlation of freeair gravity anomalies with the terrain. Journal of Geophysical research. 104 (11). 275-296.
7. von Frese, R. R. B., Jones, M., Kim, J. W. and Kim, J. H. (1997). Analysis of anomaly correlations. Geophysics. 62 (1).342-351.
8. von Frese, R. R. B., Alsdorf, D. E., Kim, J. H., Stepp, T. M., O´ Connell, D. R. H., Hayden, K. J. and Li, W. S. (1992). Regional geophysical imaging of the Antarctic lithosphere, in Recent Progress in Antarctic Earth Science, edited by Yoshida, Y., Kaminuma, K., and Shiraishi, K. pp. 465-474, TERRAPUB, Tokio, Japan.
9. von Frese, R. R. B. (1980). Lithospheric interpretation of satellite elevation gravity and magnetic anomaly data, PhD dissertation, Department of Geosciences, Purdue University, W. Lafayette, IN. USA. 1-164.
10. INGEOMINAS (1987). Recursos Minerales de Colombia. Tomos I, II, Pub.Geol.Rsp.1, Bogota, 1119 pp.
11. Kellogg J. N., Vega, V. Stalling, T. C., and Aiken, L. V. (1995). Tectonic development of Panama, Costa Rica and the Columbian Andes: Constraints from Global Position System geodetic studies and gravity, in Geol. and Tect Devel of the Caribbean Plate boundary in Southern Central America. Geol. Soc. Am. Spec. Pap. 295. 75-86.
12. Kellogg, J. N., Ogujiofor I. J. and Kansakar, D. R. (1985). Cenozoic tectonics of the Panama and North Andes blocks, Mem. Congr. Latinoam. Geol. 6. 40-59.
13. Kerr, A. C., Tarney, J., Marriner, G. F., Nivia, A., and Saunders, A. D. (1997). The Caribbean - Colombian Cretaceous Igneous Province: The Internal anatomy of an Ocean Plateau. AGU Geophysical Monograph. 100. 123-144.
14. Kim, J. W., von Frese, R. R. B., and Rae, K. H. (2000). Crustal modeling from spectrally correlated free-air and terrain gravity data A case study of Ohio. Geophysics. 65 (4). 1057-1069.
15. Lamb, S. (2004). Devil in the Mountain, A Search for the origin of the Andes. Princeton University Press, New Jersey, 335 pp.
16. Leftwich, T. E., von Frese, R. R. B., Potts, L. V., Kim, H. R., Roman, D. R., Taylor, P. T., and Barton, M. (2005). Crustal modeling of the North Atlantic from spectrally correlated free-air and terrain gravity. J. Geodynamics. 40. 23-50.
17. Lemoine, F. G., Chinn, D. S., Cox, S. M., Factor, J. K. Kenyon, S. C., Klosko, S. M., Luthcke, S. B., Olson, T. R., Pavlis, E. C., Pavlis, N. K., Rapp, R. H., Smith, W. H. F., and Sandwell, D. T. (1994). Bathymetric prediction from dense satellite altimetry and sparse shipboard bathymetry. J. Geoph. Res. 99. 21803-21824.
18. Liu M., Yang Y., Stein S., and Klosko, E. (2002). Crustal shortening and Extension in the Central Andes: Insights for a Viscoelastic model. AGU Geod. 30. 325-339.
19. MATHWORKS, inc. (2005). MATLAB version 7.0.4 352 (R14). http://www.mathworks.com/.
20. Merriam, D. F. and Sneath, P. H. A. (1966). Quantitative comparison of contour maps. J. Geophysical Research. 7. 1105-1115.
21. Roman, D. R. (1999). An integrated Geophysical Investigation of Greenlands Tectonic History. PhD Dissertation. The Ohio State University, 1- 270.
22. Romero, F. H., Schultz, R. A., and Chavez, T. (1996). Ocurrencias diamantíferas de la quebrada Alba Rosa, caño Guapito, Guainia, Colombia. Geol. Col. 20.
23. Smith, W. H. F. and Sandwell, D. T. (1994), Bathymetric prediction from dense satellite altimetry and sparse shipboard bathymetry. Journal of Geophysical research. 99. 21803- 21824.
24. Stein, S. and Freymueller, J. T. (2002). Plate Boundary Zones, editors. AGU Geodynamic Series. 30. 425 pp.
25. Stein, S. and Sella, G. F. (2002), Plate boundary zones: Concepts and approaches. AGU Geodynamics series. 30. 1-26.
26. Tan, L. and von Frese, R. R. B. (1997). Satellite geopotential models of East Asian Lithosphere (Abstract), Trans. AGU. Spring meeting supplements. 78. 114-121.
27. Trummer, I. (2002). S-Wave processing and interpretation of wide-angle seismic refraction data, Malpelo ridge, eastern Panama basin, PhD dissertation, der Mathematisch- Naturwissenschaftlichen Fakultat, der Christian Albrechts- Universitat, Zu Kiel.
28. Watts, A. B. (2001). Isostasy and flexure of the lithosphere. First edition, Cambridge University Press, New York, 458 pp.

CHAPTER 20

PLATE TECTONICS DRIVE TROPICAL REEF BIODIVERSITY DYNAMICS

Fabien Leprieur, Patrice Descombes, Théo Gaboriau, Peter F. Cowman, Valeriano Parravicini, Michel Kulbicki, Carlos J. Melián, Charles N. de Santana, Christian Heine, David Mouillot, David R. Bellwood, and Loïc Pellissier

Tropical shallow reefs support the world's greatest diversity of marine organisms[1] and have experienced a long history of climate-driven habitat change[2,3]. Sea level and temperature fluctuations during the Quaternary have strongly shaped the distribution of coral reefs and their associated fish faunas[3]. Nevertheless, most of the diversification of tropical marine faunas predates the Quaternary[4], arguing for a major role of older historical events[2,4,5]. Because plate tectonics modulate the distribution of seafloors through continental movement and collision over geological timescales[6], they are potentially one of the main determinants of tropical marine biodiversity[2,5]. The fossil record indicates that the location of the major marine biodiversity hotspot has moved across the globe during the last 50 million years[2], transitioning from the western Tethys during the Eocene (55–33 Ma), to the Arabian Peninsula and Western Indian Ocean during the Late Eocene and Oligocene (37–15 Ma), and finally to the Indo-Australian Archipelago (IAA) in the Miocene (15–5 Ma). This movement was primarily driven by successive closure of the Tethys followed by the collision of the Arabian and Indian plates with Eurasia. Yet, the role of plate tectonics in shaping diversification and dispersal processes of tropical marine biodiversity still remains elusive. Correlative approaches have had limited success in disentangling alternative hypotheses and mechanistic models may provide a clearer understanding of the processes shaping marine biogeographic patterns[7,8].

We coupled a mechanistic model of species diversification to a model of synthetic paleobathymetry estimates for the past 140 Myr. The gridded paleobathymetry model is based on global ocean paleo-age grids[6] merged with paleocoastline distributions for continental areas[9]. The model outlines the absolute position of continents over time and the associated distribution and residence time of shallow epicontinental seas. Using a novel dynamic model that simulates clade dispersal, speciation and extinction from a single ancestral species, bounded by a temporal sequence of habitat availability, we show that plate motions predict the emergence and movement of diversity hotspots on tropical reefs. Incipient species emerge within (that is, sympatric) or adjacent (that is, parapatric) to the geographical range of its ancestor[10,11] and disperse at a distance *d* during each time step. Parapatric speciation

Citation: Leprieur, F. et al. Plate tectonics drive tropical reef biodiversity dynamics. Nat. Commun. 7:11461 doi: 10.1038/ncomms11461 (2016).

arises when a species range is split into one or more distinct areas separated by a distance d_s, while sympatric speciation arises within a cell with a probability P_s. Consistent with the speciation-area hypothesis[12], species occupying a large area show higher sympatric speciation frequency $S=P_s \times A$. Regardless of the speciation mode, the models predict the emergence and movement from the western Tethys towards the IAA. The diversification model assumes that speciation and extinction are modulated exclusively by variations in habitat size and spatial distribution. The model is based on strict niche conservatism, but provides a realistic picture of the historical dynamic of marine reef fauna.

RESULTS AND DISCUSSION

Hopping Biodiversity Hotspots

We show that the spatial distribution of tropical reef habitats over time (Fig. 1;) explains the emergence and movement of coral reef biodiversity hotspots since the Cretaceous with both parapatric and sympatric speciation modes (Fig. 2). The parapatric model (d=4, d_s=5) provided better estimates of fossil biodiversity in the Eocene (40 Ma, R^2=0.26) and the Miocene (20 Ma, R^2=0.38) than the sympatric model (d=5, P_s=6e−5, 40 Ma: R^2=0.12; 20 Ma: R^2=0.13). The paleobathymetry model suggests that the Tethys contained extensive but disjointed reef habitats promoting both modes of speciation, but the biodiversity peak in the Western Tethys is most consistent with parapatric speciation fuelled by waves of disconnection and reconnection of isolated reef regions . The highly dynamic and complex tectonics of the Western Tethys[13] provided fertile ground for parapatric speciation. Modelled diversification, computed as speciation minus extinction rates, peaked at 85 Ma when fragmentation was highest, underpinning the biodiversity hotspot in Western Tethys. During the mid/late Eocene–Oligocene times, the closure of the Tethys was associated with a narrowing coastal corridor[14], that allowed Tethyan biodiversity to move eastward (Fig. 2), supporting the hypothesis of faunal migration into the IAA[2,15]. Since high faunal turnover is expected at the transition across the Cretaceous–Paleogene (K/Pg)[16,17], which is not considered in our habitat reconstructions, a sensitivity analyses demonstrated the robustness of this dynamic to a high degree of forced extinction at 66 Ma (80%,). In addition, our results showed robustness in regards to the choice of plate kinematic model used to reconstruct plate positions. Together, these results indicate that the Tethys was a major diversification and emigration hotspot explaining the Tethyan origin of most tropical shallow reef lineages, here shown for corals, but with parallels in fishes[18] and foraminifera[2].

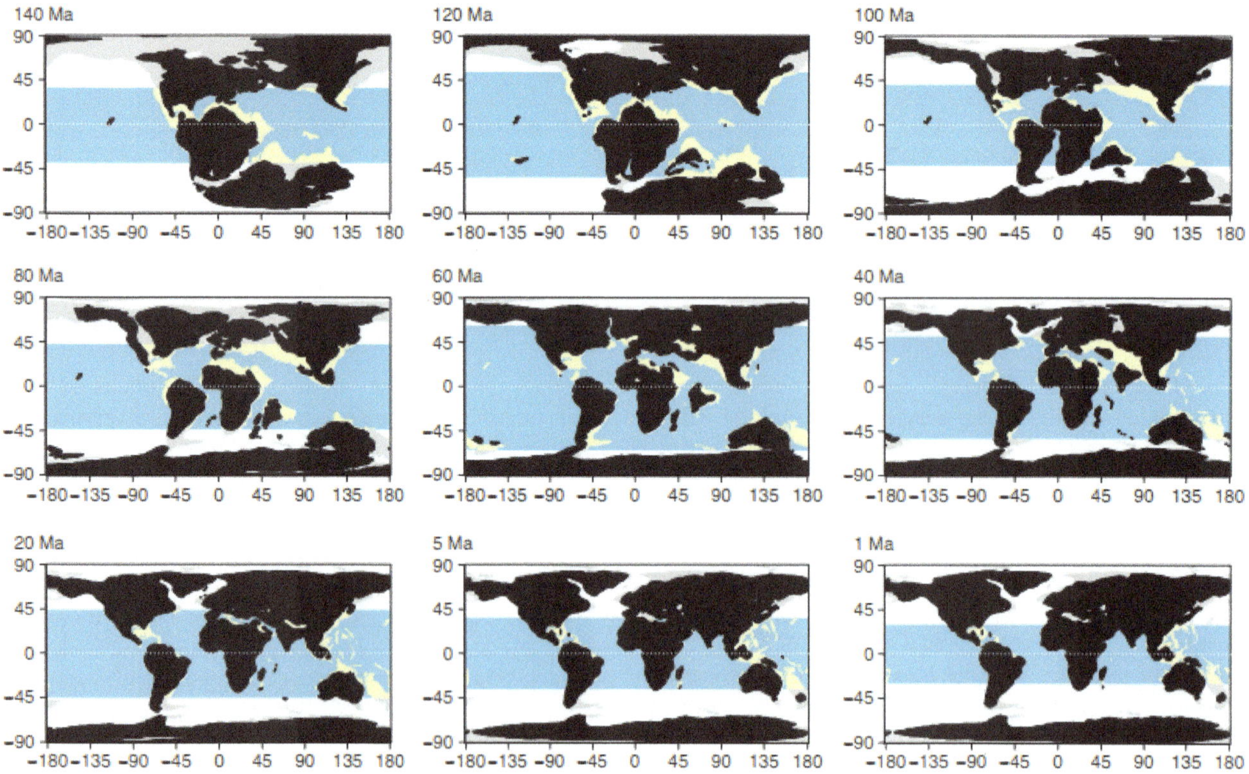

Figure 1: Distribution of shallow and deep ocean sea floor across the past 140 Myr.

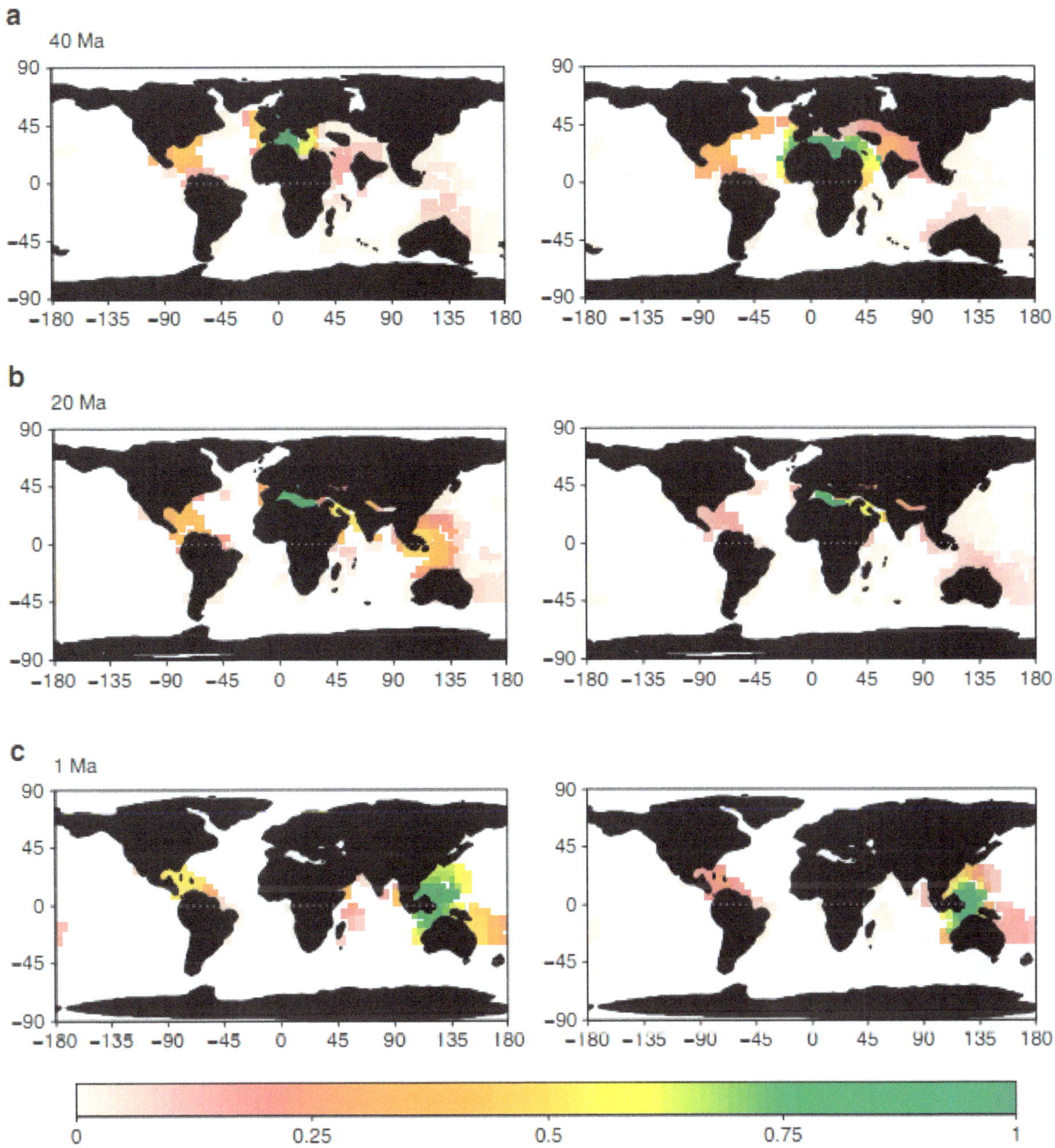

Figure 2: Hopping biodiversity hotspots.

Current Biodiversity Patterns

We examined whether the parapatric and sympatric models predicted realistic current biodiversity patterns for two globally distributed coral (Acroporidae) and fish (Labridae) clades. Parapatric (R^2=0.36 and R^2=0.42) and sympatric R^2=0.46 and R^2=0.45) models both explain global variation in coral and fish species richness, which peaks in the Central Indo-Pacific region, and more particularly in the IAA, centred between Indonesia and Papua New Guinea (Fig. 2). The lower diversity in the Atlantic is not only the result of less extensive Quaternary refugia[3], but also lower opportunity for immigration and speciation since the Eocene than in the geologically complex Central Indo-Pacific region[18,19] . The models also provided realistic values of between-assemblage (β) diversity for both parapatric (Acroporidae: mantel r_m=0.56; Labridae: r_m=0.63) and sympatric (Acroporidae: r_m=0.76; Labridae: r_m=0.77) speciation modes. Observed and predicted values of assemblage dissimilarity increase with sea distance, and the model captures the high level of faunal turnover between the Atlantic and Indo-Pacific oceans[20] , which were largely disconnected after the closure of the Tethys Sea[18]. Overall, these results demonstrate the important role played by isolation and plate tectonics in shaping patterns of speciation and dispersal, and in explaining the present-day longitudinal gradient in tropical marine shallow reef biodiversity.

Nestedness Patterns and the Merging of Lineages

The parapatric model accurately (Acroporidae, Spearman' rank correlation: r_s=0.54, P<0.001, Labridae,r_s=0.59, P<0.001) mimicked the nested structure of assemblages in the Central Indo-Pacific region[19,21](Parapatric model: $NODF_{sites}$=33.2, P<0.001; fish: $NODF_{sites}$=82.9, P<0.001;

coral: $NODF_{sites}=90.5, P<0.001$; Fig. 3). Assemblages in peripheral and species-poor cells are composed of species that constitute subsets of species that occur in successively richer cells of the IAA (Fig. 3). Furthermore, these peripheral areas display distinct tropical reef faunas, which are also mimicked by our mechanistic model. The mechanistic model predicts that two faunas previously separated, from Tethys and coastal Australia, fused when the Sunda and Sahul shelves entered in collision. Among the candidate lineages of Australian origin, a biogeographic reconstruction based on the labrid phylogeny suggests that the hypsigenyine and pseudocheiline lineages became isolated in the Central Indo-Pacific region in the Eocene before the emergence of the IAA. Those may have colonized the Australian coast across the Southern Indian Ocean. In contrast, the scarines, a lineage with earliest fossils in Tethys, represents a more recent colonization in the Miocene from Tethys. This, together with other evidence for sharks[22] and corals[23]may suggest the existence of an Australian endemic marine fauna before the Miocene. The confluence of two major faunal elements, Tethyan and Australian shaped the accumulation of biodiversity in the IAA[18], which provides a marine analogue to the juxtaposed terrestrial taxa across the Wallace line[24].

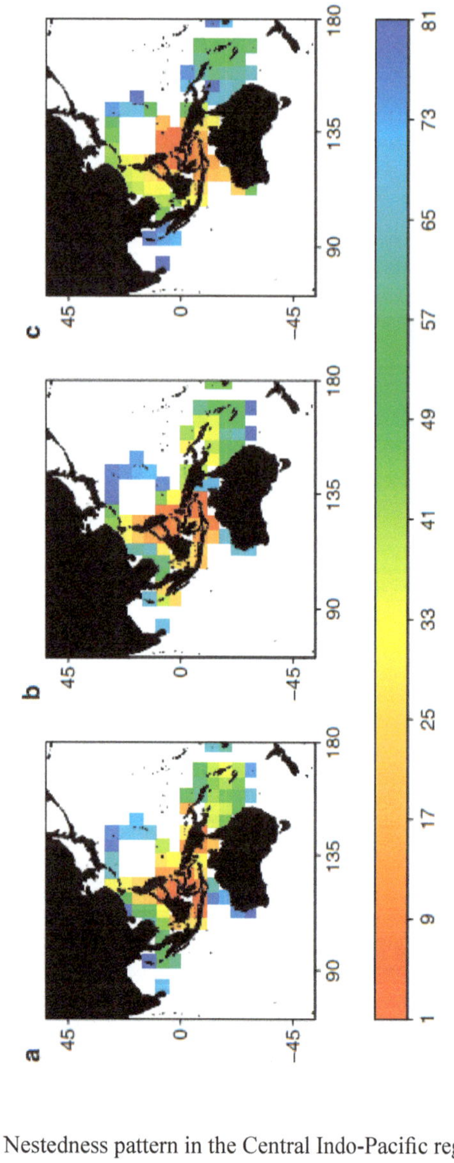

Figure 3: Nestedness pattern in the Central Indo-Pacific region.

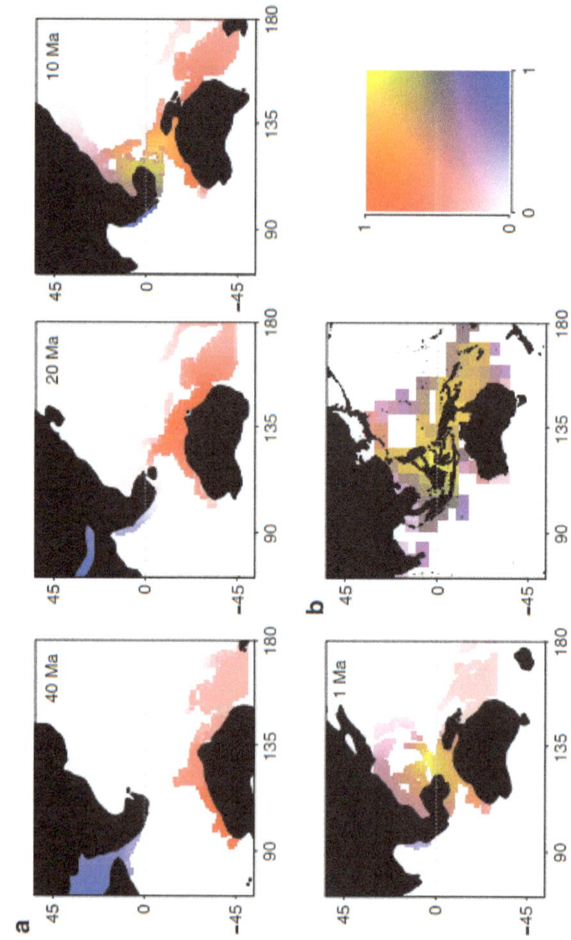

Figure 4: Biodiversity patterns through time.

Paralleling major radiations in terrestrial systems associated to island formation[25] or orogeny[26], our study highlights how the dispersal and amalgamation of continental plates since the Cretaceous has shaped the dynamics of biodiversity for two major groups of tropical marine organisms, namely corals and fishes. Both speciation modes emphasize that diversification of shallow reef biodiversity and its movement across the globe was contingent on habitat configuration (Fig. 2). Our model corroborating the nestedness pattern of assemblages suggests that the marine fauna in IAA was formed, at least in part, by the merging of Australian and Tethyan faunas. Yet, determining the exact origin of extant lineages will remain challenging given the potential for long distance dispersal in the marine realm, which, together with extinction, may blur phylogeographical inferences[18]. Together, our next-generation dynamic macroecological model provides new hypotheses with which to explore and to interpret empirical patterns, such as the fusion of the Tethyan and Australian faunas in the IAA (Figs 3 and and4).4). While our model assumes niche conservatism, ecological diversification is also expected to have driven diversification on tropical[11,27] but bounded by available tropical reef habitat through time. Incorporating ecological niche evolution together with habitat changes into a mechanistic model of species diversification would be a new avenue of research for understanding tropical marine biodiversity.

Our results remain contingent on the quality of the synthetic paleobathymetry, reconstructed from multiple data sources with their own uncertainty, propagating up from simple navigational errors for the marine surveys collecting magnetic anomaly data[6] to larger spatio-temporal uncertainties associated to the location of paleo-shorelines. We used two absolute plate kinematic models based on absolute moving hotspot reference frames, one of them corrected for the effects of true polar wander (TPW) before 100 Ma (ref. [28]). TPW is an effect that causes the whole mantle and lithosphere to rotate latitudinally with respect to the Earth's spin axis, commonly only detectable by paleomagnetic reference frames and not visible in mantle/hotspot reference frames[29]. However, the magnitude of TPW over the past 140 Ma is debated[29]. Our two input plate kinematic models result in identical paleolatitudinal positions for a hypothetical point (20° E 33° N) on the northern edge of Africa for the time from 100 Ma to the present, where most of the species diversification occurred. Comparison with a pure paleomagnetic model[29] shows that there are up to 700–800 km (at 60 Ma) of potential paleolatitudinal difference in the location of this point between our models and the paleomagnetic model. The habitat complexity shaping reef biodiversity would still remain in the tropical climate zones even considering a paleomagnetic reference frame and our results are robust to this uncertainty. Moreover, our models are corroborated by fossil records, indicating a prolonged state of tropical climatic conditions associated to habitat complexity in the Western Tethys at the end of Cretaceous to Eocene, fuelling species diversification[2,30].

Here, we highlight the major role of plate tectonics in shaping biodiversity patterns in the tropical marine realm, but we acknowledge that other shifts in oceanic conditions such as episodic high temperature, acidity levels[31], or ecological conditions such as habitat complexity[32] have also influenced biodiversity dynamics. In addition, even though our process-based model provides realistic patterns of tropical marine biodiversity dynamics, this does necessarily mean that the process is realistic, arguing for further validation. Overall, our study paves the way toward a new generation of spatially explicit diversification models[8], integrating paleo-environmental changes, providing a potential mechanistic understanding of extant biodiversity patterns.

METHODS

Bathymetry Layers and Tropical Limit Across Time

Plate tectonics drive the evolution of oceanic basins and mountain chains through the motions of continental plates across the surface of the Earth[6]. Plate kinematic models provide insights on whether and when plates are located in favourable latitudes for the establishment of tropical marine biodiversity associated with shallow water reef systems. We used two different absolute plate motion models extending back to 140 Ma, constructed using relative plate motions derived from current marine magnetic anomalies and oceanic fracture zones[33]. Different choices of reference frames determine the absolute positions of plates back in time for each model. The first model (Model 1), is based on a combined moving Indian/Atlantic Hotspot reference frame[34]back until 100 Ma, and a fixed hotspot reference frame[35] back to 140 Ma. For the Pacific Plate, we used the absolute reference frame from Wessel *et al.*[36] The second model (Model 2) uses a plate kinematic framework published by Seton *et al.*[37] This second model uses a hybrid reference frame, which merges a moving Indian/Atlantic hotspot reference frame back to 100 Ma (ref. [34]) with a paleomagnetically derived true polar wander corrected reference frame[38] back to 140 Ma.

For subducted ocean crust, the marine magnetic anomalies of *in situ* oceanic crustal remnants are utilized in combination with geological information and paleomagnetic data from accreted terranes and microcontinents along with the rules of plate tectonics to generate sets of synthetic paleo-age grids for oceanic lithosphere for the past 140 Myr. Due to the age-depth relationship related to the cooling of oceanic crust, one can compute the paleobathymetry of oceanic crust, assuming either no sedimentation or relatively simple sedimentation patterns based on present-day observations[6]. We used the GDH-1 cooling model[39] to compute the paleobathymetry grids for each time step. Estimates for continental topography, including continental margins, however, do not follow such relatively simple rules due to the effects of tectonic deformation, sediment loading and erosion. We used a set of newly digitized paleo-shorelines[9] from the compilation of Smith *et al.*[40] to generate simple continental shelf topography fringing continental plates at each reconstruction time step. For both models, the depth at the continent–ocean boundary is determined by age of the adjacent oceanic crust. We combine both data sets to generate synthetic paleobathymetry models with a 1° spatial resolution for the past 140 Myr in 1 million year time steps. Shelfal area are assigned a constant bathymetry of −200 m in the grids, but this represents a coarse depth approximation of the entire shelfal area which in reality displays a much more complex bathymetry. The bathymetric reconstructions only partially considers volcanic activity leading to shelf emergence and we did not model the biodiversity of the Tropical Eastern Pacific that contains many islands of volcanic origin.

Since the study focuses on the history of tropical shallow reef fauna and because climate strongly fluctuated in the last 140 millions of years[41], we also reconstructed the paleo-latitudes of tropical ocean limits from reef-forming coral fossil records. Reef-forming coral species can only develop at warm water temperature >25 °C (ref. [42]) and thus the paleo-distribution of this group constitutes a good indicator of the past latitudinal tropical borders. We collected fossil occurrences of scleractinian corals from paleoDB (www.paleodb.org) corresponding to 31,392 occurrences[43] from −140 Ma to the present. From the reconstructed paleo-latitude for each million year time slice, we computed the 95th percentile of the paleo-latitude at which corals were living to inform on the latitudinal border of tropical oceans across time. By combining reconstructed shelfal areas with tropical limits, we generated 1 map per million years of tropical shallow marine habitats for the last 140 Myr.

Mapping Current and Past Biodiversity Patterns

Current biodiversity patterns were extracted from a global-scale distribution database[3,44] for tropical reef fishes and from the IUCN (http://www.iucnredlist.org/technical-documents/spatial-data) for Scleractinian corals. By examining almost 500 references and extracting information from published works, regional checklists, monographs on specific families or genera, and reports, we obtained information on presence/absence for

6,316 reef fishes in grid cells of 5° × 5°, corresponding to ~555 × 555 km at the Equator. To compare model simulations and observed biodiversity gradients, we focused on one coral clade and one fish clade based on two criteria: the clade should show a global distribution and have a stem age within the time frame of the simulations. For the fishes, we selected the Labridae family (n=559 species) with a median age of 76±10 Myr obtained from the dated phylogeny. For the corals, we considered the Acroporidae family (n=274 species) also with a global distribution and with the earliest fossil in the data set dating to 131±5 Ma.

To compare simulated past biodiversity patterns to empirical data, we used fossil occurrences of corals obtained from PaleoDB (www.paleodb.org). We mapped ancient coral species richness for three epochs with sufficient fossil occurrences, the Miocene (20 Ma), Eocene (40 Ma) and late Cretaceous (70 Ma). We retained only occurrences of fossils that were identified at the species level (that is, 25,077 species occurrences). Occurrences were converted into diversity maps by computing the sum of species present in the surrounding 40° × 40° window of each 5° × 5° cell when the fossil date range overlapped with the date of interest. Diversity maps were rescaled between 0 and 1.

Landscape-based Model of Biodiversity Dynamics

The use of correlative approaches in macroecological studies does not provide a mechanistic understanding of the way historical processes have shaped the distribution of alpha- and beta-diversity. To disentangle causalities for emergent patterns, the application of dynamic simulation models provide more understanding on the processes than statistical approaches[8,19]. In addition, phylogeography and diversification models are generally calibrated without information on the distribution of paleo-habitats and biogeographic history[45,46]. Ideally, diversification analyses should consider the historical sequence of habitat distribution, which implies to explicitly integrate landscape dynamics in the model. Accounting for shifting habitat through time within a dynamic model of speciation and dispersal was proposed by Gotelli *et al.*[8] under the term general simulation model. General simulation model should track the presence (1) or absence (0) of a given species in a particular grid cell, the phylogenetic history of lineages, and historical biodiversity information which can be compared with palaeontological or phylogenetic evidences[8]. Nevertheless, this idea was so far never applied to real data, possibly because of the lack of habitat reconstruction for deep time periods[47].

Here, we develop a diversification model constrained by shallow tropical reef maps. The simulation starts with a unique ancestral species and new species originate either within (that is, sympatric) or adjacent to (that is, parapatric) the geographical range of its ancestor. In early Cretaceous times (~140 Ma), a continuous passive continental margin along the northern coast of Gondwana provided less opportunity for geographical separations with limited faunal beta-diversity[48]. We selected the simulation start that should be the most representative of all the possible lineages arising from the breakup of Gondwana. For younger clades, for example, those arising in the Miocene, it would be more appropriate to start in a subregion of an earlier time step and assess the sensitivity to the starting point. The model starts 140 Ma with one species occupying the entire available suitable cells along the coast of Gondwana which represented a well-connected habitat. Yet, other distant regions not connected to Gondwana likely harboured allopatric lineages that also contributed to extant biodiversity especially in the IAA.

The model records at any single point in time within a data frame the distribution of each species in each cell, together with its genealogy. During each million year time slice the simulation performs two phases:

The first phase is the speciation. We considered two modes of speciation, parapatric and sympatric implemented within two alternative models as well as a model combining both modes of speciation (the combined model hereafter). We refer to parapatric speciation because it is mainly dispersal limitation and not geographic isolation which drives divergence and speciation. Parapatric speciation arises when a species range is split into two or more distinct areas separated by a sea distance larger than d_s . This mode of speciation was modelled as a cluster splits using cluster optimization algorithm based on the 'dbscan' function in the fbc R library[49]. If the range of a species i is fragmented into n patches separated by a given distance threshold of d_s, the species will separate into n species with smaller ranges. The model assumes that discontinuous range after 1 million years of separation generates full speciation with no possibility of hybridization or introgression. The main process generating species with the parapatric model is the successive phases of habitat patch disconnection and reconnection. For the sympatric speciation mode, sympatric speciation arises stochastically in a cell occupied by a species i with a probability P_s . Consistent with the speciation-area-hypothesis[12], species occupying a large area, A, will show higher sympatric speciation frequency because $S=P_s \times A$.

The second phase is the dispersal. The species in the time step t are allowed to disperse to all habitat cells at the time step t+1 that are distant by a threshold value lower than d from a Weibull dispersal kernel. This threshold is randomly sampled for each species at each time step. If a given species does not find a habitat cell within the dispersal distance, it will go extinct only if all the habitat cells inhabited by the species at time $t$disappear at time t+1.

The model can account for the parapatric mode (two parameters), the sympatric mode (two parameters) or both (three parameters) and can make inferences about assemblage properties (alpha-, beta-diversity), diversification rates (that is, speciation and extinction rates) and phylogenetic branching (that is, phylogeny). We run the simulation for a range of dispersal ($d \in \{1°:40°\}$) and speciation ($P_s \in \{1e^{-5}:1e^{-4}\}$, $d_s \in \{1°:40°\}$) rates. Values beyond those ranges were unrealistic as they predict no species (full extinction), too many species (for example, >20,000, especially when $d > d_s$) or predict all species everywhere. The model was run until 1 Ma (and not to the current period) since we were interested in highlighting the geological contingencies on tropical reef assemblages. The effects of glaciations during the Quaternary (2.6 Ma to Present) have been studied elsewhere[3]. Therefore, we did not consider Quaternary climatic fluctuations in the current study. Parameter combinations (d, d_s, P_s) were run 20 times, the stochastic component in the dispersal kernel of the parapatric model and a larger stochastic component linked to cell speciation probability in the sympatric model. We ran a total of 16,000 simulations.

We highlighted the best model (BM) among all simulations across a set of empirical validations including (1) current alpha and beta-diversity of corals and fishes globally and for the Central Indo-Pacific region, (2) past species richness 70, 40 and 20 Ma based on fossil coral data. For fossil coral data, we focused on those three time periods because they were the most differentiated and contained the most available fossil information. We compared observed and predicted current beta diversity patterns using the pairwise Jaccard's dissimilarity index. We compared observed (for corals and fishes) and predicted species richness using a simple regression model from which we extracted the coefficient of determination (R^2). Because models differed in the number of estimated parameters, we also computed the Bayesian information criterion (BIC) of each simulation as follow:

$$\mathrm{BIC} = \log(n)k + n\log(\mathrm{RSS}/n) \quad (1)$$

Where RSS represent the difference between scaled observed and predicted cell values, n the number of samples and k the number of model parameters. For each of the comparisons (fossil diversity gradients, current alpha and beta diversities), we summed the BIC to rank the simulations. We obtained general models of marine faunal dynamics explaining shared properties in the historical biodiversity dynamics of both corals and fishes.

Nestedness and Species Turnover Patterns

The IAA also called the coral triangle (that is, the Indonesian archipelago, the Malaysian peninsula, the Philippines, New Guinea and northern Australia) is widely recognized as the global centre of modern marine biodiversity[50]. At the scale of the Central Indo-Pacific region, the number of reef fishes and corals was found to decrease with increasing distance from the coral triangle, with the tendency that distant assemblages being nested within those of the coral triangle[19,21,51]. We therefore evaluated if the parapatric and sympatric models faithfully reproduced this biogeographical pattern. To do so, we used the NODF index based on paired overlap and decreasing fill[52] to measure the extent to which species present in species-poor cells constitute proper subsets of those ones present at species-rich cells. More specifically, we calculated the NODF index among sites (hereafter $\mathrm{NODF}_{\mathrm{sites}}$) that ranges from 0 (cells are not nested) to 100 (cells are perfectly nested). To determine the probability that nestedness could be obtained by chance, we compared the observed $\mathrm{NODF}_{\mathrm{sites}}$ value to those obtained by a null model where species-presence matrices are randomly drawn (n=9,999) under the sequential swap algorithm[53]. Last, we extracted the rank order of sites in the maximally packed matrix (that is, maximally ordered) for simulated occurrence data under the parapatric and sympatric models and observed fish occurrence data. We assessed the level of congruency between observed and predicted rank orders using a Spearman's rank correlation test.

Because differences in species composition between places may result from both nestedness and species turnover (that is, the replacement of some species by others between assemblages), we also quantified the levels of species turnover across the Central Indo-Pacific region using a pairwise turnover metric (β_{jtu}) that is independent of species richness differences[54]. We quantified the level of congruency between observed and predicted distance matrices using a Mantel test (999 permutations). To visualize observed and predicted patterns of species turnover, we plotted the corresponding distance matrices along reduced axes using the non-metric multidimensional scaling neighbour-joining algorithm. The ordination results were then plotted and mapped by assigning a colour to each grid cell according to its position in the ordination space.

Phylogenetic Reconstruction from Molecular Sequences

We reconstructed the phylogenetic relationships of species occurring in tropical reefs of the Labridae family together with outgroups according to previous phylogenies[55]. Molecular sequences from published sources were downloaded on GenBank focusing on the most frequent loci. Those loci were then concatenated to perform a supermatrix approach for phylogenetic reconstruction with sufficient taxon overlap for each loci. We used four mitochondrial gene regions (non-coding: 12S and 16S, coding: Cytb and CO1) and three nuclear genes (RAG2, S7 II and tmo4c4), getting data for 55% of known species. Each marker data set was independently aligned using the algorithm MUSCLE with standard settings and alignments were then manually edited using MEGA 6 (ref. [56]). We tested substitution models using a maximum likelihood approach with a random starting tree for each gene alignment and we selected the best models using the BIC criterion on Jmodeltest 2.1.6 (refs [57, 58]). Alignments were afterwards concatenated in a supermatrix using SequenceMatrix[59]. We then analysed the supermatrix using a Maximum Likelihood approach via the algorithm RAxML (ref. [60]) implemented in RAxMLGUI (ref. [61]). We performed 10 parallel runs to fully explore tree space and to avoid being retained on a local optimum. We then selected the best tree using likelihood scores. To estimate node ages, we used the penalized likelihood method on the best likelihood tree. We then used the ultrametric tree as a starting tree for Bayesian inference of tree topologies and node ages using Markov chain Monte Carlo (MCMC) in BEAST v1.8.1 (ref. [62]). We ran six independent MCMC, 40×10^6 steps long under individual gene models previously selected for each reconstruction. We used the same fossil and biogeography based calibrations of Cowman and Bellwood[63] as we had no update in fossil calibrations since then, and there is no fossil data available to our knowledge for the additional families. We checked for the convergence and stationarity of each independent run using Tracer v1.6 (ref. [64]). We combined the six independent runs for each data set (after removing an appropriate burn-in) using LogCombiner v1.8.1 (ref. [62]) to reach an effective sample size above 200 for all our estimates and we extracted the maximum clade credibility tree for combined tree sets using TreeAnnotator v1.8.1 (ref. [62]). The resulting time-calibrated phylogeny is consistent with previously published phylogenies for the Labridae, with a few differences. Overall, the node age estimates appear to be 5–10 million years older than previously estimated[40] . In addition, few nodes showed low posterior probabilities.

Biogeographic Reconstructions Using the Phylogeny

Due to the lack of knowledge of the sister family of the Labridae, we confined our biogeographic analyses to within the Labridae. Based on a global-scale distribution database[3,44], we classified the distribution of the species belonging to the Labridae into three main areas, namely Atlantic, Western Indian Ocean and Central Indo-Pacific. We performed ancestral biogeographic estimation on the reconstructed empirical phylogeny using the R package BioGeoBEARS[65]. BioGeoBEARS is designed to perform inference of biogeographic history on time-calibrated phylogenies using models of dispersal, extinction and cladogenesis (DEC)[66] with the addition of several other biogeographic parameters that can be combined to form other biogeographic models (for example, DIVA, BayAREA) in a likelihood framework. In addition, it allows the inclusion of founder-event speciation scenarios through the inclusion of the 'Jump' (+J) parameter. Alternate models can be compared using Akaike information criterion (AIC). We estimated the biogeographic range evolution under a DEC model which allowed many alternative biogeographic scenarios. In addition we compared the inclusion of the '+J' parameter for the empirical data using AIC. Species ranges were categorized into three separate biogeographic regions – Atlantic, West Indian Ocean and Central Indo-Pacific. To reflect the tectonic history of marine habitat we allowed a fourth area to only exist in the past – the Tethys. Lineages were only permitted to occupy, or include the Tethys region as part of its range before 12 Ma, a time that signifies the final closure of the Tethys seaway[67]. To reflect the fossil and paleogeographic record, we constrain the deeper nodes in the tree (before 70 Ma) to include the Tethys region in the node range estimation (but not exclude other adjacent area combinations). Dispersal was only allowed between adjacent areas and a dispersal matrix was constructed to reflect the formation of land bridges at different time period (closure of the Tethys Seaway – 12 Ma; Isthmus of Panama – 3.1 Ma). Recent geologic evidence points to earlier closures of the Isthmus of Panama[68] and its extended evolutionary influence on marine and terrestrial biotas[69]. Here, we model our dispersal matrix on the final closure of the Isthmus, but earlier vicariance events are reported in the Labridae[70]. Under the DEC+J model, we identified the age of colonization of the central Indo-Pacific among the different lineages of Labridae.

Diversification Rates

To compare to the simulated diversification rate to empirical patterns, we estimated extinction and speciation rates shaping diversification for the coral family Acroporidae using fossils occurrences following the method (PyRate) proposed by Silvestro *et al.*[71] This method includes the probability of preservation and sampling of fossils to estimate the lifespan of each lineage. We first removed all undetermined taxa from the data set. Due to the lack of fossil occurrences before 60 Ma, we only estimated diversification rates during the last 60 Myr. We first ran 10,000,000 BDMCMC generations both assuming constant and Gamma distributed preservation rate to test the heterogeneity of preservation rate. We then generated 10 randomization of the age of fossils occurrences and performed 10 independent BDMCMC analyses to estimate the diversification rates through time. We examined the posterior samples using Tracer v1.6 (ref. [64]) and combined it after removing an appropriate burn-in to get a good estimation of speciation and extinction rates.

Because missing extinct lineages in the most ancient section of the phylogeny of Labridae preclude any diversification estimate in the Cretaceous from phylogenies, we computed the rate of diversification from a phylogeny containing 18 tropical reef fish families (that is, *Labridae* including *Scaridae*, *Pomacentridae*,*Chaetodontidae*, *Acanthuridae*, *Haemulidae*, *Balistidae*, *Carangidae*, *Serranidae*, *Lutjanidae*, *Sparidae*,*Caesionidae*, *Holocentridae*, *Mullidae*, *Muraenidae*, *Tetraodontidae*, *Lethrinidae* and *Siganidae*). We obtained this phylogeny by pruning a time-calibrated phylogeny for 7,822 extant fish species[55]. These families were selected as the most representative reef fish families, that is, they are abundant and speciose on tropical reefs[72]. These families are well sampled in the phylogeny of Rabosky *et al.*[55] with more than 80% of known genus. As missing species may influence the estimation of diversification rates, we grafted those species on the pruned phylogenetic tree (1,291 missing species out of 2,224) based on published phylogenies for these families, supplemented by taxonomic information from fish identification guides and FishBase (www.fishbase.org). Specifically, new tips representing unsampled species were added to direct sister species when information allowed or to the base of the clade representing its genus. Using this empirical phylogeny, we estimated diversification rates based on an evolutionary model under a birth-death-shift process[46]. The birth-death-shift model uses a likelihood approach to estimate speciation and extinction rates like in a simple birth-death process but it also allows changes of those parameters through time if we force diversification shifts to happen at given times. For a given number of shifts, we simultaneously estimated time at which shifts occur, and speciation and extinction rates within each time interval between diversification shifts. We calculated those parameters for several numbers of shifts and we compared the resulting models using the AIC. The method proposed by Stadler[46] is robust to phylogenetic uncertainty due to polytomies. Finally, we extracted the simulated values of diversification rate expected under the parapatric and sympatric models by computing for each time step the speciation rate minus the extinction rate and we compared the observed diversification rates with those obtained from the observed phylogeny for fishes and fossils for corals.

Data Availability

The R script to run the species diversification model together with the bathymetry reconstructions are available online on FigShare (DOI: 10.6084/m9.figshare.3081226). Species diversity data and fossils are made available through the IUCN http://www.iucnredlist.org/ and http://fossilworks.org/.

ACKNOWLEDGMENTS

This project was financed by the ANR-FNS project "REEFISH" n° 310030E-164294. PFC was funded by the Gaylord Donnelley Environmental Postdoctoral Fellowship administered by the Yale Institute for Biospheric Studies (YIBS).

Footnotes

Author contributions F.L. and L.P. led the study design; L.P. developed the diversification model from conceptual discussion with F.L., C.J.M. and C.N.d.S.; C.H. produced the paleobathymetry reconstruction. P.D., T.G. and P.F.C. contributed to data analyses; F.L. and L.P. wrote a first version of the manuscript and all authors contributed to its revision.

REFERENCES

1. Roberts C. M. et al. . Marine biodiversity hotspots and conservation priorities for tropical reefs.Science 295, 1280–1284 (2002) .
2. Renema W. et al. . Hopping hotspots: global shifts in marine biodiversity. Science 321, 654–657 (2008) .
3. Pellissier L. et al. . Quaternary coral reef refugia preserved fish diversity. Science 344, 1016–1020 (2014) .
4. Williams S. T. & Duda T. F. Did tectonic activity stimulate Oligo-Miocene speciation in the Indo-West Pacific? Evolution 62, 1618–1634 (2008) .
5. Wilson M. E. J. & Rosen B. R. in Biogeography and Geological Evolution of SE Asia eds. Hall R., Holloway J. D. 165–195Backhuys Publishers (1998) .
6. Müller R. D., Sdrolias M., Gaina C., Steinberger B. & Heine C. Long-term sea-level fluctuations driven by ocean basin dynamics. Science 319, 1357–1362 (2008) .
7. Muneepeerakul R. et al. . Neutral metacommunity models predict fish diversity patterns in Mississippi–Missouri basin. Nature 453, 220–223 (2008) .
8. Gotelli N. J. et al. . Patterns and causes of species richness: a general simulation model for macroecology. Ecol. Lett. 12, 873–886 (2009) .
9. Heine C., Yeo L. G. & Müller R. D. Evaluating global paleoshoreline models for the Cretaceous and Cenozoic. Aust. J. Earth Sci. 62, 275–287 (2015) .
10. Budd A. F. & Pandolfi J. M. Evolutionary novelty is concentrated at the edge of coral species distributions. Science 328, 1558–1561 (2010) .
11. Bowen B. W., Rocha L. a., Toonen R. J. & Karl S. a. The origins of tropical marine biodiversity.Trends Ecol. Evol. 28, 359–366 (2013) .
12. Losos J. B. & Schluter D. Analysis of an evolutionary species-area relationship. Nature 408, 847–850 (2000) .
13. Jolivet L. & Faccenna C. Mediterranean extension and the Africa-Eurasia collision. Nature 19, 1095–1106 (2000) .
14. van Hinsbergen D. J. J. et al. . Restoration of Cenozoic deformation in Asia and the size of Greater India. Tectonics 30, TC5003 (2011) .
15. Dornburg A., Moore J., Beaulieu J. M., Eytan R. I. & Near T. J. The impact of shifts in marine biodiversity hotspots on patterns of range evolution: Evidence from the Holocentridae (squirrelfishes and soldierfishes). Evolution 69, 146–161 (2015) .
16. Goatley C. H. R., Bellwood D. R. & Bellwood O. Fishes on coral reefs: changing roles over the past 240 million years. Paleobiology 36, 415–427 (2010) .
17. Aberhan M. & Kiessling W. Persistent ecological shifts in marine molluscan assemblages across the end-Cretaceous mass extinction. Proc. Natl. Acad. Sci. 112, 7207–7212 (2015) .
18. Cowman P. F. & Bellwood D. R. The historical biogeography of coral reef fishes: global patterns of origination and dispersal. J. Biogeogr. 40, 209–224 (2013) .
19. Keith S. A., Baird A. H., Hughes T. P., Madin J. S. & Connolly S. R. Faunal breaks and species composition of Indo-Pacific corals: the role of plate tectonics, environment and habitat distribution.Proc. R. Soc. B 280, 2013–0818 (2013) .
20. Kulbicki M. et al. . Global biogeography of reef fishes: a hierarchical quantitative delineation of regions. PLoS ONE 8, e81847 (2013) .
21. Mouillot D. et al. . The challenge of delineating biogeographical regions: nestedness matters for Indo-Pacific coral reef fishes. J. Biogeogr. 40, 2228–2237 (2013) .
22. Last P. R. & White W. T. Biogeographic patterns in the Australian chondrichthyan fauna. J. Fish Biol.79, 1193–1213 (2011) .
23. Santini F. & Winterbottom R. Historical biogeography of Indo-western Pacific coral reef biota: is the Indonesian region a centre of origin? J. Biogeogr. 29, 189–205 (2002) .
24. Barber P. H., Palumbi S. R., Erdmann M. V. & Moosa M. K. Biogeography: a marine Wallace's line?Nature 406, 692–693 (2000) .
25. Losos J. B. & Ricklefs R. E. The Theory of Island Biogeography Revisited Princeton University Press (2009) .
26. Hoorn C. et al. . Amazonia through time: Andean uplift, climate change, landscape evolution, and biodiversity. Science 330, 927–931 (2010) .
27. Mouillot D. et al. . Functional over-redundancy and high functional vulnerability in global fish faunas of tropical reefs. Proc. Natl Acad. Sci. USA 111, 13757–13762 (2014) .
28. van Hinsbergen D. J. J. et al. . A paleolatitude calculator for paleoclimate studies. PLoS One 10, e0126946 (2015) .
29. Torsvik T. H. et al. . Phanerozoic polar wander, palaeogeography and dynamics. Earth-Sci. Rev. 114, 325–368 (2012) .
30. Wallace C. C. & Rosen B. R. Diverse staghorn corals (Acropora) in high-latitude Eocene assemblages: implications for the evolution of modern diversity patterns of reef corals. Proc. Biol. Sci. 273, 975–982 (2006) .
31. Ridgwell A. & Schmidt D. N. Past constraints on the vulnerability of marine calcifiers to massive carbon dioxide release. Nat. Geosci. 3, 196–200 (2010) .
32. Gower D. et al. . Biotic evolution and environmental change in Southeast Asia Cambridge University Press

(2012) .

33. Müller R. D., Sdrolias M., Gaina C. & Roest W. R. Age, spreading rates, and spreading asymmetry of the world's ocean crust. Geochem. Geophys. Geosyst. 9, Q04006 (2008) .
34. O'Neill C., Müller D. & Steinberger B. On the uncertainties in hot spot reconstructions and the significance of moving hot spot reference frames. Geochem. Geophys. Geosyst. 6, Q04003 (2005) .
35. Müller R. D., Royer J.-Y. & Lawver L. A. Revised plate motions relative to the hotspots from combined Atlantic and Indian Ocean hotspot tracks. Geology 21, 275–278 (1993) .
36. Wessel P., Harada Y. & Kroenke L. W. Toward a self-consistent, high-resolution absolute plate motion model for the Pacific. Geochem. Geophys. Geosyst. 7, Q03L12 (2006) .
37. Seton M. et al. . Global continental and ocean basin reconstructions since 200 Ma. Earth-Sci. Rev. 113, 212–270 (2012) .
38. Steinberger B. & Torsvik T. H. Absolute plate motions and true polar wander in the absence of hotspot tracks. Nature 452, 620–623 (2008) .
39. Stein C. A. & Stein S. A model for the global variation in oceanic depth and heat flow with lithospheric age. Nature 359, 123–129 (1992) .
40. Smith A. G., Smith D. G. & Funnell B. M. Atlas of Mesozoic and Cenozoic coastlines (2004) .
41. Zachos J., Pagani M., Sloan L., Thomas E. & Billups K. Trends, rhythms, and aberrations in global climate 65 Ma to present. Science 292, 686–693 (2001) .
42. Kleypas J. A., McManus J. W. & Meñez L. A. B. Environmental limits to coral reef development: where do we draw the line? Am. Zool. 39, 146–159 (1999) .
43. Kiessling W., Hendy A. J. W., Miller A. I. & Fursich F. T. Taxonomic occurrences of Scleractinia recorded in, Fossilworks, the Evolution of Terrestrial Ecosystems database, and the Paleobiology Database. Fossilworks. Available at http://fossilworks.org (2014) .
44. Parravicini V. et al. . Global patterns and predictors of tropical reef fish species richness. Ecography36, 1254–1262 (2013) .
45. Morlon H. Phylogenetic approaches for studying diversification. Ecol. Lett. 17, 508–525 (2014) .
46. Stadler T. Mammalian phylogeny reveals recent diversification rate shifts. Proc. Natl Acad. Sci USA108, 6187–6192 (2011) .
47. Svenning J.-C., Fløjgaard C., Marske K. A., Nógues-Bravo D. & Normand S. Applications of species distribution modeling to paleobiology. Quat. Sci. Rev. 30, 2930–2947 (2011) .
48. Kauffman E. G. in *Cretaceous-Tertiary boundary events symposium* 29–37 (University of Copenhagen, 1979) .
49. Hennig C. Fixed point clusters, clusterwise regression and discriminant plots. Package fpc for R (2009) .
50. Veron J. E. N. et al. . Delineating the coral triangle. Galaxea J. Coral Reef Stud. 11, 91–100 (2009) .
51. Bellwood D. R., Hughes T. P., Connolly S. R. & Tanner J. Environmental and geometric constraints on Indo-Pacific coral reef biodiversity. Ecol. Lett. 8, 643–651 (2005) .
52. Almeida-Neto M., Guimaraes P., Guimarães P. R., Loyola R. D. & Ulrich W. A consistent metric for nestedness analysis in ecological systems: reconciling concept and measurement. Oikos 117, 1227–1239 (2008) .
53. Gotelli N. J. Null model analysis of species co-occurrence patterns. Ecology 81, 2606–2621 (2000) .
54. Baselga A. & Leprieur F. Comparing methods to separate components of beta diversity. Methods Ecol. Evol. 6, 1069–1079 (2015) .
55. Rabosky D. L. et al. . Rates of speciation and morphological evolution are correlated across the largest vertebrate radiation. Nat. Commun. 4, 1–8 (2013) .
56. Tamura K., Stecher G., Peterson D., Filipski A. & Kumar S. MEGA6: Molecular evolutionary genetics analysis version 6.0. Mol. Biol. Evol. 30, 2725–2729 (2013) .
57. Darriba D., Taboada G. L., Doallo R. & Posada D. jModelTest 2: more models, new heuristics and parallel computing. Nat. Methods 9, 772–772 (2012) .
58. Guindon S. et al. . New algorithms and methods to estimate maximim-likelihood phylogenies: assessing the performance of PhyML 3.0. Syst. Biol. 59, 307–321 (2010) .
59. Vaidya G., Lohman D. J. & Meier R. SequenceMatrix: concatenation software for the fast assembly of multi-gene datasets with character set and codon information. Cladistics 27, 171–180 (2011) .
60. Stamatakis A. RAxML version 8: a tool for phylogenetic analysis and post-analysis of large phylogenies. Bioinformatics 30, 1312–1313 (2014) .
61. Silvestro D. & Michalak I. RaxmlGUI: a graphical front-end for RAxML. Org. Divers. Evol. 12, 335–337 (2012)
62. Drummond A. J., Suchard M. a., Xie D. & Rambaut A. Bayesian phylogenetics with BEAUti and the BEAST 1.7. Mol. Biol. Evol. 29, 1969–1973 (2012) .
63. Cowman P. F. & Bellwood D. R. Coral reefs as drivers of cladogenesis: expanding coral reefs, cryptic extinction events, and the development of biodiversity hotspots. J. Evol. Biol. 24, 2543–2562 (2011) .
64. Rambaut A., Suchard M. A., Xie D. & Drummond A. J. Tracer v1.6. (2014). Available athttp://beast.bio.ed.ac.uk/Tracer.
65. Matzke N. J. BioGeoBEARS: biogeography with Bayesian (and likelihood) evolutionary analysis in R Scripts. R package, version 0.2. (2013) .
66. Ree R. H. & Smith S. a Maximum likelihood inference of geographic range evolution by dispersal, local extinction, and cladogenesis. Syst. Biol. 57, 4–14 (2008) .
67. Steininger F. F. & Rögl F. The paratethys history. A contribution towards the Neogene geodynamics of the alpine orogene. Ann. Geol. Pays Hell 3, 1153–1165 (1979) .
68. Farris D. W. et al. . Fracturing of the Panamanian Isthmus during initial collision with: South America.Geology 39,

1007–1010 (2011) .

69. Bacon C. D. et al. . Biological evidence supports an early and complex emergence of the Isthmus of Panama. Proc. Natl Acad. Sci. USA 112, 6110–6115 (2015) .

70. Cowman P. F. & Bellwood D. R. Vicariance across major marine biogeographic barriers: temporal concordance and the relative intensity of hard versus soft barriers. Proc. R. Soc. London B Biol. Sci.280, 2013–1541 (2013) .

71. Silvestro D., Salamin N. & Schnitzler J. PyRate: a new program to estimate speciation and extinction rates from incomplete fossil data. Methods Ecol. Evol. 5, 1126–1131 (2014) .

72. Choat J. H. & Bellwood D. R. in The Ecology of Fishes on Coral Reefs ed. Sale P. F. 39–66Academic Press (1991) .

INDEX

T

W